APPLIED ANALYSIS

John K. Hunter
Bruno Nachtergaele
Department of Mathematics
University of California, Davis
USA

APPLIED
ANALYSIS

 **World Scientific**

NEW JERSEY · LONDON · SINGAPORE · BEIJING · SHANGHAI · HONG KONG · TAIPEI · CHENNAI

Published by

World Scientific Publishing Co. Pte. Ltd.

5 Toh Tuck Link, Singapore 596224

USA office: 27 Warren Street, Suite 401-402, Hackensack, NJ 07601

UK office: 57 Shelton Street, Covent Garden, London WC2H 9HE

Library of Congress Cataloging-in-Publication Data
Hunter, John K.
   Applied analysis / by John K. Hunter and Bruno Nachtergaele.
      p. cm.
   Includes bibliographical references and index.
   ISBN 9810241917 (alk. paper)
     1. Mathematical analysis. I. Nachtergaele, Bruno. II. Title.

   QA300 .H93 2001
   515--dc21   00-051304

British Library Cataloguing-in-Publication Data
A catalogue record for this book is available from the British Library.

First published 2001
Reprinted 2005

Printed in Singapore.

To Shelley and Marijke, Caitlin and Lauren, and Sigrid and Shanti.

Preface

Mathematical knowledge and sophistication, computational power, and areas of application are expanding at an enormous rate. As a result, the demands on the training of applied mathematicians are increasing all the time. It is therefore not easy to decide what should constitute the core mathematical training of an applied mathematician. We take the view that every applied mathematician, whatever his or her ultimate area of interest may turn out to be, should have a grounding in the fundamentals of analysis.

The aim of this book is to supply an introduction for beginning graduate students to those parts of analysis that are most useful in applications. The material is selected for its use in applied problems, and is presented as clearly and simply as we are able, but without the sacrifice of mathematical rigor.

We focus on ideas of central importance, and attempt to avoid technicalities and detours into areas of more specialized interest. While we make every effort to motivate the ideas introduced, and include a variety of examples from different fields, this book is first and foremost about analysis.

We do not assume extensive mathematical prerequisites of the reader. The book is intended to be accessible to students from a wide variety of backgrounds, including undergraduate students entering applied mathematics from non-mathematical fields, and graduate students in the sciences and engineering who would like to learn analysis. A basic background in calculus, linear algebra, ordinary differential equations, and some familiarity with functions and sets should be sufficient. We occasionally use some elementary results from complex analysis, but we do not develop any methods from complex analysis in the text.

We provide detailed proofs for the main topics. We make no attempt to state results in maximum generality, but instead illustrate the main ideas in simple, concrete settings. We often return to the same ideas in different contexts, even if this leads to some repetition of previous definitions and results. We make extensive use of examples and exercises to illustrate the concepts introduced. The exercises are at various levels; some are elementary, although we have omitted many of the routine exercises that we assign while teaching the class, and some are harder and are an excuse to introduce new ideas or applications not covered in the main text. One area where we do not give a complete treatment is Lebesgue measure and integration. A full development of measure theory would take us too far afield, and, in any event, the Lebesgue integral is much easier to use than to construct.

In writing this book, the material has expanded beyond what can be covered in a year long course for beginning graduate students. When teaching a three-quarter course, we usually cover Chapters 1–5 in the first quarter, which provide a review of advanced calculus and discuss the basic properties of metric and normed spaces, followed by Chapters 6–9 in the second quarter, which focus on Hilbert spaces, including Fourier series and bounded linear operators. In the last quarter, we cover a selection of topics from Chapters 10–13, which discuss Green's functions, unbounded operators, distribution theory, the Fourier transform, measure theory, function spaces, and differential calculus in Banach spaces. The choice and emphasis of the topics depends on the backgrounds and interests of the students.

The material presented here is standard. Many of the sources we have drawn upon are listed in the bibliography. The bibliography is not comprehensive, however, and is limited to books that we feel will be useful to the intended audience of this text, either for background reading, or to pursue in greater depth some of the topics treated here.

We thank the students who have taken this course and contributed comments and suggestions on early drafts of the course notes. In particular, Scott Beaver, Sergio Lucero, and John Thoo were helpful in the preparation of figures and the proofreading of the manuscript.

Contents

APPLIED ANALYSIS

Chapter 1

Metric and Normed Spaces

We are all familiar with the geometrical properties of ordinary, three-dimensional Euclidean space. A persistent theme in mathematics is the grouping of various kinds of objects into abstract spaces. This grouping enables us to extend our intuition of the relationship between points in Euclidean space to the relationship between more general kinds of objects, leading to a clearer and deeper understanding of those objects.

The simplest setting for the study of many problems in analysis is that of a metric space. A metric space is a set of points with a suitable notion of the distance between points. We can use the metric, or distance function, to define the fundamental concepts of analysis, such as convergence, continuity, and compactness.

A metric space need not have any kind of algebraic structure defined on it. In many applications, however, the metric space is a linear space with a metric derived from a norm that gives the "length" of a vector. Such spaces are called normed linear spaces. For example, n-dimensional Euclidean space is a normed linear space (after the choice of an arbitrary point as the origin). A central topic of this book is the study of infinite-dimensional normed linear spaces, including function spaces in which a single point represents a function. As we will see, the geometrical intuition derived from finite-dimensional Euclidean space remains essential, although completely new features arise in the case of infinite-dimensional spaces.

In this chapter, we define and study metric spaces and normed linear spaces. Along the way, we review a number of definitions and results from real analysis.

1.1 Metrics and norms

Let X be an arbitrary nonempty set.

Definition 1.1 A *metric*, or *distance function*, on X is a function

$$d : X \times X \to \mathbb{R},$$

1

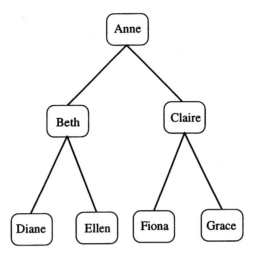

Fig. 1.1   A family tree used in the definition of the ultrametric in Example 1.3.

with the following properties:

(a) $d(x, y) \geq 0$ for all $x, y \in X$, and $d(x, y) = 0$ if and only if $x = y$;
(b) $d(x, y) = d(y, x)$, for all $x, y \in X$;
(c) $d(x, y) \leq d(x, z) + d(z, y)$, for all $x, y, z \in X$.

A *metric space* (X, d) is a set X equipped with a metric d.

When the metric d is understood from the context, we denote a metric space simply by the set X. In words, the definition states that:

(a) distances are nonnegative, and the only point at zero distance from x is x itself;
(b) the distance is a symmetric function;
(c) distances satisfy the *triangle inequality*.

For points in the Euclidean plane, the triangle inequality states that the length of one side of a triangle is less than the sum of the lengths of the other two sides.

Example 1.2 The set of real numbers $\mathbb{R}$ with the distance function $d(x, y) = |x - y|$ is a metric space. The set of complex numbers $\mathbb{C}$ with the distance function $d(z, w) = |z - w|$ is also a metric space.

Example 1.3 Let X be a set of people of the same generation with a common ancestor, for example, all the grandchildren of a grandmother (see Figure 1.1). We define the distance $d(x, y)$ between any two individuals x and y as the number of generations one has to go back along female lines to find the first common ancestor.

For example, the distance between two sisters is one. It is easy to check that d is a metric. In fact, d satisfies a stronger condition than the triangle inequality, namely

$$d(x,y) \leq \max\{d(x,z), d(z,y)\} \qquad \text{for all } x, y, z \in X. \tag{1.1}$$

A metric d which satisfies (1.1) is called an *ultrametric*. Ultrametrics have been used in taxonomy to characterize the genetic proximity of species.

Example 1.4 Let X be the set of n-letter words in a k-character alphabet $A = \{a_1, a_2, \ldots, a_k\}$, meaning that $X = \{(x_1, x_2, \ldots, x_n) \mid x_i \in A\}$. We define the distance $d(x,y)$ between two words $x = (x_1, \ldots, x_n)$ and $y = (y_1, \ldots, y_n)$ to be the number of places in which the words have different letters. That is,

$$d(x,y) = \#\{i \mid x_i \neq y_i\}.$$

Then (X, d) is a metric space.

Example 1.5 Suppose (X, d) is any metric space and Y is a subset of X. We define the distance between points of Y by restricting the metric d to Y. The resulting metric space $(Y, d|_Y)$, or (Y, d) for short, is called a *metric subspace* of (X, d), or simply a subspace when it is clear that we are talking about metric spaces. For example, $(\mathbb{R}, |\cdot|)$ is a metric subspace of $(\mathbb{C}, |\cdot|)$, and the space of rational numbers $(\mathbb{Q}, |\cdot|)$ is a metric subspace of $(\mathbb{R}, |\cdot|)$.

Example 1.6 If X and Y are sets, then the *Cartesian product* $X \times Y$ is the set of ordered pairs (x, y) with $x \in X$ and $y \in Y$. If d_X and d_Y are metrics on X and Y, respectively, then we may define a metric $d_{X \times Y}$ on the product space by

$$d_{X \times Y}((x_1, y_1), (x_2, y_2)) = d_X(x_1, x_2) + d_Y(y_1, y_2)$$

for all $x_1, x_2 \in X$ and $y_1, y_2 \in Y$.

We recall the definition of a linear, or vector, space. We consider only real or complex linear spaces.

Definition 1.7 A *linear space* X over the scalar field $\mathbb{R}$ (or $\mathbb{C}$) is a set of points, or vectors, on which are defined operations of vector addition and scalar multiplication with the following properties:

(a) the set X is a commutative group with respect to the operation $+$ of vector addition, meaning that for all $x, y, z \in X$, we have $x + y = y + x$ and $x + (y + z) = (x + y) + z$, there is a zero vector 0 such that $x + 0 = x$ for all $x \in X$, and for each $x \in X$ there is a unique vector $-x$ such that $x + (-x) = 0$;

(b) for all $x, y \in X$ and $\lambda, \mu \in \mathbb{R}$ (or $\mathbb{C}$), we have $1x = x$, $(\lambda + \mu)x = \lambda x + \mu x$, $\lambda(\mu x) = (\lambda \mu)x$, and $\lambda(x + y) = \lambda x + \lambda y$.

We assume that the reader is familiar with the elementary theory of linear spaces. Some references are given in Section 1.9.

A norm on a linear space is a function that gives a notion of the "length" of a vector.

Definition 1.8 A *norm* on a linear space X is a function $\| \cdot \| : X \to \mathbb{R}$ with the following properties:

   (a) $\|x\| \geq 0$, for all $x \in X$ (nonnegative);
   (b) $\|\lambda x\| = |\lambda| \|x\|$, for all $x \in X$ and $\lambda \in \mathbb{R}$ (or $\mathbb{C}$) (homogeneous);
   (c) $\|x + y\| \leq \|x\| + \|y\|$, for all $x, y \in X$ (triangle inequality) ;
   (d) $\|x\| = 0$ implies that $x = 0$ (strictly positive).

A *normed linear space* $(X, \| \cdot \|)$ is a linear space X equipped with a norm $\| \cdot \|$.

A normed linear space is a metric space with the metric

$$d(x, y) = \|x - y\|. \tag{1.2}$$

All the concepts we define for metric spaces therefore apply, in particular, to normed linear spaces. The metric associated with a norm in this way has the special properties of translation invariance, meaning that for all $z \in X$, $d(x + z, y + z) = d(x, y)$, and homogeneity, meaning that for all $\lambda \in \mathbb{R}$ (or $\mathbb{C}$), $d(\lambda x, \lambda y) = |\lambda| d(x, y)$.

The *closed unit ball* $\overline{B}$ of a normed linear space X is the set

$$\overline{B} = \{x \in X : \|x\| \leq 1\}.$$

A subset C of a linear space is *convex* if

$$tx + (1 - t)y \in C \tag{1.3}$$

for all $x, y \in C$ and all real numbers $0 \leq t \leq 1$, meaning that the line segment joining any two points in the set lies in the set. The triangle inequality implies that the unit ball is convex, and its shape gives a good picture of the norm's geometry.

Example 1.9 The set of real numbers $\mathbb{R}$ with the absolute value norm $\|x\| = |x|$ is a one-dimensional real normed linear space. More generally, $\mathbb{R}^n$, where $n = 1, 2, 3, \ldots$, is an n-dimensional linear space. We define the *Euclidean norm* of a point $x = (x_1, x_2, \ldots, x_n) \in \mathbb{R}^n$ by

$$\|x\| = \sqrt{x_1^2 + x_2^2 + \cdots + x_n^2},$$

and call $\mathbb{R}^n$ equipped with the Euclidean norm n-dimensional *Euclidean space*. We can also define other norms on $\mathbb{R}^n$. For example, the *sum* or *1-norm* is given by

$$\|x\|_1 = |x_1| + |x_2| + \cdots + |x_n|.$$

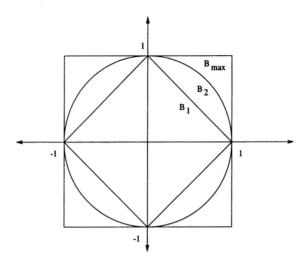

Fig. 1.2 The unit balls in $\mathbb{R}^2$ for the Euclidean norm (B_2), the sum norm (B_1), and the maximum norm ($B_{\max}$).

The *maximum norm* is given by

$$\|x\|_{\max} = \max\{|x_1|, |x_2|, \ldots, |x_n|\}.$$

We also call the maximum norm the ∞-*norm*, and denote it by $\|x\|_\infty$. The unit balls in $\mathbb{R}^2$ for each of these norms are shown in Figure 1.2. We will equip $\mathbb{R}^n$ with the Euclidean norm, unless stated otherwise.

Example 1.10 A *linear subspace* of a linear space, or simply a *subspace* when it is clear we are talking about linear spaces, is a subset that is itself a linear space. A subset M of a linear space X is a subspace if and only if $\lambda x + \mu y \in M$ for all $\lambda, \mu \in \mathbb{R}$ (or $\mathbb{C}$) and all $x, y \in M$. A subspace of a normed linear space is a normed linear space with norm given by the restriction of the norm on X to M.

We will see later on that all norms on a finite-dimensional linear space lead to exactly the same notion of convergence, so often it is not important which norm we use. Different norms on an infinite-dimensional linear space, such as a function space, may lead to completely different notions of convergence, so the specification of a norm is crucial in this case.

We will always regard a normed linear space as a metric space with the metric defined in equation (1.2), unless we explicitly state otherwise. Nevertheless, this equation is not the only way to define a metric on a normed linear space.

Example 1.11 If $(X, \|\cdot\|)$ is a normed linear space, then

$$d(x, y) = \frac{\|x - y\|}{1 + \|x - y\|} \tag{1.4}$$

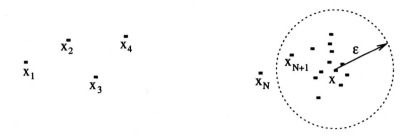

Fig. 1.3 A sequence (x_n) converging to x.

defines a nonhomogeneous, translation invariant metric on X. In this metric, the distance between two points is always less than one.

1.2   Convergence

We first consider the convergence of sequences of real numbers. A *sequence* of real numbers is a map from the natural numbers $\mathbb{N} = \{1, 2, 3, \ldots\}$ to $\mathbb{R}$. That is, with each $n \in \mathbb{N}$, we associate a real number $x_n \in \mathbb{R}$. We denote a sequence by (x_n), or $(x_n)_{n=1}^{\infty}$ when we want to indicate the range of the index n. The index n is a "dummy" index, and we may also write the sequence as (x_k) or $(x_k)_{k=1}^{\infty}$.

Another common notation for a sequence is $\{x_n\}$. This notation is a little ambiguous because a sequence is not the same thing as a set. For example,

$$(0, 1, 0, 1, 0, \ldots) \text{ and } (1, 0, 0, 0, 0, \ldots)$$

are different sequences, but the set of terms is $\{0, 1\}$ in each case.

A *subsequence* of a sequence (x_n) is a sequence of the form (x_{n_k}), where for each $k \in \mathbb{N}$ we have $n_k \in \mathbb{N}$, and $n_k < n_{k+1}$ for all k. That is, $k \mapsto n_k$ is a strictly increasing function from the set of natural numbers to itself. For example, $(1/k^2)_{k=1}^{\infty}$ is a subsequence of $(1/n)_{n=1}^{\infty}$.

The most important concept concerning sequences is convergence.

Definition 1.12 A sequence (x_n) of real numbers *converges* to $x \in \mathbb{R}$ if for every $\epsilon > 0$ there is an $N \in \mathbb{N}$ such that $|x_n - x| < \epsilon$ for all $n \geq N$. The point x is called the *limit* of (x_n).

In this definition, the integer N depends on ϵ, since smaller ϵ's usually require larger N's, and we could write $N(\epsilon)$ to make the dependence explicit. Common ways to write the convergence of (x_n) to x are

$$x_n \to x \text{ as } n \to \infty, \qquad \lim_{n \to \infty} x_n = x.$$

A sequence that does not converge is said to *diverge*. If a sequence diverges because its terms eventually become larger than any number, it is often convenient to regard the sequence as converging to ∞. That is, we say $x_n \to \infty$ if for every $M \in \mathbb{R}$ there is an $N \in \mathbb{N}$ such that $x_n > M$ for all $n \geq N$. Similarly, we say $x_n \to -\infty$ if for every $M \in \mathbb{R}$ there is an $N \in \mathbb{N}$ such that $x_n < M$ for all $n \geq N$.

Example 1.13 Here are a few examples of the limits of convergent sequences:

$$\lim_{n\to\infty} \frac{1}{n} = 0, \quad \lim_{n\to\infty} n \sin\left(\frac{1}{n}\right) = 1, \quad \lim_{n\to\infty} \left(1 + \frac{1}{n}\right)^n = e.$$

The sequence $(\log n)$ diverges because $\log n \to \infty$ as $n \to \infty$. The sequence $((-1)^n)$ diverges because its terms oscillate between -1 and 1, and it does not converge to either ∞ or $-\infty$.

A sequence is said to be *Cauchy* if its terms eventually get arbitrarily close together.

Definition 1.14 A sequence (x_n) is a *Cauchy sequence* if for every $\epsilon > 0$ there is an $N \in \mathbb{N}$ such that $|x_m - x_n| < \epsilon$ for all $m, n \geq N$.

Suppose that (x_n) converges to x. Given $\epsilon > 0$, there is an integer N such that $|x_n - x| < \epsilon/2$ when $n \geq N$. If $m, n \geq N$, then use of the triangle inequality implies that

$$|x_m - x_n| \leq |x_m - x| + |x - x_n| < \epsilon,$$

so (x_n) is Cauchy. Thus, every convergent sequence is a Cauchy sequence. For the real numbers, the converse is also true, and every Cauchy sequence is convergent. The convergence of Cauchy sequences is a fundamental defining property of the real numbers, called *completeness*. We will discuss completeness for general metric spaces in greater detail below.

Example 1.15 The sequence (x_n) with $x_n = \log n$ is not a Cauchy sequence, since $\log n \to \infty$. Nevertheless, we have

$$|x_{n+1} - x_n| = \log\left(1 + \frac{1}{n}\right) \to 0$$

as $n \to \infty$. This example shows that it is not sufficient for successive terms in a sequence to get arbitrarily close together to ensure that the sequence is Cauchy.

We can use the definition of the convergence of a sequence to define the sum of an infinite series as the limit of its sequence of partial sums. Let (x_n) be a sequence in $\mathbb{R}$. The sequence of *partial sums* (s_n) of the *series* $\sum x_n$ is defined by

$$s_n = \sum_{k=1}^{n} x_k. \tag{1.5}$$

If (s_n) converges to a limit s, then we say that the series $\sum x_n$ *converges* to s, and write

$$\sum_{n=1}^{\infty} x_n = s.$$

If the sequence of partial sums does not converge, or converges to infinity, then we say that the series *diverges*. The series $\sum x_n$ is said to be *absolutely convergent* if the series of absolute values $\sum |x_n|$ converges. Absolute convergence implies convergence, but not conversely. A useful property of an absolutely convergent series of real (or complex) numbers is that any series obtained from it by a permutation of its terms converges to the same sum as the original series.

The definitions of convergent and Cauchy sequences generalize to metric spaces in an obvious way. A sequence (x_n) in a metric space (X, d) is a map $n \mapsto x_n$ which associates a point $x_n \in X$ with each natural number $n \in \mathbb{N}$.

Definition 1.16 A sequence (x_n) in X *converges* to $x \in X$ if for every $\epsilon > 0$ there is an $N \in \mathbb{N}$ such that $d(x_n, x) < \epsilon$ for all $n \geq N$. The sequence is *Cauchy* if for every $\epsilon > 0$ there is an $N \in \mathbb{N}$ such that $d(x_m, x_n) < \epsilon$ for all $m, n \geq N$.

Figure 1.3 shows a convergent sequence in the Euclidean plane. Property (a) of the metric in Definition 1.1 implies that if a sequence converges, then its limit is unique. That is, if $x_n \to x$ and $x_n \to y$, then $x = y$. The fact that convergent sequences are Cauchy is an immediate consequence of the triangle inequality, as before. The property that every Cauchy sequence converges singles out a particularly useful class of metric spaces, called complete metric spaces.

Definition 1.17 A metric space (X, d) is *complete* if every Cauchy sequence in X converges to a limit in X. A subset Y of X is *complete* if the metric subspace $(Y, d|_Y)$ is complete. A normed linear space that is complete with respect to the metric (1.2) is called a *Banach space*.

Example 1.18 The space of rational numbers $\mathbb{Q}$ is not complete, since a sequence of rational numbers which converges in $\mathbb{R}$ to an irrational number (such as $\sqrt{2}$ or π) is a Cauchy sequence in $\mathbb{Q}$, but does not have a limit in $\mathbb{Q}$.

Example 1.19 The finite-dimensional linear space $\mathbb{R}^n$ is a Banach space with respect to the sum, maximum, and Euclidean norms defined in Example 1.9. (See Exercise 1.6.)

Series do not make sense in a general metric space, because we cannot add points together. We can, however, consider series in a normed linear space X. Just as for real numbers, if (x_n) is a sequence in X, then the series $\sum_{n=1}^{\infty} x_n$ *converges* to $s \in X$ if the sequence (s_n) of partial sums, defined in (1.5), converges to s.

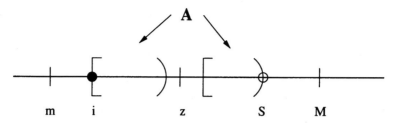

Fig. 1.4 The number M is an upper bound of A and m is a lower bound of A. The number z is neither an upper bound nor a lower bound. The number S is the supremum of A, but does not belong to A. The number i is the infimum of A, and since $i \in A$ it is also the minimum of A.

1.3 Upper and lower bounds

The real numbers have a natural ordering which we can use to define the supremum and infimum of a set of real numbers, and the lim sup and lim inf of a real sequence. Even a metric space as simple as the Euclidean plane cannot be ordered in a way that is compatible with its metric structure. Thus, the definitions in this section are restricted to real sets and sequences. We begin with the definitions of *upper bound* and *lower bound*.

Definition 1.20 Let A be a subset of $\mathbb{R}$. We say that $M \in \mathbb{R}$ is an *upper bound* of A if $x \leq M$ for all $x \in A$, and $m \in \mathbb{R}$ is a *lower bound* of A if $m \leq x$ for all $x \in A$. The set A is *bounded from above* if it has an upper bound, *bounded from below* if it has a lower bound, and *bounded* if it has both an upper and a lower bound.

If A has an upper bound M, then A has many upper bounds. For example, any number $M' \geq M$ is an upper bound.

Definition 1.21 A number M is the *supremum*, or *least upper bound*, of a set $A \subset \mathbb{R}$ if M is an upper bound of A and $M \leq M'$ for all upper bounds M' of A. A number m is the *infimum*, or *greatest lower bound*, of A if m is a lower bound of A and $m \geq m'$ for all lower bounds m' of A. We denote the supremum of A by $\sup A$, and the infimum of A by $\inf A$.

If A is given in the form $A = \{x_\alpha \mid \alpha \in \mathcal{A}\}$, where $\mathcal{A}$ is an indexing set, we also denote the supremum of A by $\sup_{\alpha \in \mathcal{A}} x_\alpha$, or $\sup x_\alpha$ for short.

The supremum and infimum are unique if they exist. For example, if M_1 and M_2 are both least upper bounds of a set A, then the definition implies that $M_1 \leq M_2$ and $M_2 \leq M_1$, so $M_1 = M_2$. The existence of the supremum of every set bounded from above, or the existence of the infimum of every set bounded from below, is a consequence of the completeness of $\mathbb{R}$, and is in fact equivalent to it.

Example 1.22 The subset $A = \{x \in \mathbb{Q} \mid x < \sqrt{2}\}$ of the rational numbers $\mathbb{Q}$ is bounded from above by $\sqrt{2}$, but has no supremum in $\mathbb{Q}$. The supremum in $\mathbb{R}$ is the irrational number $\sqrt{2}$. In this example, the supremum of A does not belong to A.

If A does not have an upper bound, we define $\sup A = \infty$, and if A does not have a lower bound, we define $\inf A = -\infty$. The convention that every number is both an upper and a lower bound of the empty set $\emptyset$ is sometimes convenient, so that $\sup \emptyset = -\infty$ and $\inf \emptyset = \infty$.

The supremum of a set A may, or may not, belong to A itself. If it does, then $\sup A$ is called the *maximum* of A, and is also denoted by $\max A$. Similarly, if the infimum belongs to A, then $\inf A$ is called the *minimum* of A, and is also denoted by $\min A$. The illustration in Figure 1.4 shows an example.

Thus, provided we allow the values $\pm\infty$, every set of real numbers has a supremum and an infimum, but it does not necessarily have a maximum or a minimum.

Next, we define the lim inf and lim sup of a real sequence. First, we consider monotone sequences. A sequence (x_n) is said to be *monotone increasing* if $x_n \leq x_{n+1}$, for every n, and *monotone decreasing* if $x_n \geq x_{n+1}$, for every n. A *monotone* sequence is a sequence that is monotone increasing or monotone decreasing. A monotone increasing sequence converges to its supremum (which could be ∞), and a monotone decreasing sequence converges to its infimum (which could be $-\infty$). Thus, provided that we allow for convergence to $\pm\infty$, all monotone sequences converge.

Now suppose that (x_n) is an arbitrary sequence of real numbers. We construct a new sequence (y_n) by taking the supremum of successively truncated "tails" of the original sequence, $y_n = \sup \{x_k \mid k \geq n\}$. The sequence (y_n) is monotone decreasing because the supremum is taken over smaller sets for larger n's. Therefore, the sequence (y_n) has a limit, which we call the lim sup of the sequence (x_n), and denote by $\limsup x_n$. Similarly, taking the infimum of the successively truncated "tails" of (x_n), we get a monotone increasing sequence. We call the limit of that sequence, the lim inf of (x_n), and denote it by $\liminf x_n$. Thus, we have the following definition.

Definition 1.23 Let (x_n) be a sequence of real numbers. Then

$$\limsup_{n\to\infty} x_n = \lim_{n\to\infty} \left[\sup \{x_k \mid k \geq n\}\right],$$
$$\liminf_{n\to\infty} x_n = \lim_{n\to\infty} \left[\inf \{x_k \mid k \geq n\}\right].$$

Another common notation for the lim sup and lim inf is

$$\limsup x_n = \overline{\lim} x_n, \qquad \liminf x_n = \underline{\lim} x_n.$$

We make the natural convention that if

$$\sup \{x_k \mid k \geq n\} = \infty, \quad \text{or} \quad \inf \{x_k \mid k \geq n\} = -\infty,$$

for every n, then $\limsup x_n = \infty$, or $\liminf x_n = -\infty$, respectively. In contrast to the limit, the lim inf and lim sup of a sequence of real numbers always exist, provided that we allow the values $\pm\infty$. The lim sup of a sequence whose terms are bounded from above is finite or $-\infty$, and the lim inf of a sequence whose terms are bounded from below is finite or ∞.

It follows from the definition that

$$\liminf_{n\to\infty} x_n \le \limsup_{n\to\infty} x_n.$$

Moreover, a sequence (x_n) converges if and only if

$$\liminf_{n\to\infty} x_n = \limsup_{n\to\infty} x_n,$$

and, in that case, the limit is the common value of $\liminf x_n$ and $\limsup x_n$.

Example 1.24 If $x_n = (-1)^n$, then

$$\liminf_{n\to\infty} x_n = -1, \qquad \limsup_{n\to\infty} x_n = 1.$$

The $\liminf$ and $\limsup$ have different values and the sequence does not have a limit.

Example 1.25 If $\{x_{n,\alpha} \in \mathbb{R} \mid n \in \mathbb{N}, \alpha \in \mathcal{A}\}$ is a set of real numbers indexed by the natural numbers $\mathbb{N}$ and an arbitrary set $\mathcal{A}$, then

$$\sup_{\alpha \in \mathcal{A}} \left[\liminf_{n\to\infty} x_{n,\alpha} \right] \le \liminf_{n\to\infty} \left[\sup_{\alpha \in \mathcal{A}} x_{n,\alpha} \right].$$

See Exercise 1.10 for the proof, and the analogous inequality with inf and lim sup.

Suppose that A is a nonempty subset of a general metric space X. The *diameter* of A is

$$\operatorname{diam} A = \sup\{d(x,y) \mid x,y \in A\}.$$

The set A is *bounded* if its diameter is finite. It follows that A is bounded if and only if there is an $M \in \mathbb{R}$ and an $x_0 \in X$ such that $d(x_0, x) \le M$ for all $x \in A$. The *distance* $d(x, A)$ of a point $x \in X$ from the set A is defined by

$$d(x, A) = \inf\{d(x,y) \mid y \in A\}.$$

The statement $d(x, A) = 0$ does not imply that $x \in A$.

We say that a function $f : X \to Y$ is *bounded* if its range $f(X)$ is bounded. For example, a real-valued function $f : X \to \mathbb{R}$ is bounded if there is a finite number M such that $|f(x)| \le M$ for all $x \in X$. We say that $f : X \to \mathbb{R}$ is *bounded from above* if there is an $M \in \mathbb{R}$ such that $f(x) \le M$ for all $x \in X$, and *bounded from below* if there is an $M \in \mathbb{R}$ such that $f(x) \ge M$ for all $x \in X$.

1.4 Continuity

A real function $f : \mathbb{R} \to \mathbb{R}$ is *continuous* at a point $x_0 \in \mathbb{R}$ if for every $\epsilon > 0$ there is a $\delta > 0$ such that $|x - x_0| < \delta$ implies $|f(x) - f(x_0)| < \epsilon$. Thus, continuity of f at x_0 is the property that the value of f at a point close to x_0 is close to the value of f at x_0. The definition of continuity for functions between metric spaces is an obvious generalization of the definition for real functions. Let (X, d_X) and (Y, d_Y) be two metric spaces.

Definition 1.26 A function $f : X \to Y$ is *continuous* at $x_0 \in X$ if for every $\epsilon > 0$ there is a $\delta > 0$ such that $d_X(x, x_0) < \delta$ implies $d_Y(f(x), f(x_0)) < \epsilon$. The function f is *continuous on* X if it is continuous at every point in X.

If f is not continuous at x, then we say that f is *discontinuous* at x. There are continuous functions on any metric space. For example, every constant function is continuous.

Example 1.27 Let $a \in X$, and define $f : X \to \mathbb{R}$ by $f(x) = d(x, a)$. Then f is continuous on X.

We can also define continuity in terms of limits. If $f : X \to Y$, we say that $f(x) \to y_0$ as $x \to x_0$, or

$$\lim_{x \to x_0} f(x) = y_0,$$

if for every $\epsilon > 0$ there is a $\delta > 0$ such that $0 < d_X(x, x_0) < \delta$ implies that $d_Y(f(x), y_0) < \epsilon$. More generally, if $f : D \subset X \to Y$ has domain D, and x_0 is a limit of points in D, then we say $f(x) \to y_0$ as $x \to x_0$ in D if for every $\epsilon > 0$ there is a $\delta > 0$ such that $0 < d_X(x, x_0) < \delta$ and $x \in D$ implies that $d_Y(f(x), y_0) < \epsilon$. A function $f : X \to Y$ is continuous at $x_0 \in X$ if

$$\lim_{x \to x_0} f(x) = f(x_0),$$

meaning that the limit of $f(x)$ as $x \to x_0$ exists and is equal to the value of f at x_0.

Example 1.28 If $f : (0, a) \to Y$ for some $a > 0$, and $f(x) \to L$ as $x \to 0$, then we write

$$\lim_{x \to 0^+} f(x) = L.$$

Similarly, if $f : (-a, 0) \to Y$, and $f(x) \to L$ as $x \to 0$, then we write

$$\lim_{x \to 0^-} f(x) = L.$$

If $f : X \to Y$ and E is a subset of X, then we say that f is *continuous on* E if it is continuous at every point $x \in E$. This property is, in general, not equivalent to the continuity of the restriction $f|_E$ of f on E.

Example 1.29 Let $f : \mathbb{R} \to \mathbb{R}$ be the characteristic function of the rationals, which is defined by

$$f(x) = \begin{cases} 1 & \text{if } x \in \mathbb{Q}, \\ 0 & \text{if } x \notin \mathbb{Q}. \end{cases}$$

The function f is discontinuous at every point of $\mathbb{R}$, but $f|_{\mathbb{Q}} : \mathbb{Q} \to \mathbb{R}$ is the constant function $f|_{\mathbb{Q}}(x) = 1$, so $f|_{\mathbb{Q}}$ is continuous on $\mathbb{Q}$.

A subtle, but important, strengthening of continuity is *uniform continuity*.

Definition 1.30 A function $f : X \to Y$ is *uniformly continuous* on X if for every $\epsilon > 0$ there is a $\delta > 0$ such that $d_X(x, y) < \delta$ implies $d_Y(f(x), f(y)) < \epsilon$ for all $x, y \in X$.

The crucial difference between Definition 1.30 and Definition 1.26 is that the value of δ does not depend on the point $x \in X$, so that $f(y)$ gets closer to $f(x)$ at a uniform rate as y gets closer to x.

In the following, we will denote all metrics by d when it is clear from the context which metric is meant.

Example 1.31 The function $r : (0, 1) \to \mathbb{R}$ defined by $r(x) = 1/x$ is continuous on $(0, 1)$ but not uniformly continuous. The function $s : \mathbb{R} \to \mathbb{R}$ defined by $s(x) = x^2$ is continuous on $\mathbb{R}$ but not uniformly continuous. If $[a, b]$ is any bounded interval, then $s|_{[a,b]}$ is uniformly continuous on $[a, b]$.

Example 1.32 A function $f : \mathbb{R}^n \to \mathbb{R}^m$ is *affine* if

$$f(tx + (1 - t)y) = tf(x) + (1 - t)f(y) \qquad \text{for all } x, y \in \mathbb{R}^n \text{ and } t \in [0, 1].$$

Every affine function is uniformly continuous. An affine function f can be written in the form $f(x) = Ax + b$, where A is a constant $m \times n$ matrix and b is a constant m-vector. Affine functions are more general than linear functions, for which $b = 0$.

There is a useful equivalent way to characterize continuous functions on metric spaces in terms of sequences.

Definition 1.33 A function $f : X \to Y$ is *sequentially continuous* at $x \in X$ if for every sequence (x_n) that converges to x in X, the sequence $(f(x_n))$ converges to $f(x)$ in Y.

Proposition 1.34 Let X, Y be metric spaces. A function $f : X \to Y$ is continuous at x if and only if it is sequentially continuous at x.

Proof. First, we show that if f is continuous, then it is sequentially continuous. Suppose that f is continuous at x, and $x_n \to x$. Let $\epsilon > 0$ be given. By the continuity of f, we can choose $\delta > 0$ so that $d(x, x_n) < \delta$ implies $d(f(x), f(x_n)) < \epsilon$. By the convergence of (x_n), we can choose N so that $n \geq N$ implies $d(x, x_n) < \delta$. Therefore, $n \geq N$ implies $d(f(x), f(x_n)) < \epsilon$, and $f(x_n) \to f(x)$.

To prove the converse, we show that if f is discontinuous, then it is not sequentially continuous. If f is discontinuous at x, then there is an $\epsilon > 0$ such that for every $n \in \mathbb{N}$ there exists $x_n \in X$ with $d(x, x_n) < 1/n$ and $d(f(x), f(x_n)) \geq \epsilon$. The sequence (x_n) converges to x but $(f(x_n))$ does not converge to $f(x)$. $\square$

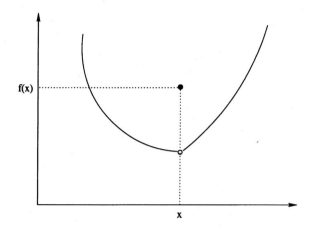

Fig. 1.5 The function f is upper semicontinuous, but not continuous, at the point x.

There are two kinds of "half-continuous" real-valued functions, defined as follows.

Definition 1.35 A function $f : X \to \mathbb{R}$ is *upper semicontinuous* on X if for all $x \in X$ and every sequence $x_n \to x$, we have

$$\limsup_{n \to \infty} f(x_n) \leq f(x).$$

A function f is *lower semicontinuous* on X if for all $x \in X$ and every sequence $x_n \to x$, we have

$$\liminf_{n \to \infty} f(x_n) \geq f(x).$$

The definition is illustrated in Figure 1.5. A function $f : X \to \mathbb{R}$ is continuous if and only if it is upper and lower semicontinuous.

1.5   Open and closed sets

Open sets provide another way to formulate the concepts of convergence and continuity. In this section, we define open sets in a metric space. We will discuss open sets in the more general context of topological spaces in Chapter 4.

Let (X, d) be a metric space. The *open ball*, $B_r(a)$, with radius $r > 0$ and center $a \in X$ is the set

$$B_r(a) = \{x \in X \mid d(x, a) < r\}.$$

The *closed ball*, $\overline{B}_r(a)$, is the set

$$\overline{B}_r(a) = \{x \in X \mid d(x, a) \leq r\}.$$

Definition 1.36 A subset G of a metric space X is *open* if for every $x \in G$ there is an $r > 0$ such that $B_r(x)$ is contained in G. A subset F of X is *closed* if its complement $F^c = X \setminus F$ is open.

For example, an open ball is an open set, and a closed ball is a closed set. The following properties of open and closed sets are easy to prove from the definition.

Proposition 1.37 Let X be a metric space.

    (a) The empty set $\emptyset$ and the whole set X are open and closed.
    (b) A finite intersection of open sets is open.
    (c) An arbitrary union of open sets is open.
    (d) A finite union of closed sets is closed.
    (e) An arbitrary intersection of closed sets is closed.

Example 1.38 The interval $I_n = (-1/n, 1)$ is open in $\mathbb{R}$ for every $n \in \mathbb{N}$, but the intersection

$$\bigcap_{n=1}^{\infty} I_n = [0, 1)$$

is not open. Thus, an infinite intersection of open sets need not be open.

Example 1.39 Let $\{q_n \mid n \in \mathbb{N}\}$ be an enumeration of the rational numbers $\mathbb{Q}$, and $\epsilon > 0$. We define the open interval I_n in $\mathbb{R}$ by

$$I_n = \left(q_n - \frac{\epsilon}{2^n}, q_n + \frac{\epsilon}{2^n} \right).$$

Then $G = \bigcup_{n=1}^{\infty} I_n$ is an open set which contains $\mathbb{Q}$. The sum of the lengths of the intervals I_n is 2ϵ, which can be made as small as we wish. Nevertheless, every interval in $\mathbb{R}$ contains infinitely many rational numbers, and therefore infinitely many intervals I_n.

A subset of $\mathbb{R}$ has *Lebesgue measure zero* if for every $\epsilon > 0$ there is a countable collection of open intervals whose union contains the subset such that the sum of the lengths of the intervals is less than ϵ. Thus, the previous example shows that the set of rational numbers $\mathbb{Q}$, or any other countable subset of $\mathbb{R}$, has measure zero. A property which holds everywhere except on a set of measure zero is said to hold *almost everywhere*, abbreviated *a.e.* For example, the function $\chi_{\mathbb{Q}} : \mathbb{R} \to \mathbb{R}$ that is one on the rational numbers and zero on the irrational numbers is zero almost everywhere.

Every open set in $\mathbb{R}$ is a countable union of disjoint open intervals. The structure of open sets in $\mathbb{R}^n$ for $n \geq 2$ may be much more complicated.

Example 1.40 We define a closed set F_1 in $\mathbb{R}$ by removing the "middle third" $(1/3, 2/3)$ of the interval $[0,1]$. That is,

$$F_1 = [0, 1/3] \cup [2/3, 1].$$

We define F_2 by removing the middle thirds of the intervals in F_1, so that

$$F_2 = [0, 1/9] \cup [2/9, 1/3] \cup [2/3, 7/9] \cup [8/9, 1].$$

Continuing this removal of middle thirds, we obtain a nested sequence of closed sets (F_n). The intersection $F = \bigcap_{n=1}^{\infty} F_n$ is a closed set called the *Cantor set*. A number $x \in [0,1]$ belongs to the Cantor set if and only if it has a base three expansion that contains no 1's. The endpoints of the closed intervals in the F_n's do not have a unique expansion. For example, we can write $1/3 \in F$ in base three as $0.1000\ldots$ and as $0.0222\ldots$. The Cantor set is an uncountable set of Lebesgue measure zero which contains no open intervals, and is a simple example of a *fractal*. Heuristically, any part of the set — for example, the left part contained in the interval $[0, 1/3]$ — is a scaled version of the whole set. The name fractal refers to the fact that, with a suitable definition of the Hausdorff dimension of a set, the Cantor set has a fractional dimension of $\log 2 / \log 3 \approx 0.631$. The Hausdorff dimension of the Cantor set lies between that of a point, which has dimension 0, and an interval, which has dimension 1.

Closed sets in a metric space can be given an alternative, sequential characterization as sets that contain their limit points.

Proposition 1.41 A subset F of a metric space is closed if and only if every convergent sequence of elements in F converges to a limit in F. That is, if $x_n \to x$ and $x_n \in F$ for all n, then $x \in F$.

Example 1.42 A subset of a complete metric space is complete if and only if it is closed.

The *closure* $\overline{A}$ of a set $A \subset X$ is the smallest closed set containing A. From property (e) of Proposition 1.37, the closure $\overline{A}$ is the intersection of all closed sets that contain A. In a metric space, the closure of a set A can also be obtained by adding to A all limits of convergent sequences of elements of A. That is,

$$\overline{A} = \{x \in X \mid \text{there exist } a_n \in A \text{ such that } a_n \to x\}. \tag{1.6}$$

The closure of the set of rational numbers $\mathbb{Q}$ in the space of real numbers $\mathbb{R}$ is the whole space $\mathbb{R}$. Sets with this property are said to be dense.

Definition 1.43 A subset A of a metric space X is *dense* in X if $\overline{A} = X$.

It follows from (1.6) that A is a dense subset of the metric space X if and only if for every $x \in X$ there is a sequence (a_n) in A such that $a_n \to x$. Thus, every point in X can be approximated arbitrarily closely by points in the dense set A. We will encounter many dense sets later on. Theorem 2.9, the Weierstrass approximation theorem, gives one example.

Definition 1.44 A metric space is *separable* if it has a countable dense subset.

For example, $\mathbb{R}$ with its usual metric is separable because $\mathbb{Q}$ is a countable dense subset. On the other hand, $\mathbb{R}$ with the discrete metric $d(x, y) = 1$ when $x \neq y$ is not separable.

Definition 1.45 Let x be a point in a metric space X. A set $U \subset X$ is a *neighborhood* of x if there is an open set $G \subset U$ with $x \in G$.

Equivalently, a set U is a neighborhood of x if U contains a ball $B_r(x)$ centered at x for some $r > 0$. Definition 1.16 for the convergence of a sequence can therefore be rephrased in the following way. A sequence (x_n) *converges* to x if for every neighborhood U of x there is an $N \in \mathbb{N}$ such that $x_n \in U$ for all $n \geq N$.

The following proposition characterizes continuous functions as functions that "pull back" open sets to open sets.

Proposition 1.46 Let X, Y be metric spaces and $f : X \to Y$. The function f is continuous on X if and only if $f^{-1}(G)$ is open in X for every open set G in Y.

Proof. Suppose that f is continuous and $G \subset Y$ is open. If $a \in f^{-1}(G)$, then there is a $b \in G$ with $b = f(a)$. Since G is open, there is an $\epsilon > 0$ with $B_\epsilon(b) \subset G$. Since f is continuous, there is a $\delta > 0$ such that $d(x, a) < \delta$ implies $d(f(x), b) < \epsilon$. It follows that $B_\delta(a) \subset f^{-1}(G)$, so $f^{-1}(G)$ is open.

Conversely, suppose that f is discontinuous at some point a in X. Then there is an $\epsilon > 0$ such that for every $\delta > 0$, there is an $x \in X$ with $d(x, a) < \delta$ and $d(f(x), f(a)) \geq \epsilon$. It follows that, although a belongs to the inverse image of the open set $B_\epsilon(f(a))$ under f, the inverse image does not contain $B_\delta(a)$ for any $\delta > 0$, so it is not open. $\square$

Example 1.47 If $s : \mathbb{R} \to \mathbb{R}$ is the function $s(x) = x^2$, then $s^{-1}((-4, 4)) = (-2, 2)$ is open, as required by continuity. On the other hand, $s((-2, 2)) = [0, 4)$ is not open. Thus, continuous functions need not map open sets to open sets.

1.6 The completion of a metric space

Working with incomplete metric spaces is very inconvenient. For example, suppose we wish to solve an equation for which we cannot write an explicit expression for the solution. We may instead construct a sequence (x_n) of approximate solutions, for example, by use of an iterative method or some kind of numerical scheme. If the approximate solutions get closer and closer together with increasing n, meaning that they form a Cauchy sequence in a metric space, then we would like to conclude that the approximate solutions have a limit, and then try to show that the limit is a solution. We cannot do this unless the metric space in which the approximations lie is complete.

In this section we explain how to extend an incomplete metric space X to a larger, complete metric space, called the completion of X. We construct the completion of X as a set of equivalence classes of Cauchy sequences in X which "ought" to converge to the same point. For a brief review of equivalence relations and equivalence classes, see Exercise 1.22. A point $x \in X$ is naturally identified with the class of Cauchy sequences in X that converge to x, while classes of Cauchy sequences that do not converge in X correspond to new points in the completion. In effect, we construct the completion by filling the "holes" in X that are detected by its Cauchy sequences.

Example 1.48 The completion of the set of rational numbers $\mathbb{Q}$ is the set of real numbers $\mathbb{R}$. A real number x is identified with the equivalence class of rational Cauchy sequences that converge to x. When we write a real number in decimal notation, we give a Cauchy sequence of rational numbers that converges to it.

In order to give a formal definition of the completion, we require the notion of an isometry between two metric spaces (X, d_X) and (Y, d_Y).

Definition 1.49 A map $\imath : X \to Y$ which satisfies

$$d_Y(\imath(x_1), \imath(x_2)) = d_X(x_1, x_2) \tag{1.7}$$

for all $x_1, x_2 \in X$ is called an *isometry* or an *isometric embedding* of X into Y. An isometry which is onto is called a *metric space isomorphism*, or an *isomorphism* when it is clear from the context that we are dealing with metric spaces. Two metric spaces X and Y are *isomorphic* if there is an isomorphism $\imath : X \to Y$.

Equation (1.7) implies that an isometry $\imath$ is one-to-one and continuous. We think of $\imath$ as "identifying" a point $x \in X$ with its image $\imath(x) \in Y$, so that $\imath(X)$ is a "copy" of X embedded in Y. Two isomorphic metric spaces are indistinguishable as metric spaces, although they may differ in other ways.

Example 1.50 The map $\imath : \mathbb{C} \to \mathbb{R}^2$ defined by $\imath(x + iy) = (x, y)$ is a metric space isomorphism between the complex numbers $(\mathbb{C}, |\cdot|)$ and the Euclidean plane $(\mathbb{R}^2, \|\cdot\|)$. In fact, since $\imath$ is linear, the spaces $\mathbb{C}$ and $\mathbb{R}^2$ are isomorphic as real normed linear spaces.

We can now define the completion of a metric space. The example of the real and rational numbers is helpful to keep in mind while reading this definition.

Definition 1.51 A metric space $(\widetilde{X}, \widetilde{d})$ is called the *completion* of (X, d) if the following conditions are satisfied:

(a) there is an isometric embedding $\imath : X \to \widetilde{X}$;
(b) the image space $\imath(X)$ is dense in $\widetilde{X}$;
(c) the space $(\widetilde{X}, \widetilde{d})$ is complete.

The main theorem about the completion of metric spaces is the following.

Theorem 1.52 Every metric space has a completion. The completion is unique up to isomorphism.

Proof. First, we prove that the completion is unique up to isomorphism, if it exists. Suppose that $(\widetilde{X}_1, \widetilde{d}_1)$ and $(\widetilde{X}_2, \widetilde{d}_2)$ are two completions of (X, d), with corresponding isometric embeddings $\imath_1 : X \to \widetilde{X}_1$ and $\imath_2 : X \to \widetilde{X}_2$. We will use $\imath_1$ to extend $\imath_2$ from X to the completion $\widetilde{X}_1$ and obtain an isomorphism $\widetilde{\imath} : \widetilde{X}_1 \to \widetilde{X}_2$.

To define $\widetilde{\imath}$ on $\widetilde{x} \in \widetilde{X}_1$, we pick a sequence (x_n) in X such that $(\imath_1(x_n))$ converges to $\widetilde{x}$ in $\widetilde{X}_1$. Such a sequence exists because $\imath_1(X)$ is dense in $\widetilde{X}_1$. The sequence $(\imath_1(x_n))$ is Cauchy because it converges. Since $\imath_1$ and $\imath_2$ are isometries, it follows that (x_n) and $(\imath_2(x_n))$ are also Cauchy. The space $\widetilde{X}_2$ is complete, hence $(\imath_2(x_n))$ converges in $\widetilde{X}_2$. We define

$$\widetilde{\imath}(\widetilde{x}) = \lim_{n \to \infty} \imath_2(x_n). \tag{1.8}$$

If (x_n') is another sequence in X such that $(\imath_1(x_n'))$ converges to $\widetilde{x}$ in $\widetilde{X}_1$, then

$$\widetilde{d}_2\left(\imath_2(x_n'), \imath_2(x_n)\right) = d\left(x_n', x_n\right) = \widetilde{d}_1\left(\imath_1(x_n'), \imath_1(x_n)\right) \to 0$$

as $n \to \infty$. Thus, $(\imath_2(x_n'))$ and $(\imath_2(x_n))$ converge to the same limit, and $\widetilde{\imath}(\widetilde{x})$ is well-defined.

If $\widetilde{x}$, $\widetilde{y}$ belong to $\widetilde{X}_2$, and

$$\widetilde{\imath}(\widetilde{x}) = \lim_{n \to \infty} \imath_2(x_n), \qquad \widetilde{\imath}(\widetilde{y}) = \lim_{n \to \infty} \imath_2(y_n),$$

then

$$\widetilde{d}_2\left(\widetilde{\imath}(\widetilde{x}), \widetilde{\imath}(\widetilde{y})\right) = \lim_{n \to \infty} \widetilde{d}_2(\imath_2(x_n), \imath_2(y_n)) = \lim_{n \to \infty} d(x_n, y_n) = \widetilde{d}_1(\widetilde{x}, \widetilde{y}).$$

Therefore $\widetilde{\imath}$ is an isometry of $\widetilde{X}_1$ into $\widetilde{X}_2$. By using constant sequences in X, we see that $\widetilde{\imath} \circ \imath_1(x) = \imath_2(x)$ for all $x \in X$, so that $\widetilde{\imath}$ identifies the image of X in $\widetilde{X}_1$ under $\imath_1$ with the image of X in $\widetilde{X}_2$ under $\imath_2$.

To show that $\widetilde{\imath}$ is onto, we observe that $\widetilde{X}_1$ contains the limit of all Cauchy sequences in $\imath_1(X)$, so the isomorphic space $\widetilde{\imath}(\widetilde{X}_1)$ contains the limit of all Cauchy sequences in $\imath_2(X)$. Therefore $\overline{\imath_2(X)} \subset \widetilde{\imath}(\widetilde{X}_1)$. By assumption, $\imath_2(X)$ is dense in $\widetilde{X}_2$, so $\overline{\imath_2(X)} = \widetilde{X}_2$, and $\widetilde{\imath}(\widetilde{X}_1) = \widetilde{X}_2$. This shows that any two completions are isomorphic.

Second, we prove the completion exists. To do this, we construct a completion from Cauchy sequences in X. We define a relation $\sim$ between Cauchy sequences $x = (x_n)$ and $y = (y_n)$ in X by

$$x \sim y \quad \text{if and only if} \quad \lim_{n\to\infty} d(x_n, y_n) = 0.$$

Two convergent Cauchy sequences x, y satisfy $x \sim y$ if and only if they have the same limit. It is straightforward to check that $\sim$ is an equivalence relation on the set $\mathcal{C}$ of Cauchy sequences in X. Let $\widetilde{X}$ be the set of equivalence classes of $\sim$ in $\mathcal{C}$. We call an element $(x_n) \in \widetilde{x}$ of an equivalence class $\widetilde{x} \in \widetilde{X}$, a *representative* of $\widetilde{x}$.

We define $\widetilde{d} : \widetilde{X} \times \widetilde{X} \to \mathbb{R}$ by

$$\widetilde{d}(\widetilde{x}, \widetilde{y}) = \lim_{n\to\infty} d(x_n, y_n), \tag{1.9}$$

where (x_n) and (y_n) are any two representatives of $\widetilde{x}$ and $\widetilde{y}$, respectively. The limit in (1.9) exists because $(d(x_n, y_n))_{n=1}^{\infty}$ is a Cauchy sequence of real numbers. For this definition to make sense, it is essential that the limit is independent of which representatives of $\widetilde{x}$ and $\widetilde{y}$ are chosen. Suppose that (x_n), (x_n') represent $\widetilde{x}$ and (y_n), (y_n') represent $\widetilde{y}$. Then, by the triangle inequality, we have

$$\begin{aligned} d(x_n, y_n) &\leq d(x_n, x_n') + d(x_n', y_n') + d(y_n', y_n), \\ d(x_n, y_n) &\geq d(x_n', y_n') - d(x_n, x_n') - d(y_n', y_n). \end{aligned}$$

Taking the limit as $n \to \infty$ of these inequalities, and using the assumption that $(x_n) \sim (x_n')$ and $(y_n) \sim (y_n')$, we find that

$$\lim_{n\to\infty} d(x_n, y_n) = \lim_{n\to\infty} d(x_n', y_n').$$

Thus, the limit in (1.9) is independent of the representatives, and $\widetilde{d}$ is well-defined. It is straightforward to check that $\widetilde{d}$ is a metric on $\widetilde{X}$.

To show that the metric space $(\widetilde{X}, \widetilde{d})$ is a completion of (X, d), we define an embedding $\imath : X \to \widetilde{X}$ as the map that takes a point $x \in X$ to the equivalence class of Cauchy sequences that contains the constant sequence (x_n) with $x_n = x$ for all n. This map is an isometric embedding, since if (x_n) and (y_n) are the constant sequences with $x_n = x$ and $y_n = y$, we have

$$\widetilde{d}(\imath(x), \imath(y)) = \lim_{n\to\infty} d(x_n, y_n) = d(x, y).$$

The image $\imath(X)$ consists of the equivalence classes in $\widetilde{X}$ which have a constant representative Cauchy sequence. To show the density of $\imath(X)$ in $\widetilde{X}$, let (x_n) be a representative of an arbitrary point $\widetilde{x} \in \widetilde{X}$. We define a sequence $(\widetilde{y}_n)$ of constant sequences by $\widetilde{y}_n = (y_{n,k})_{k=1}^{\infty}$ where $y_{n,k} = x_n$ for all $n, k \in \mathbb{N}$. From the definition of $(\widetilde{y}_n)$ and the fact that (x_n) is a Cauchy sequence, we have

$$\lim_{n\to\infty} \widetilde{d}(\widetilde{y}_n, \widetilde{x}) = \lim_{n\to\infty} \lim_{k\to\infty} d(x_n, x_k) = 0.$$

Thus, $\imath(X)$ is dense in $\widetilde{X}$.

Finally, we prove that $(\widetilde{X}, \widetilde{d})$ is complete. We will use Cantor's "diagonal" argument, which is useful in many other contexts as well. Let $(\widetilde{x}_n)$ be a Cauchy sequence in $\widetilde{X}$. In order to prove that a Cauchy sequence is convergent, it is enough to prove that it has a convergent subsequence, because the whole sequence converges to the limit of any subsequence. Picking a subsequence, if necessary, we can assume that $(\widetilde{x}_n)$ satisfies

$$\widetilde{d}\,(\widetilde{x}_m, \widetilde{x}_n) \le \frac{1}{N} \quad \text{for all } m, n \ge N. \tag{1.10}$$

For each term $\widetilde{x}_n$, we choose a representative Cauchy sequence in X, denoted by $(x_{n,k})_{k=1}^{\infty}$. Any subsequence of a representative Cauchy sequence of $\widetilde{x}_n$ is also a representative of $\widetilde{x}_n$. We can therefore choose the representative so that

$$d\,(x_{n,k}, x_{n,l}) < \frac{1}{n} \quad \text{for all } k, l \ge n. \tag{1.11}$$

We claim that the "diagonal" sequence $(x_{k,k})_{k=1}^{\infty}$ is a Cauchy sequence, and that the equivalence class $\widetilde{x}$ to which it belongs is the limit of $(\widetilde{x}_n)$ in $\widetilde{X}$. The fact that we can obtain the limit of a Cauchy sequence of sequences by taking a diagonal sequence is the key point in proving the existence of the completion.

To prove that the diagonal sequence is Cauchy, we observe that for any $i \in \mathbb{N}$,

$$d\,(x_{k,k}, x_{l,l}) \le d\,(x_{k,k}, x_{k,i}) + d\,(x_{k,i}, x_{l,i}) + d\,(x_{l,i}, x_{l,l}). \tag{1.12}$$

The definition of $\widetilde{d}$ and (1.10) imply that for all $k, l \ge N$,

$$\widetilde{d}(\widetilde{x}_k, \widetilde{x}) = \lim_{i \to \infty} d(x_{k,i}, x_{l,i}) \le \frac{1}{N}. \tag{1.13}$$

Taking the lim sup of (1.12) as $i \to \infty$, and using (1.11) and (1.13) in the result, we find that for all $k, l \ge N$,

$$d\,(x_{k,k}, x_{l,l}) \le \frac{3}{N}.$$

Therefore $(x_{k,k})$ is Cauchy.

By a similar argument, we find that for all $k, n \ge N$,

$$d\,(x_{n,k}, x_{k,k}) \le \limsup_{i \to \infty} \{ d\,(x_{n,k}, x_{n,i}) + d\,(x_{n,i}, x_{k,i}) + d\,(x_{k,i}, x_{k,k}) \} \le \frac{3}{N}.$$

Therefore, for $n \ge N$, we have

$$\widetilde{d}\,(\widetilde{x}_n, \widetilde{x}) = \lim_{k \to \infty} d(x_{n,k}, x_{k,k}) \le \frac{3}{N}.$$

Hence, the Cauchy sequence $(\widetilde{x}_n)$ converges to $\widetilde{x}$ as $n \to \infty$, and $\widetilde{X}$ is complete. $\square$

It is slightly annoying that the completion $\widetilde{X}$ is constructed as a space of equivalence classes of sequences in X, rather than as a more direct extension of X. For example, if X is a space of functions, then there is no guarantee that its completion can be identified with a space of functions that is obtained by adding more functions to the original space.

Example 1.53 Let $C([0,1])$ be the set of continuous functions $f : [0,1] \to \mathbb{R}$. We define the L^2-norm of f by

$$\|f\| = \left(\int_0^1 |f(x)|^2 \, dx \right)^{1/2}.$$

The associated metric $d(f,g) = \|f - g\|$ is a very useful one, analogous to the Euclidean metric on $\mathbb{R}^n$, but the space $C([0,1])$ is not complete with respect to it. The completion is denoted by $L^2([0,1])$, and it can nearly be identified with the space of Lebesgue measurable, square-integrable functions. More precisely, a point in $L^2([0,1])$ can be identified with an equivalence class of square-integrable functions, in which two functions that differ on a set of Lebesgue measure zero are equivalent. According to the Riesz-Fisher theorem, if (f_n) is a Cauchy sequence with respect to the L^2-norm, then there is a subsequence (f_{n_k}) that converges pointwise-a.e. to a square-integrable function, and this fact provides one way to identify an element of the completion with an equivalence class of functions. Many of the usual operations on functions can be defined on equivalence classes, independently of which representative function is chosen, but the pointwise value of an element $f \in L^2([0,1])$ cannot be defined unambiguously.

In a similar way, the space $L^2(\mathbb{R})$ of equivalence classes of Lebesgue measurable, square integrable functions on $\mathbb{R}$ is the completion of the space $C_c(\mathbb{R})$ of continuous functions on $\mathbb{R}$ with compact support (see Definition 2.6) with respect to the L^2-norm

$$\|f\| = \left(\int_{\mathbb{R}} |f(x)|^2 \, dx \right)^{1/2}.$$

We will see later on that these L^2 spaces are fundamental examples of infinite-dimensional Hilbert spaces. We discuss measure theory in greater detail in Chapter 12. We will use facts from that chapter as needed throughout the book, including Fubini's theorem for the exchange in the order of integration, and the dominated convergence theorem for passage to the limit under an integral sign.

1.7 Compactness

Compactness is one the most important concepts in analysis. A simple and useful way to define compact sets in a metric space is by means of sequences.

Definition 1.54 A subset K of a metric space X is *sequentially compact* if every sequence in K has a convergent subsequence whose limit belongs to K.

We can take $K = X$ in this definition, so that X is sequentially compact if every sequence in X has a convergent subsequence. A subset K of (X, d) is sequentially compact if and only if the metric subspace $(K, d|_K)$ is sequentially compact.

Example 1.55 The space of real numbers $\mathbb{R}$ is not sequentially compact. For example, the sequence (x_n) with $x_n = n$ has no convergent subsequence because $|x_m - x_n| \geq 1$ for all $m \neq n$. The closed, bounded interval $[0, 1]$ is a sequentially compact subset of $\mathbb{R}$, as we prove below. The half-open interval $(0, 1]$ is not a sequentially compact subset of $\mathbb{R}$, because the sequence $(1/n)$ converges to 0, and therefore has no subsequence with limit in $(0, 1]$. The limit does, however, belong to $[0, 1]$.

The full importance of compact sets will become clear only in the setting of infinite-dimensional normed spaces. It is nevertheless interesting to start with the finite-dimensional case. Compact subsets of $\mathbb{R}^n$ have a simple, explicit characterization.

Theorem 1.56 (Heine-Borel) A subset of $\mathbb{R}^n$ is sequentially compact if and only if it is closed and bounded.

The fact that closed, bounded subsets of $\mathbb{R}^n$ are sequentially compact is a consequence of the following theorem, called the Bolzano-Weierstrass theorem, even though Bolzano had little to do with its proof. We leave it to the reader to use this theorem to complete the proof of the Heine-Borel theorem.

Theorem 1.57 (Bolzano-Weierstrass) Every bounded sequence in $\mathbb{R}^n$ has a convergent subsequence.

Proof. We will construct a Cauchy subsequence from an arbitrary bounded sequence. Since $\mathbb{R}^n$ is complete, the subsequence converges.

Let (x_k) be a bounded sequence in $\mathbb{R}^n$. There is an $M > 0$ such that $x_k \in [-M, M]^n$ for all k. The set $[-M, M]^n$ is an n-dimensional cube of side $2M$. We denote this cube by C_0. We partition C_0 into 2^n cubes of side M. We denote by C_1 one of the smaller cubes that contains infinitely many terms of the sequence (x_k), meaning that $x_k \in C_1$ for infinitely many $k \in \mathbb{N}$. Such a cube exists because there is a finite number of cubes and an infinite number of terms in the sequence. Let k_1 be the smallest index such that $x_{k_1} \in C_1$. We pick x_{k_1} as the first term of the subsequence.

To choose the second term, we form a new sequence (y_k) by deleting from (x_k) the term x_{k_1} and all terms which do not belong to C_1. We repeat the procedure described in the previous paragraph, but with (x_k) replaced by (y_k), and C_0 replaced

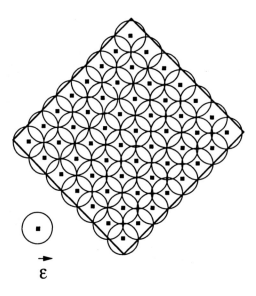

Fig. 1.6   A set with a finite ϵ-net for it.

by C_1. This procedure gives a subcube C_2 of C_1 of side $M/2$, which contains infinitely many terms of the original sequence, and an element y_{k_1}. We pick $x_{k_2} = y_{k_1}$ as the second element of the subsequence.

By repeating this procedure, we obtain a subsequence $(x_{k_i})_{i=1}^{\infty}$. We never "exhaust" the original sequence, because every cube in the construction contains infinitely many terms. We have $x_{k_i} \in C_j$ for all $i \geq j$ where C_j is a cube of side $M/2^{j-1}$. Therefore (x_{k_i}) is a Cauchy sequence, and hence it converges.   □

The following criterion for the sequential compactness of a metric space is often easier to verify than the definition. Let A be a subset of a metric space X. We say that a collection $\{G_\alpha \mid \alpha \in \mathcal{A}\}$ of subsets of X is a *cover* of A if its union contains A, meaning that

$$A \subset \bigcup_{\alpha \in \mathcal{A}} G_\alpha.$$

The number of sets in the cover is not required to be countable. If every G_α in the cover is open, then we say that $\{G_\alpha\}$ is an *open cover* of A.

Let $\epsilon > 0$. A subset $\{x_\alpha \mid \alpha \in \mathcal{A}\}$ of X is called an *ϵ-net* of the subset A if the family of open balls $\{B_\epsilon(x_\alpha) \mid \alpha \in \mathcal{A}\}$ is an open cover of A. If the set $\{x_\alpha\}$ is finite, then we say that $\{x_\alpha\}$ is a *finite ϵ-net* of A (see Figure 1.6).

Definition 1.58 A subset of a metric space is *totally bounded* if it has a finite ϵ-net for every $\epsilon > 0$.

That is, a subset A of a metric space X is totally bounded if for every $\epsilon > 0$ there is a finite set of points $\{x_1, x_2, \ldots, x_n\}$ in X such that $A \subset \bigcup_{i=1}^{n} B_\epsilon(x_i)$.

Theorem 1.59 A subset of a metric space is sequentially compact if and only if it is complete and totally bounded.

Proof. The proof that a complete, totally bounded set K is sequentially compact is the same as the proof of the Bolzano-Weierstrass theorem 1.57. Suppose that (x_n) is a sequence in K. Then, since K is totally bounded, there is a sequence of balls (B_k) such that B_k has radius $1/2^k$ and every intersection $A_k = \bigcap_{i=1}^{k} B_i$ contains infinitely many terms of the sequence. We can therefore choose a subsequence (x_{n_k}) such that $x_{n_k} \in A_k$ for every k. This subsequence is Cauchy, and, since K is complete, it converges.

To prove the converse, we show that a sequentially compact space is complete, and that a space which is not totally bounded is not sequentially compact.

If (x_n) is a Cauchy sequence in a sequentially compact space K, then it has a convergent subsequence. The whole Cauchy sequence converges to the limit of any convergent subsequence. Hence K is complete.

Now suppose that K is not totally bounded. Then there is an $\epsilon > 0$ such that K has no finite ϵ-net. For every finite subset $\{x_1, \ldots, x_n\}$ of K, there is a point $x_{n+1} \in K$ such that $x_{n+1} \notin \bigcup_{i=1}^{n} B_\epsilon(x_i)$. Consequently, we can find an infinite sequence (x_n) in K such that $d(x_m, x_n) \geq \epsilon$ for all $m \neq n$. This sequence does not contain a Cauchy subsequence, and hence has no convergent subsequence. Therefore K is not sequentially compact. $\qquad\square$

Another way to define compactness is in terms of open sets. We say that a cover $\{G_\alpha\}$ of A has a finite subcover if there is a finite subcollection of sets $\{G_{\alpha_1}, \ldots, G_{\alpha_n}\}$ such that $A \subset \bigcup_{i=1}^{n} G_{\alpha_i}$.

Definition 1.60 A subset K of a metric space X is *compact* if every open cover of K has a finite subcover.

Example 1.61 The space of real numbers $\mathbb{R}$ is not compact, since the open cover $\{(n-1, n+1) \mid n \in \mathbb{Z}\}$ of $\mathbb{R}$ has no finite subcover. The half-open interval $(0,1]$ is not compact, since the open cover $\{(1/2n, 2/n) \mid n \in \mathbb{N}\}$ has no finite subcover. If this open cover is extended to an open cover of $[0,1]$, then the extension must contain an open neighborhood of 0. This open neighborhood, together with a finite number of sets from the cover of $(0,1]$, is a finite subcover of $[0,1]$.

For metric spaces, compactness and sequential compactness are equivalent.

Theorem 1.62 A subset of a metric space is compact if and only if it is sequentially compact.

Proof. First, we prove that sequential compactness implies compactness. We will show that an arbitrary open cover of a sequentially compact set has a countable subcover, and that a countable cover has a finite subcover.

Lemma 1.63 A sequentially compact metric space is separable.

Proof. By Theorem 1.59, there is a finite $(1/n)$-net A_n of a sequentially compact space K for every $n \in \mathbb{N}$. Let $A = \bigcup_{n=1}^{\infty} A_n$. Then A is countable, because it is a countable union of finite sets, and A is dense in K by construction. $\square$

Suppose that $\{G_\alpha \mid \alpha \in \mathcal{A}\}$ is an arbitrary open cover of a sequentially compact space K. From Lemma 1.63, the space K has a countable dense subset A. Let $\mathcal{B}$ be the collection of open balls with rational radius and center in A, and let $\mathcal{C}$ be the subcollection of balls in $\mathcal{B}$ that are contained in at least one of the open sets G_α. The collection $\mathcal{B}$ is countable because it is a countable union of countable sets. Hence, the subcollection $\mathcal{C}$ is also countable.

For every $x \in K$, there is a set G_α in the open cover of K with $x \in G_\alpha$. Since G_α is open, there is an $\epsilon > 0$ such that $B_\epsilon(x) \subset G_\alpha$. Since A is dense in K, there is a point $y \in A$ such that $d(x, y) < \epsilon/3$. Then $x \in B_{\epsilon/3}(y)$, and $B_{2\epsilon/3}(y) \subset G_\alpha$. (It may help to draw a picture!) Thus, if q is a rational number with $\epsilon/3 < q < 2\epsilon/3$, then $x \in B_q(y)$ and $B_q(y) \subset G_\alpha$. It follows that $B_q(y) \in \mathcal{C}$, so any point x in K belongs to a ball in $\mathcal{C}$. Hence $\mathcal{C}$ is an open cover of K. For every $B \in \mathcal{C}$, we pick an $\alpha_B \in \mathcal{A}$ such that $B \subset G_{\alpha_B}$. Then $\{G_{\alpha_B} \mid B \in \mathcal{C}\}$ is a countable subcover of K, because $\bigcup_{B \in \mathcal{C}} G_{\alpha_B}$ contains $\bigcup_{B \in \mathcal{C}} B$, which contains K.

We will show by contradiction that a countable open cover has a finite subcover. Suppose that $\{G_n \mid n \in \mathbb{N}\}$ is a countable open cover of a sequentially compact space K that does not have a finite subcover. Then the finite union $\bigcup_{n=1}^{N} G_n$ does not contain K for any N. We can therefore construct a sequence (x_k) in K as follows. We pick a point $x_1 \in K$. Since $\{G_n\}$ covers K, there is an N_1 such that $x_1 \in G_{N_1}$. We pick $x_2 \in K$ such that $x_2 \notin \bigcup_{n=1}^{N_1} G_n$, and choose N_2 such that $x_2 \in G_{N_2}$. Then we pick $x_3 \in K$ such that $x_3 \notin \bigcup_{n=1}^{N_2} G_n$, and so on. Since

$$x_k \in G_{N_k}, \quad \text{and} \quad x_k \notin \bigcup_{n=1}^{N_{k-1}} G_n,$$

the open set G_{N_k} is not equal to G_n for any $n \leq N_{k-1}$. Thus, the sequence (N_k) is strictly increasing, and $N_k \to \infty$ as $k \to \infty$. It follows that, for any n, there is an integer K_n such that $x_k \notin G_n$ when $k \geq K_n$. If $x \in G_n$, then all points of the sequence eventually leave the open neighborhood G_n of x, so no subsequence of (x_k) can converge to x. Since the collection $\{G_n\}$ covers K, the sequence (x_n) has no subsequence that converges to a point of K. This contradicts the sequential compactness of K, and proves that sequential compactness implies compactness.

To prove the converse, we show that if a space is not sequentially compact, then it is not compact. Suppose that K has a sequence (x_n) with no convergent subsequence. Such a sequence must contain an infinite number of distinct points, so we can assume without loss of generality that $x_m \neq x_n$ for $m \neq n$.

Let $x \in K$. If the open ball $B_\epsilon(x)$ contains a point in the sequence that is distinct from x for every $\epsilon > 0$, then x is the limit of a subsequence, which contradicts the assumption that the sequence has no convergent subsequence in K. Hence, there is an $\epsilon_x > 0$ such that the open ball $B_{\epsilon_x}(x)$ contains either no points in the sequence, if x itself does not belong to the sequence, or one point, if x belongs to the sequence.

The collection of open balls $\{B_{\epsilon_x}(x) \mid x \in K\}$ is an open cover of K. Every finite subcollection of n open balls contains at most n terms of the sequence. Since the terms of the sequence are distinct, no finite subcollection covers K. Thus, K has an open cover with no finite subcover, and K is not compact. $\qquad\square$

In future, we will abbreviate "sequentially compact" to "compact" when referring to metric spaces. The following terminology is often convenient.

Definition 1.64 A subset A of a metric space X is *precompact* if its closure in X is compact.

The term "relatively compact" is frequently used instead of "precompact." This definition means that A is precompact if every sequence in A has a convergent subsequence. The limit of the subsequence can be any point in X, and is not required to belong to A. Since compact sets are closed, a set is compact if and only if it is closed and precompact. A subset of a complete metric space is precompact if and only if it is totally bounded.

Example 1.65 A subset of $\mathbb{R}^n$ is precompact if and only if it is bounded.

Continuous functions on compact sets have several nice properties. From Proposition 1.34, continuous functions preserve the convergence of sequences. It follows immediately from Definition 1.54 that continuous functions preserve compactness.

Theorem 1.66 Let $f : K \to Y$ be continuous on K, where K is a compact metric space and Y is any metric space. Then $f(K)$ is compact.

Since compact sets are bounded, continuous functions on a compact set are bounded. Moreover, continuous functions on compact sets are uniformly continuous.

Theorem 1.67 Let $f : K \to Y$ be a continuous function on a compact set K. Then f is uniformly continuous.

Proof. Suppose that f is not uniformly continuous. Then there is an $\epsilon > 0$ such that for all $\delta > 0$, there are $x, y \in X$ with $d(x,y) < \delta$ and $d(f(x), f(y)) \geq \epsilon$. Taking $\delta = 1/n$ for $n \in \mathbb{N}$, we find that there are sequences (x_n) and (y_n) in X such that

$$d(x_n, y_n) < \frac{1}{n}, \qquad d(f(x_n), f(y_n)) \geq \epsilon. \qquad (1.14)$$

Since K is compact there are convergent subsequences of (x_n) and (y_n) which, for simplicity, we again denote by (x_n) and (y_n). From (1.14), the subsequences converge to the same limit, but the sequences $(f(x_n))$ and $(f(y_n))$ either diverge or converge to different limits. This contradicts the continuity of f. $\qquad\square$

1.8   Maxima and minima

Maximum and minimum problems are of central importance in applications. For example, in many physical systems, the equilibrium state is one which minimizes energy or maximizes entropy, and in optimization problems, the desirable state of a system is one which minimizes an appropriate cost function. The mathematical formulation of these problems is the maximization or minimization of a real-valued function f on a state space X. Each point of the state space, which is often a metric space, represents a possible state of the system. The existence of a maximizing, or minimizing, point of f in X may not be at all clear; indeed, such a point may not exist. The following theorem gives sufficient conditions for the existence of maximizing or minimizing points — namely, that the function f is continuous and the state space X is compact. Although these conditions are fundamental, they are too strong to be useful in many applications. We will return to these issues later on.

Theorem 1.68 Let K be a compact metric space and $f : K \to \mathbb{R}$ a continuous, real-valued function. Then f is bounded on K and attains its maximum and minimum. That is, there are points $x, y \in K$ such that

$$f(x) = \inf_{z \in K} f(z), \qquad f(y) = \sup_{z \in K} f(z). \tag{1.15}$$

Proof.   From Theorem 1.66, the image $f(K)$ is a compact subset of $\mathbb{R}$, and therefore f is bounded by the Heine-Borel theorem in Theorem 1.56.

It is enough to prove that f attains its infimum, because the application of this result to $-f$ implies that f attains its supremum. Since f is bounded, it is bounded from below, and the infimum m of f on K is finite. By the definition of the infimum, for each $n \in \mathbb{N}$ there is an $x_n \in K$ such that

$$m \leq f(x_n) < m + \frac{1}{n}.$$

This inequality implies that

$$\lim_{n \to \infty} f(x_n) = m. \tag{1.16}$$

The sequence $(x_n)_{n=1}^{\infty}$ need not converge, but since K is compact the sequence has a convergent subsequence, which we denote by $(x_{n_k})_{k=1}^{\infty}$. We denote the limit of the subsequence by x. Then, since f is continuous, we have from (1.16) that

$$f(x) = \lim_{k \to \infty} f(x_{n_k}) = m.$$

Therefore, f attains its infimum m at x. $\square$

The strategy of this proof is typical of many compactness arguments. We construct a sequence of approximate solutions of our problem, in this case a minimizing sequence (x_n) that satisfies (1.16). We use compactness to extract a convergent sub-sequence, and show that the limit of the convergent subsequence is a solution of our problem, in this case a point where f attains its infimum. The following examples illustrate Theorem 1.68 and some possible behaviors of minimizing sequences.

Example 1.69 The function $f(x) = x^4/4 - x^2/2$ is continuous and bounded on $[-2, 2]$. It attains its minimum at $x = \pm 1$. An example of a minimizing sequence (x_n) is given by $x_n = (-1)^n$. In fact, $f(x_n) = \inf f(x)$ for all n. This minimizing sequence does not converge because its terms jump back and forth between $x = -1$ and $x = 1$. The subsequences (x_{2k+1}) and (x_{2k}) converge, to $x = -1$ and $x = 1$, respectively.

As this example shows, the compactness argument does not imply that a point where f attains its minimum is unique. There are many possible minimizing se-quences, and there may be subsequences of a given minimizing sequence that con-verge to different limits. If, however, the function f attains its minimum at a unique point, then it follows from Exercise 1.27 that every minimizing sequence must converge to that point.

Example 1.70 The function $f(x) = e^{-x}$ is continuous and bounded from below on the noncompact set $\mathbb{R}$. The infimum of f on $\mathbb{R}$ is zero, but f does not attain its infimum. An example of a minimizing sequence (x_n) is given by $x_n = n$. The terms of the minimizing sequence "escape" to infinity, and it has no convergent subsequence.

Example 1.71 The discontinuous function f on the compact set $[0, 1]$ defined by

$$f(x) = \begin{cases} \log x & \text{if } 0 < x \leq 1, \\ 0 & \text{if } x = 0, \end{cases}$$

is not bounded from below. A sequence (x_n) is a minimizing sequence if $x_n \to 0$ as $n \to \infty$. In that case, $f(x_n) \to -\infty$ as $n \to \infty$, but f is discontinuous at the limit point $x = 0$.

Some of the conclusions of Theorem 1.68 still hold for semicontinuous functions. An almost identical proof shows the following result.

Theorem 1.72 Let K be a compact metric space. If $f : K \to \mathbb{R}$ is upper semicon-tinuous, then f is bounded from above and attains its supremum. If $f : K \to \mathbb{R}$ is lower semicontinuous, then f is bounded from below and attains its infimum.

Example 1.73 We define $f, g : [0, 1] \to \mathbb{R}$ by

$$f(x) = \begin{cases} x & \text{if } 0 < x \leq 1, \\ 1 & \text{if } x = 0, \end{cases} \qquad g(x) = \begin{cases} x & \text{if } 0 < x \leq 1, \\ -1 & \text{if } x = 0. \end{cases}$$

The function f is upper semicontinuous, and does not attain its infimum, while g is lower semicontinuous and attains its minimum at $x = 0$.

1.9   References

For introductions to basic real analysis, see Marsden and Hoffman [37] or Rudin
[47]. Simmons [50] gives a clear and accessible discussion of metric, normed, and
topological spaces. For linear algebra, see Halmos [19] and Lax [30]. Two other
books with a similar purpose to this one are Naylor and Sell [40] and Stakgold [52].

1.10   Exercises

Exercise 1.1 A set A is *countably infinite* if there is a one-to-one, onto map from A
to N. A set is *countable* if it is finite or countably infinite, otherwise it is *uncountable*.

(a) Prove that the set $\mathbb{Q}$ of rational numbers is countably infinite.
(b) Prove that the set $\mathbb{R}$ of real numbers is uncountable.

Exercise 1.2 Give an ϵ-δ proof that

$$\sum_{n=0}^{\infty} x^n = \frac{1}{1-x},$$

when $|x| < 1$.

Exercise 1.3 If x, y, z are points in a metric space (X, d), show that

$$d(x,y) \geq |d(x,z) - d(y,z)|.$$

Exercise 1.4 Suppose that (X, d_X) and (Y, d_Y) are metric spaces. Prove that the
Cartesian product $Z = X \times Y$ is a metric space with metric d defined by

$$d(z_1, z_2) = d_X(x_1, x_2) + d_Y(y_1, y_2),$$

where $z_1 = (x_1, y_1)$ and $z_2 = (x_2, y_2)$.

Exercise 1.5 Suppose that $(X, \| \cdot \|)$ is a normed linear space. Prove that (1.2)
and (1.4) define metrics on X.

Exercise 1.6 Starting from the fact that $\mathbb{R}$ equipped with its usual distance function is complete, prove that $\mathbb{R}^n$ equipped with the sum, maximum, or Euclidean
norm is complete.

Exercise 1.7 Show that the series

$$\sum_{n=1}^{\infty} \frac{(-1)^n}{n}$$

is not absolutely convergent. Show that by permuting the terms of this series one
can obtain series with different limits.

Exercise 1.8 Let (x_n) be a sequence of real numbers. A point $c \in \mathbb{R} \cup \{\pm\infty\}$ is called a *cluster point* of (x_n) if there is a convergent subsequence of (x_n) with limit c. Let C denote the set of cluster points of (x_n). Prove that C is closed and

$$\limsup x_n = \max C \quad \text{and} \quad \liminf x_n = \min C.$$

Exercise 1.9 Let (x_n) be a bounded sequence of real numbers.

(a) Prove that for every $\epsilon > 0$ and every $N \in \mathbb{N}$ there are $n_1, n_2 \geq N$, such that

$$\limsup_{n \to \infty} x_n \leq x_{n_1} + \epsilon, \quad x_{n_2} - \epsilon \leq \liminf_{n \to \infty} x_n.$$

(b) Prove that for every $\epsilon > 0$ there is an $N \in \mathbb{N}$ such that

$$x_m \leq \limsup_{n \to \infty} x_n + \epsilon, \quad x_m \geq \liminf_{n \to \infty} x_n - \epsilon$$

for all $m \geq N$.

(c) Prove that (x_n) converges if and only if

$$\liminf_{n \to \infty} x_n = \limsup_{n \to \infty} x_n.$$

Exercise 1.10 Consider a family $\{x_{n,\alpha}\}$ of real numbers indexed by $n \in \mathbb{N}$ and $\alpha \in A$. Prove the following relations:

$$\limsup_{n \to \infty} \left(\inf_\alpha x_{n,\alpha} \right) \leq \inf_\alpha \left(\limsup_{n \to \infty} x_{n,\alpha} \right);$$

$$\sup_\alpha \left(\liminf_{n \to \infty} x_{n,\alpha} \right) \leq \liminf_{n \to \infty} \left(\sup_\alpha x_{n,\alpha} \right).$$

Exercise 1.11 If (x_n) is a sequence of real numbers such that

$$\lim_{n \to \infty} x_n = x,$$

and $a_n \leq x_n \leq b_n$, prove that

$$\limsup_{n \to \infty} a_n \leq x \leq \liminf_{n \to \infty} b_n.$$

Exercise 1.12 Let $(X, d_X), (Y, d_Y)$, and (Z, d_Z) be metric spaces and let $f : X \to Y$, and $g : Y \to Z$ be continuous functions. Show that the composition

$$h = g \circ f : X \to Z,$$

defined by $h(x) = g(f(x))$, is also continuous.

Exercise 1.13 A function $f : \mathbb{R} \to \mathbb{R}$ is said to be *differentiable* at x if the following limit exists and is finite:

$$f'(x) = \lim_{h \to 0} \frac{f(x + h) - f(x)}{h}.$$

(a) Prove that if f is differentiable at x, then f is continuous at x.
(b) Show that the function

$$f(x) = \begin{cases} x^2 \sin\left(1/x^2\right) & \text{if } x \neq 0, \\ 0 & \text{if } x = 0. \end{cases}$$

is differentiable at $x = 0$ but the derivative is not continuous at $x = 0$.
(c) Prove that if f is differentiable at x and has a local maximum or minimum at x, then $f'(x) = 0$.

Exercise 1.14 If $f : [a, b] \to \mathbb{R}$ is continuous on $[a, b]$ and differentiable in (a, b), then prove that there is a a $< \xi < b$ such that

$$f(b) - f(a) = f'(\xi)(b - a).$$

This result is called the *mean value theorem*. Deduce that if $f'(x) = 0$ for all $a < x < b$ then f is a constant function.

Exercise 1.15 Prove that every compact subset of a metric space is closed and bounded. Prove that a closed subset of a compact space is compact.

Exercise 1.16 Suppose that F and G are closed and open subsets of $\mathbb{R}^n$, respectively, such that $F \subset G$. Show that there is a continuous function $f : \mathbb{R}^n \to \mathbb{R}$ such that:

(a) $0 \leq f(x) \leq 1$;
(b) $f(x) = 1$ for $x \in F$;
(c) $f(x) = 0$ for $x \in G^c$.

HINT. Consider the function

$$f(x) = \frac{d(x, G^c)}{d(x, G^c) + d(x, F)}.$$

This result is called *Urysohn's lemma*.

Exercise 1.17 Let (X, d) be a complete metric space, and $Y \subset X$. Prove that (Y, d) is complete if and only if Y is a closed subset of X.

Exercise 1.18 Let (X, d) be a metric space, and let (x_n) be a sequence in X. Prove that if (x_n) has a Cauchy subsequence, then, for any decreasing sequence of positive $\epsilon_k \to 0$, there is a subsequence (x_{n_k}) of (x_n) such that

$$d(x_{n_k}, x_{n_l}) \leq \epsilon_k \qquad \text{for all } k \leq l.$$

Exercise 1.19 Following the construction of the Cantor set C by the removal of middle thirds, we define a function F on the complement of the Cantor set $[0, 1] \setminus C$ as follows. First, we define $F(x) = 1/2$ for $1/3 < x < 2/3$. Then $F(x) = 1/4$ for $1/9 < x < 2/9$ and $F(x) = 3/4$ for $7/9 < x < 8/9$, and so on. Prove that F extends to a unique continuous function $F : [0, 1] \to \mathbb{R}$. Prove that F is differentiable at every $x \in \mathbb{R} \setminus C$ and $F'(x) = 0$. This function is called the *Cantor function*. Its graph is sometimes called the *devil's staircase*.

Exercise 1.20 Let X be a normed linear space. A series $\sum x_n$ in X is *absolutely convergent* if $\sum \|x_n\|$ converges to a finite value in $\mathbb{R}$. Prove that X is a Banach space if and only if every absolutely convergent series converges.

Exercise 1.21 Suppose that X is a Banach space, and (x_{mn}) is a doubly indexed sequence in X such that

$$\sum_{m=1}^{\infty} \sum_{n=1}^{\infty} \|x_{mn}\| < \infty.$$

Prove that

$$\sum_{m=1}^{\infty} \left(\sum_{n=1}^{\infty} x_{mn} \right) = \sum_{n=1}^{\infty} \left(\sum_{m=1}^{\infty} x_{mn} \right).$$

Exercise 1.22 Let S be a set. A relation $\sim$ between points of S is called an *equivalence relation* if, for all $a, b, c \in S$, we have:

(a) $a \sim a$;
(b) $a \sim b$ implies $b \sim a$;
(c) $a \sim b$ and $b \sim c$ implies $a \sim c$.

Define the equivalence class C_a associated with $a \in S$ by

$$C_a = \{b \in S \mid a \sim b\}.$$

Prove that two equivalence classes are either disjoint or equal, so $\sim$ partitions S into a union of disjoint equivalence classes. Show that the relation $\sim$ between Cauchy sequences defined in the proof of Theorem 1.52 is an equivalence relation.

Exercise 1.23 Suppose that $f : X \to \mathbb{R}$ is lower semicontinuous and M is a real number. Define $f_M : X \to \mathbb{R}$ by

$$f_M(x) = \min \left(f(x), M \right).$$

Prove that f_M is lower semicontinuous.

Exercise 1.24 Let $f : X \to \mathbb{R}$ be a real-valued function on a set X. The *epigraph* epi f of f is the subset of $X \times \mathbb{R}$ consisting of points that lie above the graph of f:

$$\text{epi } f = \{(x, t) \in X \times \mathbb{R} \mid t \geq f(x)\}.$$

Prove that a function is lower semicontinuous if and only if its epigraph is a closed set.

Exercise 1.25 A function $f : \mathbb{R}^n \to \mathbb{R}$ is *coercive* if

$$\lim_{\|x\|\to\infty} f(x) = \infty. \tag{1.17}$$

Explicitly, this condition means that for any $M > 0$ there is an $R > 0$ such that $\|x\| \geq R$ implies $f(x) \geq M$. Prove that if $f : \mathbb{R}^n \to \mathbb{R}$ is lower semicontinuous and coercive, then f is bounded from below and attains its infimum.

Exercise 1.26 Let $p : \mathbb{R}^2 \to \mathbb{R}$ be a polynomial function of two real variables. Suppose that $p(x, y) \geq 0$ for all $x, y \in \mathbb{R}$. Does every such function attain its infimum? Prove or disprove.

Exercise 1.27 Suppose that (x_n) is a sequence in a compact metric space with the property that every convergent subsequence has the same limit x. Prove that $x_n \to x$ as $n \to \infty$.

Chapter 2

Continuous Functions

In Chapter 1, we introduced the notion of a normed linear space, with finite-dimensional Euclidean space $\mathbb{R}^n$ as the main example. In this chapter, we study linear spaces of continuous functions on a compact set equipped with the uniform norm. These function spaces are our first examples of infinite-dimensional normed linear spaces, and we explore the concepts of convergence, completeness, density, and compactness in this context. As an application of compactness, we prove an existence result for initial value problems for ordinary differential equations.

2.1 Convergence of functions

Suppose that (f_n) is a sequence of real-valued functions $f_n : X \to \mathbb{R}$ defined on a metric space X. What would we mean by $f_n \to f$? Two natural ways to answer this question are the following.

(a) The functions f_n are defined by their values, so the functions converge if the values converge. That is, we say $f_n \to f$ if $f_n(x) \to f(x)$ for all $x \in X$. This definition reduces the convergence of real-valued functions to the convergence of real numbers, with which we are already familiar. This type of convergence is called *pointwise convergence*.

(b) We define a suitable notion of the distance between functions, and say that $f_n \to f$ if the distance between f_n and f tends to zero. In this approach, we regard the functions as points in a metric space, and use *metric convergence*.

Both of these ideas are useful. It turns out, however, that they are not compatible. For most domains X — for example, any uncountable domain — pointwise convergence cannot be expressed as convergence with respect to a metric. The next example shows that pointwise convergence is not a good notion of convergence to use for continuous functions because it does not preserve continuity.

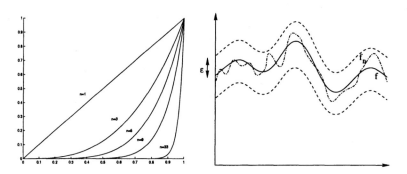

Fig. 2.1 Left: the sequence of functions $f_n(x) = x^n$ converges pointwise but not uniformly on $[0, 1]$. Right: graphically, uniform convergence means that for an arbitrarily narrow tubular neighborhood of the limiting function, the functions f_n will be contained in it for all sufficiently large n.

Example 2.1 We define $f_n : [0, 1] \to \mathbb{R}$ by

$$f_n(x) = x^n.$$

As illustrated in Figure 2.1, the sequence (f_n) converges pointwise to the function f given by

$$f(x) = \begin{cases} 0 & \text{if } 0 \le x < 1, \\ 1 & \text{if } x = 1. \end{cases}$$

The pointwise limit f is discontinuous at $x = 1$.

In view of these somewhat pathological features of pointwise convergence, we consider metric convergence. As we will see, there are many different ways to define a distance between functions, and different metrics or norms usually lead to different types of convergence. A natural norm on spaces of continuous functions is the *uniform* or *sup* norm, which is defined by

$$\|f\| = \sup_{x \in X} |f(x)|. \tag{2.1}$$

The norm $\|f\|$ is finite if and only if f is bounded. The uniform norm is often denoted by $\| \cdot \|_{\text{sup}}$ or $\| \cdot \|_\infty$. The reason for the latter notation will become clear when we study L^p spaces in Chapter 12. In this chapter, we only use the uniform norm, so we denote it by $\| \cdot \|$ without ambiguity.

As illustrated in Figure 2.1, two functions are close in the metric associated with the uniform norm if their pointwise values are uniformly close. Metric convergence with respect to the uniform norm is called *uniform convergence*.

Definition 2.2 A sequence of bounded, real-valued functions (f_n) on a metric space X *converges uniformly* to a function f if

$$\lim_{n \to \infty} \|f_n - f\| = 0,$$

where $\| \cdot \|$ is defined in (2.1).

Uniform convergence implies pointwise convergence. The sequence defined in Example 2.1 shows that the opposite implication does not hold, since $f_n \to f$ pointwise but $\|f_n - f\| = 1$ for every n. Unlike pointwise convergence, uniform convergence preserves continuity.

Theorem 2.3 Let (f_n) be a sequence of bounded, continuous, real-valued functions on a metric space (X, d). If $f_n \to f$ uniformly, then f is continuous.

Proof. In order to show that f is continuous at $x \in X$, we need to prove that for every $\epsilon > 0$ there is a $\delta > 0$ such that $d(x, y) < \delta$ implies $|f(x) - f(y)| < \epsilon$. By the triangle inequality, we have

$$|f(x) - f(y)| \leq |f(x) - f_n(x)| + |f_n(x) - f_n(y)| + |f_n(y) - f(y)|.$$

Since $f_n \to f$ uniformly, there is an n such that

$$|f(x) - f_n(x)| < \frac{\epsilon}{3}, \quad |f_n(y) - f(y)| < \frac{\epsilon}{3} \qquad \text{for all } x, y \in X.$$

Since f_n is continuous at x, there is a $\delta > 0$ such that $d(x, y) < \delta$ implies that

$$|f_n(y) - f_n(x)| < \frac{\epsilon}{3}.$$

It follows that $d(x, y) < \delta$ implies $|f(x) - f(y)| < \epsilon$, so f is continuous at x. $\qquad \square$

The "$\epsilon/3$-trick" used in this proof has many other applications. The proof fails if $f_n \to f$ pointwise but not uniformly.

2.2   Spaces of continuous functions

Let X be a metric space. We denote the set of continuous, real-valued functions $f : X \to \mathbb{R}$ by $C(X)$. The set $C(X)$ is a real linear space under the pointwise addition of functions and the scalar multiplication of functions by real numbers. That is, for $f, g \in C(X)$ and $\lambda \in \mathbb{R}$, we define

$$(f + g)(x) = f(x) + g(x), \qquad (\lambda f)(x) = \lambda(f(x)).$$

From Theorem 1.68, a continuous function f on a compact metric space K is bounded, so the uniform norm $\|f\|$ is finite for $f \in C(K)$. It is straightforward to check that $C(K)$ equipped with the uniform norm is a normed linear space. For example, the triangle inequality holds because

$$\|f + g\| = \sup_{x \in K} |f(x) + g(x)| \leq \sup_{x \in K} |f(x)| + \sup_{x \in K} |g(x)| = \|f\| + \|g\|.$$

We will always use the uniform norm on $C(K)$, unless we state explicitly otherwise. A basic property of $C(K)$ is that it is complete, and therefore a Banach space.

Theorem 2.4 Let K be a compact metric space. The space $C(K)$ is complete.

Proof. Let (f_n) be a Cauchy sequence in $C(K)$ with respect to the uniform norm. We have to show that (f_n) converges uniformly. We do this in two steps. First, we construct a candidate function f for the limit of the sequence, as the pointwise limit of the sequence. Second, we show that the sequence converges uniformly to f.

First, the fact that (f_n) is Cauchy in $C(K)$ implies that the sequence $(f_n(x))$ is Cauchy in $\mathbb{R}$ for each $x \in K$. Since $\mathbb{R}$ is complete, the sequence of pointwise values converges, and we can define a function $f : K \to \mathbb{R}$ by

$$f(x) = \lim_{n \to \infty} f_n(x).$$

For the second step, we use the fact that (f_n) is Cauchy in $C(K)$ to prove that it converges uniformly to f. Since $f_m(x) \to f(x)$ as $m \to \infty$, we have

$$
\begin{aligned}
\|f_n - f\| &= \sup_{x \in K} |f_n(x) - f(x)| \\
&= \sup_{x \in K} \lim_{m \to \infty} |f_n(x) - f_m(x)| \\
&\leq \liminf_{m \to \infty} \sup_{x \in K} |f_n(x) - f_m(x)|. \quad (2.2)
\end{aligned}
$$

The fact that (f_n) is Cauchy in the uniform norm means that for all $\epsilon > 0$ there is an N such that

$$\sup_{x \in K} |f_n(x) - f_m(x)| < \epsilon \quad \text{for all } m, n \geq N.$$

It follows from (2.2) that $\|f_n - f\| \leq \epsilon$ for $n \geq N$, which proves that $\|f_n - f\| \to 0$ as $n \to \infty$. By Theorem 2.3, the limit function f is continuous, and therefore belongs to $C(K)$. Hence, $C(K)$ is complete. □

Example 2.5 Suppose $K = \{x_1, \ldots, x_n\}$ is a finite space, with metric d defined by $d(x_i, x_j) = 1$ for $i \neq j$. A function $f : K \to \mathbb{R}$ can be identified with a point $y = (y_1, \ldots, y_n) \in \mathbb{R}^n$, where $f(x_i) = y_i$, and

$$\|f\| = \max_{1 \leq i \leq n} |y_i|.$$

Thus, the space $C(K)$ is linearly isomorphic to the finite-dimensional space $\mathbb{R}^n$ with the maximum norm, which we have already observed is a Banach space. If K contains infinitely many points, for example if $K = [0, 1]$, then $C(K)$ is an infinite-dimensional Banach space.

The same proof applies to complex-valued functions $f : K \to \mathbb{C}$, and the space of complex-valued continuous functions on a compact metric space is also a Banach space with the uniform norm (2.1).

The pointwise product of two continuous functions is continuous, so $C(K)$ has an *algebra* structure. The product is compatible with the norm, in the sense that

$$\|fg\| \leq \|f\| \|g\|. \tag{2.3}$$

We say that $C(K)$ is a *Banach algebra*. Strict inequality may occur in (2.3); for example, the product of two functions that are nonzero on disjoint sets is zero.

Equation (2.1) does not define a norm on $C(X)$ when X is not compact, since continuous funtions may be unbounded. The space $C_b(X)$ of bounded continuous functions on X is a Banach space with respect to the uniform norm.

Definition 2.6 The *support*, supp f, of a function $f : X \to \mathbb{R}$ (or $f : X \to \mathbb{C}$) on a metric space X is the closure of the set on which f is nonzero,

$$\text{supp } f = \overline{\{x \in X \mid f(x) \neq 0\}}.$$

We say that f has *compact support* if supp f is a compact subset of X, and denote the space of continuous functions on X with compact support by $C_c(X)$.

The space $C_c(X)$ is a linear subspace of $C_b(X)$, but it need not be closed, in which case it is not a Banach space. We denote the closure of $C_c(X)$ in $C_b(X)$ by $C_0(X)$. Since $C_0(X)$ is a closed linear subspace of a Banach space, it is also a Banach space. (We warn the reader that the notation $C_0(X)$ is often used to denote the space $C_c(X)$ of functions with compact support.) We have the following inclusions between these spaces of continuous functions:

$$C(X) \supset C_b(X) \supset C_0(X) \supset C_c(X).$$

If X is compact, then these spaces are equal.

Example 2.7 A function $f : \mathbb{R}^n \to \mathbb{R}$ has compact support if there is an $R > 0$ such that $f(x) = 0$ for all x with $\|x\| > R$. The space $C_0(\mathbb{R}^n)$ consists of continuous functions that vanish at infinity, meaning that for every $\epsilon > 0$ there is an $R > 0$ such that $\|x\| > R$ implies that $|f(x)| < \epsilon$. We write this condition as $\lim_{\|x\| \to \infty} f(x) = 0$.

Example 2.8 Consider real functions $f : \mathbb{R} \to \mathbb{R}$. Then $f(x) = x^2$ is in $C(\mathbb{R})$ but not $C_b(\mathbb{R})$. The constant function $f(x) = 1$ is in $C_b(\mathbb{R})$ but not $C_0(\mathbb{R})$. The function $f(x) = e^{-x^2}$ is in $C_0(\mathbb{R})$ but not $C_c(\mathbb{R})$. The function

$$f(x) = \begin{cases} 1 - x^2 & \text{if } |x| \leq 1, \\ 0 & \text{if } |x| > 1, \end{cases}$$

is in $C_c(\mathbb{R})$.

2.3 Approximation by polynomials

A *polynomial* $p : [a, b] \to \mathbb{R}$ on a closed, bounded interval $[a, b]$ is a function of the form

$$p(x) = \sum_{k=0}^{n} c_k x^k,$$

where the coefficients c_k are real numbers. If $c_n \neq 0$, the integer $n \geq 0$ is called the *degree* of p. The Weierstrass Approximation Theorem states that every continuous function $f : [a, b] \to \mathbb{R}$ can be approximated by a polynomial with arbitrary accuracy in the uniform norm.

Theorem 2.9 (Weierstrass approximation) The set of polynomials is dense in $C([a, b])$.

Proof. We need to show that for any $f \in C([a, b])$ there is a sequence of polynomials (p_n) such that $p_n \to f$ uniformly.

We first show that, by shifting and rescaling x, it is sufficient to prove the theorem in the case $[a, b] = [0, 1]$. We define $T : C([a, b]) \to C([0, 1])$ by

$$(Tf)(x) = f(a + (b - a)x).$$

Then T is linear and invertible, with inverse

$$\left(T^{-1} f\right)(x) = f\left(\frac{x - a}{b - a}\right).$$

Moreover, T is an isometry, since $\|Tf\| = \|f\|$, and for any polynomial p both Tp and $T^{-1}p$ are polynomials. If polynomials are dense in $C([0, 1])$, then for any $f \in C([a, b])$ we have polynomials p_n such that $p_n \to Tf$ in $C([0, 1])$. It follows that the polynomials $T^{-1}p_n$ converge to f in $C([a, b])$.

To show that polynomials are dense in $C([0, 1])$, we use a proof by Bernstein, which gives an explicit formula for a sequence of polynomials converging to a function f in $C([0, 1])$. These polynomials are called the Bernstein polynomials of f, and are defined by

$$B_n(x; f) = \sum_{k=0}^{n} f\left(\frac{k}{n}\right) \binom{n}{k} x^k (1 - x)^{n-k}. \tag{2.4}$$

Notice that each term $x^k(1 - x)^{n-k}$, attains its maximum at $x = k/n$. This is illustrated in Figure 2.2 for $n = 20$ and some values of k. The value of $B_n(x; f)$ for x near k/n, is therefore predominantly determined by the values of $f(x)$ near $x = k/n$. In (2.4), we use the standard notation for the binomial coefficients,

$$\binom{n}{k} = \frac{n!}{(n - k)! k!}.$$

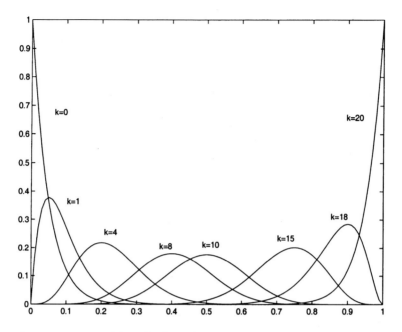

Fig. 2.2 The polynominals $x^k(1-x)^{n-k}$, for the case $n = 20$, appearing in the definition of the Bernstein polynomials (2.4). Note that they attain their maximum at $x = k/n$.

The binomial theorem implies that

$$\sum_{k=0}^{n} \binom{n}{k} x^k (1-x)^{n-k} = 1.$$

Therefore, the difference between f and its nth Bernstein polynomial can be written as

$$B_n(x; f) - f(x) = \sum_{k=0}^{n} \left[f\left(\frac{k}{n}\right) - f(x) \right] \binom{n}{k} x^k (1-x)^{n-k}. \tag{2.5}$$

Taking the supremum with respect to x of the absolute value of this equation, we get

$$\|B_n(\cdot; f) - f\| \leq \sup_{0 \leq x \leq 1} \left[\sum_{k=0}^{n} \left| f\left(\frac{k}{n}\right) - f(x) \right| \binom{n}{k} x^k (1-x)^{n-k} \right]. \tag{2.6}$$

Here, we use $B_n(x; f)$ to denote the value of the Bernstein polynomial at x, and $B_n(\cdot; f)$ to denote the corresponding polynomial function.

Let $\epsilon > 0$ be an arbitrary positive number. From Theorem 1.67, the function f is uniformly continuous, so there is a $\delta > 0$ such that

$$|x - y| < \delta \quad \text{implies} \quad |f(x) - f(y)| < \epsilon, \tag{2.7}$$

for all $x, y \in [0,1]$. To estimate the right hand side of (2.6), we divide the terms in the series into two groups. We let

$$I(x) = \{k \mid 0 \le k \le n \text{ and } |x - (k/n)| < \delta\},$$
$$J(x) = \{k \mid 0 \le k \le n \text{ and } |x - (k/n)| \ge \delta\}. \tag{2.8}$$

From (2.6), (2.7), and (2.8), we get the following estimate,

$$\|B_n(\cdot\,; f) - f\| \;\le\; \epsilon \sup_{0 \le x \le 1} \left[\sum_{k \in I(x)} \binom{n}{k} x^k (1-x)^{n-k} \right]$$

$$+ \sup_{0 \le x \le 1} \left[\sum_{k \in J(x)} \left| f\left(\frac{k}{n}\right) - f(x) \right| \binom{n}{k} x^k (1-x)^{n-k} \right]$$

$$\le \;\; \epsilon + 2\|f\| \sup_{0 \le x \le 1} \left[\sum_{k \in J(x)} \binom{n}{k} x^k (1-x)^{n-k} \right]. \tag{2.9}$$

Since $[x - (k/n)]^2 \ge \delta^2$ for $k \in J(x)$, the sum on the right hand side of (2.9) can be estimated as follows:

$$\sup_{0 \le x \le 1} \left[\sum_{k \in J(x)} \binom{n}{k} x^k (1-x)^{n-k} \right]$$

$$\le \frac{1}{\delta^2} \sup_{0 \le x \le 1} \left[\sum_{k \in J(x)} \left(x - \frac{k}{n} \right)^2 \binom{n}{k} x^k (1-x)^{n-k} \right]$$

$$\le \frac{1}{\delta^2} \sup_{0 \le x \le 1} \left[\sum_{k=0}^{n} \left(x^2 - \frac{2k}{n} x + \frac{k^2}{n^2} \right) \binom{n}{k} x^k (1-x)^{n-k} \right]$$

$$\le \frac{1}{\delta^2} \sup_{0 \le x \le 1} \left[x^2 B_n(x; 1) - 2x B_n(x; x) + B_n(x; x^2) \right]. \tag{2.10}$$

To find an expression for the Bernstein polynomials $B_n(x; 1)$, $B_n(x; x)$, and $B_n(x; x^2)$ of the polynomials 1, x, and x^2, we write out the binomial expansion of $(x + y)^n$, compute the first and second derivatives of the expansion with respect to x, and rearrange the results. This gives

$$(x + y)^n \;=\; \sum_{k=0}^{n} \binom{n}{k} x^k y^{n-k},$$

$$x(x + y)^{n-1} \;=\; \sum_{k=0}^{n} \left(\frac{k}{n}\right) \binom{n}{k} x^k y^{n-k},$$

$$\left(\frac{n-1}{n}\right) x^2 (x + y)^{n-2} + \left(\frac{1}{n}\right) x(x + y)^{n-1} \;=\; \sum_{k=0}^{n} \left(\frac{k}{n}\right)^2 \binom{n}{k} x^k y^{n-k}.$$

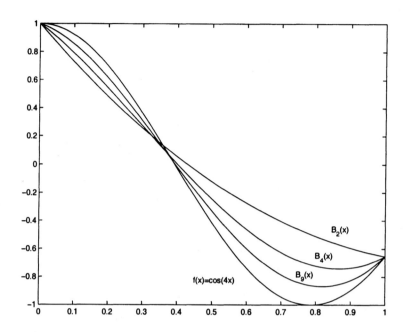

Fig. 2.3 Some approximations of the function $x \mapsto \cos(4x)$ by its Bernstein polynomials.

Evaluation of these equations at $y = 1 - x$, and the use of (2.4) gives

$$
\begin{aligned}
B_n(x; 1) &= 1, \\
B_n(x; x) &= x, \\
B_n(x; x^2) &= \left(\frac{n-1}{n}\right) x^2 + \left(\frac{1}{n}\right) x,
\end{aligned}
\tag{2.11}
$$

for all $n \geq 1$. Using (2.10) and (2.11) in (2.9), we obtain the estimate

$$
\|B_n(\cdot\,; f) - f\| \leq \epsilon + \frac{\|f\|}{2n\delta^2}.
$$

Taking the lim sup of this equation as $n \to \infty$, we get

$$
\limsup_{n \to \infty} \|B_n(\cdot\,; f) - f\| \leq \epsilon.
$$

Since ϵ is arbitrary, it follows that $\limsup_{n \to \infty} \|B_n(\cdot\,; f) - f\| = 0$, so the polynomials $B_n(\cdot\,; f)$ converge uniformly to f. $\qquad\square$

The first few approximations by Bernstein polynomials of the function $f(x) = \cos(4x)$ are graphed in Figure 2.3. Note that we could have formulated the theorem for complex-valued functions with the same proof.

The Weierstrass approximation theorem differs from Taylor's theorem, which states that a function with sufficiently many derivatives can be approximated locally by its Taylor polynomial. The Weierstrass approximation theorem applies to a continuous function, which need not be differentiable, and states that there is a global polynomial approximation of the function on the whole interval $[a, b]$.

An analogous result is the density of trigonometric polynomials in the space of periodic continuous functions on the circle, which we prove below in Theorem 7.3. Both of these theorems are special cases of the Stone-Weierstrass theorem (see Rudin [48]).

2.4   Compact subsets of $C(K)$

The proof of the Heine-Borel theorem, that a subset of $\mathbb{R}^n$ is compact if and only if it is closed and bounded, uses the finite-dimensionality of $\mathbb{R}^n$ in an essential way. Compact subsets of infinite-dimensional normed spaces are also closed and bounded, but these properties are no longer sufficient. In this section, we prove the Arzelà-Ascoli theorem, which characterizes the compact subsets of $C(K)$. To state the theorem, we introduce the notion of *equicontinuity*.

Definition 2.10 Let $\mathcal{F}$ be a family of functions from a metric space (X, d) to a metric space (Y, d). The family $\mathcal{F}$ is *equicontinuous* if for every $x \in X$ and $\epsilon > 0$ there is a $\delta > 0$ such that $d(x, y) < \delta$ implies $d(f(x), f(y)) < \epsilon$ for all $f \in \mathcal{F}$.

The crucial point in this definition is that δ does not depend on f, although it may depend on x. If δ can be chosen independent of x as well, then the family is said to be *uniformly equicontinuous*. The following theorem is a generalization of Theorem 1.67.

Theorem 2.11 An equicontinuous family of functions from a compact metric space to a metric space is uniformly equicontinuous.

Proof.   Suppose that K is a compact metric space, and $\mathcal{F}$ is a family of functions $f : K \to Y$ that is not uniformly equicontinuous. We will prove that $\mathcal{F}$ is not equicontinuous.

Since $\mathcal{F}$ is not uniformly equicontinuous, there is an $\epsilon > 0$, such that for every $n \in \mathbb{N}$ there are points $x_n, y_n \in K$ and a function $f_n \in \mathcal{F}$ with

$$d(x_n, y_n) < \frac{1}{n} \quad \text{and} \quad d(f_n(y_n), f_n(x_n)) \geq 2\epsilon. \tag{2.12}$$

Since K is compact, the sequence (x_n) has a convergent subsequence, which we also denote by (x_n). Suppose that $x_n \to x$ as $n \to \infty$. Then (2.12) implies that $y_n \to x$ as well. Hence, for all $\delta > 0$, there are points x_n, y_n such that $d(x_n, x) < \delta$ and $d(y_n, x) < \delta$. But, from (2.12), we must have either $d(f_n(x_n), f_n(x)) \geq \epsilon$ or $d(f_n(y_n), f_n(x)) \geq \epsilon$, so $\mathcal{F}$ is not equicontinuous at x. □

Next, we give necessary and sufficient conditions for compactness in $C(K)$.

Theorem 2.12 (Arzelà-Ascoli) Let K be a compact metric space. A subset of $C(K)$ is compact if and only if it is closed, bounded, and equicontinuous.

Proof. Recall that a set is precompact if its closure is compact, and that a set is compact if and only if it is closed and precompact. We will prove that a subset of $C(K)$ is precompact if and only if it is bounded and equicontinuous.

We divide the proof into three parts. First, we show that an unbounded subset is not precompact. Second, we show that a precompact subset is equicontinuous. Third, we show that a bounded, equicontinuous subset is precompact.

For the first part, suppose that $\mathcal{F}$ is an unbounded subset of $C(K)$. Then there is a sequence of functions $f_n \in \mathcal{F}$, with $\|f_{n+1}\| \geq \|f_n\| + 1$, so that $\|f_n - f_m\| \geq 1$ for all $n \neq m$. It follows that (f_n) has no Cauchy subsequence, and therefore no convergent subsequence, so $\mathcal{F}$ is not precompact.

For the second part, suppose that $\mathcal{F}$ is a precompact subset of $C(K)$. Fix $\epsilon > 0$. Since $\mathcal{F}$ is dense in $\overline{\mathcal{F}}$, we have

$$\overline{\mathcal{F}} \subset \bigcup_{f \in \mathcal{F}} B_{\epsilon/3}(f).$$

Since $\overline{\mathcal{F}}$ is compact, there is a finite subset $\{f_1, \ldots, f_k\}$ of $\mathcal{F}$ such that

$$\overline{\mathcal{F}} \subset \bigcup_{i=1}^{k} B_{\epsilon/3}(f_i).$$

Each f_i is uniformly continuous by Theorem 1.67, so there is a $\delta_i > 0$ such that $d(x,y) < \delta_i$ implies that $|f_i(x) - f_i(y)| < \epsilon/3$ for all $x, y \in K$. We define δ by

$$\delta = \min_{1 \leq i \leq k} \delta_i.$$

Since δ is the minimum of a finite set of $\delta_i > 0$, we have $\delta > 0$. For every $f \in \mathcal{F}$, there is an $1 \leq i \leq k$ such that $\|f - f_i\| < \epsilon/3$. We conclude that for $d(x,y) < \delta$

$$|f(x) - f(y)| \leq |f(x) - f_i(x)| + |f_i(x) - f_i(y)| + |f_i(y) - f(y)| < \epsilon.$$

Since ϵ is arbitrary and δ is independent of f, the set $\mathcal{F}$ is equicontinuous.

For the third part, suppose that $\mathcal{F}$ is a bounded, equicontinuous subset of $C(K)$. We will show that every sequence (f_n) in $\mathcal{F}$ has a convergent subsequence. By Lemma 1.63 there is a countable dense set $\{x_1, x_2, x_3, \ldots\}$ in the compact domain K. We choose a subsequence $(f_{1,n})$ of (f_n) such that the sequence of values $(f_{1,n}(x_1))$ converges in $\mathbb{R}$. Such a subsequence exists because $(f_n(x_1))$ is bounded in $\mathbb{R}$, since $\mathcal{F}$ is bounded in $C(K)$. We choose a subsequence $(f_{2,n})$ of $(f_{1,n})$ such that $(f_{2,n}(x_2))$ converges, which exists for the same reason. Repeating this procedure, we obtain sequences $(f_{k,n})_{n=1}^{\infty}$ for $k = 1, 2, \ldots$ such that $(f_{k+1,n})$ is a subsequence of $(f_{k,n})$, and $(f_{k,n}(x_k))$ converges as $n \to \infty$. Finally, we define a "diagonal" subsequence (g_k) by $g_k = f_{k,k}$. By construction, the sequence (g_k) is a subsequence of (f_n) with the property that $g_k(x_i)$ converges in $\mathbb{R}$ as $k \to \infty$ for all x_i in a dense subset of K.

So far, we have only used the boundedness of $\mathcal{F}$. The equicontinuity of $\mathcal{F}$ is needed to ensure the uniform convergence of (g_k). Let $\epsilon > 0$. Since $\mathcal{F}$ is equicontinuous and K is compact, Theorem 2.11 implies that $\mathcal{F}$ is uniformly equicontinuous. Consequently, there is a $\delta > 0$ such that $d(x, y) < \delta$ implies

$$|g_k(x) - g_k(y)| < \frac{\epsilon}{3}.$$

Since $\{x_i\}$ is dense in K, we have

$$K \subset \bigcup_{i=1}^{\infty} B_\delta(x_i).$$

Since K is compact, there is a finite subset of $\{x_i\}$, which we denote by $\{x_1, \ldots, x_n\}$, such that

$$K \subset \bigcup_{i=1}^{n} B_\delta(x_i).$$

The sequence $(g_k(x_i))$ is convergent for each $i = 1, \ldots, n$, and hence Cauchy, so there is an N such that

$$|g_j(x_i) - g_k(x_i)| < \frac{\epsilon}{3}$$

for all $j, k \geq N$ and $i = 1, \ldots, n$. For any $x \in K$, there is an i such that $x \in B_\delta(x_i)$. Then, for $j, k \geq N$, we have

$$|g_j(x) - g_k(x)| \leq |g_j(x) - g_j(x_i)| + |g_j(x_i) - g_k(x_i)| + |g_k(x_i) - g_k(x)| < \epsilon.$$

It follows that (g_k) is a Cauchy sequence for the uniform norm and, since $C(K)$ is complete, it converges. $\qquad\square$

In the proof of this theorem, we again used Cantor's diagonal argument, and the "$\epsilon/3$-trick."

Example 2.13 For each $n \in \mathbb{N}$, we define a function $f_n : [0, 1] \to \mathbb{R}$ by

$$f_n(x) = \begin{cases} 0 & \text{if } 0 \leq x \leq 2^{-n}, \\ 2^{n+1}(x - 2^{-n}) & \text{if } 2^{-n} \leq x \leq 3 \cdot 2^{-(n+1)}, \\ 2^{n+1}(2^{-(n-1)} - x) & \text{if } 3 \cdot 2^{-(n+1)} \leq x \leq 2^{-(n-1)}, \\ 0 & \text{if } 2^{-(n-1)} \leq x \leq 1. \end{cases} \qquad (2.13)$$

These functions consist of 'tent' functions of height one that move from right to left across the inverval $[0, 1]$, becoming narrower and steeper as they do so.

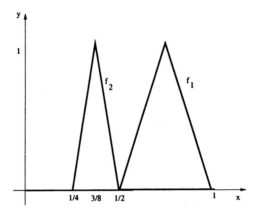

Fig. 2.4 The sequence of bounded continuous functions defined in Example 2.13 is not equicontinuous.

The first two functions are shown in Figure 2.4. Let $\mathcal{F} = \{f_n \mid n \in \mathbb{N}\}$. Then $\|f_n\| = 1$ for all $n \geq 1$, so $\mathcal{F}$ is bounded, but $\|f_m - f_n\| = 1$ for all $m \neq n$, so the sequence (f_n) does not have any convergent subsequences. Hence, the set $\mathcal{F}$ is a closed, bounded subset of $C([0,1])$ which is not compact. Note that $\mathcal{F}$ is not equicontinuous either, because the graphs of the f_n become steeper as n gets larger. The same phenomenon occurs for the set $\mathcal{F} = \{\sin(n\pi x) \mid n \in \mathbb{N}\}$.

Heuristically, a subset of an infinite-dimensional linear space is precompact if it is "almost" contained in a bounded subset of a finite-dimensional subspace. Without making this statement more precise at the moment (but see Theorem 9.17), we illustrate it with the following example.

Example 2.14 Let $\mathcal{F}$ be the subset of $C([0,1])$ that consists of functions f of the form

$$f(x) = \sum_{n=1}^{\infty} a_n \sin(n\pi x) \quad \text{with} \quad \sum_{n=1}^{\infty} n|a_n| \leq 1.$$

The series defining f converges uniformly, so f is an element of $C([0,1])$. The set $\mathcal{F}$ is bounded in $C(K)$, since for any $f \in \mathcal{F}$ we have

$$\|f\| \leq \sum_{n=1}^{\infty} |a_n| \leq \sum_{n=1}^{\infty} n|a_n| \leq 1.$$

By the mean value theorem, for any $x < y \in \mathbb{R}$ there is a $x < \xi < y$ with

$$\sin x - \sin y = (\cos \xi)(x - y).$$

Hence, for all $x, y \in \mathbb{R}$ we have

$$|\sin x - \sin y| \leq |x - y|.$$

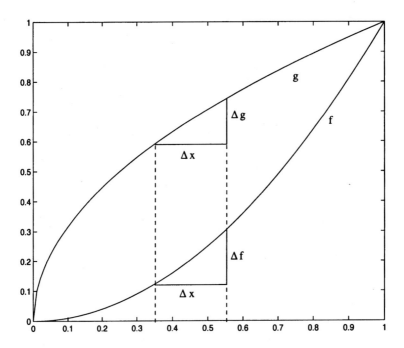

Fig. 2.5 The graph of two continuous functions on $[0,1]$: $f(x) = x^2$, and $g(x) = \sqrt{x}$. f is Lipschitz on $[0,1]$, but g is not Lipschitz at the point 0. The ratio $\Delta f/\Delta x$ is bounded for arbitrarily small Δx everywhere in $[0,1]$, but $\Delta g/\Delta x$ is unbounded for small Δx near $x = 0$.

Thus, every $f \in \mathcal{F}$ satisfies

$$|f(x) - f(y)| \leq \sum_{n=1}^{\infty} |a_n| \, |\sin(n\pi x) - \sin(n\pi y)| \leq \sum_{n=1}^{\infty} \pi n |a_n| \, |x - y| \leq \pi \, |x - y| \,.$$

Therefore, given $\epsilon > 0$, we can pick $\delta = \epsilon/\pi$, and then $|x - y| < \delta$ implies $|f(x) - f(y)| < \epsilon$ for all $f \in \mathcal{F}$. From the Arzelà-Ascoli theorem, $\mathcal{F}$ is a precompact subset of $C([0,1])$. For large N, the subset $\mathcal{F}$ is "almost" contained in the unit ball in the finite-dimensional subspace spanned by $\{\sin \pi x, \sin 2\pi x, \ldots, \sin N\pi x\}$.

The previous example illustrates a useful sufficient condition for equicontinuity, which we now describe. We begin by defining *Lipschitz continuous* functions.

Definition 2.15 A function $f : X \to \mathbb{R}$ on a metric space X is *Lipschitz continuous* on X if there is a constant $C \geq 0$ such that

$$|f(x) - f(y)| \leq C d(x,y) \qquad \text{for all } x, y \in X. \tag{2.14}$$

We will often abbreviate the term "Lipschitz continuous" to "Lipschitz." Every Lipschitz continuous function is uniformly continuous, but there are uniformly continuous functions that are not Lipschitz.

Example 2.16 As illustrated in Figure 2.5, the square function $f(x) = x^2$ is Lipschitz continuous on $[0, 1]$, but the uniformly continuous square-root function $g(x) = \sqrt{x}$ is not, because

$$\lim_{x \to 0^+} \frac{|g(x) - g(0)|}{|x - 0|} = \infty.$$

If $f : X \to \mathbb{R}$ is a Lipschitz function, then we define the *Lipschitz constant* $\mathrm{Lip}(f)$ of f by

$$\mathrm{Lip}(f) = \sup_{x \neq y} \frac{|f(x) - f(y)|}{d(x, y)}.$$

Equivalently, $\mathrm{Lip}(f)$ is the smallest constant C that works in the Lipschitz condition (2.14),

$$\mathrm{Lip}(f) = \inf \left\{ C \mid |f(x) - f(y)| \leq C d(x, y) \text{ for all } x, y \in X \right\}.$$

Suppose that K is a compact metric space and $M > 0$. We define a subset $\mathcal{F}_M$ of $C(K)$ by

$$\mathcal{F}_M = \{ f \mid f \text{ is Lipschitz on } K \text{ and } \mathrm{Lip}(f) \leq M \}. \tag{2.15}$$

The set $\mathcal{F}_M$ is equicontinuous, since if $\epsilon > 0$ and $\delta = \epsilon/M$, then

$$d(x, y) < \delta \quad \text{implies} \quad |f(x) - f(y)| < \epsilon \quad \text{for all } f \in \mathcal{F}_M.$$

The set $\mathcal{F}_M$ is closed, since if (f_n) is a sequence in $\mathcal{F}_M$ that converges uniformly to f in $C(K)$, then

$$
\begin{aligned}
\mathrm{Lip}(f) &= \sup_{x \neq y} \frac{|f(x) - f(y)|}{d(x, y)} \\
&= \sup_{x \neq y} \left[\lim_{n \to \infty} \frac{|f_n(x) - f_n(y)|}{d(x, y)} \right] \\
&\leq \liminf_{n \to \infty} \left[\sup_{x \neq y \in K} \frac{|f_n(x) - f_n(y)|}{d(x, y)} \right] \\
&\leq M.
\end{aligned}
$$

Thus, the limit f belongs to $\mathcal{F}_M$. The set $\mathcal{F}_M$ is not bounded, since the constant functions belong to $\mathcal{F}_M$ and their sup-norms are arbitrarily large. Consequently, although $\mathcal{F}_M$ itself is not compact, the Arzelà-Ascoli theorem implies that every closed, bounded subset of $\mathcal{F}_M$ is compact, and every bounded subset of $\mathcal{F}_M$ is precompact.

Example 2.17 Suppose that x_0 is a point in a compact metric space K. Let

$$\mathcal{B}_M = \{ f \in \mathcal{F}_M \mid f(x_0) = 0 \}.$$

Then $\mathcal{B}_M$ is bounded because for every $f \in \mathcal{B}_M$ we have

$$\|f\| = \sup_{x \in K} |f(x) - f(x_0)| \leq M \sup_{x \in K} |x - x_0| \leq M \operatorname{diam} K,$$

where $\operatorname{diam} K$ is finite since K is compact, and hence bounded. The set $\mathcal{B}_M$ is closed, since if $f_n(x_0) = 0$ and $f_n \to f$ in $C(K)$, then

$$f(x_0) = \lim_{n \to \infty} f_n(x_0) = 0.$$

Therefore, the set $\mathcal{B}_M$ is a compact subset of $C(K)$.

Lemma 2.18 Suppose that $f : C \to \mathbb{R}$ is a continuously differentiable function on an open, convex subset C of $\mathbb{R}^n$, and that the partial derivatives of f are bounded on C. Then, for all $x, y \in C$, we have

$$|f(x) - f(y)| \leq M \|x - y\|, \tag{2.16}$$

where $\| \cdot \|$ denotes the Euclidean norm and

$$M = \sup_{z \in C} \|\nabla f(z)\|. \tag{2.17}$$

Proof. Since C is convex, the point $tx + (1 - t)y$ lies in C for all $x, y \in C$ and $0 \leq t \leq 1$. The fundamental theorem of calculus and the chain rule imply that

$$
\begin{aligned}
f(x) - f(y) &= \int_0^1 \frac{d}{dt} f\left(tx + (1 - t)y\right) dt \\
&= \int_0^1 \nabla f\left(tx + (1 - t)y\right) \cdot (x - y) \, dt,
\end{aligned}
\tag{2.18}
$$

where

$$\nabla f = \left(\frac{\partial f}{\partial x_1}, \ldots, \frac{\partial f}{\partial x_n} \right)$$

is the gradient of f with respect to $x = (x_1, \ldots, x_n)$. We take the absolute value of equation (2.18), and estimate the resulting integral, to obtain

$$|f(x) - f(y)| \leq \sup_{0 \leq t \leq 1} \left\{ \|\nabla f\left(tx + (1 - t)y\right)\| \right\} \|x - y\|.$$

The use of (2.17) in this equation gives (2.16). $\qquad \square$

From Lemma 2.18, a continuously differentiable function with bounded partial derivatives is Lipschitz. A Lipschitz continuous function need not be differentiable everywhere, however, since its graph may have "corners."

Example 2.19 The absolute value function $f(x) = |x|$ is Lipschitz continuous, with Lipschitz constant one, because

$$|f(x) - f(y)| = |\, |x| - |y| \,| \leq |x - y|.$$

The absolute value function is not differentiable at $x = 0$.

Lemma 2.18 implies that a family of continuously differentiable functions with uniformly bounded derivatives is equicontinuous. If the family is also bounded, then it is precompact. The idea that a uniform bound on suitable norms of the derivatives of a family of functions implies that the family is precompact will reappear when we study Sobolev spaces in Chapter 12.9.

Example 2.20 Let $C^1([0,1])$ denote the space of all continuous functions f on $[0,1]$ with continuous derivative f'. For constants $M > 0$ and $N > 0$, we define the subset $\mathcal{F}$ of $C([0,1])$ by

$$\mathcal{F} = \left\{ f \in C^1([0,1]) \mid \|f\| \leq M, \|f'\| \leq N \right\},$$

where $\|\cdot\|$ denotes the sup-norm. Then $\mathcal{F}$ is precompact in $C(K)$. It is not closed, however, because the uniform limit of continuously differentiable functions need not be differentiable. Thus, $\mathcal{F}$ is not compact. Its closure in $C([0,1])$ is the compact set

$$\overline{\mathcal{F}} = \{ f \in C([0,1]) \mid \|f\| \leq M, \mathrm{Lip}(f) \leq N \}.$$

2.5 Ordinary differential equations

A differential equation is an equation that relates the values of a function and its derivatives at each point. We distinguish between ordinary differential equations (ODEs) for functions of a single variable, and partial differential equations (PDEs) for functions of several variables. In this section, we discuss the existence and uniqueness of solutions of ODEs.

To focus on the central ideas in the simplest setting, we consider a scalar, first order ODE for a real-valued function $u(t)$ of the form

$$\dot{u} = f(t, u). \tag{2.19}$$

In (2.19), we use $\dot{u}(t)$ to denote the derivative of $u(t)$ with respect to t, and $f : \mathbb{R}^2 \to \mathbb{R}$ is a given continuous function. We say that (2.19) is a *linear ODE* if $f(t, u)$ is a linear (strictly speaking, we should say "affine") function of u of the form $f(t, u) = a(t)u + b(t)$. Otherwise, we say that (2.19) is a *nonlinear ODE*.

A solution of (2.19), defined in an open interval $I \subset \mathbb{R}$, is a continuously differentiable function $u : I \to \mathbb{R}$ such that

$$\dot{u}(t) = f(t, u(t)) \qquad \text{for all } t \in I.$$

If the solution is defined on the whole of $\mathbb{R}$, then we call it a *global solution*. If the solution is defined only on a subinterval of $\mathbb{R}$, then we call it a *local solution*.

We will refer to the independent variable t in (2.19) as "time." Equation (2.19) determines the rate of change of the function u at each time in terms of the value of u. We expect that if we know the value of u at some time, then the ODE determines the values of u at nearby times, and by repetition of this process, we expect that there is a unique solution of the *initial value problem* (IVP)

$$\dot{u} = f(t, u), \tag{2.20}$$
$$u(t_0) = u_0.$$

Here, t_0 is a given initial time, and u_0 is a given initial value. As the following examples show, however, the question of the existence and uniqueness of solutions (2.20) is not always as straightforward as this naive discussion might suggest.

Example 2.21 Consider the linear initial value problem,

$$\dot{u} = au, \tag{2.21}$$
$$u(0) = u_0,$$

where $a \in \mathbb{R}$ is a constant. This initial value problem has a unique, global solution $u(t) = u_0 e^{at}$. Equation (2.21) has a simple interpretation in terms of population growth. It states that the growth rate $\dot{u}$ of a population is proportional to the population u. If the per capita growth rate $\dot{u}/u = a$ is positive, then the population grows exponentially in time, as Malthus observed in 1798.

Example 2.22 Consider the nonlinear initial value problem,

$$\dot{u} = u^2, \tag{2.22}$$
$$u(0) = u_0.$$

The unique solution is

$$u(t) = \frac{u_0}{1 - u_0 t}.$$

This solution becomes arbitrarily large as $t \to 1/u_0$. For $u_0 > 0$, the initial value problem in (2.22) has a local solution defined in the interval $-\infty < t < 1/u_0$, but it does not have a global solution. This phenomenon is called "blow-up," and is a fundamental difficulty in the study of nonlinear differential equations. When interpreted as a population model, equation (2.22) describes the growth of a population in which the per capita growth rate is equal to the population. Thus, as the population increases the growth rate increases both because the population is larger and because the per capita growth rate is larger. As a result, the solution tends to infinity in finite time.

Example 2.23 Consider the initial value problem

$$\dot{u} = \sqrt{|u|}, \tag{2.23}$$
$$u(0) = 0.$$

The zero function $u(t) = 0$ is a global solution, but it is not the only one. The following function satisfies (2.23) for any $a \geq 0$,

$$u(t) = \begin{cases} 0 & \text{if } t \leq a, \\ (t-a)^2/4 & \text{if } t > a. \end{cases}$$

In this example, the function $f(u) = \sqrt{|u|}$ is a continuous function of u, but it is not Lipschitz continuous at the initial value $u = 0$.

These examples show that the most we can hope for, if f is an arbitrary continuous function, is the existence of a local solution of the initial value problem (2.20). If f is smooth, the solution is unique, as we will see, but it may not exist globally.

For general f, we cannot prove the existence of a solution by giving an explicit analytical formula for it, as we did in the simple examples above. Instead we use a compactness argument, analogous to the one used in the proof of Theorem 1.68. We construct a family $\{u_\epsilon\}$ of functions that satisfy (2.20) in a suitable approximate sense. Since the functions are approximate solutions of the ordinary differential equation, their derivatives are uniformly bounded, and the Arzelà-Ascoli theorem implies that they form a precompact set. Consequently, there is a subsequence of approximate solutions that converges uniformly as $\epsilon \to 0$ to a function u. We then show that u is a solution of (2.20).

Theorem 2.24 Suppose that $f(t, u)$ is a continuous function on $\mathbb{R}^2$. Then, for every (t_0, u_0), there is an open interval $I \subset \mathbb{R}$ that contains t_0, and a continuously differentiable function $u : I \to \mathbb{R}$ that satisfies the initial value problem (2.20).

Proof. We say that $u_\epsilon(t)$ is an *ϵ-approximate solution* of (2.20) in an interval I containing t_0 if:

(a) $u_\epsilon(t_0) = u_0$;
(b) $u_\epsilon(t)$ is a continuous function of t that is differentiable at all but finitely many points of I;
(c) at every point $t \in I$ where $\dot{u}_\epsilon(t)$ exists, we have

$$|\dot{u}_\epsilon(t) - f(t, u_\epsilon(t))| \leq \epsilon.$$

To construct an ϵ-approximate solution u_ϵ, we first pick $T_1 > 0$, and let

$$I_1 = \{t \mid |t - t_0| \leq T_1\}.$$

We partition I_1 into $2N$ subintervals of length h, where $T_1 = Nh$, and let

$$t_k = t_0 + kh \qquad \text{for } -N \leq k \leq N.$$

We denote the values of the approximate solution at the times t_k by $u_\epsilon(t_k) = a_k$. We define these values by the following finite difference approximation of the ODE,

$$\frac{a_{k+1} - a_k}{h} = f(t_k, a_k),$$
$$a_0 = u_0.$$

This discretization of (2.20) is called the forward Euler method. It is not an accurate numerical method for the solution of (2.20), but its simplicity makes it convenient for an existence proof.

Inside the subinterval $t_k \le t \le t_{k+1}$, we define $u_\epsilon(t)$ to be the linear function of t that takes the appropriate values at the endpoints. That is,

$$u_\epsilon(t) = a_k + b_k(t - t_k) \qquad \text{for } t_k \le t \le t_{k+1},$$

where the parameters a_k and b_k are defined recursively by

$$a_0 = u_0, \qquad a_k = a_{k-1} + b_{k-1}h,$$
$$b_0 = f(t_0, u_0), \qquad b_k = f(t_k, a_k).$$

Thus, $u_\epsilon(t)$ is a continuous, piecewise linear function of t that is differentiable except possibly at the points $t = t_k$, and $\dot{u}_\epsilon(t) = b_k$ for $t_k < t < t_{k+1}$. For $t_k < t < t_{k+1}$, we have

$$|\dot{u}_\epsilon(t) - f(t, u_\epsilon(t))| = |f(t_k, a_k) - f(t, a_k + b_k(t - t_k))|, \qquad (2.24)$$
$$|t - t_k| \le h, \qquad |a_k + b_k(t - t_k) - a_k| \le |b_k| h. \qquad (2.25)$$

We choose an $L > 0$, and a $T \le T_1$ such that the graph of every u_ϵ with $|t - t_0| \le T$ is contained in the rectangle $R \subset \mathbb{R}^2$ given by

$$R = \{(t, u) \mid |t - t_0| \le T, \, |u - u_0| \le L\}.$$

To do this, we consider the closed rectangle $R_1 \subset \mathbb{R}^2$, centered at (t_0, u_0), defined by

$$R_1 = \{(t, u) \mid |t - t_0| \le T_1, \, |u - u_0| \le L\}.$$

We let

$$M = \sup\{|f(t, u)| \mid (t, u) \in R_1\}, \qquad T = \min(T_1, L/M).$$

It follows that, for $|t - t_0| \le T$, the slopes b_k of the linear segments of u_ϵ are less than or equal to M, and the graph of u_ϵ lies in the cone bounded by the lines $u - u_0 = M(t - t_0)$ and $u - u_0 = -M(t - t_0)$. Figure 2.6 shows why this is true.

Since R is compact, the function f is uniformly continuous on R. Therefore, for every $\epsilon > 0$, there is a $\delta > 0$ such that

$$|f(s, u) - f(t, v)| \le \epsilon$$

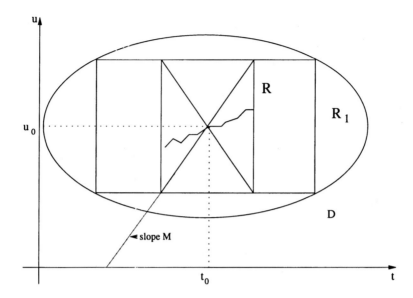

Fig. 2.6   The construction of the rectangle R used in the proof of Theorem 2.24.

for all $(s, u), (t, v) \in R$ such that $|s - t| \leq \delta$ and $|u - v| \leq \delta$. Using (2.24)–(2.25), we see that u_ϵ is an ϵ-approximate solution when $h \leq \delta$ and $Mh \leq \delta$.

Each u_ϵ is Lipschitz continuous, and its Lipschitz constant is bounded uniformly by M, independently of ϵ. We also have $u_\epsilon(t_0) = u_0$ for all ϵ. From Example 2.17, the set $\{u_\epsilon\}$ is precompact in $C([t_0 - T, t_0 + T])$. Hence there is a continuous function u and a sequence (ϵ_n) with $\epsilon_n \to 0$ as $n \to \infty$ such that $u_{\epsilon_n} \to u$ as $n \to \infty$ uniformly on $[t_0 - T, t_0 + T]$.

It remains to show that the limiting function u solves (2.20). Since u_ϵ is piecewise linear, we have

$$
\begin{aligned}
u_\epsilon(t) &= u_\epsilon(t_0) + \int_{t_0}^{t} \dot{u}_\epsilon(s) \, ds \\
&= u_0 + \int_{t_0}^{t} f(s, u_\epsilon(s)) \, ds + \int_{t_0}^{t} [\dot{u}_\epsilon - f(s, u_\epsilon(s))] \, ds. \quad (2.26)
\end{aligned}
$$

Here, $\dot{u}_\epsilon$ is not necessarily defined at the points t_k, but this does not affect the value of the integral. We set $\epsilon = \epsilon_n$ in (2.26), and let $n \to \infty$. Using Exercise 2.2 to take the limit, we find that

$$
u(t) = u_0 + \int_{t_0}^{t} f(s, u(s)) \, ds. \quad (2.27)
$$

The fundamental theorem of calculus implies that the right hand side of (2.27) is continuously differentiable. Therefore, the function u is also continuously differentiable in $|t - t_0| < T$, and $\dot{u}(t) = f(t, u(t))$.   □

More generally, the same proof applies if f is continuous only in some open set $D \subset \mathbb{R}^2$ which contains the initial point (t_0, u_0), provided we choose the rectangles R_1 and R so that they are contained in D (see Figure 2.6).

This proof shows the existence of a local solution in some interval about the initial time, but the solution need not be global. A solution has, however, a maximal open interval of existence that contains t_0.

As shown by Example 2.23, the continuity of f does not guarantee uniqueness, but if $f(t, u)$ is Lipschitz continuous in u, then the solution is unique. The condition that f is Lipschitz continuous is a mild one, and is met in nearly all applications, where f is typically a smooth function. To prove this fact, we use the following result, called *Gronwall's inequality*.

Theorem 2.25 (Gronwall's inequality) Suppose that $u(t) \geq 0$ and $\varphi(t) \geq 0$ are continuous, real-valued functions defined on the interval $0 \leq t \leq T$ and $u_0 \geq 0$ is a constant. If u satisfies the inequality

$$u(t) \leq u_0 + \int_0^t \varphi(s)u(s)\, ds \qquad \text{for } t \in [0, T], \tag{2.28}$$

then

$$u(t) \leq u_0 \exp\left(\int_0^t \varphi(s)\, ds\right) \qquad \text{for } t \in [0, T].$$

In particular, if $u_0 = 0$ then $u(t) = 0$.

Proof. Suppose first that $u_0 > 0$. Let

$$U(t) = u_0 + \int_0^t \varphi(s)u(s)\, ds.$$

Then, since $u(t) \leq U(t)$, we have that

$$\dot{U} = \varphi u \leq \varphi U, \qquad U(0) = u_0.$$

Since $U(t) > 0$, it follows that

$$\frac{d}{dt}\log U = \frac{\dot{U}}{U} \leq \varphi.$$

Hence

$$\log U(t) \leq \log u_0 + \int_0^t \varphi(s)\, ds,$$

so

$$u(t) \leq U(t) \leq u_0 \exp\left(\int_0^t \varphi(s)\, ds\right). \tag{2.29}$$

If the inequality (2.28) holds for $u_0 = 0$, then it also holds for all $u_0 > 0$, so (2.29) holds for all $u_0 > 0$. Taking the limit of (2.29) as $u_0 \to 0^+$, we conclude that $u(t) = 0$, which proves the result when $u_0 = 0$. $\qquad\square$

Theorem 2.26 Suppose that $f(t, u)$ is continuous in the rectangle

$$R = \{(t, u) \mid |t - t_0| \leq T, |u - u_0| \leq L\},$$

and that

$$|f(t, u)| \leq M \qquad \text{if } (t, u) \in R.$$

Let $\delta = \min(T, L/M)$. If $u(t)$ is any solution of (2.20), then

$$|u(t) - u_0| \leq L \qquad \text{when } |t - t_0| \leq \delta. \tag{2.30}$$

Suppose, in addition, that f is a Lipschitz continuous function of u, uniformly in t, meaning that there is a constant C such that

$$|f(t, u) - f(t, v)| \leq C|u - v| \qquad \text{for all } (t, u) \in R.$$

Then the solution of (2.20) is unique in the interval $|t - t_0| \leq \delta$.

Proof. The result in (2.30) is intuitively obvious: if a solution $u(t)$ stays inside the interval $|u(t) - u_0| \leq L$, then its derivative is bounded by M, so the solution cannot escape the interval in less time than L/M. To avoid circularity in the proof, we use a "continuous induction" argument. We consider the set D defined by

$$D = \{0 \leq \eta \leq \delta \mid |u(t) - u_0| \leq L \text{ for all } |t - t_0| \leq \eta\}.$$

Then $0 \in D$, and if $\eta \in D$, then $\eta' \in D$ for all $0 \leq \eta' \leq \eta$. Thus, D is a nonempty interval. Moreover, D is closed in $[0, \delta]$ because $u(t)$ is a continuous function of t. If $\eta \in D$ and $\eta < \delta$, then $f(t, u(t)) \leq M$ for $|t - t_0| \leq \eta$, so

$$|u(t) - u_0| \leq \left| \int_{t_0}^{t} f(s, u(s)) \, ds \right| \leq M\eta < M\delta = L.$$

Since we have strict inequality, and u is continuous, it follows that there is an $\epsilon > 0$ such that $|u(t) - u_0| \leq L$ when $|t - t_0| \leq \eta + \epsilon$. Thus, D is open in $[0, \delta]$, from which we conclude that $D = [0, \delta]$. This proves the first part of the theorem.

To prove the uniqueness part, we use a common strategy: we derive an equation for the difference of two solutions which shows that it is zero. Suppose that u and v are solutions of (2.20) on a interval I that contains t_0. Then subtraction and integration of the ODEs satisfied by u and v implies that

$$u(t) - v(t) = \int_{t_0}^{t} [f(s, u(s)) - f(s, v(s))] \, ds.$$

Taking the absolute value of this equation, and estimating the result, we find that $w = |u - v|$ satisfies the inequality

$$w(t) \leq \int_{t_0}^{t} |f(s, u(s)) - f(s, v(s))| \, ds. \tag{2.31}$$

By the first part of the theorem, the graph of any solution remains in R for $|t - t_0| \leq \delta$. The Lipschitz continuity of f in R therefore implies that

$$|f(t, u(t)) - f(t, v(t))| \leq C|u(t) - v(t)| = Cw(t).$$

The use of this inequality in (2.31) implies that $w \geq 0$ satisfies

$$w(t) \leq C \int_{t_0}^{t} w(s) \, ds.$$

Therefore, from Gronwall's inequality, we have $w = 0$, and $u = v$.   □

We will give another proof of the existence and uniqueness of solutions of the initial value problem for ODEs in the next chapter, as a consequence of the contraction mapping theorem. The existence theorem above, based on compactness, is called the *Peano existence theorem*, while the theorem in the next chapter, based on the contraction mapping theorem, is called the *Picard existence theorem*.

2.6   References

Most of the material in this chapter is also covered in Rudin [47] and Marsden and Hoffman [37]. For an introduction to the theory of ordinary differential equations, see Hirsch and Smale [21].

2.7   Exercises

Exercise 2.1 Define $f : [0, 1] \to \mathbb{R}$ by

$$f(x) = \begin{cases} x & \text{if } x \text{ is irrational,} \\ p\sin(1/q) & \text{if } x = p/q, \text{ where } p,q \text{ are relatively prime integers.} \end{cases}$$

Determine the set of points where f is continuous.

Exercise 2.2 Let $f_n \in C([a, b])$ be a sequence of functions converging uniformly to a function f. Show that

$$\lim_{n \to \infty} \int_{a}^{b} f_n(x) \, dx = \int_{a}^{b} f(x) \, dx.$$

Give a counterexample to show that the pointwise convergence of continuous functions f_n to a continuous function f does not imply the convergence of the corresponding integrals.

Exercise 2.3 Suppose that $f : G \to \mathbb{R}$ is a uniformly continuous function defined on an open subset G of a metric space X. Prove that f has a unique extension to a continuous function $\overline{f} : \overline{G} \to \mathbb{R}$ defined on the closure $\overline{G}$ of G. Show that such an extension need not exist if f is continuous but not uniformly continuous on G.

Exercise 2.4 Give a counterexample to show that $f_n \to f$ in $C([0,1])$ and f_n continuously differentiable does not imply that f is continuously differentiable.

Exercise 2.5 Consider the space of continuously differentiable functions,

$$C^1 \left([a, b]\right) = \{f : [a, b] \to \mathbb{R} \mid f, \, f' \text{ are continuous}\},$$

with the C^1-norm,

$$\|f\| = \sup_{a \leq x \leq b} |f(x)| + \sup_{a \leq x \leq b} |f'(x)|.$$

Prove that $C^1 \left([a, b]\right)$ is a Banach space.

Exercise 2.6 Show that the space $C \left([a, b]\right)$ equipped with the L^1-norm $\| \cdot \|_1$ defined by

$$\|f\|_1 = \int_a^b |f(x)| \, dx,$$

is incomplete. Show that if $f_n \to f$ with respect to the sup-norm $\| \cdot \|_\infty$, then $f_n \to f$ with respect to $\| \cdot \|_1$. Give a counterexample to show that the converse statement is false.

Exercise 2.7 Prove that the set of Lipschitz continuous functions on $[0, 1]$ with Lipschitz constant less than or equal to one and zero integral is compact in $C([0,1])$.

Exercise 2.8 Prove that $C([a, b])$ is separable.

Exercise 2.9 Let $w : [0, 1] \to \mathbb{R}$ be a nonnegative, continuous function. For $f \in C([0, 1])$, we define the *weighted supremum norm* by

$$\|f\|_w = \sup_{0 \leq x \leq 1} \{w(x)|f(x)|\}.$$

If $w(x) > 0$ for $0 < x < 1$, show that $\| \cdot \|_w$ is a norm on $C([0, 1])$. If $w(x) > 0$ for $0 \leq x \leq 1$, show that $\| \cdot \|_w$ is equivalent to the usual sup-norm, corresponding to $w = 1$. (See Definition 5.21 for the definition of equivalent norms.) Show that the norm $\| \cdot \|_x$ corresponding to $w(x) = x$ is not equivalent to the usual sup-norm. Is the space $C([0, 1])$ equipped with the weighted norm $\| \cdot \|_x$ a Banach space?

Exercise 2.10 Let $C_0(\mathbb{R}^n)$ be the closure of the space $C_c(\mathbb{R}^n)$ of continuous, compactly supported functions with respect to the uniform norm. Prove that $C_0(\mathbb{R}^n)$ is the space of continuous functions that vanish at infinity.

Exercise 2.11 Suppose $f_n \in C([0,1])$ is a monotone decreasing sequence that converges pointwise to $f \in C([0,1])$. Prove that f_n converges uniformly to f. This result is called *Dini's monotone convergence theorem*.

Exercise 2.12 Let $\{f_n \in C([0,1]) \mid n \in \mathbb{N}\}$ be equicontinuous. If $f_n \to f$ pointwise, prove that f is continuous.

Exercise 2.13 Consider the scalar initial value problem,

$$\dot{u}(t) = |u(t)|^\alpha,$$
$$u(0) = 0.$$

Show that the solution is unique if $\alpha \geq 1$, but not if $0 \leq \alpha < 1$.

Exercise 2.14 Suppose that $f(t,u)$ is a continuous function $f : \mathbb{R}^2 \to \mathbb{R}$ such that

$$|f(t,u) - f(t,v)| \leq K|u - v| \qquad \text{for all } t, u, v \in \mathbb{R}.$$

Also suppose that

$$M = \sup \{|f(t, u_0)| \mid |t - t_0| \leq T\}.$$

Prove that the solution $u(t)$ of the initial value problem

$$\dot{u} = f(t, u), \qquad u(t_0) = u_0$$

satisfies the estimate

$$|u(t) - u_0| \leq MTe^{KT} \qquad \text{for } |t - t_0| \leq T.$$

Explicitly check this estimate for the linear initial value problem

$$\dot{u} = Ku, \qquad u(t_0) = u_0.$$

Chapter 3

The Contraction Mapping Theorem

In this chapter we state and prove the *contraction mapping theorem*, which is one
of the simplest and most useful methods for the construction of solutions of linear
and nonlinear equations. We also present a number of applications of the theorem.

3.1 Contractions

Definition 3.1 Let (X, d) be a metric space. A mapping $T : X \to X$ is a *con-traction mapping*, or *contraction*, if there exists a constant c, with $0 \le c < 1$, such
that

$$d(T(x), T(y)) \le c\, d(x, y) \tag{3.1}$$

for all $x, y \in X$.

Thus, a contraction maps points closer together. In particular, for every $x \in X$,
and any $r > 0$, all points y in the ball $B_r(x)$, are mapped into a ball $B_s(Tx)$, with

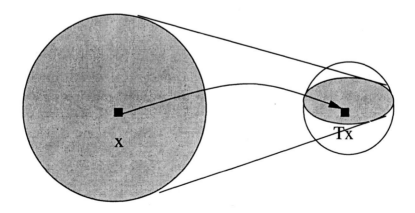

Fig. 3.1 T is a contraction.

$s < r$. This is illustrated in Figure 3.1. Sometimes a map satisfying (3.1) with $c = 1$ is also called a contraction, and then a map satisfying (3.1) with $c < 1$ is called a *strict contraction*. It follows from (3.1) that a contraction mapping is uniformly continuous.

If $T : X \to X$, then a point $x \in X$ such that

$$T(x) = x \tag{3.2}$$

is called a *fixed point* of T. The contraction mapping theorem states that a strict contraction on a complete metric space has a unique fixed point. The contraction mapping theorem is only one example of what are more generally called fixed-point theorems. There are fixed-point theorems for maps satisfying (3.1) with $c = 1$, and even for arbitrary continuous maps on certain metric spaces. For example, the *Schauder fixed point theorem* states that a continuous mapping on a convex, compact subset of a Banach space has a fixed point. The proof is topological in nature (see Kantorovich and Akilov [27]), and we will not discuss such fixed point theorems in this book.

In general, the condition that c is strictly less than one is needed for the uniqueness and the existence of a fixed point. For example, if $X = \{0, 1\}$ is the discrete metric space with metric determined by $d(0, 1) = 1$, then the map T defined by $T(0) = 1$, $T(1) = 0$ satisfies (3.1) with $c = 1$, but T does not have any fixed points. On the other hand, the identity map on any metric space satisfies (3.1) with $c = 1$, and every point is a fixed point.

It is worth noting that (3.2), and hence its solutions, do not depend on the metric d. Thus, if we can find any metric on X such that X is complete and T is a contraction on X, then we obtain the existence and uniqueness of a fixed point. It may happen that X is not complete in any of the metrics for which one can prove that T is a contraction. This can be an indication that the solution of the fixed point problem does not belong to X, but to a larger space, namely the completion of X with respect to a suitable metric d.

Theorem 3.2 (Contraction mapping) If $T : X \to X$ is a contraction mapping on a complete metric space (X, d), then there is exactly one solution $x \in X$ of (3.2).

Proof. The proof is constructive, meaning that we will explicitly construct a sequence converging to the fixed point. Let x_0 be any point in X. We define a sequence (x_n) in X by

$$x_{n+1} = T x_n \qquad \text{for } n \geq 0.$$

To simplify the notation, we often omit the parentheses around the argument of a map. We denote the nth iterate of T by T^n, so that $x_n = T^n x_0$.

First, we show that (x_n) is a Cauchy sequence. If $n \geq m \geq 1$, then from (3.1) and the triangle inequality, we have

$$
\begin{aligned}
d(x_n, x_m) &= d(T^n x_0, T^m x_0) \\
&\leq c^m d(T^{n-m} x_0, x_0) \\
&\leq c^m \left[d(T^{n-m} x_0, T^{n-m-1} x_0) + d(T^{n-m-1} x_0, T^{n-m-2} x_0) \right. \\
&\qquad \left. + \cdots + d(T x_0, x_0) \right] \\
&\leq c^m \left[\sum_{k=0}^{n-m-1} c^k \right] d(x_1, x_0) \\
&\leq c^m \left[\sum_{k=0}^{\infty} c^k \right] d(x_1, x_0) \\
&\leq \left(\frac{c^m}{1-c} \right) d(x_1, x_0),
\end{aligned}
$$

which implies that (x_n) is Cauchy. Since X is complete, (x_n) converges to a limit $x \in X$. The fact that the limit x is a fixed point of T follows from the continuity of T:

$$
Tx = T \lim_{n \to \infty} x_n = \lim_{n \to \infty} T x_n = \lim_{n \to \infty} x_{n+1} = x.
$$

Finally, if x and y are two fixed points, then

$$
0 \leq d(x, y) = d(Tx, Ty) \leq cd(x, y).
$$

Since $c < 1$, we have $d(x, y) = 0$, so $x = y$ and the fixed point is unique. □

3.2 Fixed points of dynamical systems

A dynamical system describes the evolution in time of the state of some system. Dynamical systems arise as models in many different disciplines, including physics, chemistry, engineering, biology, and economics. They also arise as an auxiliary tool for solving other problems in mathematics, and the properties of dynamical systems are of intrinsic mathematical interest.

A dynamical system is defined by a state space X, whose elements describe the different states the system can be in, and a prescription that relates the state $x_t \in X$ at time t to the state at a previous time. We call a dynamical system continuous or discrete, depending on whether the time variable is continuous or discrete. For a continuous dynamical system, the time t belongs to an interval in $\mathbb{R}$, and the dynamics of the system is typically described by an ODE of the form

$$
\dot{x} = f(x), \tag{3.3}
$$

where the dot denotes a time derivative, and f is a vector field on X. There is little loss of generality in assuming this form of equation. For example, a second order, nonautonomous ODE

$$\ddot{y} = g(t, y, \dot{y})$$

may be written as a first order, autonomous system of the form (3.3) for the state variable $x = (s, y, v)$, where $s = t$ and $v = \dot{y}$, with

$$\dot{s} = 1, \qquad \dot{y} = v, \qquad \dot{v} = g(s, y, v).$$

For a discrete dynamical system, we may take time $t = n$ to be an integer, and the dynamics is defined by a map $T : X \to X$ that relates the state x_{n+1} at time $t = n + 1$ to the state x_n at time $t = n$,

$$x_{n+1} = T x_n. \tag{3.4}$$

If T is not invertible, then the dynamics is defined only forward in time, while if T is invertible, then the dynamics is defined both backward and forward in time. A fixed point of the map T correponds to an equilibrium state of the discrete dynamical system. If the state space is a complete metric space and T is a contraction, then the contraction mapping theorem implies that there is a unique equilibrium state, and that the system approaches this state as time tends to infinity starting from any initial state. In this case, we say that the fixed point is globally asymptotically stable.

One of the simplest, and most famous, discrete dynamical systems is the *logistic equation* of population dynamics,

$$x_{n+1} = 4\mu x_n (1 - x_n), \tag{3.5}$$

where $0 \le \mu \le 1$ is a parameter, and $x_n \in [0, 1]$. This equation is of the form (3.4) where $T : [0, 1] \to [0, 1]$ is defined by

$$T x = 4\mu x(1 - x).$$

See Figure 3.2 for a plot of $x \mapsto T x$, for three different values of μ. We may interpret x_n as the population of the nth generation of a reproducing species. The linear equation $x_{n+1} = 4\mu x_n$ describes the exponential growth (if $4\mu > 1$) or decay (if $4\mu < 1$) of a population with constant birth or death rate. The nonlinearity in (3.5) provides a simple model for a species in which the effects of overcrowding lead to a decline in the birth rate as the population increases.

The logistic equation shows that the iterates of even very simple nonlinear maps can have amazingly complex behavior. Analyzing the full behavior of the logistic map is beyond the scope of the elementary application of the contraction mapping theorem we give here.

When $0 \le \mu \le 1/4$, the point 0 is the only fixed point of T in $[0, 1]$, and T is a contraction on $[0, 1]$. The proof of the contraction mapping theorem therefore

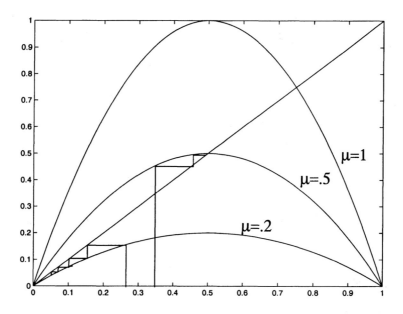

Fig. 3.2   The logistic map for three different values of the parameter μ. For $\mu = 0.2$, 0 is the only fixed point and it is stable. For $\mu = 0.5$, 0 is an unstable fixed point and there is a nonzero, stable, fixed point. For $\mu = 1$, the two fixed points are unstable and more complex (and more interesting) asymptotic behavior of the dynamical system occurs.

implies that $T^n x_0 \to 0$ for an arbitrary initial population $x_0 \in [0,1]$, meaning that the population dies out. When $1/4 < \mu \le 1$, there is a second fixed point at $x = (4\mu - 1)/4\mu$. The appearance of a new fixed point as the parameter μ varies is an example of a *bifurcation* of fixed points. As μ increases further, there is an infinite sequence of more complicated bifurcations, leading to chaotic dynamics for $\mu \ge 0.89\ldots$.

As a second application of the contraction mapping theorem, we consider the solution of an equation $f(x) = 0$. One way to obtain a solution is to recast the equation in the form of a fixed point equation $x = Tx$, and then construct approximations x_n starting from an initial guess x_0 by the iteration scheme

$$x_{n+1} = Tx_n.$$

In other words, we are attempting to find the solution as the time-asymptotic state of an associated discrete dynamical system which has the solution as a stable fixed point. Similar ideas apply in other contexts. For example, we may attempt to construct the solution of an elliptic PDE as the time-asymptotic state of an associated parabolic PDE.

There are many ways to rewrite an equation $f(x) = 0$ as a fixed point problem, some of which will work better than others. Ideally, we would like to rewrite the equation as a fixed point equation in which T is a contraction on the whole space, or at least a contraction on some set that contains the solution we seek.

To provide a simple illustration of these ideas, we prove the convergence of an algorithm to compute square roots. If $a > 0$, then $x = \sqrt{a}$ is the positive solution of the equation

$$x^2 - a = 0.$$

We rewrite this equation as the fixed point problem

$$x = \frac{1}{2}\left(x + \frac{a}{x}\right).$$

The associated iteration scheme is then

$$x_{n+1} = \frac{1}{2}\left(x_n + \frac{a}{x_n}\right),$$

corresponding to a map $T : (0, \infty) \to (0, \infty)$ given by

$$Tx = \frac{1}{2}\left(x + \frac{a}{x}\right). \tag{3.6}$$

Clearly, $x = \sqrt{a}$ is a fixed point of T. Moreover, given an approximation x_n of $\sqrt{a}$, the average of x_n and a/x_n should be a better approximation provided that x_n is not too small, so it is reasonable to expect that the sequence of approximations obtained by iteration of the fixed point equation converges to $\sqrt{a}$. We will prove that this is indeed true by finding an interval on which T is a contraction with respect to the usual absolute value metric on $\mathbb{R}$.

First, let us see whether T contracts at all. For $x_1, x_2 > 0$, we estimate that

$$\begin{aligned} |Tx_1 - Tx_2| &= \left| \frac{1}{2}\left(x_1 + \frac{a}{x_1}\right) - \frac{1}{2}\left(x_2 + \frac{a}{x_2}\right) \right| \\ &= \frac{1}{2}\left| 1 - \frac{a}{x_1 x_2} \right| |x_1 - x_2|. \end{aligned}$$

It follows that T contracts distances when $3x_1 x_2 > a$. To satisfy this condition, we need to exclude arguments x that are too small. Therefore, we consider the action of T on an interval of the form $[b, \infty)$ with $b > 0$. This is a complete metric space because $[b, \infty)$ is a closed subset of $\mathbb{R}$ and $\mathbb{R}$ is complete.

In order to make a good choice for b we first observe that

$$Tx = \sqrt{a} + \frac{(x - \sqrt{a})^2}{2x} \geq \sqrt{a} \tag{3.7}$$

for all $x > 0$. Therefore, the restriction of T to $[\sqrt{a}, \infty)$ is well-defined, since

$$T\left([\sqrt{a}, \infty)\right) \subset [\sqrt{a}, \infty),$$

and T is a contraction on $[\sqrt{a}, \infty)$ with $c = 1/2$. It follows that for any $x_0 \geq \sqrt{a}$, the sequence $x_n = T^n x_0$ converges to $\sqrt{a}$ as $n \to \infty$. Moreover, as shown in the proof of Theorem 3.2, the convergence is exponentially fast, with

$$
\begin{aligned}
|T^n x_0 - \sqrt{a}| &\le |T^n x_0 - T^m x_0| \\
&\le \frac{c^n}{1-c}|T x_0 - x_0| \\
&\le \frac{1}{2^{n-1}}\left|\frac{a}{x_0} - x_0\right|.
\end{aligned}
$$

If $0 < x_0 < \sqrt{a}$, then $x_1 > \sqrt{a}$, and subsequent iterates remain in $[\sqrt{a}, \infty)$, so the iterates converge for any starting guess $x_0 \in (0, \infty)$.

Newton's method for the solution of a nonlinear system of equations, discussed in Section 13.5, can also be formulated as a fixed point iteration.

3.3 Integral equations

A linear *Fredholm integral equation of the second kind* for an unknown function $f : [a, b] \to \mathbb{R}$ is an equation of the form

$$
f(x) - \int_a^b k(x, y) f(y)\, dy = g(x), \tag{3.8}
$$

where $k : [a, b] \times [a, b] \to \mathbb{R}$ and $g : [a, b] \to \mathbb{R}$ are given functions. A *Fredholm integral equation of the first kind* is an equation of the form

$$
\int_a^b k(x, y) f(y)\, dy = g(x).
$$

The integral equation (3.8) may be written as a fixed point equation $Tf = f$, where the map T is defined by

$$
Tf(x) = g(x) + \int_a^b k(x, y) f(y)\, dy. \tag{3.9}
$$

Theorem 3.3 Suppose that $k : [a, b] \times [a, b] \to \mathbb{R}$ is a continuous function such that

$$
\sup_{a \le x \le b} \left\{\int_a^b |k(x, y)|\, dy\right\} < 1, \tag{3.10}
$$

and $g : [a, b] \to \mathbb{R}$ is a continuous function. Then there is a unique continuous function $f : [a, b] \to \mathbb{R}$ that satisfies (3.8).

Proof. We prove this result by showing that, when (3.10) is satisfied, the map T is a contraction on the normed space $C([a, b])$ with the uniform norm $\|\cdot\|_\infty$.

From Theorem 2.4, the space $C([a, b])$ is complete. Moreover, T is a contraction since, for any $f_1, f_2 \in C([a, b])$, we have

$$
\begin{aligned}
\|Tf_1 - Tf_2\|_\infty &= \sup_{a \leq x \leq b} \left| \int_a^b k(x, y)(f_1(y) - f_2(y)) \, dy \right| \\
&\leq \sup_{a \leq x \leq b} \int_a^b |k(x, y)| \, |f_1(y) - f_2(y)| \, dy \\
&\leq \|f_1 - f_2\|_\infty \sup_{a \leq x \leq b} \left\{ \int_a^b |k(x, y)| \, dy \right\} \\
&\leq c\|f_1 - f_2\|_\infty,
\end{aligned}
$$

where

$$
c = \sup_{a \leq x \leq b} \left\{ \int_a^b |k(x, y)| \, dy \right\} < 1.
$$

The result then follows from the contraction mapping theorem. □

From the proof of the contraction mapping theorem, we can obtain the fixed point f as a limit,

$$
f = \lim_{n \to \infty} T^n f_0, \tag{3.11}
$$

for any $f_0 \in C([a, b])$. It is interesting to reinterpret this limit as a series. We define a map $K : C([a, b]) \to C([a, b])$ by

$$
Kf = \int_a^b k(x, y) f(y) \, dy.
$$

The map K is called a *Fredholm integral operator*, and the function k is called the *kernel* of K. Equation (3.8) may be written as

$$
(I - K)f = g, \tag{3.12}
$$

where I is the identity map, meaning that $If = f$. The contraction mapping T is given by $Tf = g + Kf$, which implies that

$$
\begin{aligned}
T^n f_0 &= g + K\left(g + \ldots + K\left(g + Kf_0\right)\right) \\
&= g + Kg + \ldots + K^n g + K^{n+1} f_0.
\end{aligned}
$$

Using this equation in (3.11), we find that

$$
f = \sum_{n=0}^\infty K^n g.
$$

Since $f = (I - K)^{-1}g$, we may write this equation formally as

$$(I - K)^{-1} = \sum_{n=0}^{\infty} K^n. \tag{3.13}$$

This series is called the *Neumann series*. The use of the partial sums of this series to approximate the inverse is called the *Born approximation*. Explicitly, we have

$$\left(I + K + K^2 + \ldots\right) f(x)$$
$$= f(x) + \int_a^b k(x, y) f(y) \, dy + \int_a^b \int_a^b k(x, y) k(y, z) f(z) \, dy dz + \ldots.$$

The Neumann series resembles the geometric series,

$$(1 - x)^{-1} = \sum_{n=0}^{\infty} x^n \qquad \text{for } |x| < 1.$$

In fact, (3.13) really is a geometric series that is absolutely convergent with respect to a suitable operator norm when $\|K\| < 1$ (see Exercise 5.17). This explains why we do not need a condition on g; equation (3.10) is a condition that ensures $I - K$ is invertible, and this only involves k.

3.4 Boundary value problems for differential equations

Consider a copper rod or pipe that is wrapped with imperfect insulation. We use the temperature outside the rod as the zero point of our temperature scale. We denote the spatial coordinate along the rod by x, nondimensionalized so that the length of the rod is one, and time by t. The temperature $u(x, t)$ of the rod then satisfies the following linear PDE,

$$u_t = u_{xx} - q(x)u, \tag{3.14}$$

where the subscripts denote partial derivatives,

$$u_t = \frac{\partial u}{\partial t}, \qquad u_{xx} = \frac{\partial^2 u}{\partial x^2}.$$

The lateral heat loss is proportional to the coefficient function $q(x)$ and the temperature difference u between the rod and the outside. If the rod is perfectly insulated, then $q = 0$ and (3.14) is the one-dimensional *heat* or *diffusion equation*.

Equation (3.14) does not uniquely determine u, and it has to be supplemented by an initial condition

$$u(x, 0) = u_0(x) \qquad \text{for } 0 < x < 1$$

that specifies the initial temperature $u_0(x)$ of the rod, and boundary conditions at the ends of the rod. We suppose that the ends $x = 0$ and $x = 1$ of the rod are kept at constant temperatures of T_0 and T_1, respectively. Then

$$u(0, t) = T_0, \qquad u(1, t) = T_1 \qquad \text{for all } t > 0.$$

As $t \to \infty$, the system "forgets" its initial state and approaches an equilibrium state $u = u(x)$, which satisfies the following boundary value problem (BVP) for an ODE:

$$-u''(x) + q(x)u(x) = 0, \qquad 0 < x < 1, \tag{3.15}$$
$$u(0) = T_0, \quad u(1) = T_1. \tag{3.16}$$

If q is a constant function, then (3.15) is easy to solve explicitly; but if $q(x)$ is not constant, explicit integration is in general impossible. We will use the contraction mapping theorem to show that there is a unique solution of this BVP when q is not too large.

First, we replace the nonhomogeneous boundary conditions (3.16) by the corresponding homogeneous conditions. To do this, we write u as

$$u(x) = v(x) + u_p(x),$$

where v is a new unknown function and u_p is a function that satisfies the nonhomogeneous boundary conditions. A convenient choice for u_p is the linear function

$$u_p(x) = T_0 + (T_1 - T_0)\, x. \tag{3.17}$$

The function v then satisfies

$$-v'' + q(x)v(x) = f(x), \tag{3.18}$$
$$v(0) = 0, \quad v(1) = 0, \tag{3.19}$$

where $f = -q u_p$. The transfer of nonhomogeneous terms between the boundary conditions and the differential equation is a common procedure in the analysis of linear boundary value problems. The boundary value problem (3.18)–(3.19) is an example of a Sturm-Liouville problem, which we will study in Chapter 10. We will use the following proposition to reformulate this boundary value problem as a fixed point problem for a Fredholm integral operator.

Proposition 3.4 Let $f : [0, 1] \to \mathbb{R}$ be a continuous function. The unique solution v of the boundary value problem

$$-v'' = f, \tag{3.20}$$
$$v(0) = 0, \quad v(1) = 0, \tag{3.21}$$

is given by

$$v(x) = \int_0^1 g(x, y) f(y)\, dy,$$

where

$$g(x,y) = \begin{cases} x(1-y) & \text{if } 0 \le x \le y \le 1, \\ y(1-x) & \text{if } 0 \le y \le x \le 1. \end{cases} \qquad (3.22)$$

Proof. Integrating (3.20) twice, we obtain that

$$v(x) = -\int_0^x \int_1^y f(s) \, ds \, dy + C_1 x + C_2,$$

where C_1 and C_2 are two real constants to be determined later. Integration by parts of the integral with respect to y in this equation gives

$$\begin{aligned} v(x) &= -\left[y \int_1^y f(s) \, ds \right]_0^x + \int_0^x y f(y) \, dy + C_1 x + C_2 \\ &= x \int_x^1 f(y) \, dy + \int_0^x y f(y) \, dy + C_1 x + C_2. \end{aligned}$$

We determine C_1 and C_2 from (3.21), which implies that

$$C_1 = -\int_0^1 y f(y) \, dy, \qquad C_2 = 0.$$

It follows that

$$v(x) = \int_0^x y(1-x) f(y) \, dy + \int_x^1 x(1-y) f(y) \, dy,$$

which is what we had to prove. □

The function $g(x,y)$ constructed in this proposition is called the *Green's function* of the differential operator $A = -d^2/dx^2$ in (3.20) with the Dirichlet boundary conditions (3.21). The inverse of the differential operator A is an integral operator whose kernel is the Green's function. We will study Green's functions in greater detail in Chapter 10.

Replacing f by $-qv + f$ in Proposition 3.4, we may rewrite (3.18) as an integral equation for v,

$$v(x) = -\int_0^1 g(x,y) q(y) v(y) \, dy + \int_0^1 g(x,y) f(y) \, dy.$$

This equation has the form

$$(I - K)v = h,$$

where the integral operator K and the right-hand side h are given by:

$$\begin{aligned} Kv(x) &= -\int_0^1 g(x,y) q(y) v(y) \, dy, \\ h(x) &= -\int_0^1 g(x,y) q(y) u_p(y) \, dy. \end{aligned} \qquad (3.23)$$

Theorem 3.3 now implies the following result.

Theorem 3.5 If $q : [0, 1] \to \mathbb{R}$ is continuous, and

$$\sup_{0 \leq x \leq 1} \left\{ \int_0^1 |g(x, y)q(y)| \, dy \right\} < 1, \qquad (3.24)$$

where $g(x, y)$ is defined in (3.22), then the boundary value problem (3.15)–(3.16) has a unique solution.

Using (3.22), we find the estimate

$$\sup_{0 \leq x \leq 1} \left\{ \int_0^1 |g(x, y)q(y)| \, dy \right\} \leq \frac{1}{8} \|q\|_\infty.$$

Thus, there is a unique solution of (3.15)–(3.16) for any continuous q with $\|q\|_\infty < 8$. Existence or uniqueness may break down when $\|q\|_\infty$ is sufficiently large. For example, if $q = -n^2 \pi^2$, where $n = 1, 2, 3, \ldots$, then the BVP

$$-u'' - n^2 \pi^2 u = 0,$$
$$u(0) = 0, \qquad u(1) = 0,$$

has the one-parameter family of solutions $u = c \sin n\pi x$, where c is an arbitrary constant, and no solution may exist in the case of nonzero boundary conditions. Since $\pi^2 > 8$, this result is consistent with Theorem 3.5.

The nonuniqueness breaks down only if q is negative. This is not physically realistic in the heat flow problem, where it would correspond to the flow of heat from cold to hot, but the same equation arises in many other problems with negative coefficient functions q.

If $q(x) > 0$ in $0 < x < 1$, we may prove uniqueness by a simple maximum principle argument. Suppose that $-u'' + q(x)u = 0$ and $u(0) = u(1) = 0$. Since u is continuous, it attains its maximum and minimum on the interval $[0, 1]$. If u attains its maximum at an interior point, then $u'' \leq 0$ at that point, so $u = u''/q \leq 0$. Since $u = 0$ at the endpoints, we conclude that

$$\max_{0 \leq x \leq 1} u(x) \leq 0.$$

Similarly, at an interior minimum, we have $u'' \geq 0$, so $u \geq 0$, and therefore

$$\min_{0 \leq x \leq 1} u(x) \geq 0.$$

It follows that $u = 0$, so the solution of the boundary value problem is unique. Generalizations of this maximum principle argument apply to scalar elliptic partial differential equations, such as Laplace's equation.

3.5 Initial value problems for differential equations

The contraction mapping theorem may be used to prove the existence and unique-ness of solutions of the initial value problem for ordinary differential equations. We consider a first-order system of ODEs for a function $u(t)$ that takes values in $\mathbb{R}^n$,

$$\dot{u}(t) = f\left(t, u(t)\right), \tag{3.25}$$

$$u(t_0) = u_0. \tag{3.26}$$

The function $f(t, u)$ also takes values in $\mathbb{R}^n$, and is assumed to be a continuous function of t and a Lipschitz continuous function of u on a suitable domain. The initial value problem (3.25)–(3.26) can be reformulated as an integral equation,

$$u(t) = u_0 + \int_{t_0}^{t} f\left(s, u(s)\right) ds. \tag{3.27}$$

By the fundamental theorem of calculus, a continuous solution of (3.27) is a con-tinuously differentiable solution of (3.25)–(3.26). Equation (3.27) may be written as a fixed point equation

$$u = Tu \tag{3.28}$$

for the map T defined by

$$Tu(t) = u_0 + \int_{t_0}^{t} f(s, u(s)) ds. \tag{3.29}$$

We want to find conditions which guarantee that T is a contraction on a suitable space of continuous functions. The simplest such condition is given in the following definition.

Definition 3.6 Suppose that $f : I \times \mathbb{R}^n \to \mathbb{R}^n$, where I is an interval in $\mathbb{R}$. We say that $f(t, u)$ is a *globally Lipschitz continuous function of u uniformly in t* if there is a constant $C > 0$ such that

$$\|f(t, u) - f(t, v)\| \leq C \|u - v\| \qquad \text{for all } u, v \in \mathbb{R}^n \text{ and all } t \in I. \tag{3.30}$$

Theorem 3.7 Suppose that $f : I \times \mathbb{R}^n \to \mathbb{R}^n$, where I is an interval in $\mathbb{R}$ and t_0 is a point in the interior of I. If $f(t, u)$, is a continuous function of (t, u) and a globally Lipschitz continuous function of u, uniformly in t, on $I \times \mathbb{R}^n$, then there is a unique continuously differentiable function $u : I \to \mathbb{R}^n$ that satisfies (3.25).

Proof. We will show that T is a contraction on the space of continuous functions defined on a time interval $t_0 \leq t \leq t_0 + \delta$, for sufficiently small δ. Suppose that $u, v : [t_0, t_0 + \delta] \to \mathbb{R}^n$ are two continuous functions. Then, from (3.29) and (3.30), we estimate

$$
\begin{aligned}
\|Tu - Tv\|_\infty &= \sup_{t_0 \le t \le t_0 + \delta} \|Tu(t) - Tv(t)\| \\
&= \sup_{t_0 \le t \le t_0 + \delta} \left\| \int_{t_0}^t f(s, u(s)) - f(s, v(s)) \, ds \right\| \\
&\le \sup_{t_0 \le t \le t_0 + \delta} \int_{t_0}^t \|f(s, u(s)) - f(s, v(s))\| \, ds \\
&\le \sup_{t_0 \le t \le t_0 + \delta} \int_{t_0}^t C \|u(s) - v(s)\| \, ds \\
&\le C\delta \|u - v\|_\infty.
\end{aligned}
$$

It follows that if $\delta < 1/C$, then T is a contraction on $C([t_0, t_0 + \delta])$. Therefore, there is a unique solution $u : [t_0, t_0 + \delta] \to \mathbb{R}^n$. The argument holds for any $t_0 \in I$, and by covering I with overlapping intervals of length less than $1/C$, we see that (3.25) has a unique continuous solution defined on all of I. The same proof applies for times $t_0 - \delta < t < t_0$.   □

We may have $I = \mathbb{R}$ in this theorem, in which case the solution exists globally.

Example 3.8 Linear ODEs with continuous coefficients have unique global solutions. Since higher order ODEs may be reduced to first-order systems, it is sufficient to consider a first-order linear system of the form

$$
\dot{u}(t) = A(t)u(t) + b(t),
$$
$$
u(0) = u_0,
$$

where $A : \mathbb{R} \to \mathbb{R}^{n \times n}$ is a continuous matrix-valued function (with respect to any matrix norm — see the discussion in Section 5.2 below) and $b : \mathbb{R} \to \mathbb{R}^n$ is a continuous vector-valued function. For any bounded interval $I \subset \mathbb{R}$, there exists a constant C such that

$$
\|A(t)u\| \le C\|u\| \qquad \text{for all } t \in I \text{ and } u \in \mathbb{R}^n.
$$

Therefore,

$$
\|A(t)u - A(t)v\| \le C\|u - v\|,
$$

for all $u, v \in \mathbb{R}^n$ and $t \in I$, so the hypotheses of Theorem 3.7 are satisfied, and we have a unique continuous solution on I. Since I is an arbitrary interval, we conclude that there is a unique continuous solution for all $t \in \mathbb{R}$.

The applications of Theorem 3.7 are not limited to linear ODEs.

Example 3.9 Consider the nonlinear, scalar ODE given by

$$\dot{u}(t) = \sqrt{a(t)^2 + u(t)^2},$$
$$u(0) = u_0,$$

where $a : \mathbb{R} \to \mathbb{R}$ is a continuous function. Here,

$$f(t, u) = \sqrt{a(t)^2 + u^2}.$$

This function is globally Lipschitz, with

$$|f(t, u) - f(t, v)| \le |u - v| \qquad \text{for all } t, u, v, \in \mathbb{R},$$

since

$$\left| \sqrt{a^2 + u^2} - \sqrt{a^2 + v^2} \right| = \frac{\left| (a^2 + u^2) - (a^2 + v^2) \right|}{\sqrt{a^2 + u^2} + \sqrt{a^2 + v^2}}$$
$$\le |u - v| \frac{|u| + |v|}{\sqrt{a^2 + u^2} + \sqrt{a^2 + v^2}}$$
$$\le |u - v|.$$

Theorem 3.7 implies that there is a unique global solution of this ODE.

The global Lipschitz condition (3.30) plays two roles in Theorem 3.7. First, it ensures uniqueness, which may fail if f is only a continuous function of u. Second, it implies that f does not grow faster than a linear function of u as $\|u\| \to \infty$. This is what guarantees global existence. If f is a nonlinear function, such as $f(u) = u^2$ in Example 2.22, that satisfies a local Lipschitz condition but not a global Lipschitz condition, then "blow-up" may occur, so that the solution exists only locally.

The above proof may be modified to provide a local existence result.

Theorem 3.10 (Local existence for ODEs) Let $f : I \times \overline{B}_R(u_0) \to \mathbb{R}^n$, where

$$I = \{ t \in \mathbb{R} \mid |t - t_0| \le T \}$$

is an interval in $\mathbb{R}$, and

$$\overline{B}_R(u_0) = \{ u \in \mathbb{R}^n \mid \|u - u_0\| \le R \}$$

is the closed ball of radius $R > 0$ centered at $u_0 \in \mathbb{R}^n$. Suppose that $f(t, u)$ is continuous on $I \times \overline{B}_R(u_0)$ and Lipschitz continuous with respect to u uniformly in t. Let

$$M = \sup \left\{ \|f(t, u)\| \mid t \in I \text{ and } u \in \overline{B}_R(u_0) \right\} < \infty.$$

Then the initial value problem

$$\dot{u} = f(u), \qquad u(t_0) = u_0$$

has a unique continuously differentiable local solution $u(t)$, defined in the time interval $|t - t_0| < \delta$, where $\delta = \min(T, R/M)$.

Proof.   We rewrite the initial value problem as a fixed point equation $u = Tu$, where

$$Tu(t) = u_0 + \int_{t_0}^{t} f(s, u(s)) \, ds.$$

For $0 < \eta < \delta$ we define

$$X = \left\{ u : [t_0 - \eta, t_0 + \eta] \to \overline{B}_R(u_0) \mid u \text{ is continuous} \right\},$$

where X is equipped with the sup-norm,

$$\|u\|_\infty = \sup_{|t - t_0| \leq \eta} \|u(t)\|.$$

We will show that T maps X into X, and is a contraction when η is sufficiently small.

First, if $u \in X$, then

$$\|Tu(t) - u_0\| = \left\| \int_{t_0}^{t} f(s, u(s)) \, ds \right\| \leq M\eta < R.$$

Hence $Tu \in X$ so that $T : X \to X$.

Second, we estimate

$$\begin{aligned}
\|Tu - Tv\|_\infty &= \sup_{|t - t_0| \leq \eta} \left\| \int_{t_0}^{t} [f(s, u(s)) - f(s, v(s))] \, ds \right\| \\
&\leq C\eta \|u - v\|_\infty,
\end{aligned}$$

where C is a Lipschitz constant for f. Hence if we choose $\eta = 1/(2C)$ then T is a contraction on X and it has a unique fixed point.

Since η depends only on the Lipschitz constant of f and on the distance R of the initial data from the boundary of $\overline{B}_R(u_0)$, repeated application of this result gives a unique local solution defined for $|t - t_0| < \delta$. $\square$

A significant feature of this result is that if $f(t, u)$ is continuous for all $t \in \mathbb{R}$, then the existence time δ only depends on the norm of u. Thus, the only way in which the solution of an ODE can fail to exist, assuming that the vector field f is Lipschitz continuous on any ball, is if $\|u(t)\|$ becomes unbounded. There are many functions that are bounded and continuous on an open interval which cannot be extended continuously to $\mathbb{R}$; for example, $u(t) = \sin(1/t)$ is bounded and continuous on $(-\infty, 0)$ but has no continuous extension to $(-\infty, 0]$. This kind of behavior cannot happen for solutions of ODEs with continuous right-hand sides, because the derivative of the solution cannot become large unless the solution itself becomes large. If we can prove that for every $T > 0$ any local solution satisfies an *a priori* estimate of the form

$$\|u(t)\| \leq R \qquad \text{for } |t| \leq T,$$

then the local existence theorem implies that the local solution can be extended to the interval $(-T, T)$, and hence to a global solution.

Example 3.11 A *gradient flow* is defined by a system of ODEs of the form

$$\dot{u} = -\nabla V(u), \tag{3.31}$$

where $V : \mathbb{R}^n \to \mathbb{R}$ is a smooth real-valued function of u, and ∇ denotes the gradient with respect to u. The component form of this equation is

$$\dot{u}_i = -\frac{\partial V}{\partial u_i}.$$

Solutions of a gradient system flow "down hill" in the direction of decreasing V. It follows from (3.31) and the chain rule that

$$\dot{V}(u) = \nabla V(u) \cdot \dot{u} = -\|\nabla V(u)\|^2 \leq 0.$$

Thus, if $V_0 = V(u(0))$, we have

$$V(u(t)) \leq V_0 \qquad \text{for } t > 0$$

for any local solution. Therefore, if the set $\{u \in \mathbb{R}^n \mid V(u) \leq V_0\}$ is bounded (which is the case, for example, if $V(u) \to \infty$ as $\|u\| \to \infty$), then the solution of the initial value problem for (3.31) exists for all $t > 0$. The function V is an example of a *Liapunov function*.

Most systems of ODEs cannot be written as a gradient system for any potential $V(u)$. If f is a smooth vector field on $\mathbb{R}^n$, then $f = -\nabla V$ if and only if its components f_i satisfy the integrability conditions that arise from the equality of mixed partial derivatives of V,

$$\frac{\partial f_i}{\partial u_j} = \frac{\partial f_j}{\partial u_i}.$$

With a suitable definition of the gradient of functionals on an infinite-dimensional space (see Chapter 13), a number of PDEs, such as the heat equation, can also be interpreted as gradient flows.

Example 3.12 A *Hamiltonian system* of ODEs for $q(t), p(t) \in \mathbb{R}^n$ is a system of the form

$$\dot{q} = \nabla_p H, \qquad \dot{p} = -\nabla_q H, \tag{3.32}$$

where $H(q,p)$ is a given smooth function, $H : \mathbb{R}^{2n} \to \mathbb{R}$, called the *Hamiltonian*. The chain rule implies that

$$\dot{H} = \nabla_q H \cdot \dot{q} + \nabla_p H \cdot \dot{p} = 0,$$

so H is constant on solutions. If the level sets of H are bounded, then solutions of (3.32) exist globally in time.

3.6   References

General references for this chapter are Simmons [50] and Marsden and Hoffman [37]. For more about the logistic map and chaotic dynamical systems, see Devaney [8], Guckenheimer and Holmes [18], and Schroeder [51]. Hirsch and Smale [21] discusses gradient flows.

3.7   Exercises

Exercise 3.1 Show that $T : \mathbb{R} \to \mathbb{R}$ defined by

$$T(x) = \frac{\pi}{2} + x - \tan^{-1} x$$

has no fixed point, and

$$|T(x) - T(y)| < |x - y| \qquad \text{for all } x \neq y \in \mathbb{R}.$$

Why doesn't this example contradict the contraction mapping theorem?

Exercise 3.2   (a) Show that for any $y > 0$, the convergence of $T^n y$ to $\sqrt{x}$ is in fact faster than exponential. Start from (3.7) to obtain a good estimate for $|Ty - \sqrt{x}|$.
  (b) Take $y = 1$ and $x = 2$. Find a good estimate of how large n needs to be for $T^n 1$ and $\sqrt{2}$ to have identical first d digits. For example, how many iterations of the map T are sufficient to compute the first 63 digits of $\sqrt{2}$?

Exercise 3.3 The *secant method* for solving an equation $f(x) = 0$, where $f : \mathbb{R} \to \mathbb{R}$, is a variant of Newton's method. Starting with two initial points x_0, x_1, we compute a sequence of iterates (x_n) by:

$$x_{n+1} = x_n - f(x_n) \frac{x_n - x_{n-1}}{f(x_n) - f(x_{n-1})} \qquad \text{for } n = 1, 2, 3, \ldots$$

Formulate and prove a convergence theorem for the secant method with $f(x) = x^2 - 2$.

Exercise 3.4 If T satisfies (3.1) for $c < 1$, show that we can estimate the distance of the fixed point x from the initial point x_0 of the fixed point iteration by

$$d(x, x_0) \leq \frac{1}{1 - c} d(x_1, x_0).$$

Exercise 3.5 An $n \times n$ matrix A is said to be *diagonally dominant* if for each row the sum of the absolute values of the off-diagonal terms is less than the absolute value of the diagonal term. We write $A = D - L - U$ where D is diagonal, L is lower triangular, and U is upper triangular. If A is diagonally dominant, show that

$$\|L + U\|_\infty < \|D\|_\infty,$$

where the ∞-norm $\|\cdot\|_\infty$ of a matrix is defined in (5.9). Use the contraction mapping theorem to prove that if A is diagonally dominant, then A is invertible and that the following iteration schemes converge to a solution of the equation $Ax = b$:

$$x_{n+1} = D^{-1} (L + U) x_n + D^{-1} b, \tag{3.33}$$
$$x_{n+1} = (D - L)^{-1} U x_n + (D - L)^{-1} b. \tag{3.34}$$

What can you say about the rates of convergence?

Such iterative schemes provide an efficient way to compute numerical solutions of large, sparse linear systems. The iteration (3.33) is called *Jacobi's method*, and (3.34) is called the *Gauss-Seidel method*.

Exercise 3.6 The following integral equation for $f : [-a, a] \to \mathbb{R}$ arises in a model of the motion of gas particles on a line:

$$f(x) = 1 + \frac{1}{\pi} \int_{-a}^{a} \frac{1}{1 + (x - y)^2} f(y) \, dy \qquad \text{for } -a \leq x \leq a.$$

Prove that this equation has a unique bounded, continuous solution for every $0 < a < \infty$. Prove that the solution is nonnegative. What can you say if $a = \infty$?

Exercise 3.7 Prove that there is a unique solution of the following nonlinear BVP when the constant λ is sufficiently small,

$$-u'' + \lambda \sin u = f(x),$$
$$u(0) = 0, \qquad u(1) = 0.$$

Here, $f : [0, 1] \to \mathbb{R}$ is a given continuous function. Write out the first few iterates of a uniformly convergent sequence of approximations, beginning with $u_0 = 0$. HINT. Reformulate the problem as a nonlinear integral equation.

Chapter 4

Topological Spaces

In the previous chapters, we discussed the convergence of sequences, the continuity of functions, and the compactness of sets. We expressed these properties in terms of a metric or norm. Some types of convergence, such as the pointwise convergence of real-valued functions defined on an interval, cannot be expressed in terms of a metric on a function space. Topological spaces provide a general framework for the study of convergence, continuity, and compactness. The fundamental structure on a topological space is not a distance function, but a collection of open sets; thinking directly in terms of open sets often leads to greater clarity as well as greater generality.

4.1 Topological spaces

Definition 4.1 A *topology* on a nonempty set X is a collection of subsets of X, called *open sets*, such that:

    (a) the empty set $\emptyset$ and the set X are open;
    (b) the union of an arbitrary collection of open sets is open;
    (c) the intersection of a finite number of open sets is open.

A subset A of X is a *closed set* if and only if its complement, $A^c = X \setminus A$, is open.

More formally, a collection $\mathcal{T}$ of subsets of X is a topology on X if:

    (a) $\emptyset, X \in \mathcal{T}$;
    (b) if $G_\alpha \in \mathcal{T}$ for $\alpha \in \mathcal{A}$, then $\bigcup_{\alpha \in \mathcal{A}} G_\alpha \in \mathcal{T}$;
    (c) if $G_i \in \mathcal{T}$ for $i = 1, 2 \ldots, n$, then $\bigcap_{i=1}^{n} G_i \in \mathcal{T}$.

We call the pair $(X, \mathcal{T})$ a *topological space*; if $\mathcal{T}$ is clear from the context, then we often refer to X as a topological space.

Example 4.2 Let X be a nonempty set. The collection $\{\emptyset, X\}$, consisting of the empty set and the whole set, is a topology on X, called the *trivial topology* or

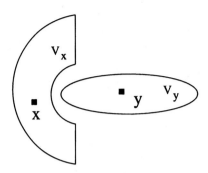

Fig. 4.1   The Hausdorff property.

indiscrete topology. The *power set* $\mathcal{P}(X)$ of X, consisting of all subsets of X, is a topology on X, called the *discrete topology.*

Example 4.3 Let (X, d) be a metric space. Then the set of all open sets defined in Definition 1.36 is a topology on X, called the *metric topology.* For instance, a subset G of $\mathbb{R}$ is open with respect to the standard, metric topology on $\mathbb{R}$ if and only if for every $x \in G$ there is an open interval I such that $x \in I$ and $I \subset G$.

Example 4.4 Let $(X, \mathcal{T})$ be a topological space and Y a subset of X. Then

$$\mathcal{S} = \{ H \subset Y \mid H = G \cap Y \text{ for some } G \in \mathcal{T} \}$$

is a topology on Y. The open sets in Y are the intersections of open sets in X with Y. This topology is called the *induced* or *relative topology* of Y in X, and $(Y, \mathcal{S})$ is called a topological subspace of $(X, \mathcal{T})$. For instance, the interval $[0, 1/2)$ is an open subset of $[0, 1]$ with respect to the induced metric topology of $[0, 1]$ in $\mathbb{R}$, since $[0, 1/2) = (-1/2, 1/2) \cap [0, 1]$.

A set $V \subset X$ is a *neighborhood* of a point $x \in X$ if there exists an open set $G \subset V$ with $x \in G$. We do not require that V itself is open. A topology $\mathcal{T}$ on X is called *Hausdorff* if every pair of distinct points $x, y \in X$ has a pair of nonintersecting neighborhoods, meaning that there are neighborhoods V_x of x and V_y of y such that $V_x \cap V_y = \emptyset$ (see Figure 4.1). When the topology is clear, we often refer to X as a Hausdorff space. Almost all the topological spaces encountered in analysis are Hausdorff. For example, all metric topologies are Hausdorff. On the other hand, if X has at least two elements, then the trivial topology on X is not Hausdorff.

We can express the notions of convergence, continuity, and compactness in terms of open sets. Let X and Y be a topological spaces.

Definition 4.5 A sequence (x_n) in X *converges* to a limit $x \in X$ if for every neighborhood V of x, there is a number N such that $x_n \in V$ for all $n \geq N$.

This definition says that the sequence eventually lies entirely in every neighborhood of x.

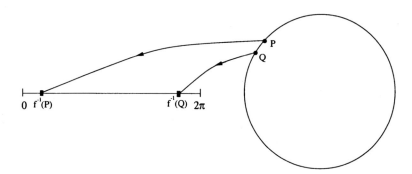

Fig. 4.2 The interval and the circle are not homeomorphic. There are arbitrarily close points on the circle, P and Q in the figure, which have inverse images near the left and right end points of the interval respectively. Hence, the inverse images are not close and the inverse map cannot be continuous.

Definition 4.6 A function $f : X \to Y$ is *continuous* at $x \in X$ if for each neighborhood W of $f(x)$ there exists a neighborhood V of x such that $f(V) \subset W$. We say that f is continuous on X if it is continuous at every $x \in X$.

Theorem 4.7 Let $(X, \mathcal{T})$ and $(Y, \mathcal{S})$ be two topological spaces and $f : X \to Y$. Then f is continuous on X if and only if $f^{-1}(G) \in \mathcal{T}$ for every $G \in \mathcal{S}$.

Thus, a continuous function is characterized by the property that the inverse image of an open set is open. We leave the proof to Exercise 4.4.

Definition 4.8 A function $f : X \to Y$ between topological spaces X and Y is a *homeomorphism* if it is a one-to-one, onto map and both f and f^{-1} are continuous. Two topological spaces X and Y are *homeomorphic* if there is a homeomorphism $f : X \to Y$.

Homeomorphic spaces are indistinguishable as topological spaces. For example, if $f : X \to Y$ is a homeomorphism, then G is open in X if and only if $f(G)$ is open in Y, and a sequence (x_n) converges to x in X if and only if the sequence $(f(x_n))$ converges to $f(x)$ in Y.

A one-to-one, onto map f always has an inverse f^{-1}, but f^{-1} need not be continuous even if f is.

Example 4.9 We define $f : [0, 2\pi) \to \mathbb{T}$ by $f(\theta) = e^{i\theta}$, where $[0, 2\pi) \subset \mathbb{R}$ with the topology induced by the usual topology on $\mathbb{R}$, and $\mathbb{T} \subset \mathbb{C}$ is the unit circle with the topology induced by the usual topology on $\mathbb{C}$. Then, as illustrated in Figure 4.2, f is continuous but f^{-1} is not.

Definition 4.10 A subset K of a topological space X is *compact* if every open cover of K contains a finite subcover.

It follows from the definition that a subset K of X is compact in the topology on X if and only if K is compact as a subset of itself with respect to the relative topology of K in X. This contrasts with the fact that a set $G \subset Y$ may be relatively open in Y, yet not be open in X. For this reason, while we define the notion of relatively open, we do not define the notion of relatively compact.

4.2   Bases of open sets

The collection of all open sets in a topological space is often huge and unwieldy. The topological properties of metric spaces can be expressed entirely in terms of open balls, which form a rather small subset of the open sets. In this section we introduce subsets of a topological space that play a similar role to open balls in a metric space.

Definition 4.11 A subset $\mathcal{B}$ of a topology $\mathcal{T}$ is a *base* for $\mathcal{T}$ if for every $G \in \mathcal{T}$ there is a collection of sets $B_\alpha \in \mathcal{B}$ such that $G = \bigcup_\alpha B_\alpha$. A collection $\mathcal{N}$ of neighborhoods of a point $x \in X$ is called a *neighborhood base* for x if for each neighborhood V of x there is a neighborhood $W \in \mathcal{N}$ such that $W \subset V$. A topological space X is *first countable* if every $x \in X$ has a countable neighborhood base, and *second countable* if X has a countable base.

Example 4.12 The collection of all open intervals (a, b) with $a, b \in \mathbb{R}$ is a base for the standard topology on $\mathbb{R}$. The collection of all open intervals $(a, b) \subset \mathbb{R}$ with rational endpoints $a, b \in \mathbb{Q}$ is a countable base for the standard topology on $\mathbb{R}$. Thus, the standard topology is second countable.

Example 4.13 Let X be a metric space and A a dense subspace of X. The set of open balls $B_{1/n}(x)$, with $n \geq 1$ and $x \in A$ is a base for the metric topology on X. A metric space is first countable, and a separable metric space is second countable.

Example 4.14 If X is topological space with the discrete topology, then the collection of open sets

$$\mathcal{B} = \{\{x\} \mid x \in X\}$$

is a base. The discrete topology is first countable, and if X is countable, then it is second countable.

It is often useful to define a topology in terms of a base.

Theorem 4.15 A collection of open sets $\mathcal{B} \subset \mathcal{T}$ is a base for the topology $\mathcal{T}$ on a set X if and only if $\mathcal{B}$ contains a neighborhood base for x for every $x \in X$.

Proof.   Suppose $\mathcal{B}$ is a base for $\mathcal{T}$. If N is a neighborhood of $x \in X$, then there is an open set $G \in \mathcal{T}$ such that $x \in G \subset N$. Since $\mathcal{B}$ is a base, there are sets $B_\alpha \in \mathcal{B}$

such that $\bigcup_\alpha B_\alpha = G$. Therefore, there is an α such that $x \in B_\alpha$ and $B_\alpha \subset N$. It follows that $\mathcal{B}$ contains a neighborhood base for x.

Conversely, if a collection of open sets $\mathcal{B}$ contains a neighborhood base for every $x \in X$, then for every open set $G \in \mathcal{T}$ and every $x \in G$ there exists $B_x \in \mathcal{B}$ such that $x \in B_x \subset G$. Therefore, $\bigcup_x B_x = G$, so $\mathcal{B}$ is a base for the topology. $\qquad \square$

Example 4.16 Suppose that X is the space of all real-valued functions on the interval $[a, b]$. We may identify a function $f : [a, b] \to \mathbb{R}$ with a point $\prod_{x\in[a,b]} f(x)$ in $\mathbb{R}^{[a,b]}$, so $X = \mathbb{R}^{[a,b]}$ is the $[a, b]$-fold Cartesian product of $\mathbb{R}$. Let $x = \{x_1, \dots, x_n\}$, where $x_i \in [a, b]$, and $y = \{y_1, \dots, y_n\}$, where $y_i \in \mathbb{R}$, be finite subsets of $[a, b]$ and $\mathbb{R}$, respectively. For $\epsilon > 0$, we define a subset $B_{x,y,\epsilon}$ of X by

$$B_{x,y,\epsilon} = \{f \in X \mid |f(x_i) - y_i| < \epsilon \text{ for } i = 1, \dots, n\}. \tag{4.1}$$

The topology of pointwise convergence is the smallest topology on X that contains the sets $B_{x,y,\epsilon}$ for all finite sets $x \subset [a, b]$, $y \subset \mathbb{R}$, and $\epsilon > 0$. We have $f_n \to f$ with respect to this topology if and only if $f_n(x) \to f(x)$ for every $x \in [a, b]$. If $f \in X$ and $y_i = f(x_i)$, then the sets $B_{x,y,\epsilon}$ form a neighborhood base for $f \in X$. This topology is not first countable.

The set $B_{x,y,\epsilon}$ in (4.1) is called a *cylinder set*. It has a rectangular base

$$(y_1 - \epsilon, y_1 + \epsilon) \times (y_2 - \epsilon, y_2 + \epsilon) \times \dots \times (y_n - \epsilon, y_n + \epsilon)$$

in the $x_1, x_2, \dots, x_n$ coordinates, and is unrestricted in the other coordinate directions. More picturesquely, $B_{x,y,\epsilon}$ is sometimes called a "slalom set," because it consists of all functions whose graphs pass through the "slalom gates" at x_i with radius ϵ and center y_i.

A base for the topology of pointwise convergence is given by all finite intersections of sets of the form $B_{x,y,\epsilon}$. In fact, it is sufficient to take the sets of the form

$$\{f \in X \mid |f(x_i) - y_i| < \epsilon_i \text{ for } i = 1, \dots, n\} \tag{4.2}$$

where $n \in \mathbb{N}$, $\{x_1, \dots, x_n\} \subset [a, b]$, $\{y_1, \dots, y_n\} \subset \mathbb{R}$, and $\epsilon_i > 0$. The sets of functions in (4.2) with intervals of variable width $\epsilon_i > 0$ generate the same topology as the sets with intervals of a fixed width because $B_{x,y,\epsilon}$ with $\epsilon = \min \epsilon_i > 0$ is contained inside the set in (4.2).

We say that a topological space $(X, \mathcal{T})$ is *metrizable* if there is a metric on X whose metric topology is $\mathcal{T}$. For a metrizable space, we can give sequential characterizations of compact sets (Theorem 1.62), closed sets (Proposition 1.41), and continuous functions (Proposition 1.34). These sequential characterizations may not apply in a nonmetrizable topological space. There is, however, a generalization of sequences, called *nets*, that can be used to express all the above properties in an analogous way [12]. We will not make use of nets in this book.

For example, the closure $\overline{A}$ of a subset A of a topological space X is the smallest closed set that contains A. If X is metrizable, then $\overline{A}$ is the set of limits of convergent sequences whose terms are in A (see Section 1.5), but if X is a not metrizable, then this procedure may fail. We call the set of limit points of sequences in A the *sequential closure* of A and denote it by $\overline{A}^S$. The sequential closure is a subset of the closure, but it may be a strict subset, as illustrated by the following example.

Example 4.17 Consider the space of all functions $f : [0,1] \to \mathbb{R}$ with the topology of pointwise convergence. For each $m, n \geq 1$, we let

$$f_{m,n}(x) = [\cos(m!\pi x)]^{2n}.$$

We define functions f_m and f by the pointwise limits,

$$f_m(x) = \lim_{n \to \infty} f_{m,n}(x) = \begin{cases} 1 & \text{if } x = k/m!, \ k = 0, \ldots, m!, \\ 0 & \text{otherwise,} \end{cases}$$

$$f(x) = \lim_{m \to \infty} f_m(x) = \begin{cases} 1 & \text{if } x \in \mathbb{Q} \cap [0,1], \\ 0 & \text{otherwise.} \end{cases}$$

Let $A = \{f_{m,n} \mid m, n \geq 1\}$. Then these limits show that

$$f_m \in \overline{A}^S, \qquad f \in \overline{\overline{A}^S}^S.$$

It is possible to show that the pointwise limit of a sequence of continuous functions on $[0,1]$ is continuous on a dense subset of $[0,1]$. Since f is nowhere continuous in $[0,1]$, it is not the pointwise limit of any subsequence of the continuous functions $f_{m,n}$. Therefore, $f \in \overline{A}$ but $f \notin \overline{A}^S$. This example shows that the topology of pointwise convergence on the real-valued functions on $[0,1]$ is not metrizable.

A linear space with a topology defined on it, which need not be derived from a norm or metric, such that the operations of vector addition and scalar multiplication are continuous is called a *topological linear space*, or a *topological vector space*. The space of real-valued functions on a set with the topology of pointwise convergence is an example of a topological linear space. Topological linear spaces, such as the Schwartz space, also arise in connection with distribution theory (see Chapter 11).

4.3   Comparing topologies

Let $\mathcal{T}_1$, $\mathcal{T}_2$ be two topologies on the same space X. Then $\mathcal{T}_2$ is said to be *finer* or *stronger* than $\mathcal{T}_1$ if $\mathcal{T}_1 \subset \mathcal{T}_2$, meaning that $\mathcal{T}_2$ has more open sets; we also say that $\mathcal{T}_1$ is *coarser* or *weaker* than $\mathcal{T}_2$. If $\mathcal{T}_1$ is stronger than $\mathcal{T}_2$, then $x_n \to x$ with respect to $\mathcal{T}_1$ implies that $x_n \to x$ with respect to $\mathcal{T}_2$. For example, the strongest topology on any set is the discrete topology, and a sequence converges with respect to the discrete topology if and only if it is eventually constant. The weakest topology on any set is the trivial topology, and every sequence converges with respect to the trivial topology. It is possible that two topologies $\mathcal{T}_1$, $\mathcal{T}_2$ are not comparable, meaning that $\mathcal{T}_1$ is neither finer nor coarser than $\mathcal{T}_2$.

Proposition 4.18 Let X and Y be two spaces, each with two topologies, $\mathcal{T}_1, \mathcal{T}_2$ and $\mathcal{S}_1, \mathcal{S}_2$ respectively. Suppose that $f : (X, \mathcal{T}_1) \to (Y, \mathcal{S}_1)$ is a map from X to Y that is continuous with respect to the indicated topologies.

    (a) If $\mathcal{T}_2$ is finer than $\mathcal{T}_1$, then $f : (X, \mathcal{T}_2) \to (Y, \mathcal{S}_1)$ is continuous.
    (b) If $\mathcal{S}_2$ is coarser than $\mathcal{S}_1$, then $f : (X, \mathcal{T}_1) \to (Y, \mathcal{S}_2)$ is continuous.

Proof. These statements are a direct consequence of the general definition of continuity in Definition 4.6. $\qquad\qquad\qquad\qquad\qquad\qquad\qquad\qquad\qquad\qquad\quad\square$

The identity map $I : (X, \mathcal{T}) \to (X, \mathcal{T})$, where $I(x) = x$, is a homeomorphism when we use the same topology $\mathcal{T}$ on the domain and range. This is not true when we use two different topologies on X. For example, the identity map from a set X containing at least two elements equipped with the trivial topology to the set X equipped with the discrete topology,

$$I : (X, \{\emptyset, X\}) \to (X, \mathcal{P}(X)),$$

is discontinuous at every point $x \in X$. As the following theorems show, the identity map is a useful tool for comparing topologies on a set.

Theorem 4.19 Let $\mathcal{T}_1$ and $\mathcal{T}_2$ be two topologies on X. Then the identity map $I : (X, \mathcal{T}_1) \to (X, \mathcal{T}_2)$ is continuous if and only if $\mathcal{T}_1$ is finer than $\mathcal{T}_2$.

Proof. This is a direct consequence of Theorem 4.7. $\qquad\qquad\qquad\qquad\qquad\quad\square$

Corollary 4.20 The identity map $I : (X, \mathcal{T}_1) \to (X, \mathcal{T}_2)$ is a homeomorphism if and only if $\mathcal{T}_1 = \mathcal{T}_2$.

Theorem 4.21 Let $\mathcal{T}_1$ and $\mathcal{T}_2$ be two topologies on X. Then the equality of $\mathcal{T}_1$ and $\mathcal{T}_2$ is equivalent to the following condition: for all topological spaces $(Y, \mathcal{S})$, a function $f : (X, \mathcal{T}_1) \to (Y, \mathcal{S})$ is continuous if and only if the function $f : (X, \mathcal{T}_2) \to (Y, \mathcal{S})$ is continuous.

Proof. If $\mathcal{T}_1 = \mathcal{T}_2$, then the condition about continuous functions $f : X \to Y$ is trivial. Conversely, taking $(Y, \mathcal{S}) = (X, \mathcal{T}_1)$ and $(Y, \mathcal{S}) = (X, \mathcal{T}_2)$, we see that the condition implies that $I : (X, \mathcal{T}_1) \to (X, \mathcal{T}_2)$ is a homeomorphism, so $\mathcal{T}_1 = \mathcal{T}_2$ from Corollary 4.20. $\qquad\qquad\qquad\qquad\qquad\qquad\qquad\qquad\qquad\qquad\qquad\qquad\quad\square$

A topology is often defined by the specification of a neighborhood base at each point. We therefore want to compare topologies in terms of their neighborhood bases.

Theorem 4.22 Let $\mathcal{T}_1$ and $\mathcal{T}_2$ be two topologies on X. Suppose that for each $x \in X$ there are neighborhood bases $\mathcal{N}_1$ and $\mathcal{N}_2$ of x for $\mathcal{T}_1$ and $\mathcal{T}_2$, respectively, such that for every $V_1 \in \mathcal{N}_1$ there is a $V_2 \in \mathcal{N}_2$ with $V_2 \subset V_1$. Then $\mathcal{T}_2$ is finer than $\mathcal{T}_1$.

Proof. The hypothesis of the theorem implies that $I : (X, \mathcal{T}_2) \to (X, \mathcal{T}_1)$ is continuous, so the result follows from Theorem 4.19.   □

Corollary 4.23 Let $\mathcal{T}_1$ and $\mathcal{T}_2$ be two topologies on X. Then $\mathcal{T}_1 = \mathcal{T}_2$ if and only if for each $x \in X$ there are neighborhood bases $\mathcal{M}_1$ and $\mathcal{M}_2$ of x for $\mathcal{T}_1$ and $\mathcal{T}_2$, respectively, such that for every $V_1 \in \mathcal{M}_1$ there is a $V_2 \in \mathcal{M}_2$ with $V_2 \subset V_1$, and there are neighborhood bases $\mathcal{N}_1$ and $\mathcal{N}_2$ of x for $\mathcal{T}_1$ and $\mathcal{T}_2$, respectively, such that for every $W_2 \in \mathcal{N}_2$ there is a $W_1 \in \mathcal{N}_1$ with $W_1 \subset W_2$.

Different metrics, or norms, on a space X can lead to the same topology. For example, this is certainly the case if d_1 and d_2 are two metrics on X such that $d_1(x, y) = 2d_2(x, y)$ for all $x, y \in X$. More generally, if two metrics lead to the same set of convergent sequences, then all their topological properties are the same.

Theorem 4.24 Two metric topologies, defined by two metrics on the same space, are equal if and only if they have the same collection of convergent sequences with the same limits.

Proof. The proof is a direct application of Corollary 4.20 and the sequential characterization of continuity on metric spaces.   □

4.4   References

In this chapter, we have limited our discussion to the basic definitions of point set topology. For more information, see Kelley [28] and Rudin [48].

4.5   Exercises

Exercise 4.1 Suppose that K is a compact subspace of a Hausdorff space. Prove that K is closed. Show that this result need not be true if X is not Hausdorff.

Exercise 4.2 If A is a subset of a topological space, then the *interior* A° of A is the union of all open sets contained in A, the *closure* $\overline{A}$ of A is the intersection of all closed sets that contain A, and the *boundary* ∂A of A is defined by $\partial A = \overline{A} \cap \overline{A^c}$. Show that a set is closed if and only if it contains its boundary, and open if and only if it is disjoint from its boundary. What are the closure, interior, and boundary of the Cantor set in $\mathbb{R}$ with its usual topology?

Exercise 4.3 Let (X, d_1) and (Y, d_2) be metric spaces. Prove that the topological definitions of convergence and continuity are equivalent to the metric space definitions in Definitions 1.12 and 1.26.

Exercise 4.4 Prove Theorem 4.7.

Exercise 4.5 A topological space is *connected* if it is not the disjoint union of two non-empty open sets.

    (a) What are the connected subsets of $\mathbb{R}$?

    (b) Show that $X \times Y$ is connected if X and Y are connected.

Exercise 4.6 Show that $\mathbb{R}$ is homeomorphic to $(0, 1)$, but not to $\mathbb{R}^2$.

HINT. Show that $\mathbb{R}^2$ remains connected when one point is removed.

Chapter 5

Banach Spaces

Many linear equations may be formulated in terms of a suitable linear operator acting on a Banach space. In this chapter, we study Banach spaces and linear operators acting on Banach spaces in greater detail. We give the definition of a Banach space and illustrate it with a number of examples. We show that a linear operator is continuous if and only if it is bounded, define the norm of a bounded linear operator, and study some properties of bounded linear operators. Unbounded linear operators are also important in applications: for example, differential operators are typically unbounded. We will study them in later chapters, in the simpler context of Hilbert spaces.

5.1 Banach spaces

A normed linear space is a metric space with respect to the metric d derived from its norm, where $d(x, y) = \|x - y\|$.

Definition 5.1 A *Banach space* is a normed linear space that is a complete metric space with respect to the metric derived from its norm.

The following examples illustrate the definition. We will study many of these examples in greater detail later on, so we do not present proofs here.

Example 5.2 For $1 \leq p < \infty$, we define the p-norm on $\mathbb{R}^n$ (or $\mathbb{C}^n$) by

$$\|(x_1, x_2, \ldots, x_n)\|_p = \left(|x_1|^p + |x_2|^p + \ldots + |x_n|^p\right)^{1/p}.$$

For $p = \infty$, we define the ∞, or maximum, norm by

$$\|(x_1, x_2, \ldots, x_n)\|_\infty = \max \left\{|x_1|, |x_2|, \ldots, |x_n|\right\}.$$

Then $\mathbb{R}^n$ equipped with the p-norm is a finite-dimensional Banach space for $1 \leq p \leq \infty$.

Example 5.3 The space $C([a,b])$ of continuous, real-valued (or complex-valued) functions on $[a,b]$ with the sup-norm is a Banach space. More generally, the space $C(K)$ of continuous functions on a compact metric space K equipped with the sup-norm is a Banach space.

Example 5.4 The space $C^k([a,b])$ of k-times continuously differentiable functions on $[a,b]$ is not a Banach space with respect to the sup-norm $\|\cdot\|_\infty$ for $k \geq 1$, since the uniform limit of continuously differentiable functions need not be differentiable. We define the C^k-norm by

$$\|f\|_{C^k} = \|f\|_\infty + \|f'\|_\infty + \ldots + \|f^{(k)}\|_\infty.$$

Then $C^k([a,b])$ is a Banach space with respect to the C^k-norm. Convergence with respect to the C^k-norm is uniform convergence of functions and their first k derivatives.

Example 5.5 For $1 \leq p < \infty$, the *sequence space* $\ell^p(\mathbb{N})$ consists of all infinite sequences $x = (x_n)_{n=1}^\infty$ such that

$$\sum_{n=1}^\infty |x_n|^p < \infty,$$

with the p-norm,

$$\|x\|_p = \left(\sum_{n=1}^\infty |x_n|^p\right)^{1/p}.$$

For $p = \infty$, the sequence space $\ell^\infty(\mathbb{N})$ consists of all bounded sequences, with

$$\|x\|_\infty = \sup\{|x_n| \mid n = 1, 2, \ldots\}.$$

Then $\ell^p(\mathbb{N})$ is an infinite-dimensional Banach space for $1 \leq p \leq \infty$. The sequence space $\ell^p(\mathbb{Z})$ of bi-infinite sequences $x = (x_n)_{n=-\infty}^\infty$ is defined in an analogous way.

Example 5.6 Suppose that $1 \leq p < \infty$, and $[a,b]$ is an interval in $\mathbb{R}$. We denote by $L^p([a,b])$ the set of Lebesgue measurable functions $f : [a,b] \to \mathbb{R}$ (or $\mathbb{C}$) such that

$$\int_a^b |f(x)|^p \, dx < \infty,$$

where the integral is a Lebesgue integral, and we identify functions that differ on a set of measure zero (see Chapter 12). We define the L^p-norm of f by

$$\|f\|_p = \left(\int_a^b |f(x)|^p \, dx\right)^{1/p}.$$

For $p = \infty$, the space $L^\infty\left([a,b]\right)$ consists of the Lebesgue measurable functions $f : [a,b] \to \mathbb{R}$ (or $\mathbb{C}$) that are *essentially bounded* on $[a,b]$, meaning that f is bounded on a subset of $[a,b]$ whose complement has measure zero. The norm on $L^\infty\left([a,b]\right)$ is the *essential supremum*

$$\|f\|_\infty = \inf \left\{ M \mid |f(x)| \leq M \text{ a.e. in } [a,b] \right\}.$$

More generally, if Ω is a measurable subset of $\mathbb{R}^n$, which could be equal to $\mathbb{R}^n$ itself, then $L^p(\Omega)$ is the set of Lebesgue measurable functions $f : \Omega \to \mathbb{R}$ (or $\mathbb{C}$) whose pth power is Lebesgue integrable, with the norm

$$\|f\|_p = \left(\int_\Omega |f(x)|^p \, dx \right)^{1/p}.$$

We identify functions that differ on a set of measure zero. For $p = \infty$, the space $L^\infty(\Omega)$ is the space of essentially bounded Lebesgue measurable functions on Ω with the essential supremum as the norm. The spaces $L^p(\Omega)$ are Banach spaces for $1 \leq p \leq \infty$.

Example 5.7 The *Sobolev spaces*, $W^{k,p}$, consist of functions whose derivatives satisfy an integrability condition. If (a,b) is an open interval in $\mathbb{R}$, then we define $W^{k,p}\left((a,b)\right)$ to be the space of functions $f : (a,b) \to \mathbb{R}$ (or $\mathbb{C}$) whose derivatives of order less than or equal to k belong to $L^p\left((a,b)\right)$, with the norm

$$\|f\|_{W^{k,p}} = \left(\sum_{j=0}^{k} \int_a^b \left| f^{(j)}(x) \right|^p dx \right)^{1/p}.$$

The derivatives $f^{(j)}$ are defined in a weak, or distributional, sense as we explain later on. More generally, if Ω is an open subset of $\mathbb{R}^n$, then $W^{k,p}(\Omega)$ is the set of functions whose partial derivatives of order less than or equal to k belong to $L^p(\Omega)$. Sobolev spaces are Banach spaces. We will give more detailed definitions of these spaces, and state some of their main properties, in Chapter 12.

A closed linear subspace of a Banach space is a Banach space, since a closed subset of a complete space is complete. Infinite-dimensional subspaces need not be closed, however. For example, infinite-dimensional Banach spaces have proper dense subspaces, something which is difficult to visualize from our intuition of finite-dimensional spaces.

Example 5.8 The space of polynomial functions is a linear subspace of $C\left([0,1]\right)$, since a linear combination of polynomials is a polynomial. It is not closed, and Theorem 2.9 implies that it is dense in $C\left([0,1]\right)$. The set $\{f \in C\left([0,1]\right) \mid f(0) = 0\}$ is a closed linear subspace of $C\left([0,1]\right)$, and is a Banach space equipped with the sup-norm.

Example 5.9 The set $\ell_c(\mathbb{N})$ of all sequences of the form $(x_1, x_2, \ldots, x_n, 0, 0, \ldots)$ whose terms vanish from some point onwards is an infinite-dimensional linear subspace of $\ell^p(\mathbb{N})$ for any $1 \le p \le \infty$. The subspace $\ell_c(\mathbb{N})$ is not closed, so it is not a Banach space. It is dense in $\ell^p(\mathbb{N})$ for $1 \le p < \infty$. Its closure in $\ell^\infty(\mathbb{N})$ is the space $c_0(\mathbb{N})$ of sequences that converge to zero.

A *Hamel basis*, or algebraic basis, of a linear space is a maximal linearly independent set of vectors. Each element of a linear space may be expressed as a unique *finite* linear combination of elements in a Hamel basis. Every linear space has a Hamel basis, and any linearly independent set of vectors may be extended to a Hamel basis by the repeated addition of linearly independent vectors to the set until none are left (a procedure which is formalized by the axiom of choice, or Zorn's lemma, in the case of infinite-dimensional spaces). A Hamel basis of an infinite-dimensional space is frequently very large. In a normed space, we have a notion of convergence, and we may therefore consider various types of topological bases in which infinite sums are allowed.

Definition 5.10 Let X be a separable Banach space. A sequence (x_n) is a *Schauder basis* of X if for every $x \in X$ there is a unique sequence of scalars (c_n) such that $x = \sum_{n=1}^{\infty} c_n x_n$.

The concept of a Schauder basis is not as straightforward as it may appear. The Banach spaces that arise in applications typically have Schauder bases, but Enflo showed in 1973 that there exist separable Banach spaces that do not have any Schauder bases. As we will see, this problem does not arise in Hilbert spaces, which always have an orthonormal basis.

Example 5.11 A Schauder basis $(f_n)_{n=0}^{\infty}$ of $C([0,1])$ may be constructed from "tent" functions. For $n = 0, 1$, we define

$$f_0(x) = 1, \qquad f_1(x) = x.$$

For $2^{k-1} < n \le 2^k$, where $k \ge 1$, we define

$$f_n(x) = \begin{cases} 2^k \left[x - \left(2^{-k}(2n-2) - 1 \right) \right] & \text{if } x \in I_n, \\ 1 - 2^k \left[x - \left(2^{-k}(2n-1) - 1 \right) \right] & \text{if } x \in J_n, \\ 0 & \text{otherwise}, \end{cases}$$

where

$$\begin{aligned} I_n &= [2^{-k}(2n-2), \ 2^{-k}(2n-1)), \\ J_n &= [2^{-k}(2n-1), \ 2^{-k}2n). \end{aligned}$$

The graphs of these functions form a sequence of "tents" of height one and width 2^{-k+1} that sweep across the interval $[0,1]$. If $f \in C([0,1])$, then we may compute the coefficients c_n in the expansion

$$f(x) = \sum_{n=0}^{\infty} c_n f_n(x)$$

by equating the values of f and the series at the points $x = 2^{-k}m$ for $k \in \mathbb{N}$ and $m = 0, 1, \ldots, 2^k$. The uniform continuity of f implies that the resulting series converges uniformly to f.

5.2 Bounded linear maps

A *linear map* or *linear operator* T between real (or complex) linear spaces X, Y is a function $T : X \to Y$ such that

$$T(\lambda x + \mu y) = \lambda Tx + \mu Ty \qquad \text{for all } \lambda, \mu \in \mathbb{R} \text{ (or } \mathbb{C}) \text{ and } x, y \in X.$$

A linear map $T : X \to X$ is called a *linear transformation of X*, or a *linear operator on X*. If $T : X \to Y$ is one-to-one and onto, then we say that T is *nonsingular* or *invertible*, and define the inverse map $T^{-1} : Y \to X$ by $T^{-1}y = x$ if and only if $Tx = y$, so that $TT^{-1} = I$, $T^{-1}T = I$. The linearity of T implies the linearity of T^{-1}.

If X, Y are normed spaces, then we can define the notion of a bounded linear map. As we will see, the boundedness of a linear map is equivalent to its continuity.

Definition 5.12 Let X and Y be two normed linear spaces. We denote both the X and Y norms by $\|\cdot\|$. A linear map $T : X \to Y$ is *bounded* if there is a constant $M \geq 0$ such that

$$\|Tx\| \leq M\|x\| \qquad \text{for all } x \in X. \tag{5.1}$$

If no such constant exists, then we say that T is *unbounded*. If $T : X \to Y$ is a bounded linear map, then we define the *operator norm* or *uniform norm* $\|T\|$ of T by

$$\|T\| = \inf\{M \mid \|Tx\| \leq M\|x\| \text{ for all } x \in X\}. \tag{5.2}$$

We denote the set of all linear maps $T : X \to Y$ by $\mathcal{L}(X, Y)$, and the set of all bounded linear maps $T : X \to Y$ by $\mathcal{B}(X, Y)$. When the domain and range spaces are the same, we write $\mathcal{L}(X, X) = \mathcal{L}(X)$ and $\mathcal{B}(X, X) = \mathcal{B}(X)$.

Equivalent expressions for $\|T\|$ are:

$$\|T\| = \sup_{x \neq 0} \frac{\|Tx\|}{\|x\|}; \quad \|T\| = \sup_{\|x\| \leq 1} \|Tx\|; \quad \|T\| = \sup_{\|x\| = 1} \|Tx\|. \tag{5.3}$$

We also use the notation $\mathbb{R}^{m \times n}$, or $\mathbb{C}^{m \times n}$, to denote the space of linear maps from $\mathbb{R}^n$ to $\mathbb{R}^m$, or $\mathbb{C}^n$ to $\mathbb{C}^m$, respectively.

Example 5.13 The linear map $A : \mathbb{R} \to \mathbb{R}$ defined by $Ax = ax$, where $a \in \mathbb{R}$, is bounded, and has norm $\|A\| = |a|$.

Example 5.14 The identity map $I : X \to X$ is bounded on any normed space X, and has norm one. If a map has norm zero, then it is the zero map $0x = 0$.

Linear maps on infinite-dimensional normed spaces need not be bounded.

Example 5.15 Let $X = C^{\infty}([0,1])$ consist of the smooth functions on $[0,1]$ that have continuous derivatives of all orders, equipped with the maximum norm. The space X is a normed space, but it is not a Banach space, since it is incomplete. The differentiation operator $Du = u'$ is an unbounded linear map $D : X \to X$. For example, the function $u(x) = e^{\lambda x}$ is an eigenfunction of D for any $\lambda \in \mathbb{R}$, meaning that $Du = \lambda u$. Thus $\|Du\|/\|u\| = |\lambda|$ may be arbitrarily large. The unboundedness of differential operators is a fundamental difficulty in their study.

Suppose that $A : X \to Y$ is a linear map between finite-dimensional real linear spaces X, Y with $\dim X = n$, $\dim Y = m$. We choose bases $\{e_1, e_2, \ldots, e_n\}$ of X and $\{f_1, f_2, \ldots, f_m\}$ of Y. Then

$$A(e_j) = \sum_{i=1}^{m} a_{ij} f_i,$$

for a suitable $m \times n$ matrix (a_{ij}) with real entries. We expand $x \in X$ as

$$x = \sum_{i=1}^{n} x_i e_i, \tag{5.4}$$

where $x_i \in \mathbb{R}$ is the ith component of x. It follows from the linearity of A that

$$A\left(\sum_{j=1}^{n} x_j e_j\right) = \sum_{i=1}^{m} y_i f_i,$$

where

$$y_i = \sum_{j=1}^{n} a_{ij} x_j.$$

Thus, given a choice of bases for X, Y we may represent A as a linear map $A : \mathbb{R}^n \to \mathbb{R}^m$ with matrix $A = (a_{ij})$, where

$$\begin{pmatrix} y_1 \\ y_2 \\ \vdots \\ y_m \end{pmatrix} = \begin{pmatrix} a_{11} & a_{12} & \cdots & a_{1n} \\ a_{21} & a_{22} & \cdots & a_{2n} \\ \vdots & \vdots & \ddots & \vdots \\ a_{m1} & a_{m2} & \cdots & a_{mn} \end{pmatrix} \begin{pmatrix} x_1 \\ x_2 \\ \vdots \\ x_n \end{pmatrix}. \tag{5.5}$$

We will often use the same notation A to denote a linear map on a finite-dimensional space and its associated matrix, but it is important not to confuse the geometrical notion of a linear map with the matrix of numbers that represents it.

Each pair of norms on $\mathbb{R}^n$ and $\mathbb{R}^m$ induces a corresponding operator, or matrix, norm on A. We first consider the *Euclidean norm*, or 2-norm, $\|A\|_2$ of A. The Euclidean norm of a vector x is given by $\|x\|_2^2 = (x, x)$, where $(x, y) = x^T y$. From (5.3), we may compute the Euclidean norm of A by maximizing the function $\|Ax\|_2^2$ on the unit sphere $\|x\|_2^2 = 1$. The maximizer x is a critical point of the function

$$f(x, \lambda) = (Ax, Ax) - \lambda \{(x, x) - 1\},$$

where λ is a Lagrange multiplier. Computing ∇f and setting it equal to zero, we find that x satisfies

$$A^T A x = \lambda x. \tag{5.6}$$

Hence, x is an eigenvector of the matrix $A^T A$ and λ is an eigenvalue. The matrix $A^T A$ is an $n \times n$ symmetric matrix, with real, nonnegative eigenvalues. At an eigenvector x of $A^T A$ that satisfies (5.6), normalized so that $\|x\|_2 = 1$, we have $(Ax, Ax) = \lambda$. Thus, the maximum value of $\|Ax\|_2^2$ on the unit sphere is the maximum eigenvalue of $A^T A$.

We define the *spectral radius* $r(B)$ of a matrix B to be the maximum absolute value of its eigenvalues. It follows that the Euclidean norm of A is given by

$$\|A\|_2 = \sqrt{r(A^T A)}. \tag{5.7}$$

In the case of linear maps $A : \mathbb{C}^n \to \mathbb{C}^m$ on finite dimensional complex linear spaces, equation (5.7) holds with $A^T A$ replaced by $A^* A$, where A^* is the Hermitian conjugate of A. Proposition 9.7 gives a formula for the spectral radius of a bounded operator in terms of the norms of its powers.

To compute the maximum norm of A, we observe from (5.5) that

$$
\begin{aligned}
|y_i| &\leq |a_{i1}||x_1| + |a_{i2}||x_2| + \ldots + |a_{in}||x_n| \\
&\leq (|a_{i1}| + |a_{i2}| + \ldots + |a_{in}|) \|x\|_\infty.
\end{aligned}
$$

Taking the maximum of this equation with respect to i and comparing the result with the definition of the operator norm, we conclude that

$$\|A\|_\infty \leq \max_{1 \leq i \leq m} (|a_{i1}| + |a_{i2}| + \ldots + |a_{in}|).$$

Conversely, suppose that the maximum on the right-hand side of this equation is attained at $i = i_0$. Let x be the vector with components $x_j = \operatorname{sgn} a_{i_0 j}$, where sgn is the *sign function*,

$$\operatorname{sgn} x = \begin{cases} 1 & \text{if } x > 0, \\ 0 & \text{if } x = 0, \\ -1 & \text{if } x < 0. \end{cases} \tag{5.8}$$

Then, if A is nonzero, we have $\|x\|_\infty = 1$, and

$$\|Ax\|_\infty = |a_{i_01}| + |a_{i_02}| + \ldots + |a_{i_0n}|.$$

Since $\|A\|_\infty \geq \|Ax\|_\infty$, we obtain that

$$\|A\|_\infty \geq \max_{1 \leq i \leq m} \left(|a_{i1}| + |a_{i2}| + \ldots + |a_{in}| \right).$$

Therefore, we have equality, and the maximum norm of A is given by the maximum row sum,

$$\|A\|_\infty = \max_{1 \leq i \leq m} \left\{ \sum_{j=1}^{n} |a_{ij}| \right\}. \tag{5.9}$$

A similar argument shows that the sum norm of A is given by the maximum column sum

$$\|A\|_1 = \max_{1 \leq j \leq n} \left\{ \sum_{i=1}^{m} |a_{ij}| \right\}.$$

For $1 < p < \infty$, one can show (see Kato [26]) that the p-matrix norm satisfies

$$\|A\|_p \leq \|A\|_1^{1/p} \|A\|_\infty^{1-1/p}.$$

There are norms on the space $\mathcal{B}(\mathbb{R}^n, \mathbb{R}^m) = \mathbb{R}^{m \times n}$ of $m \times n$ matrices that are not associated with any vector norms on $\mathbb{R}^n$ and $\mathbb{R}^m$. An example is the *Hilbert-Schmidt* norm

$$\|A\| = \left(\sum_{i=1}^{m} \sum_{j=1}^{n} |a_{ij}|^2 \right)^{1/2}.$$

Next, we give some examples of linear operators on infinite-dimensional spaces.

Example 5.16 Let $X = \ell^\infty(\mathbb{N})$ be the space of bounded sequences $\{(x_1, x_2, \ldots)\}$ with the norm

$$\|(x_1, x_2, \ldots)\|_\infty = \sup_{i \in \mathbb{N}} |x_i|.$$

A linear map $A : X \to X$ is represented by an infinite matrix $(a_{ij})_{i,j=1}^\infty$, where

$$(Ax)_i = \sum_{j=1}^{\infty} a_{ij} x_j.$$

In order for this sum to converge for any $x \in \ell^\infty(\mathbb{N})$, we require that

$$\sum_{j=1}^{\infty} |a_{ij}| < \infty$$

for each $i \in \mathbb{N}$, and in order for Ax to belong to $\ell^\infty(\mathbb{N})$, we require that

$$\sup_{i \in \mathbb{N}} \left\{ \sum_{j=1}^{\infty} |a_{ij}| \right\} < \infty.$$

Then A is a bounded linear operator on $\ell^\infty(\mathbb{N})$, and its norm is the maximum row sum,

$$\|A\|_\infty = \sup_{i \in \mathbb{N}} \left\{ \sum_{j=1}^{\infty} |a_{ij}| \right\}.$$

Example 5.17 Let $X = C([0,1])$ with the maximum norm, and

$$k : [0,1] \times [0,1] \to \mathbb{R}$$

be a continuous function. We define the linear Fredholm integral operator $K : X \to X$ by

$$Kf(x) = \int_0^1 k(x,y)f(y) \, dy.$$

Then K is bounded and

$$\|K\| = \max_{0 \le x \le 1} \left\{ \int_0^1 |k(x,y)| \, dy \right\}.$$

This expression is the "continuous" analog of the maximum row sum for the ∞-norm of a matrix.

For linear maps, boundedness is equivalent to continuity.

Theorem 5.18 A linear map is bounded if and only if it is continuous.

Proof. First, suppose that $T : X \to Y$ is bounded. Then, for all $x, y \in X$, we have

$$\|Tx - Ty\| = \|T(x - y)\| \le M\|x - y\|,$$

where M is a constant for which (5.1) holds. Therefore, we can take $\delta = \epsilon/M$ in the definition of continuity, and T is continuous.

Second, suppose that T is continuous at 0. Since T is linear, we have $T(0) = 0$. Choosing $\epsilon = 1$ in the definition of continuity, we conclude that there is a $\delta > 0$ such that $\|Tx\| \le 1$ whenever $\|x\| \le \delta$. For any $x \in X$, with $x \ne 0$, we define $\tilde{x}$ by

$$\tilde{x} = \delta \frac{x}{\|x\|}.$$

Then $\|\tilde{x}\| \leq \delta$, so $\|T\tilde{x}\| \leq 1$. It follows from the linearity of T that

$$\|Tx\| = \frac{\|x\|}{\delta}\|T\tilde{x}\| \leq M\|x\|,$$

where $M = 1/\delta$. Thus T is bounded.   □

The proof shows that if a linear map is continuous at zero, then it is continuous at every point. A nonlinear map may be bounded but discontinuous, or continuous at zero but discontinuous at other points.

The following theorem, sometimes called the BLT theorem for "bounded linear transformation" has many applications in defining and studying linear maps.

Theorem 5.19 (Bounded linear transformation) Let X be a normed linear space and Y a Banach space. If M is a dense linear subspace of X and

$$T : M \subset X \to Y$$

is a bounded linear map, then there is a unique bounded linear map $\overline{T} : X \to Y$ such that $\overline{T}x = Tx$ for all $x \in M$. Moreover, $\|\overline{T}\| = \|T\|$.

Proof.   For every $x \in X$, there is a sequence (x_n) in M that converges to x. We define

$$\overline{T}x = \lim_{n \to \infty} Tx_n.$$

This limit exists because (Tx_n) is Cauchy, since T is bounded and (x_n) Cauchy, and Y is complete. We claim that the value of the limit does not depend on the sequence in M that is used to approximate x. Suppose that (x_n) and (x_n') are any two sequences in M that converge to x. Then

$$\|x_n - x_n'\| \leq \|x_n - x\| + \|x - x_n'\|,$$

and, taking the limit of this equation as $n \to \infty$, we see that

$$\lim_{n \to \infty} \|x_n - x_n'\| = 0.$$

It follows that

$$\|Tx_n - Tx_n'\| \leq \|T\|\,\|x_n - x_n'\| \to 0 \qquad \text{as } n \to \infty.$$

Hence, (Tx_n) and (Tx_n') converge to the same limit.

The map $\overline{T}$ is an extension of T, meaning that $\overline{T}x = Tx$, for all $x \in M$, because if $x \in M$, we can use the constant sequence with $x_n = x$ for all n to define $\overline{T}x$. The linearity of $\overline{T}$ follows from the linearity of T.

The fact that $\overline{T}$ is bounded follows from the inequality

$$\|\overline{T}x\| = \lim_{n \to \infty} \|Tx_n\| \leq \lim_{n \to \infty} \|T\|\,\|x_n\| = \|T\|\,\|x\|.$$

It also follows that $\|\overline{T}\| \leq \|T\|$. Since $\overline{T}x = Tx$ for $x \in M$, we have $\|\overline{T}\| = \|T\|$.

Finally, we show that $\overline{T}$ is the unique bounded linear map from X to Y that coincides with T on M. Suppose that $\widetilde{T}$ is another such map, and let x be any point in X, We choose a sequence (x_n) in M that converges to x. Then, using the continuity of $\widetilde{T}$, the fact that $\widetilde{T}$ is an extension of T, and the definition of $\overline{T}$, we see that

$$\widetilde{T}x = \lim_{n \to \infty} \widetilde{T}x_n = \lim_{n \to \infty} Tx_n = \overline{T}x.$$

$\square$

We can use linear maps to define various notions of equivalence between normed linear spaces.

Definition 5.20 Two linear spaces X, Y are *linearly isomorphic* if there is a one-to-one, onto linear map $T : X \to Y$. If X and Y are normed linear spaces and T, T^{-1} are bounded linear maps, then X and Y are *topologically isomorphic*. If T also preserves norms, meaning that $\|Tx\| = \|x\|$ for all $x \in X$, then X, Y are *isometrically isomorphic*.

When we say that two normed linear spaces are "isomorphic" we will usually mean that they are topologically isomorphic. We are often interested in the case when we have two different norms defined on the same space, and we would like to know if the norms define the same topologies.

Definition 5.21 Let X be a linear space. Two norms $\| \cdot \|_1$ and $\| \cdot \|_2$ on X are *equivalent* if there are constants $c > 0$ and $C > 0$ such that

$$c\|x\|_1 \leq \|x\|_2 \leq C\|x\|_1 \qquad \text{for all } x \in X. \tag{5.10}$$

Theorem 5.22 Two norms on a linear space generate the same topology if and only if they are equivalent.

Proof. Let $\| \cdot \|_1$ and $\| \cdot \|_2$ be two norms on a linear space X. We consider the identity map

$$I : (X, \| \cdot \|_1) \to (X, \| \cdot \|_2).$$

From Corollary 4.20, the topologies generated by the two norms are the same if and only if I and I^{-1} are continuous. Since I is linear, it is continuous if and only if it is bounded. The boundedness of the identity map and its inverse is equivalent to the existence of constants c and C such that (5.10) holds. $\square$

Geometrically, two norms are equivalent if the unit ball of either one of the norms is contained in a ball of finite radius of the other norm.

We end this section by stating, without proof, a fundamental fact concerning linear operators on Banach spaces.

Theorem 5.23 (Open mapping) Suppose that $T : X \to Y$ is a one-to-one, onto bounded linear map between Banach spaces X, Y. Then $T^{-1} : Y \to X$ is bounded.

This theorem states that the existence of the inverse of a continuous linear map between Banach spaces implies its continuity. Contrast this result with Example 4.9.

5.3   The kernel and range of a linear map

The kernel and range are two important linear subspaces associated with a linear map.

Definition 5.24 Let $T : X \to Y$ be a linear map between linear spaces X, Y. The *null space* or *kernel* of T, denoted by $\ker T$, is the subset of X defined by

$$\ker T = \{x \in X \mid Tx = 0\}.$$

The *range* of T, denoted by $\operatorname{ran} T$, is the subset of Y defined by

$$\operatorname{ran} T = \{y \in Y \mid \text{there exists } x \in X \text{ such that } Tx = y\}.$$

The word "kernel" is also used in a completely different sense to refer to the kernel of an integral operator. A map $T : X \to Y$ is one-to-one if and only if $\ker T = \{0\}$, and it is onto if and only if $\operatorname{ran} T = Y$.

Theorem 5.25 Suppose that $T : X \to Y$ is a linear map between linear spaces X, Y. The kernel of T is a linear subspace of X, and the range of T is a linear subspace of Y. If X and Y are normed linear spaces and T is bounded, then the kernel of T is a closed linear subspace.

Proof. If $x_1, x_2 \in \ker T$ and $\lambda_1, \lambda_2 \in \mathbb{R}$ (or $\mathbb{C}$), then the linearity of T implies that

$$T(\lambda_1 x_1 + \lambda_2 x_2) = \lambda_1 T x_1 + \lambda_2 T x_2 = 0,$$

so $\lambda_1 x_1 + \lambda_2 x_2 \in \ker T$. Therefore, $\ker T$ is a linear subspace. If $y_1, y_2 \in \operatorname{ran} T$, then there are $x_1, x_2 \in X$ such that $Tx_1 = y_1$ and $Tx_2 = y_2$. Hence

$$T(\lambda_1 x_1 + \lambda_2 x_2) = \lambda_1 T x_1 + \lambda_2 T x_2 = \lambda_1 y_1 + \lambda_2 y_2,$$

so $\lambda_1 y_1 + \lambda_2 y_2 \in \operatorname{ran} T$. Therefore, $\operatorname{ran} T$ is a linear subspace.

Now suppose that X and Y are normed spaces and T is bounded. If (x_n) is a sequence of elements in $\ker T$ with $x_n \to x$ in X, then the continuity of T implies that

$$Tx = T\left(\lim_{n \to \infty} x_n\right) = \lim_{n \to \infty} Tx_n = 0,$$

so $x \in \ker T$, and $\ker T$ is closed. $\qquad\qquad\qquad\qquad\qquad\qquad\square$

The *nullity* of T is the dimension of the kernel of T, and the *rank* of T is the dimension of the range of T. We now consider some examples.

Example 5.26 The right shift operator S on $\ell^\infty(\mathbb{N})$ is defined by

$$S(x_1, x_2, x_3, \ldots) = (0, x_1, x_2, \ldots),$$

and the left shift operator T by

$$T(x_1, x_2, x_3, \ldots) = (x_2, x_3, x_4, \ldots).$$

These maps have norm one. Their matrices are the infinite-dimensional Jordan blocks,

$$[S] = \begin{pmatrix} 0 & 0 & 0 & \cdots \\ 1 & 0 & 0 & \cdots \\ 0 & 1 & 0 & \cdots \\ \vdots & \vdots & \vdots & \ddots \end{pmatrix}, \quad [T] = \begin{pmatrix} 0 & 1 & 0 & \cdots \\ 0 & 0 & 1 & \cdots \\ 0 & 0 & 0 & \cdots \\ \vdots & \vdots & \vdots & \ddots \end{pmatrix}.$$

The kernel of S is $\{0\}$ and the range of S is the subspace

$$\operatorname{ran} S = \{(0, x_2, x_3, \ldots) \in \ell^\infty(\mathbb{N})\}.$$

The range of T is the whole space $\ell^\infty(\mathbb{N})$, and the kernel of T is the one-dimensional subspace

$$\ker T = \{(x_1, 0, 0, \ldots) \mid x_1 \in \mathbb{R}\}.$$

The operator S is one-to-one but not onto, and T is onto but not one-to-one. This cannot happen for linear maps $T : X \to X$ on a finite-dimensional space X, such as $X = \mathbb{R}^n$. In that case, $\ker T = \{0\}$ if and only if $\operatorname{ran} T = X$.

Example 5.27 An integral operator $K : C([0,1]) \to C([0,1])$

$$Kf(x) = \int_0^1 k(x, y) f(y) \, dy$$

is said to be *degenerate* if $k(x, y)$ is a finite sum of separated terms of the form

$$k(x, y) = \sum_{i=1}^n \varphi_i(x) \psi_i(y),$$

where $\varphi_i, \psi_i : [0, 1] \to \mathbb{R}$ are continuous functions. We may assume without loss of generality that $\{\varphi_1, \ldots, \varphi_n\}$ and $\{\psi_1, \ldots, \psi_n\}$ are linearly independent. The range of K is the finite-dimensional subspace spanned by $\{\varphi_1, \varphi_2, \ldots, \varphi_n\}$, and the kernel of K is the subspace of functions $f \in C([0,1])$ such that

$$\int_0^1 f(y) \psi_i(y) \, dy = 0 \qquad \text{for } i = 1, \ldots, n.$$

Both the range and kernel are closed linear subspaces of $C([0,1])$.

Example 5.28 Let $X = C([0,1])$ with the maximum norm. We define the integral operator $K : X \to X$ by

$$Kf(x) = \int_0^x f(y)\,dy. \qquad (5.11)$$

An integral operator like this one, with a variable range of integration, is called a *Volterra integral operator*. Then K is bounded, with $\|K\| \le 1$, since

$$\|Kf\| \le \sup_{0 \le x \le 1} \int_0^x |f(y)|\,dy \le \int_0^1 |f(y)|\,dy \le \|f\|.$$

In fact, $\|K\| = 1$, since $K(1) = x$ and $\|x\| = \|1\|$. The range of K is the set of continuously differentiable functions on $[0,1]$ that vanish at $x = 0$. This is a linear subspace of $C([0,1])$ but it is not closed. The lack of closure of the range of K is due to the "smoothing" effect of K, which maps continuous functions to differentiable functions. The problem of inverting integral operators with similar properties arises in a number of inverse problems, where one wants to reconstruct a source distribution from remotely sensed data. Such problems are ill-posed and require special treatment.

Example 5.29 Consider the operator $T = I + K$ on $C([0,1])$, where K is defined in (5.11), which is a perturbation of the identity operator by K. The range of T is the whole space $C([0,1])$, and is therefore closed. To prove this statement, we observe that $g = Tf$ if and only if

$$f(x) + \int_0^x f(y)\,dy = g(x).$$

Writing $F(x) = \int_0^x f(y)\,dy$, we have $F' = f$ and

$$F' + F = g, \qquad F(0) = 0.$$

The solution of this initial value problem is

$$F(x) = \int_0^x e^{-(x-y)} g(y)\,dy.$$

Differentiating this expression with respect to x, we find that f is given by

$$f(x) = g(x) - \int_0^x e^{-(x-y)} g(y)\,dy.$$

Thus, the operator $T = I + K$ is invertible on $C([0,1])$ and

$$(I + K)^{-1} = I - L,$$

where L is the Volterra integral operator

$$Lg(x) = \int_0^x e^{-(x-y)} g(y)\,dy.$$

The following result provides a useful way to show that an operator T has closed range. It states that T has closed range if one can estimate the norm of the solution x of the equation $Tx = y$ in terms of the norm of the right-hand side y. In that case, it is often possible to deduce the existence of solutions (see Theorem 8.18).

Proposition 5.30 Let $T : X \to Y$ be a bounded linear map between Banach spaces X, Y. The following statements are equivalent:

(a) there is a constant $c > 0$ such that

$$c\|x\| \leq \|Tx\| \qquad \text{for all } x \in X;$$

(b) T has closed range, and the only solution of the equation $Tx = 0$ is $x = 0$.

Proof. First, suppose that T satisfies (a). Then $Tx = 0$ implies that $\|x\| = 0$, so $x = 0$. To show that $\operatorname{ran} T$ is closed, suppose that (y_n) is a convergent sequence in $\operatorname{ran} T$, with $y_n \to y \in Y$. Then there is a sequence (x_n) in X such that $Tx_n = y_n$. The sequence (x_n) is Cauchy, since (y_n) is Cauchy and

$$\|x_n - x_m\| \leq \frac{1}{c}\|T(x_n - x_m)\| = \frac{1}{c}\|y_n - y_m\|.$$

Hence, since X is complete, we have $x_n \to x$ for some $x \in X$. Since T is bounded, we have

$$Tx = \lim_{n\to\infty} Tx_n = \lim_{n\to\infty} y_n = y,$$

so $y \in \operatorname{ran} T$, and $\operatorname{ran} T$ is closed.

Conversely, suppose that T satisfies (b). Since $\operatorname{ran} T$ is closed, it is a Banach space. Since $T : X \to Y$ is one-to-one, the operator $T : X \to \operatorname{ran} T$ is a one-to-one, onto map between Banach spaces. The open mapping theorem, Theorem 5.23, implies that $T^{-1} : \operatorname{ran} T \to X$ is bounded, and hence that there is a constant $C > 0$ such that

$$\|T^{-1}y\| \leq C\|y\| \qquad \text{for all } y \in \operatorname{ran} T.$$

Setting $y = Tx$, we see that $c\|x\| \leq \|Tx\|$ for all $x \in X$, where $c = 1/C$. $\square$

Example 5.31 Consider the Volterra integral operator $K : C([0,1]) \to C([0,1])$ defined in (5.11). Then

$$K[\cos n\pi x] = \int_0^x \cos n\pi y \, dy = \frac{\sin n\pi x}{n\pi}.$$

We have $\|\cos n\pi x\| = 1$ for every $n \in \mathbb{N}$, but $\|K[\cos n\pi x]\| \to 0$ as $n \to \infty$. Thus, it is not possible to estimate $\|f\|$ in terms of $\|Kf\|$, consistent with the fact that the range of K is not closed.

5.4   Finite-dimensional Banach spaces

In this section, we prove that every finite-dimensional (real or complex) normed linear space is a Banach space, that every linear operator on a finite-dimensional space is continuous, and that all norms on a finite-dimensional space are equivalent. None of these statements is true for infinite-dimensional linear spaces. As a result, topological considerations can often be neglected when dealing with finite-dimensional spaces but are of crucial importance when dealing with infinite-dimensional spaces.

We begin by proving that the components of a vector with respect to any basis of a finite-dimensional space can be bounded by the norm of the vector.

Lemma 5.32 Let X be a finite-dimensional normed linear space with norm $\| \cdot \|$, and $\{e_1, e_2, \ldots, e_n\}$ any basis of X. There are constants $m > 0$ and $M > 0$ such that if $x = \sum_{i=1}^n x_i e_i$, then

$$m \sum_{i=1}^n |x_i| \le \|x\| \le M \sum_{i=1}^n |x_i|. \tag{5.12}$$

Proof.   By the homogeneity of the norm, it suffices to prove (5.12) for $x \in X$ such that $\sum_{i=1}^n |x_i| = 1$. The "cube"

$$C = \left\{ (x_1, \ldots, x_n) \in \mathbb{R}^n \,\middle|\, \sum_{i=1}^n |x_i| = 1 \right\}$$

is a closed, bounded subset of $\mathbb{R}^n$, and is therefore compact by the Heine-Borel theorem. We define a function $f : C \to X$ by

$$f\left((x_1, \ldots, x_n)\right) = \sum_{i=1}^n x_i e_i.$$

For $(x_1, \ldots, x_n) \in \mathbb{R}^n$ and $(y_1, \ldots, y_n) \in \mathbb{R}^n$, we have

$$\|f\left((x_1, \ldots, x_n)\right) - f\left((y_1, \ldots, y_n)\right)\| \le \sum_{i=1}^n |x_i - y_i| \|e_i\|,$$

so f is continuous. Therefore, since $\| \cdot \| : X \to \mathbb{R}$ is continuous, the map

$$(x_1, \ldots, x_n) \mapsto \|f\left((x_1, \ldots, x_n)\right)\|$$

is continuous. Theorem 1.68 implies that $\|f\|$ is bounded on C and attains its infimum and supremum. Denoting the minimum by $m \ge 0$ and the maximum by $M \ge m$, we obtain (5.12). Let $(\bar{x}_1, \ldots, \bar{x}_n)$ be a point in C where $\|f\|$ attains its minimum, meaning that

$$\|\bar{x}_1 e_1 + \ldots + \bar{x}_n e_n\| = m.$$

The linear independence of the basis vectors $\{e_1, \ldots, e_n\}$ implies that $m \ne 0$, so $m > 0$. $\square$

This result is not true in an infinite-dimensional space because, if a basis consists of vectors that become "almost" parallel, then the cancellation in linear combinations of basis vectors may lead to a vector having large components but small norm.

Theorem 5.33 Every finite-dimensional normed linear space is a Banach space.

Proof. Suppose that $(x_k)_{k=1}^\infty$ is a Cauchy sequence in a finite-dimensional normed linear space X. Let $\{e_1, \ldots, e_n\}$ be a basis of X. We expand x_k as

$$x_k = \sum_{i=1}^n x_{i,k} e_i,$$

where $x_{i,k} \in \mathbb{R}$. For $1 \leq i \leq n$, we consider the real sequence of ith components, $(x_{i,k})_{k=1}^\infty$. Equation (5.12) implies that

$$|x_{i,j} - x_{i,k}| \leq \frac{1}{m} \|x_j - x_k\|,$$

so $(x_{i,k})_{k=1}^\infty$ is Cauchy. Since $\mathbb{R}$ is complete, there is a $y_i \in \mathbb{R}$, such that

$$\lim_{k \to \infty} x_{i,k} = y_i.$$

We define $y \in X$ by

$$y = \sum_{i=1}^k y_i e_i.$$

Then, from (5.12),

$$\|x_k - y\| \leq M \sum_{i=1}^n |x_{i,k} - y_i| \, \|e_i\|,$$

and hence $x_k \to y$ as $k \to \infty$. Thus, every Cauchy sequence in X converges, and X is complete. $\quad\square$

Since a complete space is closed, we have the following corollary.

Corollary 5.34 Every finite-dimensional linear subspace of a normed linear space is closed.

In Section 5.2, we proved explicitly the boundedness of linear maps on finite-dimensional linear spaces with respect to certain norms. In fact, linear maps on finite-dimensional spaces are always bounded.

Theorem 5.35 Every linear operator on a finite-dimensional linear space is bounded.

Proof. Suppose that $A : X \to Y$ is a linear map and X is finite dimensional. Let $\{e_1, \ldots, e_n\}$ be a basis of X. If $x = \sum_{i=1}^{n} x_i e_i \in X$, then (5.12) implies that

$$\|Ax\| \leq \sum_{i=1}^{n} |x_i| \, \|Ae_i\| \leq \max_{1 \leq i \leq n} \{\|Ae_i\|\} \sum_{i=1}^{n} |x_i| \leq \frac{1}{m} \max_{1 \leq i \leq n} \{\|Ae_i\|\} \, \|x\|,$$

so A is bounded. $\qquad\square$

Finally, we show that although there are many different norms on a finite-dimensional linear space they all lead to the same topology and the same notion of convergence. This fact follows from Theorem 5.22 and the next result.

Theorem 5.36 Any two norms on a finite-dimensional space are equivalent.

Proof. Let $\|\cdot\|_1$ and $\|\cdot\|_2$ be two norms on a finite-dimensional space X. We choose a basis $\{e_1, e_2, \ldots, e_n\}$ of X. Then Lemma 5.32 implies that there are strictly positive constants m_1, m_2, M_1, M_2 such that if $x = \sum_{i=1}^{n} x_i e_i$, then

$$m_1 \sum_{i=1}^{n} |x_i| \leq \|x\|_1 \leq M_1 \sum_{i=1}^{n} |x_i|,$$

$$m_2 \sum_{i=1}^{n} |x_i| \leq \|x\|_2 \leq M_2 \sum_{i=1}^{n} |x_i|.$$

Equation (5.10) then follows with $c = m_2/M_1$ and $C = M_2/m_1$. $\qquad\square$

5.5 Convergence of bounded operators

The set $\mathcal{B}(X, Y)$ of bounded linear maps from a normed linear space X to a normed linear space Y is a linear space with respect to the natural pointwise definitions of vector addition and scalar multiplication:

$$(S + T)x = Sx + Tx, \qquad (\lambda T)x = \lambda(Tx).$$

It is straightforward to check that the operator norm in Definition 5.12,

$$\|T\| = \sup_{x \neq 0} \frac{\|Tx\|}{\|x\|},$$

defines a norm on $\mathcal{B}(X, Y)$, so that $\mathcal{B}(X, Y)$ is a normed linear space.

The composition of two linear maps is linear, and the following theorem states that the composition of two bounded linear maps is bounded.

Theorem 5.37 Let X, Y, and Z be normed linear spaces. If $T \in \mathcal{B}(X, Y)$ and $S \in \mathcal{B}(Y, Z)$, then $ST \in \mathcal{B}(X, Z)$, and

$$\|ST\| \leq \|S\| \, \|T\|. \tag{5.13}$$

Proof. For all $x \in X$ we have

$$\|STx\| \leq \|S\| \|Tx\| \leq \|S\| \|T\| \|x\|. \qquad \square$$

For example, if $T \in \mathcal{B}(X)$, then $T^n \in \mathcal{B}(X)$ and $\|T^n\| \leq \|T\|^n$. It may well happen that we have strict inequality in (5.13).

Example 5.38 Consider the linear maps A, B on $\mathbb{R}^2$ with matrices

$$A = \begin{pmatrix} \lambda & 0 \\ 0 & 0 \end{pmatrix}, \qquad B = \begin{pmatrix} 0 & 0 \\ 0 & \mu \end{pmatrix}.$$

These matrices have the Euclidean (or sum, or maximum) norms $\|A\| = |\lambda|$ and $\|B\| = |\mu|$, but $\|AB\| = 0$.

A linear space with a product defined on it is called an *algebra*. The composition of maps defines a product on the space $\mathcal{B}(X)$ of bounded linear maps on X into itself, so $\mathcal{B}(X)$ is an algebra. The algebra is associative, meaning that $(RS)T = R(ST)$, but is not commutative, since in general ST is not equal to TS. If $S, T \in \mathcal{B}(X)$, we define the *commutator* $[S, T] \in \mathcal{B}(X)$ of S and T by

$$[S, T] = ST - TS.$$

If $ST = TS$, or equivalently if $[S, T] = 0$, then we say that S and T *commute*.

The convergence of operators in $\mathcal{B}(X, Y)$ with respect to the operator norm is called uniform convergence.

Definition 5.39 If (T_n) is a sequence of operators in $\mathcal{B}(X, Y)$ and

$$\lim_{n \to \infty} \|T_n - T\| = 0$$

for some $T \in \mathcal{B}(X, Y)$, then we say that T_n *converges uniformly* to T, or that T_n converges to T in the uniform, or operator norm, topology on $\mathcal{B}(X, Y)$.

Example 5.40 Let $X = C([0, 1])$ equipped with the supremum norm. For $k_n(x, y)$ is a real-valued continuous function on $[0, 1] \times [0, 1]$, we define $K_n \in \mathcal{B}(X)$ by

$$K_n f(x) = \int_0^1 k_n(x, y) f(y) \, dy. \qquad (5.14)$$

Then $K_n \to 0$ uniformly as $n \to \infty$ if

$$\|K_n\| = \max_{x \in [0,1]} \left\{ \int_0^1 |k_n(x, y)| \, dy \right\} \to 0 \qquad \text{as } n \to \infty. \qquad (5.15)$$

An example of functions k_n satisfying (5.15) is $k_n(x, y) = xy^n$.

A basic fact about a space of bounded linear operators that take values in a Banach space is that it is itself a Banach space.

Theorem 5.41 If X is a normed linear space and Y is a Banach space, then $\mathcal{B}(X,Y)$ is a Banach space with respect to the operator norm.

Proof. We have to prove that $\mathcal{B}(X,Y)$ is complete. Let (T_n) be a Cauchy sequence in $\mathcal{B}(X,Y)$. For each $x \in X$, we have

$$\|T_n x - T_m x\| \le \|T_n - T_m\| \, \|x\|,$$

which shows that $(T_n x)$ is a Cauchy sequence in Y. Since Y is complete, there is a $y \in Y$ such that $T_n x \to y$. It is straightforward to check that $Tx = y$ defines a linear map $T : X \to Y$. We show that T is bounded. For any $\epsilon > 0$, let N_ϵ be such that $\|T_n - T_m\| < \epsilon/2$ for all $n, m \ge N_\epsilon$. Take $n \ge N_\epsilon$. Then for each $x \in X$, there is an $m(x) \ge N_\epsilon$ such that $\|T_{m(x)} x - Tx\| \le \epsilon/2$. If $\|x\| = 1$, we have

$$\|T_n x - Tx\| \le \|T_n x - T_{m(x)} x\| + \|T_{m(x)} x - Tx\| \le \epsilon. \tag{5.16}$$

It follows that if $n \ge N_\epsilon$, then

$$\|Tx\| \le \|T_n x\| + \|Tx - T_n x\| \le \|T_n\| + \epsilon$$

for all x with $\|x\| = 1$, so T is bounded. Finally, from (5.16) it follows that $\lim_{n \to \infty} \|T_n - T\| = 0$. Hence, $T_n \to T$ in the uniform norm. $\square$

A particularly important class of bounded operators is the class of compact operators.

Definition 5.42 A linear operator $T : X \to Y$ is *compact* if $T(B)$ is a precompact subset of Y for every bounded subset B of X.

An equivalent formulation is that T is compact if and only if every bounded sequence (x_n) in X has a subsequence (x_{n_k}) such that (Tx_{n_k}) converges in Y. We do not require that the range of T be closed, so $T(B)$ need not be compact even if B is a closed bounded set. We leave the proof of the following properties of compact operators as an exercise.

Proposition 5.43 Let X, Y, Z be Banach spaces. (a) If $S, T \in \mathcal{B}(X,Y)$ are compact, then any linear combination of S and T is compact. (b) If (T_n) is a sequence of compact operators in $\mathcal{B}(X,Y)$ converging uniformly to T, then T is compact. (c) If $T \in \mathcal{B}(X,Y)$ has finite-dimensional range, then T is compact. (d) Let $S \in \mathcal{B}(X,Y)$, $T \in \mathcal{B}(Y,Z)$. If S is bounded and T is compact, or S is compact and T is bounded, then $TS \in \mathcal{B}(X,Z)$ is compact.

It follows from parts (a)–(b) of this proposition that the space $\mathcal{K}(X,Y)$ of compact linear operators from X to Y is a closed linear subspace of $\mathcal{B}(X,Y)$. Part (d) implies that $\mathcal{K}(X)$ is a two-sided ideal of $\mathcal{B}(X)$, meaning that if $K \in \mathcal{K}(X)$, then $AK \in \mathcal{K}(X)$ and $KA \in \mathcal{K}(X)$ for all $A \in \mathcal{B}(X)$.

From parts (b)–(c), an operator that is the uniform limit of operators with finite rank, that is with finite-dimensional range, is compact. The converse is also true for compact operators on many Banach spaces, including all Hilbert spaces, although there exist separable Banach spaces on which some compact operators cannot be approximated by finite-rank operators. As a result, compact operators on infinite-dimensional spaces behave in many respects like operators on finite-dimensional spaces. We will discuss compact operators on a Hilbert space in greater detail in Chapter 9.

Another type of convergence of linear maps is called strong convergence.

Definition 5.44 A sequence (T_n) in $\mathcal{B}(X, Y)$ *converges strongly* if

$$\lim_{n \to \infty} T_n x = T x \qquad \text{for every } x \in X.$$

Thus, strong convergence of linear maps is convergence of their pointwise values with respect to the norm on Y. The terminology here is a little inconsistent: strong and norm convergence mean the same thing for vectors in a Banach space, but different things for operators on a Banach space. The associated strong topology on $\mathcal{B}(X, Y)$ is distinct from the uniform norm topology whenever X is infinite-dimensional, and is not derived from a norm. We leave the proof of the following theorem as an exercise.

Theorem 5.45 If $T_n \to T$ uniformly, then $T_n \to T$ strongly.

The following examples show that strong convergence does not imply uniform convergence.

Example 5.46 Let $X = \ell^2(\mathbb{N})$, and define the projection $P_n : X \to X$ by

$$P_n(x_1, x_2, \ldots, x_n, x_{n+1}, x_{n+2}, \ldots) = (x_1, x_2, \ldots, x_n, 0, 0, \ldots).$$

Then $\|P_n - P_m\| = 1$ for $n \neq m$, so (P_n) does not converge uniformly. Nevertheless, if $x \in \ell^2(\mathbb{N})$ is any fixed vector, we have $P_n x \to x$ as $n \to \infty$. Thus, $P_n \to I$ strongly.

Example 5.47 Let $X = C([0, 1])$, and consider the sequence of continuous linear functionals $K_n : X \to \mathbb{R}$, given by

$$K_n f = \int_0^1 \sin(n\pi x) \, f(x) \, dx.$$

If p is a polynomial, then an integration by parts implies that

$$K_n p = \frac{p(0) - \cos(n\pi)p(1)}{n\pi} + \frac{1}{n\pi} \int_0^1 \cos(n\pi x) \, p'(x) \, dx.$$

Hence, $K_n p \to 0$ as $n \to \infty$. If $f \in C([0,1])$, then by Theorem 2.9 for any $\epsilon > 0$ there is a polynomial p such that $\|f - p\| < \epsilon/2$, and there is an N such that $|K_n p| < \epsilon/2$ for $n \geq N$. Since $\|K_n\| \leq 1$ for all n, it follows that

$$|K_n f| \leq \|K_n\| \, \|f - p\| + |K_n p| < \epsilon$$

when $n \geq N$. Thus, $K_n f \to 0$ as $n \to \infty$ for every $f \in C([0,1])$. This result is a special case of the Riemann-Lebesgue lemma, which we prove in Theorem 11.34. On the other hand, if $f_n(x) = \sin(n\pi x)$, then $\|f_n\| = 1$ and $\|K_n f_n\| = 1/2$, which implies that $\|K_n\| \geq 1/2$. (In fact, $\|K_n\| = 2/\pi$ for each n.) Hence, $K_n \to 0$ strongly, but not uniformly.

A third type of convergence of operators, *weak convergence*, may be defined using the notion of weak convergence in a Banach space, given in Definition 5.59 below. We say that T_n *converges weakly* to T in $\mathcal{B}(X,Y)$ if the pointwise values $T_n x$ converge weakly to Tx in Y. We will not consider the weak convergence of operators in this book.

We end this section with two applications of operator convergence. First we define the *exponential* of an operator, and use it to solve a linear evolution equation. If $A : X \to X$ is a bounded linear operator on a Banach space X, then, by analogy with the power series expansion of e^a, we define

$$e^A = I + A + \frac{1}{2!}A^2 + \frac{1}{3!}A^3 + \ldots + \frac{1}{n!}A^n + \ldots . \tag{5.17}$$

A comparison with the convergent real series

$$e^{\|A\|} = 1 + \|A\| + \frac{1}{2!}\|A\|^2 + \frac{1}{3!}\|A\|^3 + \ldots + \frac{1}{n!}\|A\|^n + \ldots ,$$

implies that the series on the right hand side of (5.17) is absolutely convergent in $\mathcal{B}(X)$, and hence norm convergent. It also follows that

$$\|e^A\| \leq e^{\|A\|}.$$

If A and B commute, then multiplication and rearrangement of the series for the exponentials implies that

$$e^A e^B = e^{A+B}.$$

The solution of the initial value problem for the linear, scalar ODE $x_t = ax$ with $x(0) = x_0$ is given by $x(t) = x_0 e^{at}$. This result generalizes to a linear system,

$$x_t = Ax, \qquad x(0) = x_0, \tag{5.18}$$

where $x : \mathbb{R} \to X$, with X a Banach space, and $A : X \to X$ is a bounded linear operator on X. The solution of (5.18) is given by

$$x(t) = e^{tA} x_0.$$

This is a solution because

$$\frac{d}{dt}e^{tA} = Ae^{tA},$$

where the derivative is given by the uniformly convergent limit,

$$\frac{d}{dt}e^{tA} = \lim_{h\to 0}\left(\frac{e^{A(t+h)} - e^{tA}}{h}\right)$$

$$= e^{tA}\lim_{h\to 0}\left(\frac{e^{Ah} - I}{h}\right)$$

$$= Ae^{tA}\lim_{h\to 0}\sum_{n=0}^{\infty}\frac{1}{(n+1)!}A^n h^n$$

$$= Ae^{tA}.$$

An important application of this result is to linear systems of ODEs when $x(t) \in \mathbb{R}^n$ and A is an $n \times n$ matrix, but it also applies to linear equations on infinite-dimensional spaces.

Example 5.48 Suppose that $k : [0,1] \times [0,1] \to \mathbb{R}$ is a continuous function, and $K : C([0,1]) \to C([0,1])$ is the integral operator

$$Ku(x) = \int_0^1 k(x,y)u(y)\,dy.$$

The solution of the initial value problem

$$u_t(x,t) + \lambda u(x,t) = \int_0^1 k(x,y)u(y,t)\,dy, \qquad u(x,0) = u_0(x),$$

with $u(\cdot, t) \in C([0,1])$, is $u = e^{(K-\lambda I)t}u_0$.

The one-parameter family of operators $T(t) = e^{tA}$ is called the *flow* of the evolution equation (5.18). The operator $T(t)$ maps the solution at time 0 to the solution at time t. We leave the proof of the following properties of the flow as an exercise.

Theorem 5.49 If $A : X \to X$ is a bounded linear operator and $T(t) = e^{tA}$ for $t \in \mathbb{R}$, then:

(a) $T(0) = I$;
(b) $T(s)T(t) = T(s+t)$ for $s, t \in \mathbb{R}$;
(c) $T(t) \to I$ uniformly as $t \to 0$.

A family of bounded linear operators $\{T(t) \mid t \in \mathbb{R}\}$ that satisfies the properties (a)–(c) in this theorem is called a one-parameter *uniformly continuous group*. Properties (a)–(b) imply that the operators form a commutative group under composition, while (c) states that $T : \mathbb{R} \to \mathcal{B}(X)$ is continuous with respect to the uniform, or norm, topology on $\mathcal{B}(X)$ at $t = 0$. The group property implies that T is uniformly continuous on $\mathbb{R}$, meaning that $\|T(t) - T(t_0)\| \to 0$ as $t \to t_0$ for any $t_0 \in \mathbb{R}$.

Any one-parameter uniformly continuous group of operators can be written as $T(t) = e^{tA}$ for a suitable operator A, called the *generator* of the group. The generator A may be recovered from the operators $T(t)$ by

$$A = \lim_{t \to 0} \left(\frac{T(t) - I}{t} \right). \tag{5.19}$$

Many linear partial differential equations can be written as evolution equations of the form (5.18) in which A is an unbounded operator. Under suitable conditions on A, there exist solution operators $T(t)$, which may be defined only for $t \geq 0$, and which are strongly continuous functions of t, rather than uniformly continuous. The solution operators are then said to form a C_0-*semigroup*. For an example, see the discussion of the heat equation in Section 7.3.

As a second application of operator convergence, we consider the convergence of approximation schemes. Suppose we want to solve an equation of the form

$$Au = f, \tag{5.20}$$

where $A : X \to Y$ is a nonsingular linear operator between Banach spaces and $f \in Y$ is given. Suppose we can approximate (5.20) by an equation

$$A_\epsilon u_\epsilon = f_\epsilon, \tag{5.21}$$

whose solution u_ϵ can be computed more easily. We assume that $A_\epsilon : X \to Y$ is a nonsingular linear operator with a bounded inverse. We call the family of equations (5.21) an *approximation scheme* for (5.20). For instance, if (5.20) is a differential equation, then (5.21) may be obtained by a finite difference or finite element approximation, where ϵ is a grid spacing. One complication is that a numerical approximation A_ϵ may act on a different space X_ϵ than the space X. For simplicity, we suppose that the approximations A_ϵ may be defined on the same space as A. The primary requirement of an approximation scheme is that it is convergent.

Definition 5.50 The approximation scheme (5.21) is *convergent* to (5.20) if $u_\epsilon \to u$ as $\epsilon \to 0$ whenever $f_\epsilon \to f$.

We make precise the idea that A_ϵ approximates A in the following definition of consistency.

Definition 5.51 The approximation scheme (5.21) is *consistent* with (5.20) if $A_\epsilon v \to Av$ as $\epsilon \to 0$ for each $v \in X$.

In other words, the approximation scheme is consistent if A_ϵ converges strongly to A as $\epsilon \to 0$. Consistency on its own is not sufficient to guarantee convergence. We also need a second property called *stability*.

Definition 5.52 The approximation scheme (5.21) is *stable* if there is a constant M, independent of ϵ, such that

$$\|A_\epsilon^{-1}\| \leq M.$$

Consistency relates the operators A_ϵ to A, while stability is a property of the approximate operators A_ϵ alone. The Lax equivalence theorem states that an approximation scheme is convergent if and only if it is consistent and stable. Stability plays a crucial role in convergence, because it prevents the amplification of errors in the approximate solutions as $\epsilon \to 0$.

Theorem 5.53 (Lax equivalence) An approximation scheme is convergent if and only if it is consistent and stable.

Proof. If $Au = f$ and $A_\epsilon u_\epsilon = f_\epsilon$, then

$$u - u_\epsilon = A_\epsilon^{-1} \left(A_\epsilon u - Au + f - f_\epsilon \right).$$

Taking the norm of this equation, using the definition of the operator norm, and the triangle inequality, we find that

$$\|u - u_\epsilon\| \leq \|A_\epsilon^{-1}\| \left(\|A_\epsilon u - Au\| + \|f - f_\epsilon\| \right). \tag{5.22}$$

If the scheme is stable, then

$$\|u - u_\epsilon\| \leq M \left(\|A_\epsilon u - Au\| + \|f - f_\epsilon\| \right),$$

and if the scheme is consistent, then $A_\epsilon u \to Au$ as $\epsilon \to 0$. It follows that $u_\epsilon \to u$ if $f_\epsilon \to f$, and the scheme is convergent.

Conversely, if the approximation scheme is not convergent, then there are $f_\epsilon \in Y$ such that (f_ϵ) converges to f as $\epsilon \to 0$ but (u_ϵ) does not converge to u. Equation (5.22) then implies that either $\|A_\epsilon^{-1}\|$ is unbounded as $\epsilon \to 0$, so the scheme is unstable, or $A_\epsilon u$ does not converge to Au, so the scheme is inconsistent. $\square$

An analogous result holds for linear evolution equations of the form (5.18) (see Strikwerder [53], for example). There is, however, no general criterion for the convergence of approximation schemes for nonlinear equations.

5.6 Dual spaces

The dual space of a linear space consists of the scalar-valued linear maps on the space. Duality methods play a crucial role in many parts of analysis. In this section, we consider real linear spaces for definiteness, but all the results hold for complex linear spaces.

Definition 5.54 A scalar-valued linear map from a linear space X to $\mathbb{R}$ is called a *linear functional* or *linear form* on X. The space of linear functionals on X is called the *algebraic dual space* of X, and the space of continuous linear functionals on X is called the *topological dual space* of X.

In terms of the notation in Definition 5.12, the algebraic dual space of X is $\mathcal{L}(X, \mathbb{R})$, and the topological dual space is $\mathcal{B}(X, \mathbb{R})$. A linear functional $\varphi : X \to \mathbb{R}$ is bounded if there is a constant M such that

$$|\varphi(x)| \leq M \|x\| \qquad \text{for all } x \in X,$$

and then we define $\|\varphi\|$ by

$$\|\varphi\| = \sup_{x \neq 0} \frac{|\varphi(x)|}{\|x\|}. \tag{5.23}$$

If X is infinite dimensional, then $\mathcal{L}(X, \mathbb{R})$ is much larger than $\mathcal{B}(X, \mathbb{R})$, as we illustrate in Example 5.57 below. Somewhat confusingly, both dual spaces are commonly denoted by X^*. We will use X^* to denote the topological dual space of X. Either dual space is itself a linear space under the operations of pointwise addition and scalar multiplication of maps, and the topological dual is a Banach space, since $\mathbb{R}$ is complete.

If X is finite dimensional, then $\mathcal{L}(X, \mathbb{R}) = \mathcal{B}(X, \mathbb{R})$, so there is no need to distinguish between the algebraic and topological dual spaces. Moreover, the dual space X^* of a finite-dimensional space X is linearly isomorphic to X. To show this, we pick a basis $\{e_1, e_2, \ldots, e_n\}$ of X. The map $\omega_i : X \to \mathbb{R}$ defined by

$$\omega_i \left(\sum_{j=1}^{n} x_j e_j \right) = x_i \tag{5.24}$$

is an element of the algebraic dual space X^*. The linearity of ω_i is obvious.

For example, if $X = \mathbb{R}^n$ and

$$e_1 = (1, 0, \ldots, 0), \quad e_2 = (0, 1, \ldots, 0), \ldots, e_n = (0, 0, \ldots, 1),$$

are the coordinate basis vectors, then

$$\omega_i : (x_1, x_2, \ldots, x_n) \mapsto x_i$$

is the map that takes a vector to its ith coordinate.

The action of a general element φ of the dual space, $\varphi : X \to \mathbb{R}$, on a vector $x \in X$ is given by a linear combination of the components of x, since

$$\varphi \left(\sum_{i=1}^{n} x_i e_i \right) = \sum_{i=1}^{n} \varphi_i x_i,$$

where $\varphi_i = \varphi(e_i) \in \mathbb{R}$. It follows that, as a map,

$$\varphi = \sum_{i=1}^{n} \varphi_i \omega_i.$$

Thus, $\{\omega_1, \omega_2, \ldots, \omega_n\}$ is a basis of X^*, called the *dual basis* of $\{e_1, e_2, \ldots, e_n\}$, and both X and X^* are linearly isomorphic to $\mathbb{R}^n$. The dual basis has the property that

$$\omega_i(e_j) = \delta_{ij},$$

where δ_{ij} is the *Kronecker delta function*, defined by

$$\delta_{ij} = \begin{cases} 1 & \text{if } i = j, \\ 0 & \text{if } i \neq j. \end{cases} \tag{5.25}$$

Although a finite-dimensional space is linearly isomorphic with its dual space, there is no canonical way to identify the space with its dual; there are many isomorphisms, depending on an arbitrary choice of a basis. In the following chapters, we will study Hilbert spaces, and show that the topological dual space of a Hilbert space can be identified with the original space in a natural way through the inner product (see the Riesz representation theorem, Theorem 8.12). The dual of an infinite-dimensional Banach space is, in general, different from the original space.

Example 5.55 In Section 12.8, we will see that for $1 \leq p < \infty$ the dual of $L^p(\Omega)$ is $L^{p'}(\Omega)$, where $1/p + 1/p' = 1$. The Hilbert space $L^2(\Omega)$ is self-dual.

Example 5.56 Consider $X = C([a, b])$. For any $\rho \in L^1([a, b])$, the following formula defines a continuous linear functional φ on X:

$$\varphi(f) = \int_a^b f(x)\rho(x)\, dx. \tag{5.26}$$

Not all continuous linear functionals are of the form (5.26). For example, if $x_0 \in [a, b]$, then the evaluation of f at x_0 is a continuous linear functional. That is, if we define $\delta_{x_0} : C([a, b]) \to \mathbb{R}$ by

$$\delta_{x_0}(f) = f(x_0),$$

then δ_{x_0} is a continuous linear functional on $C([a, b])$. A full description of the dual space of $C([a, b])$ is not so simple: it may be identified with the space of Radon measures on $[a, b]$ (see [12], for example).

One way to obtain a linear functional on a linear space is to start with a linear functional defined on a subspace, extend a Hamel basis of the subspace to a Hamel basis of the whole space and extend the functional to the whole space, by use of linearity and an arbitrary definition of the functional on the additional basis elements. The next example uses this procedure to obtain a discontinuous linear functional on $C([0, 1])$.

Example 5.57 Let $M = \{x^n \mid n = 0, 1, 2, \ldots\}$ be the set of monomials in $C([0,1])$. The set M is linearly independent, so it may be extended to a Hamel basis H. Each $f \in C([0,1])$ can be written uniquely as

$$f = c_1 h_1 + \cdots + c_N h_N, \tag{5.27}$$

for suitable basis functions $h_i \in H$ and nonzero scalar coefficients c_i. For each $n = 0, 1, 2, \ldots$, we define $\varphi_n(f)$ by

$$\varphi_n(f) = \begin{cases} c_i & \text{if } h_i = x^n, \\ 0 & \text{otherwise.} \end{cases}$$

Due to the uniqueness of the decomposition in (5.27), the functional φ_n is well-defined. We define a linear functional φ on $C([0,1])$ by

$$\varphi(f) = \sum_{n=1}^{\infty} n\varphi_n(f).$$

For each f, only a finite number of terms in this sum are nonzero, so φ is a well-defined linear functional on $C([0,1])$. The functional is unbounded, since for each $n = 0, 1, 2, \ldots$ we have $\|x^n\| = 1$ and $|\varphi(x^n)| = n$.

A similar construction shows that every infinite-dimensional linear space has discontinuous linear functionals defined on it. On the other hand, Theorem 5.35 implies that all linear functionals on a finite-dimensional linear space are bounded.

It is not obvious that this extension procedure can be used to obtain bounded linear functionals on an infinite-dimensional linear space, or even that there are any nonzero bounded linear functionals at all, because the extension need not be bounded. In fact, it is possible to maintain boundedness of an extension by a suitable choice of its values off the original subspace, as stated in the following version of the Hahn-Banach theorem.

Theorem 5.58 (Hahn-Banach) If Y is a linear subspace of a normed linear space X and $\psi : Y \to \mathbb{R}$ is a bounded linear functional on Y with $\|\psi\| = M$, then there is a bounded linear functional $\varphi : X \to \mathbb{R}$ on X such that φ restricted to Y is equal to ψ and $\|\varphi\| = M$.

We omit the proof here. One consequence of this theorem is that there are enough bounded linear functionals to separate X, meaning that if $\varphi(x) = \varphi(y)$ for all $\varphi \in X^*$, then $x = y$ (see Exercise 5.6).

Since X^* is a Banach space, we can form its dual space X^{**}, called the *bidual* of X. There is no natural way to identify an element of X with an element of the dual X^*, but we can naturally identify an element of X with an element of the bidual X^{**}. If $x \in X$, then we define $F_x \in X^{**}$ by evaluation at x:

$$F_x(\varphi) = \varphi(x) \qquad \text{for every } \varphi \in X^*. \tag{5.28}$$

Thus, we may regard X as a subspace of X^{**}. If all continuous linear functionals on X^* are of the form (5.28), then $X = X^{**}$ under the identification $x \mapsto F_x$, and we say that X is *reflexive*.

Linear functionals may be used to define a notion of convergence that is weaker than norm, or strong, convergence on an infinite-dimensional Banach space.

Definition 5.59 A sequence (x_n) in a Banach space X *converges weakly* to x, denoted by $x_n \rightharpoonup x$ as $n \to \infty$, if

$$\varphi(x_n) \to \varphi(x) \qquad \text{as } n \to \infty,$$

for every bounded linear functional φ in X^*.

If we think of a linear functional $\varphi : X \to \mathbb{R}$ as defining a coordinate $\varphi(x)$ of x, then weak convergence corresponds to coordinate-wise convergence. Strong convergence implies weak convergence: if $x_n \to x$ in norm and φ is a bounded linear functional, then

$$|\varphi(x_n) - \varphi(x)| = |\varphi(x_n - x)| \leq \|\varphi\| \|x_n - x\| \to 0.$$

Weak convergence does not imply strong convergence on an infinite-dimensional space, as we will see in Section 8.6.

If X^* is the dual of a Banach space X, then we can define another type of weak convergence on X^*, called weak-$*$ convergence, pronounced "weak star."

Definition 5.60 Let X^* be the dual of a Banach space X. We say $\varphi \in X^*$ is the *weak-$*$* limit of a sequence (φ_n) in X^* if

$$\varphi_n(x) \to \varphi(x) \qquad \text{as } n \to \infty,$$

for every $x \in X$. We denote weak-$*$ convergence by

$$\varphi_n \overset{*}{\rightharpoonup} \varphi.$$

By contrast, weak convergence of (φ_n) in X^* means that

$$F(\varphi_n) \to F(\varphi) \qquad \text{as } n \to \infty,$$

for every $F \in X^{**}$. If X is reflexive, then weak and weak-$*$ convergence in X^* are equivalent because every bounded linear functional on X^* is of the form (5.28). If X^* is the dual space of a nonreflexive space X, then weak and weak-$*$ convergence are different, and it is preferable to use weak-$*$ convergence in X^* instead of weak convergence.

One reason for the importance of weak-$*$ convergence is the following compactness result, called the Banach-Alaoglu theorem.

Theorem 5.61 (Banach-Alaoglu) Let X^* be the dual space of a Banach space X. The closed unit ball in X^* is weak-$*$ compact.

We will not prove this result here, but we prove a special case of it in Theorem 8.45.

5.7 References

For more on linear operators in Banach spaces, see Kato [26]. For proofs of the Hahn-Banach, open mapping, and Banach-Alaoglu theorems, see Folland [12], Reed and Simon [45], or Rudin [48]. The use of linear and Banach spaces in optimization theory is discussed in [34]. Applied functional analysis is discussed in Lusternik and Sobolev [33]. For an introduction to semigroups associated with evolution equations, see [4]. For more on matrices, see [24]. An introduction to the numerical aspects of matrices and linear algebra is in [54]. For more on the stability, consistency, and convergence of finite difference schemes for partial differential equations, see Strikwerder [53].

5.8 Exercises

Exercise 5.1 Prove that the expressions in (5.2) and (5.3) for the norm of a bounded linear operator are equivalent.

Exercise 5.2 Suppose that $\{e_1, e_2, \ldots, e_n\}$ and $\{\widetilde{e}_1, \widetilde{e}_2, \ldots, \widetilde{e}_n\}$ are two bases of the n-dimensional linear space X, with

$$\widetilde{e}_i = \sum_{j=1}^n L_{ij} e_j, \qquad e_i = \sum_{j=1}^n \widetilde{L}_{ij} \widetilde{e}_j,$$

where (L_{ij}) is an invertible matrix with inverse $\left(\widetilde{L}_{ij}\right)$, meaning that

$$\sum_{j=1}^n L_{ij} \widetilde{L}_{jk} = \delta_{ik}.$$

Let $\{\omega_1, \omega_2, \ldots, \omega_n\}$ and $\{\widetilde{\omega}_1, \widetilde{\omega}_2, \ldots, \widetilde{\omega}_n\}$ be the associated dual bases of X^*.

(a) If $x = \sum x_i e_i = \sum \widetilde{x}_i \widetilde{e}_i \in X$, then prove that the components of x transform under a change of basis according to

$$\widetilde{x}_i = \widetilde{L}_{ij} x_j. \tag{5.29}$$

(b) If $\varphi = \sum \varphi_i \omega_i = \sum \widetilde{\varphi}_i \widetilde{\omega}_i \in X^*$, then prove that the components of φ transform under a change of basis according to

$$\widetilde{\varphi}_i = L_{ji} \varphi_j. \tag{5.30}$$

Exercise 5.3 Let $\delta : C([0,1]) \to \mathbb{R}$ be the linear functional that evaluates a function at the origin: $\delta(f) = f(0)$. If $C([0,1])$ is equipped with the sup-norm,

$$\|f\|_\infty = \sup_{0 \le x \le 1} |f(x)|,$$

show that δ is bounded and compute its norm. If $C([0,1])$ is equipped with the one-norm,

$$\|f\|_1 = \int_0^1 |f(x)|\,dx,$$

show that δ is unbounded.

Exercise 5.4 Consider the 2×2 matrix

$$A = \begin{pmatrix} 0 & a^2 \\ b^2 & 0 \end{pmatrix},$$

where $a > b > 0$. Compute the spectral radius $r(A)$ of A. Show that the Euclidean norms of powers of the matrix are given by

$$\|A^{2n}\| = a^{2n}b^{2n}, \qquad \|A^{2n+1}\| = a^{2n+2}b^{2n}.$$

Verify that $r(A) = \lim_{n\to\infty} \|A^n\|^{1/n}$.

Exercise 5.5 Define $K : C([0,1]) \to C([0,1])$ by

$$Kf(x) = \int_0^1 k(x,y)f(y)\,dy,$$

where $k : [0,1] \times [0,1] \to \mathbb{R}$ is continuous. Prove that K is bounded and

$$\|K\| = \max_{0 \le x \le 1} \left\{ \int_0^1 |k(x,y)|\,dy \right\}.$$

Exercise 5.6 Let X be a normed linear space. Use the Hahn-Banach theorem to prove the following statements.

(a) For any nonzero $x \in X$, there is a bounded linear functional $\varphi \in X^*$ such that $\|\varphi\| = 1$ and $\varphi(x) = \|x\|$.
(b) If $x, y \in X$ and $\varphi(x) = \varphi(y)$ for all $\varphi \in X^*$, then $x = y$.

Exercise 5.7 Find the kernel and range of the linear operator $K : C([0,1]) \to C([0,1])$ defined by

$$Kf(x) = \int_0^1 \sin \pi(x - y)f(y)\,dy.$$

Exercise 5.8 Prove that equivalent norms on a normed linear space X lead to equivalent norms on the space $\mathcal{B}(X)$ of bounded linear operators on X.

Exercise 5.9 Prove Proposition 5.43.

Exercise 5.10 Suppose that $k : [0,1] \times [0,1] \to \mathbb{R}$ is a continuous function. Prove that the integral operator $K : C([0,1]) \to C([0,1])$ defined by

$$Kf(x) = \int_0^1 k(x,y)f(y)\,dy$$

is compact.

Exercise 5.11 Prove that if $T_n \to T$ uniformly, then $\|T_n\| \to \|T\|$.

Exercise 5.12 Prove that if T_n converges to T uniformly, then T_n converges to T strongly.

Exercise 5.13 Suppose that Λ is the diagonal $n \times n$ matrix and N is the $n \times n$ *nilpotent matrix* (meaning that $N^k = 0$ for some k)

$$
\Lambda = \begin{pmatrix} \lambda_1 & 0 & \cdots & 0 \\ 0 & \lambda_2 & \cdots & 0 \\ \vdots & \vdots & \ddots & \vdots \\ 0 & 0 & \cdots & \lambda_n \end{pmatrix}, \qquad
N = \begin{pmatrix} 0 & 1 & 0 & \cdots & 0 \\ 0 & 0 & 1 & \cdots & 0 \\ \vdots & \vdots & \vdots & \ddots & \vdots \\ 0 & 0 & 0 & \cdots & 1 \\ 0 & 0 & 0 & \cdots & 0 \end{pmatrix}.
$$

(a) Compute the two-norms and spectral radii of Λ and N.
(b) Compute $e^{\Lambda t}$ and e^{Nt}.

Exercise 5.14 Suppose that A is an $n \times n$ matrix. For $t \in \mathbb{R}$ we define $f(t) = \det e^{tA}$.

(a) Show that

$$\lim_{t\to 0} \frac{f(t)-1}{t} = \operatorname{tr} A,$$

where $\operatorname{tr} A$ is the trace of the matrix A, that is the sum of its diagonal elements.
(b) Deduce that $f : \mathbb{R} \to \mathbb{R}$ is differentiable, and is a solution of the ODE $\dot{f} = (\operatorname{tr} A)f$.
(c) Show that

$$\det e^A = e^{\operatorname{tr} A}.$$

Exercise 5.15 Suppose that A and B are bounded linear operators on a Banach space.

(a) If A and B commute, then prove that $e^A e^B = e^{A+B}$.

(b) If $[A, [A, B]] = [B, [A, B]] = 0$, then prove that

$$e^A e^B = e^{A+B+[A,B]/2}.$$

This result is called the *Baker-Campbell-Hausdorff formula*.

Exercise 5.16 Suppose that A and B are, possibly noncommuting, bounded operators on a Banach space. Show that

$$\lim_{t \to 0} \frac{e^{t(A+B)} - e^{tA}e^{tB}}{t^2} = -\frac{1}{2}[A, B],$$

$$\lim_{t \to 0} \frac{e^{t(A+B)} - e^{tA/2}e^{tB}e^{tA/2}}{t^2} = 0.$$

Show that for small t the function $e^{tA/2}e^{tB}e^{tA/2}x(0)$ provides a better approximation to the solution of the equation $x_t = (A + B)x$ than the function $e^{tA}e^{tB}x(0)$. The approximation $e^{t(A+B)} \approx e^{tA/2}e^{tB}e^{tA/2}$, called *Strang splitting*, is useful in the numerical solution of evolution equations by *fractional step methods*.

Exercise 5.17 Suppose that $K : X \to X$ is a bounded linear operator on a Banach space X with $\|K\| < 1$. Prove that $I - K$ is invertible and

$$(I - K)^{-1} = I + K + K^2 + K^3 + \dots,$$

where the series on the right hand side converges uniformly in $\mathcal{B}(X)$.

Chapter 6

Hilbert Spaces

So far, in increasing order of specialization, we have studied topological spaces, metric spaces, normed linear spaces, and Banach spaces. Hilbert spaces are Banach spaces with a norm that is derived from an inner product, so they have an extra feature in comparison with arbitrary Banach spaces, which makes them still more special. We can use the inner product to introduce the notion of orthogonality in a Hilbert space, and the geometry of Hilbert spaces is in almost complete agreement with our intuition of linear spaces with an arbitrary (finite or infinite) number of orthogonal coordinate axes. By contrast, the geometry of infinite-dimensional Banach spaces can be surprisingly complicated and quite different from what naive extrapolations of the finite-dimensional situation would suggest.

6.1 Inner products

To be specific, we consider complex linear spaces throughout this chapter. We use a bar to denote the complex conjugate of a complex number. The corresponding results for real linear spaces are obtained by replacing $\mathbb{C}$ by $\mathbb{R}$ and omitting the complex conjugates.

Definition 6.1 An *inner product* on a complex linear space X is a map

$$(\,\cdot\,,\cdot\,) : X \times X \to \mathbb{C}$$

such that, for all $x, y, z \in X$ and $\lambda, \mu \in \mathbb{C}$:

(a) $(x, \lambda y + \mu z) = \lambda(x, y) + \mu(x, z)$ (linear in the second argument);
(b) $(y, x) = \overline{(x, y)}$ (Hermitian symmetric);
(c) $(x, x) \geq 0$ (nonnegative);
(d) $(x, x) = 0$ if and only if $x = 0$ (positive definite).

We call a linear space with an inner product an *inner product space* or a *pre-Hilbert space*.

From (a) and (b) it follows that $(\cdot, \cdot)$ is *antilinear*, or *conjugate linear*, in the first argument, meaning that

$$(\lambda x + \mu y, z) = \overline{\lambda}(x, z) + \overline{\mu}(y, z).$$

If X is real, then $(\cdot, \cdot)$ is bilinear, meaning that it is a linear function of each argument. If X is complex, then $(\cdot, \cdot)$ is said to be *sesquilinear*, a name that literally means "one-and-half" linear.

There are two conventions for the linearity of the inner product. In most of the mathematically oriented literature $(\cdot, \cdot)$ is linear in the first argument. We adopt the convention that the inner product is linear in the second argument, which is more common in applied mathematics and physics.

If X is a linear space with an inner product $(\cdot, \cdot)$, then we can define a norm on X by

$$\|x\| = \sqrt{(x, x)}. \tag{6.1}$$

Thus, any inner product space is a normed linear space. We will always use the norm defined in (6.1) on an inner product space.

Definition 6.2 A *Hilbert space* is a complete inner product space.

In particular, every Hilbert space is a Banach space with respect to the norm in (6.1).

Example 6.3 The standard inner product on $\mathbb{C}^n$ is given by

$$(x, y) = \sum_{j=1}^{n} \overline{x_j}\, y_j,$$

where $x = (x_1, \ldots, x_n)$ and $y = (y_1, \ldots, y_n)$, with $x_j, y_j \in \mathbb{C}$. This space is complete, and therefore it is a finite-dimensional Hilbert space.

Example 6.4 Let $C([a, b])$ denote the space of all complex-valued continuous functions defined on the interval $[a, b]$. We define an inner product on $C([a, b])$ by

$$(f, g) = \int_a^b \overline{f(x)} g(x)\, dx,$$

where $f, g : [a, b] \to \mathbb{C}$ are continuous functions. This space is not complete, so it is not a Hilbert space. The completion of $C([a, b])$ with respect to the associated norm,

$$\|f\| = \left(\int_a^b |f(x)|^2\, dx \right)^{1/2},$$

is denoted by $L^2([a, b])$. The spaces $L^p([a, b])$, defined in Example 5.6, are Banach spaces but they are not Hilbert spaces when $p \neq 2$.

Similarly, if $\mathbb{T}$ is the circle, then $L^2(\mathbb{T})$ is the Hilbert space of square-integrable functions $f : \mathbb{T} \to \mathbb{C}$ with the inner product

$$(f,g) = \int_{\mathbb{T}} \overline{f(x)} g(x) \, dx.$$

Example 6.5 We define the Hilbert space $\ell^2(\mathbb{Z})$ of bi-infinite complex sequences by

$$\ell^2(\mathbb{Z}) = \left\{ (z_n)_{n=-\infty}^{\infty} \;\middle|\; \sum_{n=-\infty}^{\infty} |z_n|^2 < \infty \right\}.$$

The space $\ell^2(\mathbb{Z})$ is a complex linear space, with the obvious operations of addition and multiplication by a scalar. An inner product on it is given by

$$(x,y) = \sum_{n=-\infty}^{\infty} \overline{x_n} y_n.$$

The name "ℓ^2" is pronounced "little ell two" to distinguish it from L^2 or "ell two" in the previous example. The space $\ell^2(\mathbb{N})$ of square-summable sequences $(z_n)_{n=1}^{\infty}$ is defined in an analogous way.

Example 6.6 Let $\mathbb{C}^{m \times n}$ denote the space of all $m \times n$ matrices with complex entries. We define an inner product on $\mathbb{C}^{m \times n}$ by

$$(A,B) = \operatorname{tr}(A^*B),$$

where tr denotes the trace and * denotes the Hermitian conjugate of a matrix — that is, the complex-conjugate transpose. In components, if $A = (a_{ij})$ and $B = (b_{ij})$, then

$$(A,B) = \sum_{i=1}^{m} \sum_{j=1}^{n} \overline{a}_{ij} b_{ij}.$$

This inner product is equal to the one obtained by identification of a matrix in $\mathbb{C}^{m \times n}$ with a vector in $\mathbb{C}^{mn}$. The corresponding norm,

$$\|A\| = \left(\sum_{i=1}^{m} \sum_{j=1}^{n} |a_{ij}|^2 \right)^{1/2},$$

is called the *Hilbert-Schmidt norm*.

Example 6.7 Let $C^k([a,b])$ be the space of functions with k continuous derivatives on $[a,b]$. We define an inner product on $C^k([a,b])$ by

$$(f,g) = \sum_{j=0}^{k} \int_{a}^{b} \overline{f^{(j)}(x)} g^{(j)}(x) \, dx,$$

where $f^{(j)}$ denotes the jth derivative of f. The corresponding norm is

$$\|f\| = \left(\sum_{j=0}^{k} \int_{a}^{b} |f^{(j)}(x)|^2 \, dx \right)^{1/2}. \tag{6.2}$$

The space $C^k([a,b])$ is an inner product space, but it is not complete. The Hilbert space obtained by completion of $C^k([a,b])$ with respect to the norm $\|\cdot\|$ is a *Sobolev space*, denoted by $H^k((a,b))$. In the notation of Example 5.7, we have

$$H^k((a,b)) = W^{k,2}((a,b)).$$

The following fundamental inequality on an inner product space is called the *Cauchy-Schwarz inequality*.

Theorem 6.8 (Cauchy-Schwarz) If $x, y \in X$, where X is an inner product space, then

$$|(x,y)| \leq \|x\| \|y\|, \tag{6.3}$$

where the norm $\|\cdot\|$ is defined in (6.1).

Proof. By the nonnegativity of the inner product, we have

$$0 \leq (\lambda x - \mu y, \lambda x - \mu y)$$

for all $x, y \in X$ and $\lambda, \mu \in \mathbb{C}$. Expansion of the inner product, and use of (6.1), implies that

$$\overline{\lambda}\mu(x,y) + \lambda\overline{\mu}(y,x) \leq |\lambda|^2 \|x\|^2 + |\mu|^2 \|y\|^2.$$

If $(x,y) = re^{i\varphi}$, where $r = |(x,y)|$ and $\varphi = \arg(x,y)$, then we choose

$$\lambda = \|y\|e^{i\varphi}, \qquad \mu = \|x\|.$$

It follows that

$$2\|x\| \|y\| |(x,y)| \leq 2\|x\|^2 \|y\|^2,$$

which proves the result. $\square$

An inner product space is a normed space with respect to the norm defined in (6.1). The converse question of when a norm is derived from an inner product in this way is answered by the following theorem.

Theorem 6.9 A normed linear space X is an inner product space with a norm derived from the inner product by (6.1) if and only if

$$\|x+y\|^2 + \|x-y\|^2 = 2\|x\|^2 + 2\|y\|^2 \qquad \text{for all } x, y \in X. \tag{6.4}$$

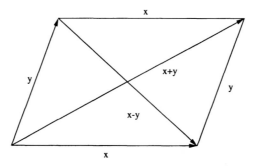

Fig. 6.1   The geometric interpretation of the parallelogram law (6.4).

Proof.   Use of (6.1) to write norms in terms of inner products, and expansion of the result, implies that (6.4) holds for any norm that is derived from an inner product. Conversely, if a norm satisfies (6.4), then the equation

$$(x,y) = \frac{1}{4}\left\{\|x+y\|^2 - \|x-y\|^2 - i\|x+iy\|^2 + i\|x-iy\|^2\right\} \qquad (6.5)$$

defines an inner product on X. We leave a detailed verification of this fact to the reader. □

The relation (6.4) is called the *parallelogram law*. Its geometrical interpretation is that the sum of the squares of the sides of a parallelogram is equal to the sum of the squares of the diagonals (see Figure 6.1). As the *polarization formula* (6.5) shows, an inner product is uniquely determined by its values on the diagonal, that is, by its values when the first and second arguments are equal.

Let $(X, (\,\cdot\,,\,\cdot\,)_X)$ and $(Y, (\,\cdot\,,\,\cdot\,)_Y)$ be two inner product spaces. Then there is a natural inner product on the Cartesian product space

$$X \times Y = \{(x,y) \mid x \in X, y \in Y\}$$

given by

$$((x_1, y_1), (x_2, y_2))_{X \times Y} = (x_1, x_2)_X + (y_1, y_2)_Y.$$

The associated norm on $X \times Y$ is

$$\|(x,y)\| = \sqrt{\|x\|^2 + \|y\|^2}$$

Unless stated otherwise, we will use this inner product and norm on the Cartesian product of two inner product spaces.

Theorem 6.10 Let X be an inner product space. The inner product is a continuous map from $X \times X \to \mathbb{C}$.

Proof. For all $x_1, x_2, y_1, y_2 \in X$, the Cauchy-Schwarz inequality implies that

$$
\begin{aligned}
|(x_1, y_1) - (x_2, y_2)| &= |(x_1 - x_2, y_1) + (x_2, y_1 - y_2)| \\
&\leq \|x_1 - x_2\| \|y_1\| + \|x_2\| \|y_1 - y_2\|.
\end{aligned}
$$

This estimate implies the continuity of the inner product. $\square$

6.2 Orthogonality

Let $\mathcal{H}$ be a Hilbert space. We denote its inner product by $\langle \cdot, \cdot \rangle$, which is another common notation for inner products that is often reserved for Hilbert spaces. The inner product structure of a Hilbert space allows us to introduce the concept of orthogonality, which makes it possible to visualize vectors and linear subspaces of a Hilbert space in a geometric way.

Definition 6.11 If x, y are vectors in a Hilbert space $\mathcal{H}$, then we say that x and y are *orthogonal*, written $x \perp y$, if $\langle x, y \rangle = 0$. We say that subsets A and B are orthogonal, written $A \perp B$, if $x \perp y$ for every $x \in A$ and $y \in B$. The *orthogonal complement* $A^\perp$ of a subset A is the set of vectors orthogonal to A,

$$
A^\perp = \{ x \in \mathcal{H} \mid x \perp y \text{ for all } y \in A \}.
$$

Theorem 6.12 The orthogonal complement of a subset of a Hilbert space is a closed linear subspace.

Proof. Let $\mathcal{H}$ be a Hilbert space and A a subset of $\mathcal{H}$. If $y, z \in A^\perp$ and $\lambda, \mu \in \mathbb{C}$, then the linearity of the inner product implies that

$$
\langle x, \lambda y + \mu z \rangle = \lambda \langle x, y \rangle + \mu \langle x, z \rangle = 0 \qquad \text{for all } x \in A.
$$

Therefore, $\lambda y + \mu z \in A^\perp$, so $A^\perp$ is a linear subspace.

To show that $A^\perp$ is closed, we show that if (y_n) is a convergent sequence in $A^\perp$, then the limit y also belongs to $A^\perp$. Let $x \in A$. From Theorem 6.10, the inner product is continuous and therefore

$$
\langle x, y \rangle = \langle x, \lim_{n \to \infty} y_n \rangle = \lim_{n \to \infty} \langle x, y_n \rangle = 0,
$$

since $\langle x, y_n \rangle = 0$ for every $x \in A$ and $y_n \in A^\perp$. Hence, $y \in A^\perp$. $\square$

The following theorem expresses one of the fundamental geometrical properties of Hilbert spaces. While the result may appear obvious (see Figure 6.2), the proof is not trivial.

Theorem 6.13 (Projection) Let $\mathcal{M}$ be a closed linear subspace of a Hilbert space $\mathcal{H}$.

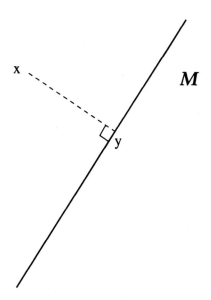

Fig. 6.2 y is the point in $\mathcal{M}$ closest to x.

(a) For each $x \in \mathcal{H}$ there is a unique closest point $y \in \mathcal{M}$ such that

$$\|x - y\| = \min_{z \in \mathcal{M}} \|x - z\|. \tag{6.6}$$

(b) The point $y \in \mathcal{M}$ closest to $x \in \mathcal{H}$ is the unique element of $\mathcal{M}$ with the property that $(x - y) \perp \mathcal{M}$.

Proof.   Let d be the distance of x from $\mathcal{M}$,

$$d = \inf \{\|x - z\| \mid z \in \mathcal{M}\}. \tag{6.7}$$

First, we prove that there is a closest point $y \in \mathcal{M}$ at which this infimum is attained, meaning that $\|x - y\| = d$. From the definition of d, there is a sequence of elements $y_n \in \mathcal{M}$ such that

$$\lim_{n \to \infty} \|x - y_n\| = d.$$

Thus, for all $\epsilon > 0$, there is an N such that

$$\|x - y_n\| \le d + \epsilon \qquad \text{when } n \ge N.$$

We show that the sequence (y_n) is Cauchy. From the parallelogram law, we have

$$\|y_m - y_n\|^2 + \|2x - y_m - y_n\|^2 = 2\|x - y_m\|^2 + 2\|x - y_n\|^2.$$

Since $(y_m + y_n)/2 \in \mathcal{M}$, equation (6.7) implies that

$$\|x - (y_m + y_n)/2\| \ge d.$$

Combining these equations, we find that for all $m, n \geq N$,

$$
\begin{aligned}
\|y_m - y_n\|^2 &= 2\|x - y_m\|^2 + 2\|x - y_n\|^2 - \|2x - y_m - y_n\|^2 \\
&\leq 4(d + \epsilon)^2 - 4d^2 \\
&\leq 4\epsilon(2d + \epsilon).
\end{aligned}
$$

Therefore, (y_n) is Cauchy. Since a Hilbert space is complete, there is a y such that $y_n \to y$, and, since M is closed, we have $y \in M$. The norm is continuous, so $\|x - y\| = \lim_{n \to \infty} \|x - y_n\| = d$.

Second, we prove the uniqueness of a vector $y \in M$ that minimizes $\|x - y\|$. Suppose y and y' both minimize the distance to x, meaning that

$$
\|x - y\| = d, \qquad \|x - y'\| = d.
$$

Then the parallelogram law implies that

$$
2\|x - y\|^2 + 2\|x - y'\|^2 = \|2x - y - y'\|^2 + \|y - y'\|^2.
$$

Hence, since $(y + y')/2 \in M$,

$$
\|y - y'\|^2 = 4d^2 - 4\|x - (y + y')/2\|^2 \leq 0.
$$

Therefore, $\|y - y'\| = 0$ so that $y = y'$.

Third, we show that the unique $y \in M$ found above satisfies the condition that the "error" vector $x - y$ is orthogonal to M. Since y minimizes the distance to x, we have for every $\lambda \in \mathbb{C}$ and $z \in M$ that

$$
\|x - y\|^2 \leq \|x - y + \lambda z\|^2.
$$

Expanding the right-hand side of this equation, we obtain that

$$
2\operatorname{Re} \lambda \langle x - y, z \rangle \leq |\lambda|^2 \|z\|^2.
$$

Suppose that $\langle x - y, z \rangle = |\langle x - y, z \rangle| e^{i\varphi}$. Choosing $\lambda = \epsilon e^{-i\varphi}$, where $\epsilon > 0$, and dividing by ϵ, we get

$$
2|\langle x - y, z \rangle| \leq \epsilon \|z\|^2.
$$

Taking the limit as $\epsilon \to 0^+$, we find that $\langle x - y, z \rangle = 0$, so $(x - y) \perp M$.

Finally, we show that y is the only element in M such that $x - y \perp M$. Suppose that y' is another such element in M. Then $y - y' \in M$, and, for any $z \in M$, we have

$$
\langle z, y - y' \rangle = \langle z, x - y' \rangle - \langle z, x - y \rangle = 0.
$$

In particular, we may take $z = y - y'$, and therefore we must have $y = y'$. $\square$

The proof of part (a) applies if M is any closed convex subset of $\mathcal{H}$ (see Exercise 6.1). Theorem 6.13 can also be stated in terms of the decomposition of $\mathcal{H}$ into an *orthogonal direct sum* of closed subspaces.

Definition 6.14 If $\mathcal{M}$ and $\mathcal{N}$ are orthogonal closed linear subspaces of a Hilbert space, then we define the *orthogonal direct sum*, or simply the *direct sum*, $\mathcal{M} \oplus \mathcal{N}$ of $\mathcal{M}$ and $\mathcal{N}$ by

$$\mathcal{M} \oplus \mathcal{N} = \{y + z \mid y \in \mathcal{M} \text{ and } z \in \mathcal{N}\}.$$

We may also define the orthogonal direct sum of two Hilbert spaces that are not subspaces of the same space (see Exercise 6.4).

Theorem 6.13 states that if $\mathcal{M}$ is a closed subspace, then any $x \in \mathcal{H}$ may be uniquely represented as $x = y + z$, where $y \in \mathcal{M}$ is the best approximation to x, and $z \perp \mathcal{M}$. We therefore have the following corollary

Corollary 6.15 If $\mathcal{M}$ is a closed subspace of a Hilbert space $\mathcal{H}$, then $\mathcal{H} = \mathcal{M} \oplus \mathcal{M}^{\perp}$.

Thus, every closed subspace $\mathcal{M}$ of a Hilbert space has a closed complementary subspace $\mathcal{M}^{\perp}$. If $\mathcal{M}$ is not closed, then we may still decompose $\mathcal{H}$ as $\mathcal{H} = \overline{\mathcal{M}} \oplus \mathcal{M}^{\perp}$. In a general Banach space, there may be no element of a closed subspace that is closest to a given element of the Banach space (see Exercise 6.2), and a closed linear subspace of a Banach space may have no closed complementary subspace. These facts are one indication of the much murkier geometrical properties of infinite-dimensional Banach spaces in comparison with Hilbert spaces.

6.3 Orthonormal bases

A subset U of nonzero vectors in a Hilbert space $\mathcal{H}$ is *orthogonal* if any two distinct elements in U are orthogonal. A set of vectors U is *orthonormal* if it is orthogonal and $\|u\| = 1$ for all $u \in U$, in which case the vectors u are said to be *normalized*. An *orthonormal basis* of a Hilbert space is an orthonormal set such that every vector in the space can be expanded in terms of the basis, in a way that we make precise below. In this section, we show that every Hilbert space has an orthonormal basis, which may be finite, countably infinite, or uncountable. Two Hilbert spaces whose orthonormal bases have the same cardinality are isomorphic — any linear map that identifies basis elements is an isomorphism — but many different concrete realizations of a given abstract Hilbert space arise in applications. The most important case in practice is that of a *separable* Hilbert space, which has a finite or countably infinite orthonormal basis. As shown in Exercise 6.10, this condition is equivalent to the separability of the Hilbert space as a metric space, meaning that it contains a countable dense subset.

Before studying orthonormal bases in general Hilbert spaces, we give some examples.

Example 6.16 A set of vectors $\{e_1, \ldots, e_n\}$ is an orthonormal basis of the finite-dimensional Hilbert spaces $\mathbb{C}^n$ if:

(a) $\langle e_j, e_k \rangle = \delta_{jk}$ for $1 \leq j, k \leq n$;

(b) for all $x \in \mathbb{C}^n$ there are unique coordinates $x_k \in \mathbb{C}$ such that

$$x = \sum_{k=1}^{n} x_k e_k, \tag{6.8}$$

where δ_{jk} is the Kronecker delta defined in (5.25). The orthonormality of the basis implies that $x_k = \langle e_k, x \rangle$. For example, the standard orthonormal basis of $\mathbb{C}^n$ consists of the vectors

$$e_1 = (1, 0, \ldots, 0), \quad e_2 = (0, 1, \ldots, 0), \ldots, \quad e_n = (0, 0, \ldots, 1).$$

Example 6.17 Consider the Hilbert space $\ell^2(\mathbb{Z})$ defined in Example 6.5. An orthonormal basis of $\ell^2(\mathbb{Z})$ is the set of coordinate basis vectors $\{e_n \mid n \in \mathbb{Z}\}$ given by

$$e_n = (\delta_{kn})_{k=-\infty}^{\infty}.$$

For example,

$$e_{-1} = (\ldots, 0, 1, 0, 0, 0, \ldots), \ e_0 = (\ldots, 0, 0, 1, 0, 0, \ldots), \ e_1 = (\ldots, 0, 0, 0, 1, 0, \ldots).$$

Example 6.18 The set of functions $\{e_n(x) \mid n \in \mathbb{Z}\}$, given by

$$e_n(x) = \frac{1}{\sqrt{2\pi}} e^{inx},$$

is an orthonormal basis of the space $L^2(\mathbb{T})$ of 2π-periodic functions, called the *Fourier basis*. We will study it in detail in the next chapter. As we will see, the inverse Fourier transform $\mathcal{F}^{-1} : \ell^2(\mathbb{Z}) \to L^2(\mathbb{T})$, defined by

$$\mathcal{F}^{-1}(c_k) = \frac{1}{\sqrt{2\pi}} \sum_{k=-\infty}^{\infty} c_k e^{ikx},$$

is a Hilbert space isomorphism between $\ell^2(\mathbb{Z})$ and $L^2(\mathbb{T})$. Both Hilbert spaces are separable with a countably infinite basis.

Example 6.19 A function that is a sum of finitely many periodic functions is said to be *quasiperiodic*. If the ratios of the periods of the terms in the sum are rational, then the sum is itself periodic, but if at least one of the ratios is irrational, then the sum is not periodic. For example,

$$f(t) = e^{it} + e^{i\pi t}$$

is quasiperiodic but not periodic. Let X be the space of quasiperiodic functions $f : \mathbb{R} \to \mathbb{C}$ of the form

$$f(t) = \sum_{k=1}^{n} a_k e^{i\omega_k t},$$

where $n \in \mathbb{N}$, $a_k \in \mathbb{C}$, and $\omega_k \in \mathbb{R}$ are arbitrary constants. We may think of t as a time variable, in which case f is a sum of time-harmonic functions with amplitudes $|a_k|$, phases $\arg a_k$, and frequencies ω_k. When some of the frequencies are incommensurable, the function f "almost" repeats itself, but it is not periodic with any period, although it is bounded.

We define an inner product on X by means of the time average,

$$\langle f, g \rangle = \lim_{T \to \infty} \frac{1}{2T} \int_{-T}^{T} \overline{f(t)} g(t) \, dt.$$

If $f(t) = \sum_{k=1}^{n} a_k e^{i\omega_k t}$ and $g(t) = \sum_{k=1}^{n} b_k e^{i\omega_k t}$, where $\omega_j \neq \omega_k$ for $j \neq k$, then

$$\langle f, g \rangle = \sum_{k=1}^{n} \overline{a_k} b_k.$$

The inner product may also be written as

$$\langle f, g \rangle = \lim_{T \to \infty} \frac{1}{T} \int_{t_0}^{t_0+T} \overline{f(t)} g(t) \, dt,$$

where t_0 is any fixed time independent of T. The set of functions

$$\left\{ e^{i\omega t} \mid \omega \in \mathbb{R} \right\} \tag{6.9}$$

is an orthonormal set in X. The space X is an inner product space, but it is not complete. We call the completion of X the space of L^2-*almost periodic functions*. This space consists of equivalence classes of functions of the form

$$f(t) = \sum_{k=1}^{\infty} a_k e^{i\omega_k t}, \tag{6.10}$$

where $\sum_{k=1}^{\infty} |a_k|^2 < \infty$. The sum converges in norm, meaning that for any $t_0 \in \mathbb{R}$,

$$\lim_{T \to \infty} \frac{1}{2T} \int_{-T}^{T} \left| f(t) - \sum_{k=1}^{n} a_k e^{i\omega_k t} \right|^2 dt \to 0 \quad \text{as } n \to \infty.$$

The set in (6.9) is an uncountable orthonormal basis of this Hilbert space, so the space is not separable.

Although in the future we will mainly consider separable Hilbert spaces, it is worth postponing this restriction for a little while. First, we say what we mean by a sum with a possibly uncountable number of terms. This definition also clarifies the sense in which our infinite sums converge, which is stronger than the sense in which infinite series converge.

Definition 6.20 Let $\{x_\alpha \in X \mid \alpha \in I\}$ be an indexed set in a Banach space X, where the index set I may be countable or uncountable. For each finite subset J of I, we define the partial sum S_J by

$$S_J = \sum_{\alpha \in J} x_\alpha.$$

The *unordered sum* of the indexed set $\{x_\alpha \mid \alpha \in I\}$ converges to $x \in X$, written

$$x = \sum_{\alpha \in I} x_\alpha, \tag{6.11}$$

if for every $\epsilon > 0$ there is a finite subset J^ϵ of I such that $\|S_J - x\| < \epsilon$ for all finite subsets J of I that contain J^ϵ. An unordered sum is said to *converge unconditionally*.

All the sums in this chapter are to be interpreted as unordered sums. The convergence of finite partial sums S_J, indexed by finite subsets J of I, is a special case of the convergence of nets [12]. It is easy to see that an unordered sum converges if and only if any permutation of its terms converges, and the sum is independent of the ordering of its terms.

A sum $\sum_{\alpha \in I} x_\alpha$ is said to *converge absolutely* if the sum $\sum_{\alpha \in I} \|x_\alpha\|$ of non-negative numbers converges unconditionally. The unordered sum of a sequence of real or complex numbers exists if and only if the corresponding series is absolutely convergent. An absolutely convergent sum in an infinite-dimensional Banach space converges unconditionally, but an unconditionally convergent sum need not converge absolutely (see Exercise 6.8 for an example).

If an unordered sum $\sum_{\alpha \in I} x_\alpha$ converges to x, then for each $n \in \mathbb{N}$, there is a finite subset $J_n \subset I$ such that for all J containing J_n, one has $\|S_J - x\| \leq 1/n$. It follows that $x_\alpha = 0$ if $\alpha \notin \bigcup_{n \in \mathbb{N}} J_n$, so a convergent unordered sum has only countably many nonzero terms. Moreover, there is a sequence (S_{J_n}) of finite partial sums that converges to x as $n \to \infty$. The continuity of the inner product implies that

$$\left\langle \sum_{\alpha \in I} x_\alpha, \sum_{\beta \in J} y_\beta \right\rangle = \sum_{(\alpha,\beta) \in I \times J} \langle x_\alpha, y_\beta \rangle.$$

There is a generalization of the Cauchy criterion for the convergence of series to unordered sums.

Definition 6.21 An unordered sum $\sum_{\alpha \in I} x_\alpha$ is *Cauchy* if for every $\epsilon > 0$ there is a finite set $J^\epsilon \subset I$ such that $\|S_K\| < \epsilon$ for every finite set $K \subset I \setminus J^\epsilon$.

Proposition 6.22 An unordered sum in a Banach space converges if and only if it is Cauchy.

Proof. First, suppose that the unordered sum $\sum_{\alpha \in I} x_\alpha$ converges to x. Let $\epsilon > 0$. By the definition of convergence, there is a finite set J^ϵ such that $\|S_J - x\| < \epsilon/2$ for all finite sets J that contain J^ϵ. If K is any finite subset of $I \setminus J^\epsilon$, then we let $J = J^\epsilon \cup K$. Since J contains J^ϵ, we have

$$\|S_K\| = \|S_J - S_{J^\epsilon}\| \le \|S_J - x\| + \|x - S_{J^\epsilon}\| < \epsilon.$$

Hence, the sequence is Cauchy.

Conversely, suppose that an unordered sum is Cauchy. Let J_n be finite subsets of I such that $\|S_K\| < 1/n$ for every $K \subset I \setminus J_n$. Without loss of generality, we may assume that $J_n \subset J_{n+1}$ for all n. It follows that for all $n \le m$ we have $\|S_{J_m} - S_{J_n}\| < 1/n$, which shows that the sequence (S_{J_n}) is Cauchy; hence, since a Banach space is complete, it converges to a point x. To complete the proof, we show that the unordered sum converges to x. Given $\epsilon > 0$, we pick n such that $1/n < \epsilon/2$ and put $J^\epsilon = J_n$. If J is a finite set that contains J^ϵ, then the Cauchy criterion for the set J_n and the convergence of the sequence (S_{J_n}) to x imply that

$$\|S_J - x\| \le \|S_J - S_{J_n}\| + \|S_{J_n} - x\| < \frac{2}{n} < \epsilon. \qquad \square$$

We may use the Cauchy criterion to give a simple necessary and sufficient condition for the unconditional convergence of a sum of orthogonal terms in a Hilbert space.

Lemma 6.23 Let $U = \{u_\alpha \mid \alpha \in I\}$ be an indexed, orthogonal subset of a Hilbert space $\mathcal{H}$. The sum $\sum_{\alpha \in I} u_\alpha$ converges unconditionally if and only if $\sum_{\alpha \in I} \|u_\alpha\|^2 < \infty$, and, in that case,

$$\left\| \sum_{\alpha \in I} u_\alpha \right\|^2 = \sum_{\alpha \in I} \|u_\alpha\|^2. \tag{6.12}$$

Proof. For any finite set J we have

$$\left\| \sum_{\alpha \in J} u_\alpha \right\|^2 = \sum_{\alpha, \beta \in J} \langle u_\alpha, u_\beta \rangle = \sum_{\alpha \in J} \langle u_\alpha, u_\alpha \rangle = \sum_{\alpha \in J} \|u_\alpha\|^2.$$

It follows that the Cauchy criterion is satisfied for $\sum_{\alpha \in I} u_\alpha$ if and only if it is satisfied for $\sum_{\alpha \in I} \|u_\alpha\|^2$. Thus, one of the sums converges unconditionally if and only if the other does. Equation (6.12) follows because the sum is the limit of a sequence of finite partial sums and the norm is a continuous function. $\qquad \square$

When combined with the following basic estimate, this lemma will imply that every element of a Hilbert space can be expanded with respect to an orthonormal basis.

Theorem 6.24 (Bessel's inequality) Let $U = \{u_\alpha \mid \alpha \in I\}$ be an orthonormal set in a Hilbert space $\mathcal{H}$ and $x \in \mathcal{H}$. Then:

(a) $\sum_{\alpha \in I} |\langle u_\alpha, x \rangle|^2 \leq \|x\|^2$;
(b) $x_U = \sum_{\alpha \in I} \langle u_\alpha, x \rangle u_\alpha$ is a convergent sum;
(c) $x - x_U \in U^\perp$.

Proof.   We begin by computing $\|x - \sum_{\alpha \in J} \langle u_\alpha, x \rangle u_\alpha\|$ for any finite subset $J \subset I$:

$$
\left\| x - \sum_{\alpha \in J} \langle u_\alpha, x \rangle u_\alpha \right\|^2 = \left\langle \left(x - \sum_{\alpha \in J} \langle u_\alpha, x \rangle u_\alpha \right), \left(x - \sum_{\beta \in J} \langle u_\beta, x \rangle u_\beta \right) \right\rangle
$$

$$
= \langle x, x \rangle - \sum_{\beta \in J} \langle u_\beta, x \rangle \langle x, u_\beta \rangle - \sum_{\alpha \in J} \overline{\langle u_\alpha, x \rangle} \langle u_\alpha, x \rangle
$$

$$
+ \sum_{\alpha, \beta \in J} \overline{\langle u_\alpha, x \rangle} \langle u_\beta, x \rangle \langle u_\alpha, u_\beta \rangle
$$

$$
= \|x\|^2 - \sum_{\alpha \in J} |\langle u_\alpha, x \rangle|^2.
$$

Hence

$$
\sum_{\alpha \in J} |\langle u_\alpha, x \rangle|^2 = \|x\|^2 - \left\| x - \sum_{\alpha \in J} \langle u_\alpha, x \rangle u_\alpha \right\|^2 \leq \|x\|^2.
$$

Since $\sum_{\alpha \in I} |\langle u_\alpha, x \rangle|^2$ is a sum of nonnegative numbers that is bounded from above by $\|x\|^2$, it is Cauchy. Therefore the sum converges and satisfies (a). The convergence claimed in (b) follows from an application of Lemma 6.23.

In order to prove (c), we consider any $u_{\alpha_0} \in U$. Using the orthonormality of U and the continuity of the inner product, we find that

$$
\left\langle x - \sum_{\alpha \in I} \langle u_\alpha, x \rangle u_\alpha, u_{\alpha_0} \right\rangle = \langle x, u_{\alpha_0} \rangle - \sum_{\alpha \in I} \overline{\langle u_\alpha, x \rangle} \langle u_\alpha, u_{\alpha_0} \rangle
$$

$$
= \langle x, u_{\alpha_0} \rangle - \langle x, u_{\alpha_0} \rangle = 0.
$$

Hence, $x - \sum_{\alpha \in I} \langle u_\alpha, x \rangle u_\alpha \in U^\perp$. $\square$

Given a subset U of $\mathcal{H}$, we define the closed linear span $[U]$ of U by

$$
[U] = \left\{ \sum_{u \in U} c_u u \ \middle| \ c_u \in \mathbb{C} \text{ and } \sum_{u \in U} c_u u \text{ converges unconditionally} \right\}. \qquad (6.13)
$$

Equivalently, $[U]$ is the smallest closed linear subspace that contains U. We leave the proof of the following lemma to the reader.

Lemma 6.25 If $U = \{ u_\alpha \mid \alpha \in I \}$ is an orthonormal set in a Hilbert space $\mathcal{H}$, then

$$
[U] = \left\{ \sum_{\alpha \in I} c_\alpha u_\alpha \ \middle| \ c_\alpha \in \mathbb{C} \text{ such that } \sum_{\alpha \in I} |c_\alpha|^2 < \infty \right\}.
$$

By combining Theorem 6.13 and Theorem 6.24 we see that x_U, defined in part (b) of Theorem 6.24, is the unique element of $[U]$ satisfying

$$\|x - x_U\| = \min_{u \in [U]} \|x - u\|.$$

In particular, if $[U] = \mathcal{H}$, then $x_U = x$, and every $x \in \mathcal{H}$ may be expanded in terms of elements of U. The following theorem gives equivalent conditions for this property of U, called *completeness*.

Theorem 6.26 If $U = \{u_\alpha \mid \alpha \in I\}$ is an orthonormal subset of a Hilbert space $\mathcal{H}$, then the following conditions are equivalent:

(a) $\langle u_\alpha, x \rangle = 0$ for all $\alpha \in I$ implies $x = 0$;
(b) $x = \sum_{\alpha \in I} \langle u_\alpha, x \rangle u_\alpha$ for all $x \in \mathcal{H}$;
(c) $\|x\|^2 = \sum_{\alpha \in I} |\langle u_\alpha, x \rangle|^2$ for all $x \in \mathcal{H}$;
(d) $[U] = \mathcal{H}$;
(e) U is a maximal orthonormal set.

Proof. We prove that (a) implies (b), (b) implies (c), (c) implies (d), (d) implies (e), and (e) implies (a). The condition in (a) states that $U^\perp = \{0\}$. Part (c) of Theorem 6.24 then implies (b). The fact that (b) implies (c) follows from Lemma 6.23. To prove that (c) implies (d), we observe that (c) implies that $U^\perp = \{0\}$, which implies that $[U]^\perp = \{0\}$, so $[U] = \mathcal{H}$. Condition (e) means that if V is a subset of $\mathcal{H}$ that contains U and is strictly larger than U, then V is not orthonormal. To prove that (d) implies (e), we note from (d) that any $v \in \mathcal{H}$ is of the form $v = \sum_{\alpha \in I} c_\alpha u_\alpha$, where $c_\alpha = \langle u_\alpha, v \rangle$. Therefore, if $v \perp U$ then $c_\alpha = 0$ for all α, and hence $v = 0$, so $U \cup \{v\}$ is not orthonormal. Finally, (e) implies (a), since (a) is just a reformulation of (e). $\quad\square$

In view of this theorem, we can make the following definition.

Definition 6.27 An orthonormal subset $U = \{u_\alpha \mid \alpha \in I\}$ of a Hilbert space $\mathcal{H}$ is *complete* if it satisfies any of the equivalent conditions (a)–(e) in Theorem 6.26. A complete orthonormal subset of $\mathcal{H}$ is called an *orthonormal basis* of $\mathcal{H}$.

Condition (a) is often the easiest condition to verify. Condition (b) is the property that is used most often. Condition (c) is called *Parseval's identity*. Condition (d) simply expresses completeness of the basis, and condition (e) will be used in the proof of the existence of an orthonormal basis in an arbitrary Hilbert space (see Theorem 6.29).

The following generalization of Parseval's identity shows that a Hilbert space $\mathcal{H}$ with orthonormal basis $\{u_\alpha \mid \alpha \in I\}$ is isomorphic to the sequence space $\ell^2(I)$.

Theorem 6.28 (Parseval's identity) Suppose that $U = \{u_\alpha \mid \alpha \in I\}$ is an orthonormal basis of a Hilbert space $\mathcal{H}$. If $x = \sum_{\alpha \in I} a_\alpha u_\alpha$ and $y = \sum_{\alpha \in I} b_\alpha u_\alpha$, where $a_\alpha = \langle u_\alpha, x \rangle$ and $b_\alpha = \langle u_\alpha, y \rangle$, then

$$\langle x, y \rangle = \sum_{\alpha \in I} \overline{a_\alpha} \, b_\alpha.$$

To show that every Hilbert space has an orthonormal basis, we use *Zorn's lemma*, which states that a nonempty partially ordered set with the property that every totally ordered subset has an upper bound has a maximal element [49].

Theorem 6.29 Every Hilbert space $\mathcal{H}$ has an orthonormal basis. If U is an orthonormal set, then $\mathcal{H}$ has an orthonormal basis containing U.

Proof. If $\mathcal{H} = \{0\}$, then the statement is trivially true with $U = \emptyset$, so we assume that $\mathcal{H} \neq \{0\}$. We introduce a partial ordering $\leq$ on orthonormal subsets of $\mathcal{H}$ by inclusion, so that $U \leq V$ if and only if $U \subset V$. If $\{U_\alpha \mid \alpha \in \mathcal{A}\}$ is a totally ordered family of orthonormal sets, meaning that for any $\alpha, \beta \in \mathcal{A}$ we have either $U_\alpha \leq U_\beta$ or $U_\beta \leq U_\alpha$, then $\bigcup_{\alpha \in \mathcal{A}} U_\alpha$ is an orthonormal set and is an upper bound, in the sense of inclusion, of the family $\{U_\alpha \mid \alpha \in \mathcal{A}\}$. Zorn's Lemma implies that the family of all orthonormal sets in $\mathcal{H}$ has a maximal element. This element satisfies (e) in Theorem 6.26, and hence is a basis. To prove that any orthonormal set U can be extended to an orthonormal basis of $\mathcal{H}$, we apply the same argument to the family of all orthonormal sets containing U. $\qquad \square$

The existence of orthonormal bases would not be useful if we did not have a means of constructing them. The Gram-Schmidt orthonormalization procedure is an algorithm for the construction of an orthonormal basis from any countable linearly independent set whose linear span is dense in $\mathcal{H}$.

Let V be a countable set of linearly independent vectors in a Hilbert space $\mathcal{H}$. The Gram-Schmidt orthonormalization procedure is a method of constructing an orthonormal set U such that $[U] = [V]$, where the closed linear span $[V]$ of V is defined in (6.13). We denote the elements of V by v_n. The orthonormal set $U = \{u_n\}$ is then constructed inductively by setting $u_1 = v_1 / \|u_1\|$, and

$$u_{n+1} = c_{n+1} \left(v_{n+1} - \sum_{k=1}^{n} \langle u_k, v_{n+1} \rangle u_k \right)$$

for all $n \geq 1$. Here $c_{n+1} \in \mathbb{C}$ is chosen so that $\|u_{n+1}\| = 1$. It is straightforward to check that $[\{v_1, \ldots, v_n\}] = [\{u_1, \ldots, u_n\}]$ for all $n \geq 1$, and hence that

$$[V] = \overline{\bigcup_n [\{v_1, \ldots, v_n\}]} = \overline{\bigcup_n [\{u_1, \ldots, u_n\}]} = [U].$$

Example 6.30 Let $(a, b) \subset \mathbb{R}$ be a finite or infinite interval and $w : (a, b) \to \mathbb{R}$ a continuous function such that $w(x) > 0$ for $a < x < b$. We define a weighted inner product on

$$C_w([a, b]) = \left\{ f : [a, b] \to \mathbb{C} \mid f \text{ continuous and } \int_a^b w(x) |f(x)|^2 \, dx < \infty \right\}$$

by

$$\langle f, g \rangle = \int_a^b w(x)\overline{f(x)}g(x)\,dx.$$

Let $L_w^2([a,b])$ be the Hilbert space obtained by the completion of $C_w([a,b])$ with respect to the norm derived from this inner product. The Gram-Schmidt procedure applied to the set of monomials $\{x^n \mid n \geq 0\}$ gives an orthonormal basis of polynomials for this Hilbert space. The simplest case is that of the space $L^2([-1,1])$, with the usual unweighted inner product, which leads to the Legendre polynomials (see Exercise 6.12). The Tchebyschev polynomials are obtained from Gram-Schmidt orthonormalization of the monomials in $L_w^2([-1,1])$ where $w(x) = \left(1 - x^2\right)^{1/2}$ (see Exercise 6.13). The Hermite polynomials are obtained by Gram-Schmidt orthonormalization of the monomials in the space $L_w^2(\mathbb{R})$ with the Gaussian weight function $w(x) = e^{-x^2/2}$ (see Exercise 6.14). For a description of other polynomials that arise in this way, such as the Jacobi and Laguerre polynomials, see [5].

6.4 Hilbert spaces in applications

In this section, we describe several applications in which Hilbert spaces arise naturally.

The first is quantum mechanics. The introduction of quantum mechanics in the 1920s represents one of the most profound shifts in history of our understanding of the physical world. The theory developed at a feverish pace, and people hardly had time to pause to think about the mathematical structures they were inventing and using. Only later was it realized, by von Neumann, that Hilbert spaces are the natural setting for quantum mechanics.

One of the simplest quantum mechanical systems consists of a particle, such as an electron, confined to move in a straight line between two parallel walls: the "particle in a box." Quantum effects are important when the kinetic energy of the particle is comparable with $E = \hbar^2/(2mL^2)$, where m is the mass of the particle, $\hbar$ is Planck's constant, and L is the distance between the walls. Planck's constant has the dimensions of action, or energy times time, so E has the dimensions of energy.

In quantum mechanics, the state of the particle at each instant in time t is described by an element $\psi(\cdot, t) \in L^2([0, L])$, that is, a vector in the Hilbert space of square-integrable, complex-valued functions on the interval $[0, L]$. The function ψ is called the wavefunction of the particle. This description contrasts with classical, Newtonian mechanics, where the state of the particle is described by just two numbers: the position $0 \leq x \leq L$ and the velocity $v \in \mathbb{R}$. The physical interpretation of the wavefunction is that $|\psi|^2$ is a probability density. If the position x of the particle is measured at some time t, then the probability of observing the particle in some interval $[a, b]$, where $0 \leq a < b \leq L$, is given by

$$\Pr\left[\text{particle is in the interval } [a, b] \text{ at time } t\right] = \frac{\int_a^b |\psi(x, t)|^2 \, dx}{\int_0^L |\psi(x, t)|^2 \, dx}.$$

The dynamics of the quantum mechanical particle is described by a partial differential equation for the wavefunction, called the Schrödinger equation. For the particle in a box, the Schrödinger equation is

$$i\hbar\psi_t = -\frac{\hbar^2}{2m}\psi_{xx}, \quad x \in [0, L], \ t \in \mathbb{R}, \tag{6.14}$$

with the boundary conditions $\psi(0, t) = \psi(L, t) = 0$ for all $t \in \mathbb{R}$.

A second way in which L^2-spaces arise naturally is as "energy" spaces. The quantity

$$\int |f(x)|^2 \, dx \tag{6.15}$$

often represents the total energy of a physical system, or some other fundamental quantity, and one often wants to restrict attention to systems for which this quantity is finite. For example, in fluid mechanics, if $\mathbf{u}(\mathbf{x})$ is the velocity of a fluid at the point $\mathbf{x}$, then

$$\int_V |\mathbf{u}(\mathbf{x})|^2 \, d\mathbf{x},$$

where $|\cdot|$ denotes the Euclidean norm of a vector, is proportional to the kinetic energy of the fluid in V. This energy should be finite for any region V with finite volume. An electromagnetic field is described by two vector fields, the electric field $\mathbf{E}$ and the magnetic field $\mathbf{B}$. In suitable units, the energy of the electromagnetic field in a region V is given by

$$\int_V \left\{ |\mathbf{E}(\mathbf{x})|^2 + |\mathbf{B}(\mathbf{x})|^2 \right\} \, d\mathbf{x}.$$

The requirement of finite energy leads naturally to the requirement that $\mathbf{E}$ and $\mathbf{B}$ belong to appropriate L^2-spaces.

A third area in which Hilbert spaces arise naturally is in probability theory. As we discuss in greater detail in Chapter 12, a random experiment is modeled mathematically by a space Ω, called the sample space, and a probability measure P on Ω. Each point $\omega \in \Omega$ corresponds to a possible outcome of the experiment. An event A is a measurable subset of Ω. The probability measure P associates with each event A a probability $P(A)$, where $0 \leq P(A) \leq 1$ and $P(\Omega) = 1$.

A *random variable* X is a measurable function $X : \Omega \to \mathbb{C}$, which associates a number $X(\omega)$ with each possible outcome $\omega \in \Omega$. The *expected value* $\mathbb{E}X$ of a random variable X is the mean, or integral, of the random variable X with respect to the probability measure P,

$$\mathbb{E}X = \int_\Omega X(\omega)\,dP(\omega).$$

A random variable X is said to be *second-order* if

$$\mathbb{E}|X|^2 < \infty.$$

The set of second-order random variables forms a Hilbert space with respect to the inner product

$$\langle X,Y\rangle = \mathbb{E}\left[\overline{X}Y\right],$$

where we identify random variables that are equal almost surely. Here, "almost surely" is the probabilistic terminology for "almost everywhere," so that two random variables are equal almost surely if they are equal on a subset of Ω which has probability one. The space of second-order random variables may be identified with the space $L^2(\Omega, P)$ of square-integrable functions on (Ω, P), with the inner product

$$\langle X,Y\rangle = \int_\Omega \overline{X(\omega)}Y(\omega)\,dP(\omega).$$

The Cauchy-Schwarz inequality and the fact that $\mathbb{E}1 = 1$ imply that a second-order random variable has finite mean, since

$$|\mathbb{E}X| = |\langle 1, X\rangle| \le \mathbb{E}\left[|X|^2\right]^{1/2}.$$

Thus, the Hilbert space of second-order random variables consists of the random variables with finite mean and finite variance, where the *variance* Var X of a random variable X is defined by

$$\mathrm{Var}\,X = \mathbb{E}\left[|X - \mathbb{E}X|^2\right].$$

Two random variables X, Y are *uncorrelated* if

$$\mathbb{E}[XY] = \mathbb{E}X\,\mathbb{E}Y.$$

In particular, two random variables with zero mean are uncorrelated if and only if they are orthogonal.

6.5   References

The material of this chapter's introduction to Hilbert space is covered in Chapter 4 of Rudin [49], and also in Simmons [50]. Halmos [20] contains a large number of problems on Hilbert spaces, together with hints and solutions. For an introduction to probability theory, see Grimmett and Stirzaker [17].

6.6   Exercises

Exercise 6.1 Prove that a closed, convex subset of a Hilbert space has a unique point of minimum norm.

Exercise 6.2 Consider $C([0,1])$ with the sup-norm. Let

$$N = \left\{ f \in C([0,1]) \;\Big|\; \int_0^1 f(x)\,dx = 0 \right\}$$

be the closed linear subspace of $C([0,1])$ of functions with zero mean. Let

$$X = \left\{ f \in C([0,1]) \;\Big|\; f(0) = 0 \right\}$$

and define $M = N \cap X$, meaning that

$$M = \left\{ f \in C([0,1]) \;\Big|\; f(0) = 0, \;\; \int_0^1 f(x)\,dx = 0 \right\}.$$

(a) If $u \in C([0,1])$, prove that

$$d(u, N) = \inf_{n \in N} \|u - n\| = |\bar{u}|\,,$$

where $|\bar{u}| = \int_0^1 u(x)\,dx$ is the mean of u, so the infimum is attained when $n = u - \bar{u} \in N$.

(b) If $u(x) = x \in X$, show that

$$d(x, M) = \inf_{m \in M} \|u - m\| = 1/2,$$

but that the infimum is not attained for any $m \in M$.

Exercise 6.3 If A is a subset of a Hilbert space, prove that

$$A^{\perp} = \overline{A}^{\perp},$$

where $\overline{A}$ is the closure of A. If $\mathcal{M}$ is a linear subspace of a Hilbert space, prove that

$$\mathcal{M}^{\perp\perp} = \overline{\mathcal{M}}.$$

Exercise 6.4 Suppose that $\mathcal{H}_1$ and $\mathcal{H}_2$ are two Hilbert spaces. We define

$$\mathcal{H}_1 \oplus \mathcal{H}_2 = \{(x_1, x_2) \mid x_1 \in \mathcal{H}_1,\ x_2 \in \mathcal{H}_2\}$$

with the inner product

$$\langle (x_1, x_2), (y_1, y_2) \rangle_{\mathcal{H}_1 \oplus \mathcal{H}_2} = \langle x_1, y_1 \rangle_{\mathcal{H}_1} + \langle x_2, y_2 \rangle_{\mathcal{H}_2}.$$

Prove that $\mathcal{H}_1 \oplus \mathcal{H}_2$ is a Hilbert space. Find the orthogonal complement of the subspace $\{(x_1, 0) \mid x_1 \in \mathcal{H}_1\}$.

Exercise 6.5 Suppose that $\{\mathcal{H}_n \mid n \in \mathbb{N}\}$ is a set of orthogonal closed subspaces of a Hilbert space $\mathcal{H}$. We define the infinite direct sum

$$\bigoplus_{n=1}^{\infty} \mathcal{H}_n = \left\{ \sum_{n=1}^{\infty} x_n \;\middle|\; x_n \in \mathcal{H}_n \text{ and } \sum_{n=1}^{\infty} \|x_n\|^2 < \infty \right\}.$$

Prove that $\bigoplus_{n=1}^{\infty} \mathcal{H}_n$ is a closed linear subspace of $\mathcal{H}$.

Exercise 6.6 Prove that the vectors in an orthogonal set are linearly independent.

Exercise 6.7 Let $\{x_\alpha\}_{\alpha \in I}$ be a family of nonnegative real numbers. Prove that

$$\sum_{\alpha \in I} x_\alpha = \sup \left\{ \sum_{\alpha \in J} x_\alpha \;\middle|\; J \subset I \text{ and } J \text{ is finite} \right\}.$$

Exercise 6.8 Let $\{x_n \mid n \in \mathbb{N}\}$ be an orthonormal set in a Hilbert space. Show that the sum $\sum_{n=1}^{\infty} x_n/n$ converges unconditionally but not absolutely.

Exercise 6.9 Prove Lemma 6.25.

Exercise 6.10 Prove that a Hilbert space is a separable metric space if and only if it has a countable orthonormal basis.

Exercise 6.11 Prove that if $\mathcal{M}$ is a dense linear subspace of a separable Hilbert space $\mathcal{H}$, then $\mathcal{H}$ has an orthonormal basis consisting of elements in $\mathcal{M}$. Does the same result hold for arbitrary dense subsets of $\mathcal{H}$?

Exercise 6.12 Define the *Legendre polynomials* P_n by

$$P_n(x) = \frac{1}{2^n n!} \frac{d^n}{dx^n} \left(x^2 - 1\right)^n.$$

(a) Compute the first few Legendre polynomials, and compare with what you get by Gram-Schmidt orthogonalization of the monomials $\{1, x, x^2, \ldots\}$ in $L^2([-1, 1])$.

(b) Show that the Legendre polynomials are orthogonal in $L^2([-1, 1])$, and that they are obtained by Gram-Schmidt orthogonalization of the monomials.

(c) Show that

$$\int_{-1}^{1} P_n(x)^2 \, dx = \frac{2}{2n + 1}.$$

(d) Prove that the Legendre polynomials form an orthogonal basis of $L^2([-1, 1])$. Suppose that $f \in L^2([-1, 1])$ is given by

$$f(x) = \sum_{n=0}^{\infty} c_n P_n(x).$$

Compute c_n and say explicitly in what sense the series converges.

(e) Prove that the Legendre polynomial P_n is an eigenfunction of the differential operator

$$L = -\frac{d}{dx}\left(1 - x^2\right)\frac{d}{dx}$$

with eigenvalue $\lambda_n = n(n+1)$, meaning that

$$LP_n = \lambda_n P_n.$$

Exercise 6.13 Let $\mathcal{H}$ be the Hilbert space of functions $f : [-1, 1] \to \mathbb{C}$ such that

$$\int_{-1}^{1} \frac{|f(x)|^2}{\sqrt{1 - x^2}}\, dx < \infty,$$

with the inner-product

$$\langle f, g \rangle = \int_{-1}^{1} \frac{\overline{f(x)}g(x)}{\sqrt{1 - x^2}}\, dx.$$

Show that the *Tchebyshev polynomials,*

$$T_n(x) = \cos(n\theta) \qquad \text{where } \cos\theta = x \text{ and } 0 \le \theta \le \pi,$$

$n = 0, 1, 2, \ldots$, form an orthogonal set in $\mathcal{H}$, and

$$\|T_0\| = \sqrt{\pi}, \qquad \|T_n\| = \sqrt{\frac{\pi}{2}} \quad n \ge 1.$$

Exercise 6.14 Define the Hermite polynomials H_n by

$$H_n(x) = (-1)^n e^{x^2} \frac{d^n}{dx^n}\left(e^{-x^2}\right).$$

(a) Show that

$$\varphi_n(x) = e^{-x^2/2} H_n(x)$$

is an orthogonal set in $L^2(\mathbb{R})$.

(b) Show that the nth Hermite function φ_n is an eigenfunction of the linear operator

$$H = -\frac{d^2}{dx^2} + x^2$$

with eigenvalue

$$\lambda_n = 2n + 1.$$

HINT: Let

$$A = \frac{d}{dx} + x, \qquad A^* = -\frac{d}{dx} + x.$$

Show that

$$A\varphi_n = 2n\varphi_{n-1}, \quad A^*\varphi_n = \varphi_{n+1}, \quad H = AA^* - 1.$$

In quantum mechanics, H is the *Hamiltonian operator* of a simple harmonic oscillator, and A^* and A are called *creation* and *annihilation*, or *ladder*, operators.

Chapter 7

Fourier Series

What makes Hilbert spaces so powerful in many applications is the possibility of expressing a problem in terms of a suitable orthonormal basis. In this chapter, we study *Fourier series*, which correspond to the expansion of periodic functions with respect to an orthonormal basis of trigonometric functions. We explore a variety of applications of Fourier series, and introduce an important related class of orthonormal bases, called *wavelets*.

7.1 The Fourier basis

A function $f : \mathbb{R} \to \mathbb{C}$ is 2π-periodic if

$$f(x + 2\pi) = f(x) \qquad \text{for all } x \in \mathbb{R}.$$

The choice of 2π for the period is simply for convenience; different periods may be reduced to this case by rescaling the independent variable. A 2π-periodic function on $\mathbb{R}$ may be identified with a function on the circle, or one-dimensional torus, $\mathbb{T} = \mathbb{R}/(2\pi\mathbb{Z})$, which we define by identifying points in $\mathbb{R}$ that differ by $2\pi n$ for some $n \in \mathbb{Z}$. We could instead represent a 2π-periodic function $f : \mathbb{R} \to \mathbb{C}$ by a function on a closed interval $f : [a, a + 2\pi] \to \mathbb{C}$ such that $f(a) = f(a + 2\pi)$, but the choice of a here is arbitrary, and it is clearer to think of the function as defined on the circle, rather than an interval.

The space $C(\mathbb{T})$ is the space of continuous functions from $\mathbb{T}$ to $\mathbb{C}$, and $L^2(\mathbb{T})$ is the completion of $C(\mathbb{T})$ with respect to the L^2-norm,

$$\|f\| = \left(\int_{\mathbb{T}} |f(x)|^2 \, dx \right)^{1/2}.$$

Here, the integral over $\mathbb{T}$ is an integral with respect to x taken over any interval of length 2π. An element $f \in L^2(\mathbb{T})$ can be interpreted concretely as an equivalence class of Lebesgue measurable, square integrable functions from $\mathbb{T}$ to $\mathbb{C}$ with respect to the equivalence relation of almost-everywhere equality. The space $L^2(\mathbb{T})$ is a

Hilbert space with the inner product

$$\langle f, g \rangle = \int_{\mathbb{T}} \overline{f(x)} g(x) \, dx.$$

The Fourier basis elements are the functions

$$e_n(x) = \frac{1}{\sqrt{2\pi}} e^{inx}. \tag{7.1}$$

Our first objective is to prove that $\{e_n \mid n \in \mathbb{Z}\}$ is an orthonormal basis of $L^2(\mathbb{T})$. The orthonormality of the functions e_n is a simple computation:

$$\begin{aligned}
\langle e_m, e_n \rangle &= \int_{\mathbb{T}} \overline{\frac{1}{\sqrt{2\pi}} e^{imx}} \frac{1}{\sqrt{2\pi}} e^{inx} \, dx \\
&= \frac{1}{2\pi} \int_0^{2\pi} e^{i(n-m)x} \, dx \\
&= \begin{cases} 1 & \text{if } m = n, \\ 0 & \text{if } m \neq n. \end{cases}
\end{aligned}$$

Thus, the main result we have to prove is the completeness of $\{e_n \mid n \in \mathbb{Z}\}$. We denote the set of all finite linear combinations of the e_n by $\mathcal{P}$. Functions in $\mathcal{P}$ are called *trigonometric polynomials*. We will prove that any continuous function on $\mathbb{T}$ can be approximated uniformly by trigonometric polynomials, a result which is closely related to the Weierstrass approximation theorem in Theorem 2.9. Since uniform convergence on $\mathbb{T}$ implies L^2-convergence, and continuous functions are dense in $L^2(\mathbb{T})$, it follows that the trigonometric polynomials are dense in $L^2(\mathbb{T})$, so $\{e_n\}$ is a basis.

The idea behind the completeness proof is to obtain a trigonometric polynomial approximation of a continuous function f by taking the convolution of f with an approximate identity that is a trigonometric polynomial. Convolutions and approximate identities are useful in many other contexts, so we begin by describing them.

The *convolution* of two continuous functions $f, g : \mathbb{T} \to \mathbb{C}$ is the continuous function $f * g : \mathbb{T} \to \mathbb{C}$ defined by the integral

$$(f * g)(x) = \int_{\mathbb{T}} f(x - y) g(y) \, dy. \tag{7.2}$$

By changing variables $y \to x - y$, we may also write

$$(f * g)(x) = \int_{\mathbb{T}} f(y) g(x - y) \, dy,$$

so that $f * g = g * f$.

Definition 7.1 A family of functions $\{\varphi_n \in C(\mathbb{T}) \mid n \in \mathbb{N}\}$ is an *approximate identity* if:

(a) $\quad \varphi_n(x) \geq 0;$ (7.3)

(b) $\quad \displaystyle\int_{\mathbb{T}} \varphi_n(x)\, dx = 1 \qquad$ for every $n \in \mathbb{N};$ (7.4)

(c) $\quad \displaystyle\lim_{n\to\infty} \int_{\delta \leq |x| \leq \pi} \varphi_n(x)\, dx = 0 \qquad$ for every $\delta > 0.$ (7.5)

In (7.5), we identify $\mathbb{T}$ with the interval $[-\pi, \pi]$.

Thus, each function φ_n has unit area under its graph, and the area concentrates closer to the origin as n increases. For large n, the convolution of a function f with φ_n therefore gives a local average of f.

Theorem 7.2 If $\{\varphi_n \in C(\mathbb{T}) \mid n \in \mathbb{N}\}$ is an approximate identity and $f \in C(\mathbb{T})$, then $\varphi_n * f$ converges uniformly to f as $n \to \infty$.

Proof. From (7.4), we have

$$f(x) = \int_{\mathbb{T}} \varphi_n(y) f(x)\, dy.$$

We also have that

$$(\varphi_n * f)(x) = \int_{\mathbb{T}} \varphi_n(y) f(x - y)\, dy.$$

We may therefore write

$$(\varphi_n * f)(x) - f(x) = \int_{\mathbb{T}} \varphi_n(y) \left[f(x - y) - f(x) \right] dy. \tag{7.6}$$

To show that the integral on the right-hand side of this equation is small when n is large, we consider the integrand separately for y close to zero and y bounded away from zero. The contribution to the integral from values of y close to zero is small because f is continuous, and the contribution to the integral from values of y bounded away from zero is small because the integral of φ_n is small.

More precisely, suppose that $\epsilon > 0$. Since f is continuous on the compact set $\mathbb{T}$, it is bounded and uniformly continuous. Therefore, there is an M such that $|f(x)| \leq M$ for all $x \in \mathbb{T}$, and there is a $\delta > 0$ such that $|f(x) - f(y)| \leq \epsilon$ whenever $|x - y| < \delta$. Then, estimating the integral in (7.6), we obtain that

$$
|(\varphi_n * f)(x) - f(x)| \quad \leq \quad \int_{-\pi}^{\pi} \varphi_n(y) \, |f(x - y) - f(x)| \, dy
$$

$$
\leq \quad \int_{|y| < \delta} \varphi_n(y) \, |f(x - y) - f(x)| \, dy
$$

$$+ \int_{|y| \geq \delta} \varphi_n(y) \, |f(x-y) - f(x)| \, dy$$

$$\leq \quad \epsilon \int_{|y| < \delta} \varphi_n(y) \, dy + \int_{|y| \geq \delta} \varphi_n(y) \, [|f(x-y)| + |f(x)|] \, dy$$

$$\leq \quad \epsilon + 2M \int_{|y| \geq \delta} \varphi_n(y) \, dy.$$

Taking the sup of this inequality over x, the lim sup as $n \to \infty$, and using (7.5), we find that

$$\limsup_{n \to \infty} \|\varphi_n * f - f\|_\infty \leq \epsilon.$$

Since $\epsilon > 0$ is arbitrary, it follows that $\varphi_n * f \to f$ uniformly in $C(\mathbb{T})$. □

Theorem 7.3 The trigonometric polynomials are dense in $C(\mathbb{T})$ with respect to the uniform norm.

Proof. For each $n \in \mathbb{N}$, we define the function $\varphi_n \geq 0$ by

$$\varphi_n(x) = c_n \left(1 + \cos x\right)^n. \tag{7.7}$$

We choose the constant c_n so that

$$\int_{\mathbb{T}} \varphi_n(x) \, dx = 1. \tag{7.8}$$

Since $1 + \cos x$ has a strict maximum at $x = 0$, the graph of φ_n is sharply peaked at $x = 0$ for large n, and the area under the graph concentrates near $x = 0$. In particular, $\{\varphi_n\}$ satisfies (7.5) (see Exercise 7.1). It follows that $\{\varphi_n\}$ is an approximate identity, and hence $\varphi_n * f$ converges uniformly to f from Theorem 7.2.

To complete the proof, we show that $\varphi_n * f$ is a trigonometric polynomial for any continuous function f. First, φ_n is a trigonometric polynomial; in fact,

$$\varphi_n(x) = \sum_{k=-n}^{n} a_{nk} e^{ikx}, \quad \text{where} \quad a_{nk} = 2^{-n} c_n \binom{2n}{n+k}.$$

Therefore,

$$\begin{aligned}
\varphi_n * f(x) &= \int_{\mathbb{T}} \sum_{k=-n}^{n} a_{nk} e^{ik(x-y)} f(y) \, dy \\
&= \sum_{k=-n}^{n} a_{nk} e^{ikx} \int_{\mathbb{T}} e^{-iky} f(y) \, dy \\
&= \sum_{k=-n}^{n} b_k e^{ikx},
\end{aligned}$$

where

$$b_k = a_{nk} \int_{\mathbb{T}} e^{-iky} f(y)\, dy.$$

Thus, $\varphi_n * f$ is a trigonometric polynomial. □

From the completeness of the Fourier basis, it follows that any function $f \in L^2(\mathbb{T})$ may be expanded in a Fourier series as

$$f(x) = \frac{1}{\sqrt{2\pi}} \sum_{n=-\infty}^{\infty} \widehat{f}_n e^{inx},$$

where the equality means convergence of the partial sums to f in the L^2-norm, or

$$\lim_{N \to \infty} \int_{\mathbb{T}} \left| \frac{1}{\sqrt{2\pi}} \sum_{n=-N}^{N} \widehat{f}_n e^{inx} - f(x) \right|^2 dx = 0.$$

From orthonormality, the *Fourier coefficients* $\widehat{f}_n \in \mathbb{C}$ of f are given by $\widehat{f}_n = \langle e_n, f \rangle$, or

$$\widehat{f}_n = \frac{1}{\sqrt{2\pi}} \int_{\mathbb{T}} f(x) e^{-inx}\, dx.$$

Moreover, Parseval's identity implies that

$$\int_{\mathbb{T}} \overline{f(x)} g(x)\, dx = \sum_{n=-\infty}^{\infty} \overline{\widehat{f}_n} \widehat{g}_n.$$

In particular, the L^2-norm of a function can be computed either in terms of the function or its Fourier coefficients, since

$$\int_{\mathbb{T}} |f(x)|^2\, dx = \sum_{n=-\infty}^{\infty} \left| \widehat{f}_n \right|^2. \tag{7.9}$$

Thus, the periodic Fourier transform $\mathcal{F} : L^2(\mathbb{T}) \to \ell^2(\mathbb{Z})$ that maps a function to its sequence of Fourier coefficients, by

$$\mathcal{F}f = \left(\widehat{f}_n \right)_{n=-\infty}^{\infty},$$

is a Hilbert space isomorphism between $L^2(\mathbb{T})$ and $\ell^2(\mathbb{Z})$. The projection theorem, Theorem 6.13, implies that the partial sum

$$f_N(x) = \frac{1}{\sqrt{2\pi}} \sum_{n=-N}^{N} \widehat{f}_n e^{inx}$$

is the best approximation of f by a trigonometric polynomial of degree N in the sense of the L^2-norm.

An important property of the Fourier transform is that it maps the convolution of two functions to the pointwise product of their Fourier coefficients. The convolution of two L^2-functions may either be defined by (7.2), where the integral is a Lebesgue integral, or by a density argument using continuous functions, as in the following proposition.

Proposition 7.4 If $f, g \in L^2(\mathbb{T})$, then $f * g \in C(\mathbb{T})$ and

$$\|f * g\|_\infty \leq \|f\|_2 \|g\|_2. \tag{7.10}$$

Proof. If $f, g \in C(\mathbb{T})$, then application of the Cauchy-Schwarz inequality to (7.2) implies that

$$|f * g(x)| \leq \|f\|_2 \|g\|_2.$$

Taking the supremum of this equation with respect to x, we get (7.10). If $f, g \in L^2(\mathbb{T})$, then there are sequences (f_k) and (g_k) of continuous functions such that $\|f - f_k\|_2 \to 0$ and $\|g - g_k\|_2 \to 0$ as $k \to \infty$. The convolutions $f_k * g_k$ are continuous functions. Moreover, they form a Cauchy sequence with respect to the sup-norm since, from (7.10),

$$\begin{aligned}
\|f_j * g_j - f_k * g_k\|_\infty &\leq \|(f_j - f_k) * g_j\|_\infty + \|f_k * (g_j - g_k)\|_\infty \\
&\leq \|f_j - f_k\|_2 \|g_j\|_2 + \|f_k\|_2 \|g_j - g_k\|_2 \\
&\leq M \left(\|f_j - f_k\|_2 + \|g_j - g_k\|_2 \right).
\end{aligned}$$

Here, we use the fact that $\|f_j\|_2 \leq M$ and $\|g_k\|_2 \leq M$ for some constant M because the sequences converge in $L^2(\mathbb{T})$. By the completeness of $C(\mathbb{T})$, the sequence $(f_k * g_k)$ converges uniformly to a continuous function $f * g$. This limit is independent of the sequences used to approximate f and g, and it satisfies (7.10).   □

The inequality (7.10) is a special case of *Young's inequality* for convolutions (see Theorem 12.58).

Theorem 7.5 (Convolution) If $f, g \in L^2(\mathbb{T})$, then

$$\widehat{(f * g)}_n = \sqrt{2\pi} \hat{f}_n \hat{g}_n. \tag{7.11}$$

Proof. Because of the density of $C(\mathbb{T})$ in $L^2(\mathbb{T})$, and the continuity of the Fourier transform and the convolution with respect to L^2-convergence, it is sufficient to prove (7.11) for continuous functions f, g. In that case, we may exchange the order of integration in the following computation:

$$\begin{aligned}
\widehat{(f * g)}_n &= \frac{1}{\sqrt{2\pi}} \int_{\mathbb{T}} f * g(x) e^{-inx} \, dx \\
&= \frac{1}{\sqrt{2\pi}} \int_{\mathbb{T}} \left(\int_{\mathbb{T}} f(x - y) g(y) \, dy \right) e^{-inx} \, dx
\end{aligned}$$

$$= \int_\mathbb{T} \frac{1}{\sqrt{2\pi}} \left(\int_\mathbb{T} f(x-y) e^{-in(x-y)} \, dx \right) g(y) e^{-iny} \, dy$$

$$= \hat{f}_n \int_\mathbb{T} g(y) e^{-iny} \, dy$$

$$= \sqrt{2\pi} \hat{f}_n \hat{g}_n.$$

This proves the theorem. □

Alternatively, we may prove Theorem 7.5 directly for $f, g \in L^1(\mathbb{T})$. The exchange in the order of integration is justified by Fubini's theorem, Theorem 12.41.

The L^2-convergence of Fourier series is particularly simple. It is nevertheless interesting to ask about other types of convergence. For example, the Fourier series of a function $f \in L^2(\mathbb{T})$ also converges pointwise a.e. to f. This result was proved by Carleson, only as recently as 1966. An analysis of the pointwise convergence of Fourier series is very subtle, and the proof is beyond the scope of this book. For smooth functions, such as continuously differentiable functions, the convergence of the partial sums is uniform, as we will show in Section 7.2 below.

The behavior of the partial sums near a point of discontinuity of a piecewise smooth function is interesting. The sums do not converge uniformly; instead the partial sums oscillate in an interval that contains the point of discontinuity. The width of the interval where the oscillations occur shrinks to zero as $N \to 0$, but the size of the oscillations does not — in fact, for large N, the magnitude of the oscillations is approximately 9% of the jump in f at the jump discontinuity. This behavior is called the *Gibbs phenomenon*. As a result, care is required when one uses Fourier series to represent discontinuous functions.

It is often convenient to modify the orthonormal basis $\{e_n(x)\}$ in (7.1) slightly. First, if we use the non-normalized orthogonal basis $\{e^{inx}\}$, then the Fourier expansion of $f \in L^2(\mathbb{T})$ is

$$f(x) = \sum_{n=-\infty}^{\infty} \hat{f}_n e^{inx},$$

where

$$\hat{f}_n = \frac{1}{2\pi} \int_\mathbb{T} f(x) e^{-inx} \, dx.$$

Second, the real-valued functions

$$\{1, \cos nx, \sin nx \mid n = 1, 2, 3, \ldots\}$$

also form an orthogonal basis, since

$$\cos nx = \frac{e^{inx} + e^{-inx}}{2}, \qquad \sin nx = \frac{e^{inx} - e^{-inx}}{2i}.$$

The corresponding Fourier expansion of $f \in L^2(\mathbb{T})$ is

$$f(x) = \frac{1}{2}a_0 + \sum_{n=1}^{\infty} \{a_n \cos nx + b_n \sin nx\},$$

where

$$a_0 = \frac{1}{\pi} \int_{\mathbb{T}} f(x)\,dx, \quad a_n = \frac{1}{\pi} \int_{\mathbb{T}} f(x) \cos nx\,dx, \quad b_n = \frac{1}{\pi} \int_{\mathbb{T}} f(x) \sin nx\,dx.$$

This basis has the advantage that a real-valued function has real Fourier coefficients a_n, b_n. A second useful property of this basis is that its elements are even or odd. A function f is *even* if $f(-x) = f(x)$ for all x, and *odd* if $f(-x) = -f(x)$ for all x. Even functions f have a *Fourier cosine expansion* of the form

$$f(x) = \frac{1}{2}a_0 + \sum_{n=1}^{\infty} a_n \cos(nx),$$

while odd functions f have a *Fourier sine expansion* of the form

$$f(x) = \sum_{n=1}^{\infty} b_n \sin(nx).$$

If a function is defined on the interval $[0, \pi]$, then we may extend it to an even or an odd 2π periodic function on $\mathbb{R}$. The original function may therefore be represented by a Fourier cosine or sine expansion on $[0, \pi]$ (see Exercise 7.3). The quality of the approximation of a function by the partial sums of a Fourier series sometimes depends significantly on the basis used for the expansion. This is illustrated in Figure 7.1.

Fourier series of multiply periodic functions are defined in an entirely analogous way. A function $f : \mathbb{R}^d \to \mathbb{C}$ is 2π-periodic in each variable if

$$f(x_1, x_2, \ldots, x_i + 2\pi, \ldots, x_d) = f(x_1, x_2, \ldots, x_i, \ldots, x_d) \qquad \text{for } i = 1, \ldots, d.$$

We may regard a multiply periodic function as a function on the d-dimensional torus $\mathbb{T}^d = \mathbb{R}^d/(2\pi\mathbb{Z})^d$, which is the Cartesian product of d circles. An orthonormal basis of $L^2(\mathbb{T}^d)$ consists of the functions

$$e_{\mathbf{n}}(\mathbf{x}) = \frac{1}{(2\pi)^{d/2}} e^{i\mathbf{n}\cdot\mathbf{x}},$$

where $\mathbf{x} = (x_1, \ldots, x_d) \in \mathbb{T}^d$, $\mathbf{n} = (n_1, \ldots, n_d) \in \mathbb{Z}^d$, and

$$\mathbf{n} \cdot \mathbf{x} = n_1 x_1 + \cdots + n_d x_d.$$

The Fourier series expansion of a function $f \in L^2(\mathbb{T}^d)$ is

$$f(\mathbf{x}) = \frac{1}{(2\pi)^{d/2}} \sum_{\mathbf{n} \in \mathbb{Z}^d} \widehat{f}_{\mathbf{n}} e^{i\mathbf{n}\cdot\mathbf{x}},$$

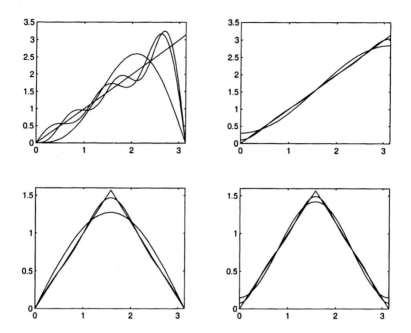

Fig. 7.1 Fourier sine (left) and cosine (right) series for two piecewise linear functions. Note the difference in the quality of the approximations.

where the series converges unconditionally with respect to the L^2-norm, and

$$\widehat{f}_{\mathbf{n}} = \frac{1}{(2\pi)^{d/2}} \int_{\mathbb{T}^d} f(\mathbf{x}) e^{-i\mathbf{n}\cdot\mathbf{x}} \, d\mathbf{x}.$$

7.2   Fourier series of differentiable functions

There is an important connection between the smoothness of a function and the rate of decay of its Fourier coefficients: the smoother a function (that is, the more times it is differentiable), the faster its Fourier coefficients decay. Heuristically, a smooth function contains a small amount of high frequency components.

If $f \in C^1(\mathbb{T})$ is continuously differentiable, then we can relate the Fourier coefficients of f' to those of f using an integration by parts:

$$
\begin{aligned}
\widehat{f'}_n &= \frac{1}{\sqrt{2\pi}} \int_0^{2\pi} e^{-inx} f'(x) \, dx \\
&= \frac{1}{\sqrt{2\pi}} [f(2\pi) - f(0)] - \frac{1}{\sqrt{2\pi}} \int_0^{2\pi} (-in) e^{-inx} f(x) \, dx \\
&= in \widehat{f}_n.
\end{aligned}
\tag{7.12}
$$

Thus, differentiation of a function corresponds to multiplication of its Fourier coefficients by in. It follows by induction that if $f \in C^k(\mathbb{T})$, then

$$\widehat{f_n^{(k)}} = (in)^k \widehat{f}_n.$$

Equation (7.12) may be used to define the notion of the derivative of a function whose derivative is square integrable, but need not be continuous. Such a derivative is called a *weak derivative*. The space of functions in L^2 whose weak derivatives are in L^2 is denoted by H^1, and is an example of a Sobolev space.

Definition 7.6 The *Sobolev space* $H^1(\mathbb{T})$ consists of all functions

$$f(x) = \frac{1}{\sqrt{2\pi}} \sum_{n=-\infty}^{\infty} \widehat{f}_n e^{inx} \in L^2(\mathbb{T})$$

such that

$$\sum_{n=-\infty}^{\infty} n^2 \left| \widehat{f}_n \right|^2 < \infty.$$

The weak L^2-derivative $f' \in L^2(\mathbb{T})$ of $f \in H^1(\mathbb{T})$ is defined by the L^2-convergent Fourier series

$$f'(x) = \frac{1}{\sqrt{2\pi}} \sum_{n=-\infty}^{\infty} in\widehat{f}_n e^{inx}.$$

The space $H^1(\mathbb{T})$ is a Hilbert space with respect to the inner product

$$\langle f, g \rangle_{H^1} = \int_{\mathbb{T}} \left\{ \overline{f(x)} g(x) + \overline{f'(x)} g'(x) \right\} \, dx.$$

By Parseval's theorem, the H^1-inner product of two functions may be written in terms of their Fourier coefficients as

$$\langle f, g \rangle_{H^1} = \sum_{n=-\infty}^{\infty} \left(1 + n^2\right) \overline{\widehat{f}_n} \widehat{g}_n.$$

Convergence with respect to the associated H^1-norm corresponds to mean-square convergence of functions and their derivatives.

A continuously differentiable function belongs to $H^1(\mathbb{T})$ and its weak derivative is equal to the usual pointwise derivative. It follows from the density of $C(\mathbb{T})$ in $L^2(\mathbb{T})$ that $H^1(\mathbb{T})$ is the completion of the space $C^1(\mathbb{T})$ of continuously differentiable functions (or the space of trigonometric polynomials) with respect to the H^1-norm.

If $f, g \in H^1(\mathbb{T})$, then the definition of f' and Parseval's theorem imply that

$$\langle f', g \rangle_{L^2} = \sum_{n=-\infty}^{\infty} \overline{in\widehat{f}_n} \widehat{g}_n = - \sum_{n=-\infty}^{\infty} \overline{\widehat{f}_n} in\widehat{g}_n = -\langle f, g' \rangle_{L^2}.$$

After replacing f by $\overline{f}$, we see that weak derivatives satisfy integration by parts:

$$\int_{\mathbb{T}} f'g\,dx = -\int_{\mathbb{T}} fg'\,dx. \tag{7.13}$$

There are no boundary terms in the integration by parts formula for periodic functions. We can use (7.13) to give an equivalent definition of the weak L^2-derivative of a function in terms of its integral against a smooth *test function*. If $f \in H^1(\mathbb{T})$, then the linear functional $F : C^1(\mathbb{T}) \subset L^2(\mathbb{T}) \to \mathbb{C}$ defined by

$$F(\varphi) = -\int_{\mathbb{T}} f\varphi'\,dx, \qquad \varphi \in C^1(\mathbb{T}),$$

is bounded. Conversely, if F is bounded for a given function $f \in L^2(\mathbb{T})$, then, since $C^1(\mathbb{T})$ is dense in $H^1(\mathbb{T})$, the bounded linear transformation theorem, Theorm 5.19, implies that F extends to a unique bounded linear functional on $L^2(\mathbb{T})$. The Riesz representation theorem (see Theorem 8.12 below) therefore implies that there is a unique function $f' \in L^2(\mathbb{T})$ such that $F(\varphi) = \langle \overline{f'}, \varphi \rangle$ holds for all $\varphi \in C^1(\mathbb{T})$. This leads to the following alternative definition of a weak L^2-derivative.

Definition 7.7 A function $f \in L^2(\mathbb{T})$ belongs to $H^1(\mathbb{T})$ if there is a constant M such that

$$\left| \int_{\mathbb{T}} f\varphi'\,dx \right| \leq M\|\varphi\|_{L^2} \qquad \text{for all } \varphi \in C^1(\mathbb{T}).$$

If $f \in H^1(\mathbb{T})$, then the *weak derivative* f' of f is the unique element of $L^2(\mathbb{T})$ such that

$$\int_{\mathbb{T}} f'\varphi\,dx = -\int_{\mathbb{T}} f\varphi'\,dx \qquad \text{for all } \varphi \in C^1(\mathbb{T}).$$

More generally, for any $k \geq 0$, we define the Sobolev space

$$H^k(\mathbb{T}) = \left\{ f \in L^2(\mathbb{T}) \,\middle|\, f(x) = \sum_{n=-\infty}^{\infty} c_n e^{inx}, \quad \sum_{n=-\infty}^{\infty} |n|^{2k}|c_n|^2 < \infty \right\}.$$

If k is a natural number, the space H^k consists of functions with k square-integrable weak derivatives, but the Fourier series definition makes sense even when k is not a natural number.

Lemma 7.8 Suppose that $f \in H^k(\mathbb{T})$ for $k > 1/2$. Let

$$S_N(x) = \frac{1}{\sqrt{2\pi}} \sum_{n=-N}^{N} \widehat{f}_n e^{inx} \tag{7.14}$$

be the Nth partial sum of the Fourier series of f, and define

$$\left\| f^{(k)} \right\| = \left(\sum_{n=-\infty}^{\infty} |n|^{2k} \left| \widehat{f}_n \right|^2 \right)^{1/2}.$$

Then there is a constant C_k, independent of f, such that

$$\|S_N - f\|_\infty \le \frac{C_k}{N^{k-1/2}} \left\| f^{(k)} \right\|,$$

and (S_N) converges uniformly to f as $N \to \infty$.

Proof. Since $S_N \in C(\mathbb{T})$ for every $N \in \mathbb{N}$ and $C(\mathbb{T})$ is complete with respect to the supremum norm, it is sufficient to prove that for all $M > N$,

$$\|S_N - S_M\|_\infty \le \frac{C_k}{N^{k-1/2}} \left\| f^{(k)} \right\|.$$

This equation follows from (7.12) and the Cauchy-Schwarz inequality:

$$
\begin{aligned}
\|S_N - S_M\|_\infty &\le \frac{1}{\sqrt{2\pi}} \sum_{N < |n| \le M} \left| \hat{f}_n \right| \\
&= \frac{1}{\sqrt{2\pi}} \sum_{N < |n| \le M} |n|^k \left| \hat{f}_n \right| \frac{1}{|n|^k} \\
&\le \frac{1}{\sqrt{2\pi}} \left[\sum_{N < |n| \le M} |n|^{2k} \left| \hat{f}_n \right|^2 \right]^{1/2} \left[\sum_{N < |n| \le M} \frac{1}{|n|^{2k}} \right]^{1/2} \\
&\le \frac{1}{\sqrt{2\pi}} \left\| f^{(k)} \right\| \left[\int_N^\infty \frac{dr}{r^{2k}} \right]^{1/2} \\
&\le \frac{1}{\sqrt{2\pi(2k-1)}} \frac{1}{N^{k-1/2}} \left\| f^{(k)} \right\|,
\end{aligned}
$$

which proves the result with

$$C_k = \frac{1}{\sqrt{2\pi(2k-1)}}.$$ □

A corollary of this lemma is a special case of the *Sobolev embedding theorem*, which implies, in particular, that if a function on $\mathbb{T}$ has a square-integrable weak derivative, then it is continuous.

Theorem 7.9 (Sobolev embedding) If $f \in H^k(\mathbb{T})$ for $k > 1/2$, then $f \in C(\mathbb{T})$.

Proof. From Lemma 7.8, the partial sums of the Fourier series of f converge uniformly, so the limit is continuous. □

More generally, if $f \in H^k(\mathbb{T})$, then the Fourier series for the derivatives $f^{(j)}$ converge uniformly when $k > j + 1/2$, so $f \in C^\ell(\mathbb{T})$, where ℓ is the greatest integer strictly less than $k - 1/2$. For functions of several variables, one finds that $f \in H^k(\mathbb{T}^d)$ is continuous when $k > d/2$, and j-times continuously differentiable when $k > j + d/2$ (see Exercise 7.5). Roughly speaking, there is a "loss" of slightly more than one-half a derivative per space dimension in passing from L^2 derivatives to continuous derivatives.

7.3 The heat equation

Fourier series are an essential tool for the study of a wide variety of problems in engineering and science. In this section, we use Fourier series to solve the heat, or diffusion, equation which models the flow of heat in a conducting body. This was the original problem that led Jean-Baptiste Fourier to develop the series expansion named after him, although similar ideas had been suggested earlier by Daniel Bernoulli. The same equation also describes the diffusion of a dye or pollutant in a fluid.

We consider a thin ring made of a heat conducting material. In a one-dimensional approximation, we can represent the ring by a circle. We choose units of space and time so that the length of the ring is 2π and the thermal conductivity of the material is equal to one. The temperature $u(x,t)$ at time $t \geq 0$ and position $x \in \mathbb{T}$ along the ring satisfies the *heat* or *diffusion equation*,

$$u_t = u_{xx}, \tag{7.15}$$
$$u(x,0) = f(x),$$

where $f : \mathbb{T} \to \mathbb{R}$ is a given function describing the initial temperature in the ring.

If $u(x,t)$ is a smooth solution of the heat equation, then, multiplying the equation by $2u$ and rearranging the result, we get that

$$\left(u^2\right)_t = (2uu_x)_x - 2u_x^2.$$

Integration of this equation over $\mathbb{T}$, and use of the periodicity of u, implies that

$$\frac{d}{dt} \int_{\mathbb{T}} u^2(x,t)\,dx = -2 \int_{\mathbb{T}} |u_x(x,t)|^2\,dx \leq 0.$$

Therefore, $\|u(\cdot,t)\| \leq \|f(\cdot)\|$, where $\|\cdot\|$ denotes the L^2-norm in x, so it is reasonable to look for solutions $u(\cdot,t)$ that belong to $L^2(\mathbb{T})$ for all $t \geq 0$.

To make the notion of solutions that belong to L^2 more precise, let us first suppose that the initial data f is a trigonometric polynomial,

$$f(x) = \sum_{n=-N}^{N} f_n e^{inx}. \tag{7.16}$$

We look for a solution

$$u(x,t) = \sum_{n=-N}^{N} u_n(t) e^{inx}, \tag{7.17}$$

that is also a trigonometric polynomial, with coefficients $u_n(t)$ that are continuously differentiable functions of t. Using (7.17) in the heat equation, computing the t and x derivatives, and equating Fourier coefficients, we find that $u_n(t)$ satisfies

$$\dot{u}_n + n^2 u_n = 0, \tag{7.18}$$

$$u_n(0) = f_n,$$

where the dot denotes a derivative with respect to t. Thus, the PDE (7.15) reduces to a decoupled system of ODEs. The solutions of (7.18) are

$$u_n(t) = f_n e^{-n^2 t}.$$

Therefore, the solution of the heat equation with the initial data (7.16) is given by

$$u(x, t) = \sum_{n=-N}^{N} f_n e^{-n^2 t} e^{inx}. \tag{7.19}$$

We may write this solution more abstractly as

$$u(\cdot, t) = T(t) f(\cdot),$$

where $T(t) : \mathcal{P} \to \mathcal{P}$ is the linear operator on the space $\mathcal{P}$ of trigonometric polynomials defined by

$$T(t) \left[\sum_{n=-N}^{N} f_n e^{inx} \right] = \sum_{n=-N}^{N} f_n e^{-n^2 t} e^{inx}.$$

Parseval's theorem (7.9) implies that, for $t \geq 0$,

$$\|T(t) f\|^2 = \sum_{n=-N}^{N} |f_n|^2 e^{-2n^2 t} \leq \sum_{n=-N}^{N} |f_n|^2 = \|f\|^2.$$

Thus, the solution operator $T(t)$ is bounded with respect to the L^2-operator norm when $t \geq 0$. The operator $T(t)$ is unbounded when $t < 0$.

By the bounded linear transformation theorem, Theorem 5.19, there is a unique bounded extension of $T(t)$ from $\mathcal{P}$ to $L^2(\mathbb{T})$, which we still denote by $T(t)$. Explicitly, if

$$f(x) = \sum_{n=-\infty}^{\infty} f_n e^{inx} \in L^2(\mathbb{T})$$

then $u(\cdot, t) = T(t) f \in L^2(\mathbb{T})$ is given by

$$u(x, t) = \sum_{n=-\infty}^{\infty} f_n e^{-n^2 t} e^{inx}. \tag{7.20}$$

We may regard this equation as defining the exponential

$$T(t) = e^{t \partial^2 / \partial x^2}$$

of the unbounded operator $A = \partial^2 / \partial x^2$ with periodic boundary conditions. Rather than consider in detail when this Fourier series converges to a continuously differentiable, or *classical*, solution of the heat equation that satisfies the initial condition pointwise, we will simply say that the function u obtained in this way is a *weak solution* of the heat equation (7.15). In Chapter 11, we will see that this point of view corresponds to interpreting the derivatives in (7.15) in a distributional sense.

The operators $T(t)$ have the following properties:

(a) $T(0) = I$;
(b) $T(s)T(t) = T(s+t)$ for $s, t \geq 0$;
(c) $T(t)f \to f$ as $t \to 0^+$ for each $f \in L^2(\mathbb{T})$.

In particular, $T(t)$ converges strongly, but not uniformly, to I as $t \to 0^+$. We say that $\{T(t) \mid t \geq 0\}$ is a C_0-*semigroup*, in contrast with the uniformly continuous group of operators with a bounded generator defined in Theorem 5.49.

The action of the solution operator $T(t)$ on a function is given by multiplication of the function's nth Fourier coefficient by $e^{-n^2 t}$. The convolution theorem, Theorem 7.5, implies that for $t > 0$ the operator has the spatial representation $T(t)f = g^t * f$ of convolution with a function g^t, called the *Green's function*, where

$$g^t(x) = \frac{1}{2\pi} \sum_{n=-\infty}^{\infty} e^{inx - n^2 t}. \tag{7.21}$$

Using the Poisson summation formula in (11.43), we can write this series as an infinite, periodic sum of Gaussians,

$$g^t(x) = \frac{1}{\sqrt{4\pi t}} \sum_{n=-\infty}^{\infty} e^{-(x - 2\pi n)^2 / 4t}.$$

We can immediately read off from (7.20) several important qualitative properties of the heat equation. The first is the *smoothing property*. For every $t > 0$, we have $u(\cdot, t) \in C^\infty(\mathbb{T})$, because the Fourier coefficients decay exponentially quickly as $n \to \infty$. This holds even if the initial condition has a discontinuity, as illustrated in Figure 7.2 for the case of a step function. In more detail, we have

$$\sum_{n=-\infty}^{\infty} |n|^{2k} \left| f_n e^{-n^2 t} \right|^2 \leq \max_{n \in \mathbb{N}} \left\{ n^{2k} e^{-n^2 t} \right\} \sum_m |f_m|^2 < \infty$$

for each $k \geq 0$, so $u(\cdot, t) \in H^k(\mathbb{T})$ for every $k \in \mathbb{N}$. The Sobolev embedding theorem in Theorem 7.9 implies that $u(\cdot, t) \in C^{k-1}(\mathbb{T})$ for every $k \in \mathbb{N}$, and therefore $u(x, t)$ has continuous partial derivatives with respect to x of all orders. It then follows from the heat equation that u has continuous partial derivatives with respect to t of all orders for $t > 0$.

The second property is *irreversibility*. A solution may not exist for $t < 0$, even if the "final data" f is C^∞. For example, if

$$f(x) = \sum_{n=-\infty}^{\infty} e^{-|n|} e^{inx} \in C^\infty(\mathbb{T}),$$

then the Fourier series solution (7.20) diverges for any $t < 0$. Equivalently, letting $t \to -t$, we see that the initial value problem for the *backwards heat equation*,

$$u_t = -u_{xx}, \qquad u(x,0) = f(x),$$

may not have a solution for $t > 0$, and is said to be *ill-posed*. This ill-posedness reflects the impossibility of determining the temperature distribution that led to a given observed temperature distribution because of the rapid damping of temperature variations that fluctuate rapidly in space.

The third property is the *exponential decay* of solutions to an equilibrium state as $t \to +\infty$. It follows from the heat equation that the mean temperature,

$$\langle u \rangle = \frac{1}{2\pi} \int_{\mathbb{T}} f(x) \, dx,$$

is independent of time since

$$\frac{d}{dt} \int_{\mathbb{T}} u(x,t) \, dx = \int_{\mathbb{T}} u_t(x,t) \, dx = \int_{\mathbb{T}} u_{xx}(x,t) \, dx = 0.$$

The solution $u(x,t)$ converges exponentially quickly to its mean value, because

$$\sup_{x \in \mathbb{T}} \left| \sum_{n=-\infty}^{\infty} f_n e^{-n^2 t} e^{inx} - \langle u \rangle \right| \leq \sum_{n \neq 0} |f_n| e^{-n^2 t}$$

$$\leq \left[\sum_{n \neq 0} |f_n|^2 \right]^{1/2} \left[2 \sum_{n=1}^{\infty} e^{-2n^2 t} \right]^{1/2}$$

$$\leq C \|f\|_{L^2} e^{-t},$$

where C is a suitable constant. The exponential decay is a consequence of a *spectral gap* between the lowest eigenvalue, zero, of the operator $\partial^2/\partial x^2$ on $\mathbb{T}$ and the rest of its spectrum.

Heat diffusion on a ring leads to periodic boundary conditions in x. Other types of problems may be analyzed in an analogous way. An interesting example is the modeling of seasonal temperature variations in the earth as a function of depth. If we neglect daily fluctuations, a reasonable assumption is that the surface temperature of the earth is a periodic function of time with period equal to one year, and that the temperature at a depth x below the surface is also a periodic function of time. We further require that the temperature be bounded at large depths.

We choose a time unit so that $1 \, \text{year} = 2\pi$, and a length unit so that the thermal conductivity of the earth, assumed constant, is equal to one. The temperature $u(x,t)$ then satisfies the following problem in $x > 0$, $t > 0$:

$$u_t = u_{xx},$$

$$u(0,t) = f(t),$$

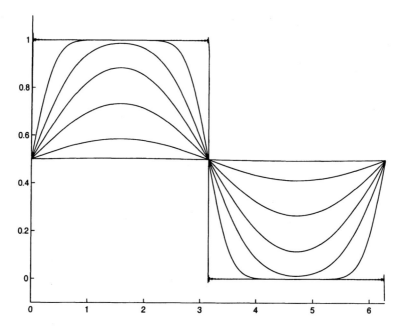

Fig. 7.2 The time evolution of the temperature distribution on a ring. The initial distribution is a step function. A truncated Fourier series is used to approximate the step function, and the Gibbs phenomenon can be seen near the points of discontinuity. The final distribution is uniform.

$$u(\cdot, t) \in L^{\infty}([0, \infty)),$$
$$u(x, t) = u(x, t + 2\pi).$$

Here, $f(t)$ is a given real-valued, 2π-periodic function that describes the seasonal temperature variations at the earth's surface.

We expand the temperature $u(x, t)$ at depth x in a Fourier series in t,

$$u(x, t) = \sum_{n=-\infty}^{\infty} u_n(x)e^{int}.$$

Use of this expansion in the heat equation implies that the coefficients $u_n(x)$ satisfy

$$-u_n'' + inu_n = 0, \tag{7.22}$$
$$u_n(0) = f_n,$$
$$u_n \in L^{\infty},$$

where the prime denotes a derivative with respect to x, and f_n is the nth Fourier coefficient of f,

$$f_n = \frac{1}{2\pi} \int_{\mathbb{T}} f(t)e^{-int} \, dt.$$

The solution of (7.22) is

$$
u_n(x) = \begin{cases} f_n \exp(\pm\sqrt{n}(1+i)x/\sqrt{2}) & \text{if } n > 0 \\ f_0 & \text{if } n = 0 \\ f_n \exp(\pm\sqrt{|n|}(1-i)x/\sqrt{2}) & \text{if } n < 0 \end{cases}.
$$

The solutions with the plus sign in the exponent are excluded because they are unbounded as $x \to \infty$. The solution for $u(x,t)$ is therefore

$$
u(x,t) = f_0 + \sum_{n=1}^{\infty} f_n e^{-|n/2|^{1/2}x} e^{i(nt - |n/2|^{1/2}x)}
$$

$$
+ \sum_{n=-\infty}^{-1} f_n e^{-|n/2|^{1/2}x} e^{i(nt + |n/2|^{1/2}x)}. \tag{7.23}
$$

For example, suppose that the surface temperature is given by a simple harmonic function

$$
u(0,t) = a + b\sin t.
$$

Then (7.23) may be written as

$$
u(x,t) = a + b\exp\left(-\frac{x}{\sqrt{2}}\right)\sin\left(t - \frac{x}{\sqrt{2}}\right).
$$

See Figure 7.3 for a graph of this solution. The exponential damping factor in front of the sine function describes a reduction in the magnitude of the variations in the earth's temperature below the surface. The argument of the sine function indicates that there is a depth-dependent phase shift in the temperature variations. At a depth $x = \sqrt{2}\pi$, the variations are reduced by a factor of $e^{-\pi} \approx 0.04$, and are opposite in phase to the surface temperature. For realistic numerical values of the thermal conductivity of the soil, this happens at a depth of about 13 feet. Thus, 13 feet below the surface the maximum temperature is reached in winter and minimum in summer! At this depth, the difference between winter and summer temperatures is reduced by a factor of about 25, as compared with the temperature difference at the surface. This reduction explains the usefulness of wine cellars, since it is important to store wine at a cool, uniform temperature.

7.4 Other partial differential equations

Fourier series may be used to study periodic solutions of any linear, constant coefficient partial differential equation. In this section, we consider a number of examples, including the wave equation and Laplace's equation, the two other classical linear partial differential equations of applied mathematics, in addition to the heat equation. The Fourier series may be interpreted either as classical solutions if they

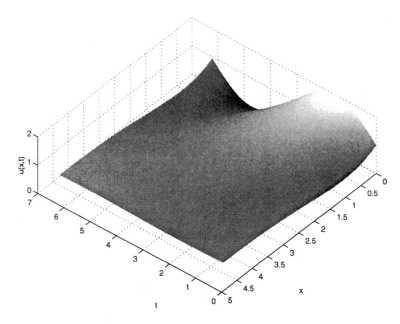

Fig. 7.3   The temperature of the earth as a function of time and depth. The time unit is one year divided by 2π. The unit of depth is roughly 1 meter. The temperature unit is arbitrary. The maximum and minimum temperature at the surface ($x = 0$) represent the maximum and minimum mean soil temperatures attained during summer and winter respectively.

converge sufficiently quickly to have continuous derivatives, or as weak solutions if they do not.

The one-dimensional *wave equation* is

$$u_{tt} = c^2 u_{xx}. \tag{7.24}$$

This equation describes the propagation of waves with a constant speed c, such as waves on an elastic string, sound waves, or light waves in a vacuum. The wave equation (7.24) is second order in time t, so we expect that two initial conditions are required to specify a unique solution. The initial value problem for wave propagation on a circle is

$$u_{tt} - c^2 u_{xx} = 0,$$
$$u(x, 0) = f(x),$$
$$u_t(x, 0) = g(x),$$

where $f, g \in L^2(\mathbb{T})$ are given functions.

The separated solutions of the wave equation (7.24), proportional to e^{inx}, are

$$u(x, t) = \left(ae^{inct} + be^{-inct}\right) e^{inx}$$

for $n \neq 0$, and

$$u(x,t) = a + bt$$

for $n = 0$. Superposing these solutions, the general solution is of the form

$$u(x,t) = a_0 + b_0 t + \sum_{n \neq 0} \left\{ a_n e^{in(x+ct)} + b_n e^{in(x-ct)} \right\}. \tag{7.25}$$

The constants a_n and b_n can be determined from the initial conditions as

$$a_0 = f_0, \quad b_0 = g_0, \quad a_n = \frac{1}{2} \left(f_n - \frac{i}{nc} g_n \right), \quad b_n = \frac{1}{2} \left(f_n + \frac{i}{nc} g_n \right),$$

where f_n and g_n are given by

$$f_n = \frac{1}{2\pi} \int_{\mathbb{T}} f(x) e^{-inx}, \quad g_n = \frac{1}{2\pi} \int_{\mathbb{T}} g(x) e^{-inx}.$$

In contrast with the heat equation, the solution exists for both $t > 0$ and $t < 0$, there is no smoothing of the initial data, and the solution does not converge to a stationary solution as $t \to \infty$.

The two-dimensional *Laplace equation* is

$$u_{xx} + u_{yy} = 0. \tag{7.26}$$

We will use Fourier series to solve a boundary value problem for Laplace's equation in the unit disc

$$\Omega = \left\{ (x,y) \mid x^2 + y^2 < 1 \right\}.$$

The Dirichlet problem consists of (7.26) in Ω with the boundary condition

$$u = f \qquad \text{on } \partial \Omega, \tag{7.27}$$

where $f : \mathbb{T} \to \mathbb{R}$ is a given function. In polar coordinates (r, θ) we may write (7.26)–(7.27) for $u(r, \theta)$ as

$$\frac{1}{r} (r u_r)_r + \frac{1}{r^2} u_{\theta\theta} = 0 \qquad \text{in } r < 1,$$
$$u(1, \theta) = f(\theta).$$

The Laplace equation in polar coordinates has the separated solutions

$$u(r, \theta) = \left(a r^n + b r^{-n} \right) e^{in\theta} \qquad \text{for } n \in \mathbb{Z}.$$

The general solution of Laplace's equation that is bounded inside the unit disc is therefore

$$u(r, \theta) = \sum_{n=-\infty}^{\infty} a_n r^{|n|} e^{in\theta}. \tag{7.28}$$

The boundary condition implies that

$$a_n = \frac{1}{2\pi} \int_{\mathbb{T}} f(\theta)e^{-in\theta}\,d\theta.$$

Using the convolution theorem, we may write (7.28) in $r < 1$ as

$$u(r,\theta) = (g^r * f)(\theta),$$

where $g^r : \mathbb{T} \to \mathbb{R}$ is the *Poisson kernel*,

$$g^r(\theta) = \frac{1}{2\pi} \sum_{n=-\infty}^{\infty} r^{|n|}e^{in\theta}.$$

The geometric series for $n > 0$ and $n < 0$ may be summed to give

$$g^r(\theta) = \frac{1}{2\pi} \frac{1-r^2}{1 - 2r\cos\theta + r^2}.$$

The series in (7.28) converges to an infinitely differentiable — in fact, analytic — function in $r < 1$ for any $f \in L^2(\mathbb{T})$, so the Laplace equation smoothes the boundary data.

In 1895 Korteweg and de Vries introduced a nonlinear PDE to describe water waves in shallow channels:

$$u_t = uu_x + u_{xxx}. \tag{7.29}$$

This *KdV equation* has exact localized traveling wave solutions called solitary waves, or solitons. A remarkable fact is that, in spite of its nonlinearity, the KdV equation can be solved exactly by the inverse scattering method introduced by Gardner, Greene, Kruskal, and Miura in 1967. This method depends on a surprising connection between the nonlinear KdV equation and a spectral problem for an associated linear operator (see Exercise 9.15). We will not discuss the inverse scattering method here, but we will use Fourier analysis to describe the *dispersive* property of the KdV equation.

If u is sufficiently small, then we do not expect the nonlinear term uu_x to influence the solution significantly, so we omit it in a first approximation. We therefore consider the linearized KdV equation,

$$u_t = u_{xxx}. \tag{7.30}$$

The general solution that is a 2π-periodic function of x is

$$u(x,t) = \sum_{n=-\infty}^{\infty} a_n e^{in(x-n^2 t)}.$$

Notice that the speed of propagation of e^{inx} depends on n, that is, on the wavelength. Since the components in the wave with different wavelengths propagate at

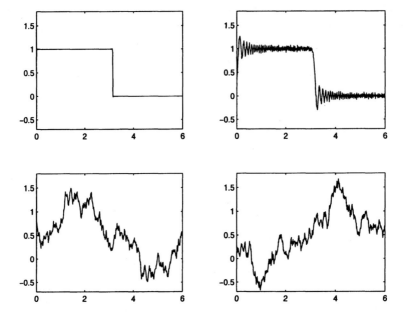

Fig. 7.4   The effect of dispersion is illustrated here with the solution (7.30) of the linearized KdV equation on a ring for times $t = 0, e^{-10}, e^{-2}$, and $t = e$. The initial condition is a step function.

different speeds, a wave generally spreads out or disperses; hence the name *dispersive* waves. In particular, a wave front does not maintain its shape while propagating. See Figure 7.4 for an illustration. Contrast this with the solution of the wave equation (7.25), where different Fourier components propagate at the same speed. The wave equation is said to be *nondispersive*. Another example of a dispersive wave equation, the Schrödinger equation from quantum mechanics, is discussed in Exercise 7.13.

7.5   More applications of Fourier series

The use of Fourier series is not restricted to differential equations. In this section, we consider two other applications.

The first is a solution of the isoperimetric problem, which states that of all closed curves of a given length, a circle encloses the maximum area. This result can also be stated as an inequality: for any closed curve of length L enclosing an area A, we have

$$4\pi A \le L^2, \tag{7.31}$$

with equality if and only if the curve is a circle. Equation (7.31) is called the *isoperimetric inequality*. There are many different proofs of this result; the one we give, using Fourier series, is due to Hurwitz.

In order to state and prove a precise result, we reformulate the problem analytically. Without loss of generality, we consider curves whose lengths are normalized to 2π, and that are parametrized by arclength, s, positively oriented in the counterclockwise direction. We may represent such a smooth, closed curve Γ in the plane $\mathbb{R}^2$ by

$$(x, y) = (f(s), g(s)),\qquad(7.32)$$

where $f, g : \mathbb{T} \to \mathbb{R}$ are continuously differentiable functions such that

$$\dot{f}(s)^2 + \dot{g}(s)^2 = 1.\qquad(7.33)$$

Here, the dot denotes a derivative with respect to s.

Green's theorem states that if Ω is a region in the plane with a smooth, positively oriented boundary $\partial\Omega$ and $u, v : \overline{\Omega} \to \mathbb{R}$ are continuously differentiable functions, then

$$\int_{\Omega} \{u_x + v_y\}\, dx dy = \int_{\partial\Omega} \{u\, dy - v\, dx\}.$$

If Γ does not intersect itself, then the use of Green's theorem with $u = x/2$ and $v = y/2$ implies that the area A enclosed by Γ is given by

$$A = \frac{1}{2} \int_{\mathbb{T}} \left\{ f(s)\dot{g}(s) - g(s)\dot{f}(s) \right\}\, ds.\qquad(7.34)$$

The expressions in (7.33) and (7.34) make sense for general functions $f, g \in H^1(\mathbb{T})$. Thus, an analytical formulation of the isoperimetric problem is to find functions $f, g \in H^1(\mathbb{T})$ that maximize the area functional A in (7.34) subject to the constraint (7.33).

Theorem 7.10 Suppose that a curve Γ is given by $x = f(s)$, $y = g(s)$, where $f, g \in H^1(\mathbb{T})$ are real-valued functions that satisfy (7.33), and the area A enclosed by Γ is given by (7.34). Then $A \leq \pi$, with equality if and only if Γ is a circle.

Proof. We Fourier expand f and g as

$$f(s) = \frac{1}{\sqrt{2\pi}} \sum_{n=-\infty}^{\infty} \hat{f}_n e^{ins}, \qquad g(s) = \frac{1}{\sqrt{2\pi}} \sum_{n=-\infty}^{\infty} \hat{g}_n e^{ins}.\qquad(7.35)$$

Since f and g are real valued, we have $\hat{f}_{-n} = \overline{\hat{f}_n}$ and $\hat{g}_{-n} = \overline{\hat{g}_n}$ for all n. Integration of (7.33) over $\mathbb{T}$ gives

$$2\pi = \int_{\mathbb{T}} \left\{ \dot{f}(s)^2 + \dot{g}(s)^2 \right\}\, ds.$$

From Parseval's theorem, this equation implies that

$$2\pi = \sum_{n=-\infty}^{\infty} n^2 \left\{ \left|\hat{f}_n\right|^2 + |\hat{g}_n|^2 \right\},\qquad(7.36)$$

and equation (7.34) implies that

$$2A = \sum_{n=-\infty}^{\infty} in \left\{ \overline{\hat{f}_n} \hat{g}_n - \hat{f}_n \overline{\hat{g}_n} \right\}.$$

Subtracting these series and rearranging the result, we find that

$$2\pi - 2A = \frac{1}{2} \sum_{n \neq 0} \left\{ \left| n\hat{f}_n - i\hat{g}_n \right|^2 + \left| n\hat{g}_n + i\hat{f}_n \right|^2 + (n^2 - 1)\left(\left| \hat{f}_n \right|^2 + |\hat{g}_n|^2 \right) \right\}.$$

Since the terms in the series on the right hand side of this equation are nonnegative, it follows that $A \leq \pi$. Moreover, we have equality if and only if $\hat{f}_n = \hat{g}_n = 0$ for $n \geq 2$, and $\hat{f}_1 = i\hat{g}_1$. Equation (7.36) implies that $|\hat{f}_1| = \sqrt{\pi/2}$, so that

$$\hat{f}_1 = \sqrt{\frac{\pi}{2}} e^{i\delta}, \qquad \hat{g}_1 = -i\sqrt{\frac{\pi}{2}} e^{i\delta},$$

for some $\delta \in \mathbb{R}$. Writing $\hat{f}_0 = \sqrt{2\pi} x_0$ and $\hat{g}_0 = \sqrt{2\pi} y_0$, where $x_0, y_0 \in \mathbb{R}$, we find from (7.35) that

$$f(s) = x_0 + \cos(s + \delta), \qquad g(s) = y_0 + \sin(s + \delta).$$

Thus, if $A = \pi$, the curve $x = f(s)$, $y = g(s)$ is a circle.   □

Our final application is an *ergodic theorem* for one of the simplest dynamical systems one can imagine, namely, rotations of the circle. We will prove another ergodic theorem for more general dynamical systems later on, in Theorem 8.37.

Let $\gamma \in \mathbb{R}$. We define a map $F_\gamma : \mathbb{T} \to \mathbb{T}$ on the circle $\mathbb{T}$ by

$$F_\gamma(x) = x + 2\pi\gamma. \tag{7.37}$$

This map is called the *circle map* or the *rotation map*. For every $x_0 \in \mathbb{T}$, the iterated application of F_γ generates a sequence of points $(x_n)_{n=0}^{\infty}$, where $x_n = F_\gamma^n(x_0)$. The set $\{x_n\}$ is called the *orbit* or *trajectory* of x_0 under F_γ. If γ is rational, then these points eventually repeat, and each orbit contains finitely many distinct points. If γ is irrational, then $x_m \neq x_n$ for $m \neq n$, and there are infinitely many points in each orbit (see Figure 7.5).

If $f : \mathbb{T} \to \mathbb{C}$ is a continuous function on $\mathbb{T}$, we define two averages of f, a *time average*

$$\langle f \rangle_t(x_0) = \lim_{N \to \infty} \frac{1}{N+1} \sum_{n=0}^{N} f(x_n),$$

and a *phase-space average*,

$$\langle f \rangle_{\mathrm{ph}} = \frac{1}{2\pi} \int_{\mathbb{T}} f(x) \, dx.$$

This phase-space average may be regarded as a probabilistic average with respect to a uniform probability measure on $\mathbb{T}$. The following ergodic theorem, proved by Weyl in 1916, states that time averages and phase-space averages are equal when γ is irrational. This result is false when γ is rational.

Theorem 7.11 (Weyl ergodic) If γ is irrational, then

$$\langle f \rangle_t(x_0) = \langle f \rangle_{\text{ph}} \tag{7.38}$$

for all $f \in C(\mathbb{T})$ and all $x_0 \in \mathbb{T}$.

Proof. First, we show that (7.38) holds for the functions e^{imx} for each $m \in \mathbb{Z}$. If $m = 0$, then both averages are equal to 1. If $m \neq 0$, then $\langle e^{imx} \rangle_{\text{ph}} = 0$, and the time average may be explicitly computed as follows:

$$
\begin{aligned}
\langle e^{imx} \rangle_t &= \lim_{N \to \infty} \frac{1}{N+1} \sum_{n=0}^{N} e^{im(x_0 + 2\pi n\gamma)} \\
&= \lim_{N \to \infty} \frac{1}{N+1} e^{imx_0} \sum_{n=0}^{N} [e^{2\pi im\gamma}]^n \\
&= \lim_{N \to \infty} \frac{1}{N+1} e^{imx_0} \left(\frac{1 - [e^{2\pi im\gamma}]^{N+1}}{1 - e^{2\pi im\gamma}} \right) \\
&= 0,
\end{aligned}
$$

where we use the fact that $e^{2\pi im\gamma} \neq 1$ for irrational γ. Since both averages are linear in f, it follows that (7.38) holds for all trigonometric polynomials.

The trigonometric polynomials are dense in $C(\mathbb{T})$. Therefore, if $f \in C(\mathbb{T})$ and $\epsilon > 0$, then there is a trigonometric polynomial p such that $\|f - p\|_\infty \leq \epsilon$, and

$$\left| \frac{1}{N+1} \sum_{n=0}^{N} f(x_n) - \frac{1}{2\pi} \int_0^{2\pi} f(x) \, dx \right| \leq 2\epsilon + \left| \frac{1}{N+1} \sum_{n=0}^{N} p(x_n) - \frac{1}{2\pi} \int_0^{2\pi} p(x) \, dx \right|.$$

Taking the $\limsup$ of this equation as $N \to \infty$, we obtain that

$$\limsup_{N \to \infty} \left| \frac{1}{N+1} \sum_{n=0}^{N} f(x_n) - \langle f \rangle_{\text{ph}} \right| \leq 2\epsilon.$$

Since $\epsilon > 0$ is arbitrary, this proves (7.38) for all $f \in C(\mathbb{T})$ and all $x_0 \in \mathbb{T}$. $\quad\square$

A consequence of this ergodic theorem is the following result, which says that the points in an orbit $\{x_n \mid n \geq 0\}$ are uniformly distributed on the circle.

Corollary 7.12 Suppose that γ is irrational and I is an interval in $\mathbb{T}$ of length λ. Then

$$\lim_{N \to \infty} \frac{\#\{n \mid 0 \leq n \leq N, \, x_n \in I\}}{N+1} = \frac{\lambda}{2\pi}, \tag{7.39}$$

where $\#S$ denotes the number of points in the set S.

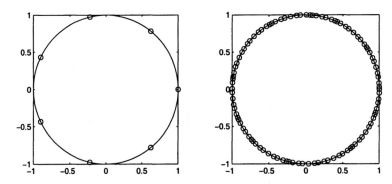

Fig. 7.5 The repeated images of the origin under the circle map, $F_\gamma^n(0)$, for $1 \leq n \leq 100$. On the left, $\gamma = 2/7$ is rational. On the right, $\gamma = (\sqrt{5}-1)/2$ is the golden ratio, which is irrational.

Proof. Let χ_I be the characteristic function of the interval I. Then (7.39) is equivalent to the statement that

$$\langle \chi_I \rangle_t = \langle \chi_I \rangle_{\text{ph}}. \tag{7.40}$$

This equation does not follow directly from Theorem 7.11 because χ_I is not continuous. We therefore approximate χ_I by continuous functions. We choose sequences (f_k) and (g_k) of nonnegative, continuous functions such that $f_k \leq \chi_I \leq g_k$ and

$$\int_{\mathbb{T}} f_k(x)\,dx \to \int_{\mathbb{T}} \chi_I(x)\,dx, \quad \int_{\mathbb{T}} g_k(x)\,dx \to \int_{\mathbb{T}} \chi_I(x)\,dx \quad \text{as } k \to \infty.$$

We leave it to the reader to construct such sequences. Since $f_k \leq \chi_I \leq g_k$,

$$\frac{1}{N+1}\sum_{n=0}^{N} f_k(x_n) \leq \frac{1}{N+1}\sum_{n=0}^{N} \chi_I(x_n) \leq \frac{1}{N+1}\sum_{n=0}^{N} g_k(x_n).$$

Taking the limit as $N \to \infty$ of this equation, and applying Theorem 7.11 to the functions f_k and g_k, we obtain that

$$\frac{1}{2\pi}\int_{\mathbb{T}} f_k(x)\,dx \;\leq\; \liminf_{N\to\infty} \frac{1}{N+1}\sum_{n=0}^{N}\chi_I(x_n)$$

$$\leq\; \limsup_{N\to\infty} \frac{1}{N+1}\sum_{n=0}^{N}\chi_I(x_n) \leq \frac{1}{2\pi}\int_{\mathbb{T}} g_k(x)\,dx.$$

Letting $k \to \infty$, we find that

$$\frac{1}{2\pi}\int_{\mathbb{T}} \chi_I(x)\,dx \;\leq\; \liminf_{N\to\infty} \frac{1}{N+1}\sum_{n=0}^{N}\chi_I(x_n)$$

$$\leq\; \limsup_{N\to\infty} \frac{1}{N+1}\sum_{n=0}^{N}\chi_I(x_n)$$

$$\leq \frac{1}{2\pi} \int_{\mathbb{T}} \chi_I(x)\, dx.$$

It follows that the limit defining the time average of χ_I exists and satisfies equation (7.40) for all $x_0 \in \mathbb{T}$. $\qquad\square$

Theorem 7.11 actually holds for every $f \in L^1(\mathbb{T})$, except possibly for a set of initial points x_0 in $\mathbb{T}$ with zero Lebesgue measure. The proof, however, requires additional results from measure theory.

One application of the ergodic theorem is to the numerical integration of functions by the *Monte Carlo method*, in which one approximates the phase average, or integral, of f by a time average. This method is not required in the simple case of functions defined on a circle, but it is useful for the numerical integration of functions that depend on a large number of independent variables, where standard numerical integration formulae may become prohibitively expensive.

7.6 Wavelets

In this section, we introduce a special class of orthonormal bases of $L^2([0,1])$ and $L^2(\mathbb{R})$, called *wavelets*. These bases have proved to be very useful in signal analysis and data compression. With this application in mind, we will refer to the independent variable as a "time" variable. Wavelet bases in several independent variables are equally useful in image compression and many other applications.

Fourier expansions provide an efficient representation of stationary functions whose properties are invariant under translations in time. They are not as efficient in representing other types of functions, such as transient functions that vanish on most of their domain, or functions which vary much more rapidly at some times than at others. In the case of periodic functions, Parseval's identity in Theorem 6.26,

$$\|f\|^2 = \sum_n \left|\widehat{f}_n\right|^2,$$

suggests that a large number of terms in a Fourier series expansion of f is needed if the quantity $\|f\|$ is distributed over a large number of coefficients $\widehat{f}_n$ which are not too small. For example, from Lemma 7.8, this happens when f is discontinuous, so that its Fourier coefficients decay slowly as $n \to \infty$. Signals with sharp, or almost discontinuous, transitions and transient signals supported on a relatively small portion of the relevant time interval, such as the short beeps transmitted by a modem, are very common.

It is often useful to compress a signal before transmission or storage. To represent a function $f(t)$ accurately on the interval $0 \leq t \leq 1$ by storing a finite number of values $f(n\Delta t)$, where $n = 0, \ldots, N$ with $N = 1/\Delta t$, we need to choose Δt small enough that all rapid transitions can be reconstructed from this list of values. If the changes are rapid, then Δt has to be small and N has to be large. Suppose, however, that we have an orthonormal basis of $L^2([0, 1])$ with the property that a finite linear combination of basis elements with M terms, where M is much smaller than N, yields a good approximation of the function f. Then we can store or transmit the function with M instead of N numbers without significant loss of information. Roughly speaking, we would then have compressed the data with a compression ratio of $N : M$. One reason for the use of wavelets in representing signals, or images, is that they allow for large compression ratios. There are many different kinds of wavelets, but all of them share the property that they describe a function at a sequence of different time, or length, scales. This allows us to represent a function efficiently by using wavelets whose local rate of variation is adapted to that of the function. We begin by describing a simple example, the *Haar wavelets*.

We define the *Haar scaling function* $\varphi \in L^2(\mathbb{R})$ by

$$\varphi(x) = \begin{cases} 1 & \text{if } 0 \leq x < 1, \\ 0 & \text{otherwise.} \end{cases} \tag{7.41}$$

The function φ is the characteristic function of the interval $[0, 1)$, and is often referred to as a "box" function because of the shape of its graph. The basic *Haar wavelet*, or *mother wavelet*, $\psi \in L^2(\mathbb{R})$ is given by

$$\psi(x) = \begin{cases} 1 & \text{if } 0 \leq x < 1/2, \\ -1 & \text{if } 1/2 \leq x < 1, \\ 0 & \text{otherwise.} \end{cases} \tag{7.42}$$

These functions satisfy the scaling relations

$$\varphi(x) = \varphi(2x) + \varphi(2x - 1), \tag{7.43}$$

$$\psi(x) = \varphi(2x) - \varphi(2x - 1). \tag{7.44}$$

For $n, k \in \mathbb{Z}$, we define scaled translates $\varphi_{n,k}, \psi_{n,k} \in L^2(\mathbb{R})$ of φ, ψ by

$$\varphi_{n,k}(x) = 2^{n/2}\varphi(2^n x - k), \quad \psi_{n,k}(x) = 2^{n/2}\psi(2^n x - k). \tag{7.45}$$

First, consider the Hilbert space $L^2([0, 1])$ with its usual inner product. For $n = 0, 1, 2, \ldots$, let V_n be the finite-dimensional subspace

$$V_n = \{f \mid f \text{ is constant on } [k/2^n, (k+1)/2^n) \text{ for } k = 0, \ldots, 2^n - 1\}. \tag{7.46}$$

Elements of V_n are step functions that are constant on intervals of length 2^{-n}. The value of $f \in V_n$ at the right endpoint $x = 1$ is irrelevant, since functions in $L^2([0, 1])$ that are equal a.e. are equivalent. Clearly, we have $V_n \subset V_{n+1}$.

The function $\varphi_{n,k}$ is the characteristic function of the interval $[k2^{-n}, (k+1)2^{-n})$. The set

$$A_n = \{\varphi_{n,k} \mid 0 \leq k \leq 2^n - 1\}$$

is therefore a basis of V_n for each $n \geq 0$. Since $A_{n+1} \supset A_n$, the sets A_n are not disjoint, and we cannot form a basis of $\bigcup_{n \in \mathbb{N}} V_n$ by taking their union. Instead, for

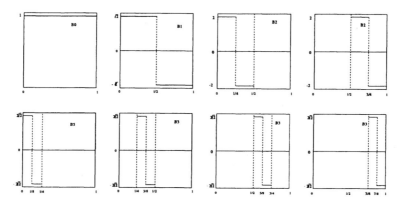

Fig. 7.6   Some members of the sets of functions B_n defined in (7.47).

$n = 0, 1, 2, \ldots$, we define subsets B_n of V_n by

$$B_0 = \{\varphi_{0,0}\}, \quad B_{n+1} = \{\psi_{n,k} \mid k = 0, 1, \ldots, 2^n - 1\}. \tag{7.47}$$

The subsets B_m and B_n are disjoint for $n \neq m$. The union of these sets, $B = \bigcup_{n=0}^{\infty} B_n$, or

$$B = \{\varphi_{0,0}\} \cup \{\psi_{n,k} \mid n = 0, 1, 2, \ldots, k = 0, 1, \ldots, 2^n - 1\}, \tag{7.48}$$

is called the *Haar wavelet basis* of $L^2([0,1])$. Some of these basis functions are illustrated in Figure 7.6.

Solving (7.43)–(7.44) for $\varphi(2x)$ and $\varphi(2x - 1)$, we get

$$\varphi(2x) = \frac{1}{2}\left(\varphi(x) + \psi(x)\right),$$

$$\varphi(2x - 1) = \frac{1}{2}\left(\varphi(x) - \psi(x)\right).$$

It follows by induction from dyadic dilations $x \mapsto 2x$ of these equations that $\varphi_{n,k}$ is a linear combination of $\varphi_{0,0}$ and $\psi_{m,k}$ with $m < n$. Hence the linear span of B contains bases A_n of V_n for every $n \in \mathbb{N}$. Using this fact, we can prove that B is a basis of $L^2([0,1])$.

Lemma 7.13 The set B in (7.48) is an orthonormal basis of $L^2([0,1])$.

Proof.   It follows from Exercise 7.17 that B is an orthonormal set, so we just have to show that it is complete. Suppose that $f \in C([0,1])$ and $\epsilon > 0$. By Theorem 1.67, f is uniformly continuous, so there is an n such that $|f(x) - f(y)| \leq \epsilon$ for all $x, y \in [0,1]$ with $|x - y| \leq 2^{-n}$. We define the step function approximation $g \in V_n$ of f by

$$g(x) = \sum_{k=0}^{2^n - 1} f\left(k 2^{-n}\right) \varphi_{n,k}(x).$$

Then g is in the linear span of $\bigcup_{m=0}^{n-1} B_m$, and

$$\sup_{0 \leq x < 1} |f(x) - g(x)| \leq \epsilon. \tag{7.49}$$

Thus any $f \in C([0,1])$ is the uniform limit of finite linear combinations of functions in B. Since the continuous functions are dense in $L^2([0,1])$, and the sup-norm is stronger than the L^2-norm, the orthonormal set B is complete in $L^2([0,1])$.   □

We define the Haar wavelet basis B of $L^2(\mathbb{R})$ in a similar way, as

$$B = \{\psi_{n,k} \mid n \in \mathbb{Z}, k \in \mathbb{Z}\}, \tag{7.50}$$

where $\psi_{n,k}$ is defined in (7.45). This basis includes wavelets supported on intervals of arbitrarily large length, when n is large and negative, as well as on intervals of arbitrarily small length, when n is large and positive. The wavelet basis of $L^2(\mathbb{R})$ does not include a scaling function φ, in contrast with the wavelet basis (7.48) of $L^2([0,1])$.

Lemma 7.14 The set B in (7.50) is an orthonormal basis of $L^2(\mathbb{R})$.

Proof.   The set B is orthonormal, so we just have to show that it is complete. Suppose that $f \in L^2(\mathbb{R})$ is orthogonal to B. Then f is orthogonal to all wavelets $\psi_{n,k}$ that are supported on any compact interval $[-2^N, 2^N]$. Since we can transform the interval $[-2^N, 2^N]$ to $[0,1]$ by a translation $x \mapsto x + 2^N$ and a dyadic dilation $x \mapsto 2^{-(N+1)}x$, and the basis B is invariant under such translations and dilations, it follows from Lemma 7.13 that f is constant on every compact interval $[-2^N, 2^N]$. Therefore f is constant on $\mathbb{R}$. Since the nonzero constant functions do not belong to $L^2(\mathbb{R})$, we conclude that $f = 0$, so B is complete.   □

The Haar wavelets are very simple, compactly supported, orthonormal, step functions that take only three different values. Each wavelet is obtained by dilation and translation of a single basic wavelet ψ, derived from a scaling function φ. These properties make the Haar wavelets especially suitable for the representation of localized functions, as well as functions that vary on different lengthscales at different locations, and functions with a self-similar, fractal structure.

A drawback of the Haar wavelets is that they are discontinuous, so the partial sums approximating a continuous function are also discontinuous. It is often desirable to have continuous approximations of continuous functions, and C^p approximations of C^p functions. This is one motivation for the introduction of other wavelet bases. For definiteness, we consider wavelet bases of $L^2(\mathbb{R})$. The following *axioms of multiresolution analysis* capture the essential properties of the Haar wavelet basis that we want to generalize.

Definition 7.15 (Multiresolution analysis) A family $\{V_n \mid n \in \mathbb{Z}\}$ of closed linear subspaces of $L^2(\mathbb{R})$ and a function $\varphi \in L^2(\mathbb{R})$ are called a *multiresolution* *analysis* of $L^2(\mathbb{R})$ if the following properties hold:

(a) $f(x) \in V_n$ if and only if $f(2x) \in V_{n+1}$ for all $n \in \mathbb{Z}$ (scaling);   (7.51)

(b) $V_n \subset V_{n+1}$ for all $n \in \mathbb{Z}$ (inclusion);   (7.52)

(c) $\overline{\bigcup_{n \in \mathbb{Z}} V_n} = L^2(\mathbb{R})$ (density);   (7.53)

(d) $\bigcap_{n \in \mathbb{Z}} V_n = \{0\}$ (maximality);   (7.54)

(e) there is a function $\varphi \in L^2(\mathbb{R}) \cap L^1(\mathbb{R})$ such that $\{\varphi(x - k) \mid k \in \mathbb{Z}\}$

is an orthonormal basis of V_0 (basis).   (7.55)

The five properties required in this definition are not independent. One can prove that (d) follows from (a), (b), and (e), and that, under the assumption that (a), (b), and (e) hold, (c) is equivalent to the property that $\widehat{\varphi}(0) \neq 0$, where $\widehat{\varphi}$ is the Fourier transform of φ. For brevity, we do not prove these statements here.

The spaces V_n defined in (7.46) and the function φ defined in (7.41) satisfy these axioms. We call φ the *scaling function* of the multiresolution analysis. We will explain how to obtain an orthonormal wavelet basis of $L^2(\mathbb{R})$ from this structure. When the scaling function φ is the box function, defined in (7.41), this procedure will reproduce the Haar wavelets, but other scaling functions lead to different orthonormal wavelet bases.

From (7.51) and (7.55) it follows that

$$A_n = \{2^{n/2}\varphi(2^n x - k) \mid k \in \mathbb{Z}\}$$

is an orthonormal basis of V_n for each $n \in \mathbb{Z}$. Since $V_n \subset V_{n+1}$, each function in A_n is a linear combination of functions of A_{n+1}, so the sets A_n are not linearly independent. To obtain linearly independent sets of functions, we define closed linear subspaces W_n of V_{n+1} by

$$V_{n+1} = V_n \oplus W_n.$$

The subspaces W_n are called *wavelet subspaces*. From their definition and the inclusion property (7.52), we see that W_m and W_n are orthogonal subspaces for $m \neq n$. Moreover, from Exercise 7.16, properties (7.52)–(7.54) imply that

$$\bigoplus_{n \in \mathbb{Z}} W_n = L^2(\mathbb{R}). \tag{7.56}$$

Now suppose that we have a function $\psi \in L^2(\mathbb{R})$, called a *wavelet*, such that

$$\{\psi(x - k) \mid k \in \mathbb{Z}\} \text{ is an orthonormal basis of } W_0.$$

Equation (7.58) below shows how the wavelet ψ is obtained from the scaling function φ. It then follows from the scaling axiom (7.51) that for each $n \in \mathbb{Z}$ the set

$$B_n = \left\{2^{n/2}\psi(2^n x - k) \mid k \in \mathbb{Z}\right\}$$

is an orthonormal basis of W_n so, from (7.56), their union

$$B = \left\{ 2^{n/2} \psi(2^n x - k) \mid n, k \in \mathbb{Z} \right\}$$

is an orthonormal basis of $L^2(\mathbb{R})$.

The axioms of multiresolution analysis impose severe restrictions on the scaling function. Translates of the scaling function must be orthogonal, and the function must be a linear combination of scaled translates of itself, meaning that there are constants c_k such that

$$\varphi(x) = \sum_{k \in \mathbb{Z}} c_k \varphi(2x - k). \tag{7.57}$$

For example, the Haar scaling function satisfies (7.43), so in that case $c_0 = c_1 = 1$ and $c_k = 0$ otherwise. For simplicity, we assume that $c_k \in \mathbb{R}$ and all but finitely many of the coefficients c_k are zero.

The basic wavelet ψ belongs to V_1 and is orthogonal to V_0. The following function satisfies these conditions:

$$\psi(x) = \sum_{k \in \mathbb{Z}} (-1)^k c_{1-k} \varphi(2x - k). \tag{7.58}$$

For example, in the case of the Haar wavelets, this equation gives (7.44). The function ψ in (7.58) clearly belongs to V_1, since it is a linear combination of the orthonormal basis elements $2^{1/2} \varphi(2x - k)$ of V_1. Moreover, for $j \in \mathbb{Z}$, we find from (7.57) and (7.58) that

$$\langle \varphi(x - j), \psi(x) \rangle = \frac{1}{2} \sum_{k \in \mathbb{Z}} (-1)^k c_{k-2j} c_{1-k}.$$

This sum is zero for every $j \in \mathbb{Z}$, since the change of summation variable from $k - 2j$ to $1 - k$ implies that

$$\sum_{k \in \mathbb{Z}} (-1)^k c_{k-2j} c_{1-k} = \sum_{k \in \mathbb{Z}} (-1)^{2j+1-k} c_{1-k} c_{k-2j} = -\sum_{k \in \mathbb{Z}} (-1)^k c_{k-2j} c_{1-k}.$$

Hence ψ is orthogonal to V_0. The translates $\{\psi(x - k) \mid k \in \mathbb{Z}\}$ form a basis of W_0, but we omit a proof of this fact here.

Next, we derive restrictions on the coefficients c_k in the scaling equation (7.57). We assume that the integral of φ is nonzero. It can, in fact, be shown that this is necessarily the case. By rescaling φ and x, we may assume without loss of generality that

$$\int_{\mathbb{R}} \varphi(x) \, dx = 1, \qquad \int_{\mathbb{R}} |\varphi(x)|^2 \, dx = 1.$$

Changing variables $x \to 2x$, we see that

$$\int_{\mathbb{R}} \varphi(2x) \, dx = \frac{1}{2}, \qquad \int_{\mathbb{R}} |\varphi(2x)|^2 \, dx = \frac{1}{2}.$$

Integration of (7.57) over $\mathbb{R}$ therefore implies that

$$\sum_{k \in \mathbb{Z}} c_k = 2. \tag{7.59}$$

Since $\{\varphi(2x - k) | k \in \mathbb{Z}\}$ is an orthogonal set, an application of Parseval's identity to (7.57) implies that

$$\sum_{k \in \mathbb{Z}} c_k^2 = 2. \tag{7.60}$$

The orthogonality of $\varphi(x - j)$ and $\varphi(x)$ for $j \neq 0$, together with (7.57), further imply that

$$\sum_{k \in \mathbb{Z}} c_{k-2j} c_k = 0 \qquad \text{for } j \in \mathbb{Z} \text{ and } j \neq 0. \tag{7.61}$$

Finally, it is often useful to require that several moments of the wavelet ψ vanish, meaning that

$$\int_{\mathbb{R}} x^m \psi(x) \, dx = 0 \qquad \text{for } m = 0, 1, \ldots, p - 1. \tag{7.62}$$

The scaling coefficients c_k and the wavelet ψ must therefore satisfy (7.59)–(7.61). For example, the Haar wavelet coefficients $c_0 = 1$, $c_1 = 1$, and $c_k = 0$ for $k \neq 0, 1$ satisfy these conditions, and (7.62) with $p = 1$, but there are many other possible choices of the scaling coefficients.

One interesting choice, that satisfies (7.62) with $p = 2$, is due to Daubechies:

$$c_0 = \frac{1}{4}(1 + \sqrt{3}), \qquad c_1 = \frac{1}{4}(3 + \sqrt{3}),$$
$$c_2 = \frac{1}{4}(3 - \sqrt{3}), \qquad c_3 = \frac{1}{4}(1 - \sqrt{3}),$$

and $c_k = 0$ otherwise. We call the corresponding wavelet the D_4 wavelet. We can find the scaling function φ by regarding (7.57) as a fixed point equation and solving it iteratively, starting with the box function, for example, as an initial guess:

$$\varphi_{n+1}(x) = \sum_{k \in \mathbb{Z}} a_k \varphi_n(2x - k), \qquad n \geq 0,$$
$$\varphi_0(x) = \chi_{[0,1)}(x).$$

It is possible to show that φ_n converges to a continuous function φ whose support is the interval $[0, 3]$. There is no explicit analytical expression for φ, which is shown in Figure 7.7. As suggested by this figure, the D_4 scaling and wavelet functions φ and ψ are Hölder continuous (see Definition 12.72) but not differentiable.

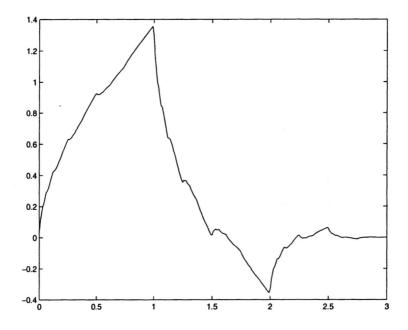

Fig. 7.7   The scaling function φ for the D_4 wavelets.

7.7   References

Beals [3] gives an elegant discussion of Fourier series, Hilbert spaces, and distributions. Rauch [44] discusses Fourier solutions of linear constant coefficient PDEs in more detail. See Whitham [56] for more on dispersive and nondispersive waves. Dym and McKean [10] contains a discussion of the Gibbs phenomenon, a proof of the isoperimetric inequality, and much more besides. Körner [29] is a wide-ranging introduction to the theory and applications of Fourier methods. In particular, it has a discussion of the Monte Carlo integration techniques mentioned in Section 7.5. There are many accounts of wavelets: for example, see Mallet [35]. Some algorithms for the numerical implementation of wavelets are described in [43].

7.8   Exercises

Exercise 7.1 Let φ_n be the function defined in (7.7).

    (a) Prove (7.5).

    (b) Prove that if the set $\mathcal{P}$ of trigonometric polynomials is dense in the space of periodic continuous functions on $\mathbb{T}$ with the uniform norm, then $\mathcal{P}$ is dense in the space of all continuous functions on $\mathbb{T}$ with the L^2-norm.

    (c) Is $\mathcal{P}$ dense in the space of all continuous functions on $[0, 2\pi]$ with the uniform norm?

Exercise 7.2 Suppose that $f : \mathbb{T} \to \mathbb{C}$ is a continuous function, and

$$S_N = \frac{1}{\sqrt{2\pi}} \sum_{n=-N}^{N} \widehat{f}_n e^{inx}$$

is the Nth partial sum of its Fourier series.

(a) Show that $S_N = D_N * f$, where D_N is the *Dirichlet kernel*

$$D_N(x) = \frac{1}{2\pi} \frac{\sin\left[(N+1/2)\,x\right]}{\sin(x/2)}.$$

(b) Let T_N be the mean of the first $N+1$ partial sums,

$$T_N = \frac{1}{N+1} \{S_0 + S_1 + \ldots + S_N\}.$$

Show that $T_N = F_N * f$, where F_N is the *Fejér kernel*

$$F_N(x) = \frac{1}{2\pi(N+1)} \left(\frac{\sin\left[(N+1)x/2\right]}{\sin(x/2)} \right)^2.$$

(c) Which of the families (D_N) and (F_N) are approximate identities as $N \to \infty$? What can you say about the uniform convergence of the partial sums S_N and the averaged partial sums T_N to f?

Exercise 7.3 Prove that the sets $\{e_n \mid n \geq 1\}$ defined by

$$e_n(x) = \sqrt{\frac{2}{\pi}} \sin nx,$$

and $\{f_n \mid n \geq 0\}$ defined by

$$f_0(x) = \sqrt{\frac{1}{\pi}}, \quad f_n(x) = \sqrt{\frac{2}{\pi}} \cos nx \quad \text{for } n \geq 1,$$

are both orthonormal bases of $L^2([0, \pi])$.

Exercise 7.4 Let $T, S \in L^2(\mathbb{T})$ be the triangular and square wave, respectively, defined by

$$T(x) = |x|, \quad \text{if } |x| \leq \pi, \qquad S(x) = \begin{cases} 1 & \text{if } 0 < x < \pi, \\ -1 & \text{if } -\pi < x < 0. \end{cases}$$

(a) Compute the Fourier series of T and S.
(b) Show that $T \in H^1(\mathbb{T})$ and $T' = S$.
(c) Show that $S \notin H^1(\mathbb{T})$.

Exercise 7.5 Consider $f : \mathbb{T}^d \to \mathbb{C}$ defined by

$$f(\mathbf{x}) = \sum_{\mathbf{n} \in \mathbb{Z}^d} a_{\mathbf{n}} e^{i\mathbf{n} \cdot \mathbf{x}},$$

where $\mathbf{x} = (x_1, x_2, \ldots, x_d)$, $\mathbf{n} = (n_1, n_2, \ldots, n_d)$, and $\mathbf{n} \cdot \mathbf{x} = n_1 x_1 + n_2 x_2 + \ldots + n_d x_d$. Prove that if

$$\sum_{\mathbf{n} \in \mathbb{Z}^d} |\mathbf{n}|^{2k} |a_{\mathbf{n}}|^2 < \infty$$

for some $k > d/2$, then f is continuous.

Exercise 7.6 Suppose that $f \in H^1([a, b])$ and $f(a) = f(b) = 0$. Prove the *Poincaré inequality*

$$\int_a^b |f(x)|^2 \, dx \leq \frac{(b - a)^2}{\pi^2} \int_a^b |f'(x)|^2 \, dx.$$

Exercise 7.7 Solve the following initial-boundary value problem for the heat equation,

$$u_t = u_{xx},$$
$$u(0, t) = 0, \quad u(L, t) = 0 \qquad \text{for } t > 0,$$
$$u(x, 0) = f(x) \qquad\qquad\quad \text{for } 0 \leq x \leq L.$$

Exercise 7.8 Find sufficient conditions on the coefficients a_n and b_n in the solution (7.25) of the wave equation so that $u(x, t)$ is a twice continuously differentiable function of x for all $t \in \mathbb{R}$.

Exercise 7.9 Suppose that $u(x, t)$ is a smooth solution of the one-dimensional wave equation,

$$u_{tt} - c^2 u_{xx} = 0.$$

Prove that

$$\left(u_t^2 + c^2 u_x^2 \right)_t - \left(2c^2 u_t u_x \right)_x = 0.$$

If $u(0, t) = u(1, t) = 0$ for all t, deduce that

$$\int_0^1 |u_t(x, t)|^2 + c^2 |u_x(x, t)|^2 \, dx = \text{constant}.$$

Exercise 7.10 Show that

$$u(x, t) = f(x + ct) + g(x - ct)$$

is a solution of the one-dimensional wave equation,

$$u_{tt} - c^2 u_{xx} = 0,$$

for arbitrary functions f and g, This solution is called *d'Alembert's solution.*

Exercise 7.11 Let $\Omega = \{(r, \theta) \mid r < 1\}$ be the unit disc in the plane, where (r, θ) are polar coordinates. The boundary of Ω is the unit circle $\mathbb{T}$. Let $u(r, \theta)$ be a solution of Laplace's equation in Ω,

$$\frac{1}{r}(ru_r)_r + \frac{1}{r^2}u_{\theta\theta} = 0 \qquad r < 1,$$

and define $f, g \in L^2(\mathbb{T})$ by $f(\theta) = u_\theta(1, \theta)$, $g(\theta) = u_r(1, \theta)$. Show that $g = \mathbb{H}f$ where $\mathbb{H}$ is the periodic Hilbert transform, defined in Example 8.32.

Exercise 7.12 Show that there is initial data $f \in C^\infty(\mathbb{T})$ for which the initial value problem for Laplace's equation,

$$u_{tt} + u_{xx} = 0,$$
$$u(x, 0) = f(x),$$
$$u_t(x, 0) = 0,$$

has no solution with $u(\cdot, t) \in L^2(\mathbb{T})$ in any interval $|t| < \delta$, where $\delta > 0$. (The initial-value problem for Laplace's equation is therefore ill-posed.)

Exercise 7.13 Use Fourier series to solve the following initial-boundary value problem for the Schrödinger equation (6.14), that describes a quantum mechanical particle in a box:

$$iu_t = -u_{xx} \tag{7.63}$$
$$u(0, t) = u(1, t) = 0 \quad \text{for all } t, \tag{7.64}$$
$$u(x, 0) = f(x). \tag{7.65}$$

Derive the following two conservation laws from your Fourier series solution and directly from the PDE:

$$\frac{d}{dt}\int_0^1 |u(x, t)|^2 \, dx = 0, \qquad \frac{d}{dt}\int_0^1 |u_x|^2 \, dx = 0.$$

Exercise 7.14 Consider the logistic map

$$x_{n+1} = 4\mu x_n(1 - x_n),$$

where $x_n \in [0, 1]$, and $\mu = 1$. Show that the solutions may be written as $x_n = \sin^2\theta_n$ where $\theta^n \in \mathbb{T}$, and

$$\theta_{n+1} = 2\theta_n.$$

What can you say about the orbits of the logistic map, the existence of an invariant measure, and the validity of an ergodic theorem?

Exercise 7.15 Consider the dynamical system on $\mathbb{T}$,

$$x_{n+1} = \alpha x_n \pmod 1,$$

where $\alpha = (1 + \sqrt{5})/2$ is the golden ratio. Show that the orbit with initial value $x_0 = 1$ is not equidistributed on the circle, meaning that it does not satisfy (7.39). HINT: Show that

$$u_n = \left(\frac{1 + \sqrt{5}}{2}\right)^n + \left(\frac{1 - \sqrt{5}}{2}\right)^n$$

satisfies the difference equation

$$u_{n+1} = u_n + u_{n-1}$$

and hence is an integer for every $n \in \mathbb{N}$. Then use the fact that

$$\left(\frac{1 - \sqrt{5}}{2}\right)^n \to 0 \quad \text{as } n \to \infty.$$

Exercise 7.16 If $\{V_n \mid n \in \mathbb{Z}\}$ is a family of closed subspaces of $L^2(\mathbb{R})$ that satisfies the axioms of multiresolution analysis, and $V_{n+1} = V_n \oplus W_n$, prove that

$$L^2(\mathbb{R}) = \bigoplus_{n \in \mathbb{Z}} W_n.$$

(See Exercise 6.5 for the definition of an infinite direct sum.)

Exercise 7.17 Let B_n and V_n be as defined in (7.46) and (7.47). Prove that $\bigcup_{n=0}^{N} B_n$ is an orthonormal basis of V_N.
HINT. Prove that the set is orthonormal and count its elements.

Exercise 7.18 Suppose that $B = \{e_n(x)\}_{n=1}^{\infty}$ is an orthonormal basis for $L^2([0,1])$. Prove the following:

(a) For any $a \in \mathbb{R}$, the set $B_a = \{e_n(x - a)\}_{n=1}^{\infty}$ is an orthonormal basis for $L^2([a, a + 1])$.

(b) For any $c > 0$, the set $B^c = \{\sqrt{c}\, e_n(cx)\}_{n=1}^{\infty}$ is an orthonormal basis for $L^2([0, c^{-1}])$.

(c) With the convention that functions are extended to a larger domain than their original domain by setting them equal to 0, prove that $B \cup B_1$ is an orthonormal basis for $L^2([0, 2])$.

(d) Prove that $\bigcup_{k \in \mathbb{Z}} B_k$ is an orthonormal basis for $L^2(\mathbb{R})$.

Chapter 8

Bounded Linear Operators on a Hilbert Space

In this chapter we describe some important classes of bounded linear operators on Hilbert spaces, including projections, unitary operators, and self-adjoint operators. We also prove the Riesz representation theorem, which characterizes the bounded linear functionals on a Hilbert space, and discuss weak convergence in Hilbert spaces.

8.1 Orthogonal projections

We begin by describing some algebraic properties of projections. If M and N are subspaces of a linear space X such that every $x \in X$ can be written uniquely as $x = y + z$ with $y \in M$ and $z \in N$, then we say that $X = M \oplus N$ is the *direct sum* of M and N, and we call N a *complementary subspace* of M in X. The decomposition $x = y + z$ with $y \in M$ and $z \in N$ is unique if and only if $M \cap N = \{0\}$. A given subspace M has many complementary subspaces. For example, if $X = \mathbb{R}^3$ and M is a plane through the origin, then any line through the origin that does not lie in M is a complementary subspace. Every complementary subspace of M has the same dimension, and the dimension of a complementary subspace is called the *codimension* of M in X.

If $X = M \oplus N$, then we define the projection $P : X \to X$ of X onto M along N by $Px = y$, where $x = y + z$ with $y \in M$ and $z \in N$. This projection is linear, with $\operatorname{ran} P = M$ and $\ker P = N$, and satisfies $P^2 = P$. As we will show, this property characterizes projections, so we make the following definition.

Definition 8.1 A *projection* on a linear space X is a linear map $P : X \to X$ such that

$$P^2 = P. \qquad (8.1)$$

Any projection is associated with a direct sum decomposition.

Theorem 8.2 Let X be a linear space.

(a) If $P : X \to X$ is a projection, then $X = \operatorname{ran} P \oplus \ker P$.
(b) If $X = M \oplus N$, where M and N are linear subpaces of X, then there is a projection $P : X \to X$ with $\operatorname{ran} P = M$ and $\ker P = N$.

Proof. To prove (a), we first show that $x \in \operatorname{ran} P$ if and only if $x = Px$. If $x = Px$, then clearly $x \in \operatorname{ran} P$. If $x \in \operatorname{ran} P$, then $x = Py$ for some $y \in X$, and since $P^2 = P$, it follows that $Px = P^2y = Py = x$.

If $x \in \operatorname{ran} P \cap \ker P$ then $x = Px$ and $Px = 0$, so $\operatorname{ran} P \cap \ker P = \{0\}$. If $x \in X$, then we have

$$x = Px + (x - Px),$$

where $Px \in \operatorname{ran} P$ and $(x - Px) \in \ker P$, since

$$P(x - Px) = Px - P^2 x = Px - Px = 0.$$

Thus $X = \operatorname{ran} P \oplus \ker P$.

To prove (b), we observe that if $X = M \oplus N$, then $x \in X$ has the unique decomposition $x = y + z$ with $y \in M$ and $z \in N$, and $Px = y$ defines the required projection. □

When using Hilbert spaces, we are particularly interested in orthogonal subspaces. Suppose that M is a closed subspace of a Hilbert space $\mathcal{H}$. Then, by Corollary 6.15, we have $\mathcal{H} = M \oplus M^{\perp}$. We call the projection of $\mathcal{H}$ onto M along $M^{\perp}$ the *orthogonal projection* of $\mathcal{H}$ onto M. If $x = y + z$ and $x' = y' + z'$, where $y, y' \in M$ and $z, z' \in M^{\perp}$, then the orthogonality of M and $M^{\perp}$ implies that

$$\langle Px, x' \rangle = \langle y, y' + z' \rangle = \langle y, y' \rangle = \langle y + z, y' \rangle = \langle x, Px' \rangle. \tag{8.2}$$

This equation states that an orthogonal projection is self-adjoint (see Section 8.4). As we will show, the properties (8.1) and (8.2) characterize orthogonal projections. We therefore make the following definition.

Definition 8.3 An *orthogonal projection* on a Hilbert space $\mathcal{H}$ is a linear map $P : \mathcal{H} \to \mathcal{H}$ that satisfies

$$P^2 = P, \qquad \langle Px, y \rangle = \langle x, Py \rangle \quad \text{for all } x, y \in \mathcal{H}.$$

An orthogonal projection is necessarily bounded.

Proposition 8.4 If P is a nonzero orthogonal projection, then $\|P\| = 1$.

Proof. If $x \in \mathcal{H}$ and $Px \neq 0$, then the use of the Cauchy-Schwarz inequality implies that

$$\|Px\| = \frac{\langle Px, Px \rangle}{\|Px\|} = \frac{\langle x, P^2 x \rangle}{\|Px\|} = \frac{\langle x, Px \rangle}{\|Px\|} \leq \|x\|.$$

Therefore $\|P\| \leq 1$. If $P \neq 0$, then there is an $x \in \mathcal{H}$ with $Px \neq 0$, and $\|P(Px)\| = \|Px\|$, so that $\|P\| \geq 1$. □

There is a one-to-one correspondence between orthogonal projections P and closed subspaces $\mathcal{M}$ of $\mathcal{H}$ such that $\operatorname{ran} P = \mathcal{M}$. The kernel of the orthogonal projection is the orthogonal complement of $\mathcal{M}$.

Theorem 8.5 Let $\mathcal{H}$ be a Hilbert space.

(a) If P is an orthogonal projection on $\mathcal{H}$, then $\operatorname{ran} P$ is closed, and

$$\mathcal{H} = \operatorname{ran} P \oplus \ker P$$

is the orthogonal direct sum of $\operatorname{ran} P$ and $\ker P$.

(b) If $\mathcal{M}$ is a closed subspace of $\mathcal{H}$, then there is an orthogonal projection P on $\mathcal{H}$ with $\operatorname{ran} P = \mathcal{M}$ and $\ker P = \mathcal{M}^\perp$.

Proof. To prove (a), suppose that P is an orthogonal projection on $\mathcal{H}$. Then, by Theorem 8.2, we have $\mathcal{H} = \operatorname{ran} P \oplus \ker P$. If $x = Py \in \operatorname{ran} P$ and $z \in \ker P$, then

$$\langle x, z \rangle = \langle Py, z \rangle = \langle y, Pz \rangle = 0,$$

so $\operatorname{ran} P \perp \ker P$. Hence, we see that $\mathcal{H}$ is the orthogonal direct sum of $\operatorname{ran} P$ and $\ker P$. It follows that $\operatorname{ran} P = (\ker P)^\perp$, so $\operatorname{ran} P$ is closed.

To prove (b), suppose that $\mathcal{M}$ is a closed subspace of $\mathcal{H}$. Then Corollary 6.15 implies that $\mathcal{H} = \mathcal{M} \oplus \mathcal{M}^\perp$. We define a projection $P : \mathcal{H} \to \mathcal{H}$ by

$$Px = y, \qquad \text{where } x = y + z \text{ with } y \in \mathcal{M} \text{ and } z \in \mathcal{M}^\perp.$$

Then $\operatorname{ran} P = \mathcal{M}$, and $\ker P = \mathcal{M}^\perp$. The orthogonality of P was shown in (8.2) above. $\qquad\square$

If P is an orthogonal projection on $\mathcal{H}$, with range $\mathcal{M}$ and associated orthogonal direct sum $\mathcal{H} = \mathcal{M} \oplus \mathcal{N}$, then $I - P$ is the orthogonal projection with range $\mathcal{N}$ and associated orthogonal direct sum $\mathcal{H} = \mathcal{N} \oplus \mathcal{M}$.

Example 8.6 The space $L^2(\mathbb{R})$ is the orthogonal direct sum of the space $\mathcal{M}$ of even functions and the space $\mathcal{N}$ of odd functions. The orthogonal projections P and Q of $\mathcal{H}$ onto $\mathcal{M}$ and $\mathcal{N}$, respectively, are given by

$$Pf(x) = \frac{f(x) + f(-x)}{2}, \qquad Qf(x) = \frac{f(x) - f(-x)}{2}.$$

Note that $I - P = Q$.

Example 8.7 Suppose that A is a measurable subset of $\mathbb{R}$ — for example, an interval — with characteristic function

$$\chi_A(x) = \begin{cases} 1 & \text{if } x \in A, \\ 0 & \text{if } x \notin A. \end{cases}$$

Then

$$P_A f(x) = \chi_A(x) f(x)$$

is an orthogonal projection of $L^2(\mathbb{R})$ onto the subspace of functions with support contained in $\overline{A}$.

A frequently encountered case is that of projections onto a one-dimensional subspace of a Hilbert space $\mathcal{H}$. For any vector $u \in \mathcal{H}$ with $\|u\| = 1$, the map P_u defined by

$$P_u x = \langle u, x \rangle u$$

projects a vector orthogonally onto its component in the direction u. Mathematicians use the tensor product notation $u \otimes u$ to denote this projection. Physicists, on the other hand, often use the "bra-ket" notation introduced by Dirac. In this notation, an element x of a Hilbert space is denoted by a "bra" $\langle x|$ or a "ket" $|x\rangle$, and the inner product of x and y is denoted by $\langle x \mid y \rangle$. The orthogonal projection in the direction u is then denoted by $|u\rangle\langle u|$, so that

$$(|u\rangle\langle u|)\, |x\rangle = \langle u \mid x \rangle |u\rangle.$$

Example 8.8 If $\mathcal{H} = \mathbb{R}^n$, the orthogonal projection $P_{\mathbf{u}}$ in the direction of a unit vector $\mathbf{u}$ has the rank one matrix $\mathbf{u}\mathbf{u}^T$. The component of a vector $\mathbf{x}$ in the direction $\mathbf{u}$ is $P_{\mathbf{u}}\mathbf{x} = (\mathbf{u}^T\mathbf{x})\mathbf{u}$.

Example 8.9 If $\mathcal{H} = l^2(\mathbb{Z})$, and $u = e_n$, where

$$e_n = (\delta_{k,n})_{k=-\infty}^{\infty},$$

and $x = (x_k)$, then $P_{e_n} x = x_n e_n$.

Example 8.10 If $\mathcal{H} = L^2(\mathbb{T})$ is the space of 2π-periodic functions and $u = 1/\sqrt{2\pi}$ is the constant function with norm one, then the orthogonal projection P_u maps a function to its mean: $P_u f = \langle f \rangle$, where

$$\langle f \rangle = \frac{1}{2\pi} \int_0^{2\pi} f(x)\, dx.$$

The corresponding orthogonal decomposition,

$$f(x) = \langle f \rangle + f'(x),$$

decomposes a function into a constant mean part $\langle f \rangle$ and a fluctuating part f' with zero mean.

8.2   The dual of a Hilbert space

A *linear functional* on a complex Hilbert space $\mathcal{H}$ is a linear map from $\mathcal{H}$ to $\mathbb{C}$. A linear functional φ is bounded, or continuous, if there exists a constant M such that

$$|\varphi(x)| \le M\|x\| \quad \text{for all } x \in \mathcal{H}. \tag{8.3}$$

The norm of a bounded linear functional φ is

$$\|\varphi\| = \sup_{\|x\|=1} |\varphi(x)|. \tag{8.4}$$

If $y \in \mathcal{H}$, then

$$\varphi_y(x) = \langle y, x \rangle \tag{8.5}$$

is a bounded linear functional on $\mathcal{H}$, with $\|\varphi_y\| = \|y\|$.

Example 8.11 Suppose that $\mathcal{H} = L^2(\mathbb{T})$. Then, for each $n \in \mathbb{Z}$, the functional $\varphi_n : L^2(\mathbb{T}) \to \mathbb{C}$,

$$\varphi_n(f) = \frac{1}{\sqrt{2\pi}} \int_{\mathbb{T}} f(x) e^{-inx}\, dx,$$

that maps a function to its nth Fourier coefficient is a bounded linear functional. We have $\|\varphi_n\| = 1$ for every $n \in \mathbb{Z}$.

One of the fundamental facts about Hilbert spaces is that all bounded linear functionals are of the form (8.5).

Theorem 8.12 (Riesz representation) If φ is a bounded linear functional on a Hilbert space $\mathcal{H}$, then there is a unique vector $y \in \mathcal{H}$ such that

$$\varphi(x) = \langle y, x \rangle \qquad \text{for all } x \in \mathcal{H}. \tag{8.6}$$

Proof. If $\varphi = 0$, then $y = 0$, so we suppose that $\varphi \neq 0$. In that case, $\ker \varphi$ is a proper closed subspace of $\mathcal{H}$, and Theorem 6.13 implies that there is a nonzero vector $z \in \mathcal{H}$ such that $z \perp \ker \varphi$. We define a linear map $P : \mathcal{H} \to \mathcal{H}$ by

$$Px = \frac{\varphi(x)}{\varphi(z)}\, z.$$

Then $P^2 = P$, so Theorem 8.2 implies that $\mathcal{H} = \operatorname{ran} P \oplus \ker P$. Moreover,

$$\operatorname{ran} P = \{\alpha z \mid \alpha \in \mathbb{C}\}, \qquad \ker P = \ker \varphi,$$

so that $\operatorname{ran} P \perp \ker P$. It follows that P is an orthogonal projection, and

$$\mathcal{H} = \{\alpha z \mid \alpha \in \mathbb{C}\} \oplus \ker \varphi$$

is an orthogonal direct sum. We can therefore write $x \in \mathcal{H}$ as

$$x = \alpha z + n, \qquad \alpha \in \mathbb{C} \text{ and } n \in \ker \varphi.$$

Taking the inner product of this decomposition with z, we get

$$\alpha = \frac{\langle z, x \rangle}{\|z\|^2},$$

and evaluating φ on $x = \alpha z + n$, we find that

$$\varphi(x) = \alpha\varphi(z).$$

The elimination of α from these equations, and a rearrangement of the result, yields

$$\varphi(x) = \langle y, x\rangle,$$

where

$$y = \frac{\overline{\varphi(z)}}{\|z\|^2}\, z.$$

Thus, every bounded linear functional is given by the inner product with a fixed vector.

We have already seen that $\varphi_y(x) = \langle y, x\rangle$ defines a bounded linear functional on $\mathcal{H}$ for every $y \in \mathcal{H}$. To prove that there is a unique y in $\mathcal{H}$ associated with a given linear functional, suppose that $\varphi_{y_1} = \varphi_{y_2}$. Then $\varphi_{y_1}(y) = \varphi_{y_2}(y)$ when $y = y_1 - y_2$, which implies that $\|y_1 - y_2\|^2 = 0$, so $y_1 = y_2$. $\qquad\qquad\square$

The map $J : \mathcal{H} \to \mathcal{H}^*$ given by $Jy = \varphi_y$ therefore identifies a Hilbert space $\mathcal{H}$ with its dual space $\mathcal{H}^*$. The norm of φ_y is equal to the norm of y (see Exercise 8.7), so J is an isometry. In the case of complex Hilbert spaces, J is antilinear, rather than linear, because $\varphi_{\lambda y} = \overline{\lambda}\varphi_y$. Thus, Hilbert spaces are *self-dual*, meaning that $\mathcal{H}$ and $\mathcal{H}^*$ are isomorphic as Banach spaces, and anti-isomorphic as Hilbert spaces. Hilbert spaces are special in this respect. The dual space of an infinite-dimensional Banach space, such as an L^p-space with $p \neq 2$ or $C([a, b])$, is in general not isomorphic to the original space.

Example 8.13 In quantum mechanics, the *observables* of a system are represented by a space $\mathcal{A}$ of linear operators on a Hilbert space $\mathcal{H}$. A *state* ω of a quantum mechanical system is a linear functional ω on the space $\mathcal{A}$ of observables with the following two properties:

$$\omega(A^*A) \geq 0 \quad \text{for all } A \in \mathcal{A}, \tag{8.7}$$
$$\omega(I) = 1. \tag{8.8}$$

The number $\omega(A)$ is the expected value of the observable A when the system is in the state ω. Condition (8.7) is called positivity, and condition (8.8) is called normalization. To be specific, suppose that $\mathcal{H} = \mathbb{C}^n$ and $\mathcal{A}$ is the space of all $n \times n$ complex matrices. Then $\mathcal{A}$ is a Hilbert space with the inner product given by

$$\langle A, B\rangle = \operatorname{tr} A^*B.$$

By the Riesz representation theorem, for each state ω there is a unique $\rho \in \mathcal{A}$ such that

$$\omega(A) = \operatorname{tr} \rho^*A \quad \text{for all } A \in \mathcal{A}.$$

The conditions (8.7) and (8.8) translate into $\rho \geq 0$, and $\operatorname{tr} \rho = 1$, respectively.

Another application of the Riesz representation theorem is given in Section 12.11, where we use it to prove the existence and uniqueness of weak solutions of Laplace's equation.

8.3 The adjoint of an operator

An important consequence of the Riesz representation theorem is the existence of the *adjoint* of a bounded operator on a Hilbert space. The defining property of the adjoint $A^* \in \mathcal{B}(\mathcal{H})$ of an operator $A \in \mathcal{B}(\mathcal{H})$ is that

$$\langle x, Ay \rangle = \langle A^*x, y \rangle \qquad \text{for all } x, y \in \mathcal{H}. \tag{8.9}$$

The uniqueness of A^* follows from Exercise 8.14. The definition implies that

$$(A^*)^* = A, \qquad (AB)^* = B^*A^*.$$

To prove that A^* exists, we have to show that for every $x \in \mathcal{H}$, there is a vector $z \in \mathcal{H}$, depending linearly on x, such that

$$\langle z, y \rangle = \langle x, Ay \rangle \qquad \text{for all } y \in \mathcal{H}. \tag{8.10}$$

For fixed x, the map φ_x defined by

$$\varphi_x(y) = \langle x, Ay \rangle$$

is a bounded linear functional on $\mathcal{H}$, with $\|\varphi_x\| \leq \|A\|\|x\|$. By the Riesz representation theorem, there is a unique $z \in \mathcal{H}$ such that $\varphi_x(y) = \langle z, y \rangle$. This z satisfies (8.10), so we set $A^*x = z$. The linearity of A^* follows from the uniqueness in the Riesz representation theorem and the linearity of the inner product.

Example 8.14 The matrix of the adjoint of a linear map on $\mathbb{R}^n$ with matrix A is A^T, since

$$\mathbf{x} \cdot (A\mathbf{y}) = (A^T\mathbf{x}) \cdot \mathbf{y}.$$

In component notation, we have

$$\sum_{i=1}^{n} x_i \left(\sum_{j=1}^{n} a_{ij} y_j \right) = \sum_{j=1}^{n} \left(\sum_{i=1}^{n} a_{ij} x_i \right) y_j.$$

The matrix of the adjoint of a linear map on $\mathbb{C}^n$ with complex matrix A is the Hermitian conjugate matrix,

$$A^* = \overline{A^T}.$$

Example 8.15 Suppose that S and T are the right and left shift operators on the sequence space $\ell^2(\mathbb{N})$, defined by

$$S(x_1, x_2, x_3, \ldots) = (0, x_1, x_2, x_3, \ldots), \quad T(x_1, x_2, x_3, \ldots) = (x_2, x_3, x_4, \ldots).$$

Then $T = S^*$, since

$$\langle x, Sy \rangle = \overline{x_2} y_1 + \overline{x_3} y_2 + \overline{x_4} y_3 + \ldots = \langle Tx, y \rangle.$$

Example 8.16 Let $K : L^2([0,1]) \to L^2([0,1])$ be an integral operator of the form

$$Kf(x) = \int_0^1 k(x, y) f(y)\, dy,$$

where $k : [0,1] \times [0,1] \to \mathbb{C}$. Then the adjoint operator

$$K^* f(x) = \int_0^1 \overline{k(y, x)} f(y)\, dy$$

is the integral operator with the complex conjugate, transpose kernel.

The adjoint plays a crucial role in studying the solvability of a linear equation

$$Ax = y, \tag{8.11}$$

where $A : \mathcal{H} \to \mathcal{H}$ is a bounded linear operator. Let $z \in \mathcal{H}$ be any solution of the homogeneous adjoint equation,

$$A^* z = 0.$$

We take the inner product of (8.11) with z. The inner product on the left-hand side vanishes because

$$\langle Ax, z \rangle = \langle x, A^* z \rangle = 0.$$

Hence, a necessary condition for a solution x of (8.11) to exist is that $\langle y, z \rangle = 0$ for all $z \in \ker A^*$, meaning that $y \in (\ker A^*)^\perp$. This condition on y is not always sufficient to guarantee the solvability of (8.11); the most we can say for general bounded operators is the following result.

Theorem 8.17 If $A : \mathcal{H} \to \mathcal{H}$ is a bounded linear operator, then

$$\overline{\operatorname{ran} A} = (\ker A^*)^\perp, \qquad \ker A = (\operatorname{ran} A^*)^\perp. \tag{8.12}$$

Proof. If $x \in \operatorname{ran} A$, there is a $y \in \mathcal{H}$ such that $x = Ay$. For any $z \in \ker A^*$, we then have

$$\langle x, z \rangle = \langle Ay, z \rangle = \langle y, A^* z \rangle = 0.$$

This proves that $\operatorname{ran} A \subset (\ker A^*)^\perp$. Since $(\ker A^*)^\perp$ is closed, it follows that $\overline{\operatorname{ran} A} \subset (\ker A^*)^\perp$. On the other hand, if $x \in (\operatorname{ran} A)^\perp$, then for all $y \in \mathcal{H}$ we have

$$0 = \langle Ay, x \rangle = \langle y, A^* x \rangle.$$

Therefore $A^* x = 0$. This means that $(\operatorname{ran} A)^\perp \subset \ker A^*$. By taking the orthogonal complement of this relation, we get

$$(\ker A^*)^\perp \subset (\operatorname{ran} A)^{\perp\perp} = \overline{\operatorname{ran} A},$$

which proves the first part of (8.12). To prove the second part, we apply the first part to A^*, instead of A, use $A^{**} = A$, and take orthogonal complements. $\square$

An equivalent formulation of this theorem is that if A is a bounded linear operator on $\mathcal{H}$, then $\mathcal{H}$ is the orthogonal direct sum

$$\mathcal{H} = \overline{\operatorname{ran} A} \oplus \ker A^*.$$

If A has closed range, then we obtain the following necessary and sufficient condition for the solvability of (8.11).

Theorem 8.18 Suppose that $A : \mathcal{H} \to \mathcal{H}$ is a bounded linear operator on a Hilbert space $\mathcal{H}$ with closed range. Then the equation $Ax = y$ has a solution for x if and only if y is orthogonal to $\ker A^*$.

This theorem provides a useful general method of proving existence from uniqueness: if A has closed range, and the solution of the adjoint problem $A^* x = y$ is unique, then $\ker A^* = \{0\}$, so every y is orthogonal to $\ker A^*$. Hence, a solution of $Ax = y$ exists for every $y \in \mathcal{H}$. The condition that A has closed range is implied by an estimate of the form $c\|x\| \leq \|Ax\|$, as shown in Proposition 5.30.

A commonly occurring dichotomy for the solvability of a linear equation is summarized in the following Fredholm alternative.

Definition 8.19 A bounded linear operator $A : \mathcal{H} \to \mathcal{H}$ on a Hilbert space $\mathcal{H}$ satisfies the *Fredholm alternative* if one of the following two alternatives holds:

(a) either $Ax = 0$, $A^* x = 0$ have only the zero solution, and the equations $Ax = y$, $A^* x = y$ have a unique solution $x \in \mathcal{H}$ for every $y \in \mathcal{H}$;
(b) or $Ax = 0$, $A^* x = 0$ have nontrivial, finite-dimensional solution spaces of the same dimension, $Ax = y$ has a (nonunique) solution if and only if $y \perp z$ for every solution z of $A^* z = 0$, and $A^* x = y$ has a (nonunique) solution if and only if $y \perp z$ for every solution z of $Az = 0$.

Any linear operator $A : \mathbb{C}^n \to \mathbb{C}^n$ on a finite-dimensional space, associated with an $n \times n$ system of linear equations $Ax = y$, satisfies the Fredholm alternative. The ranges of A and A^* are closed because they are finite-dimensional. From linear algebra, the rank of A^* is equal to the rank of A, and therefore the nullity of A is equal to the nullity of A^*. The Fredholm alternative then follows from Theorem 8.18.

Two things can go wrong with the Fredholm alternative in Definition 8.19 for bounded operators A on an infinite-dimensional space. First, $\operatorname{ran} A$ need not be closed; and second, even if $\operatorname{ran} A$ is closed, it is not true, in general, that $\ker A$ and $\ker A^*$ have the same dimension. As a result, the equation $Ax = y$ may be solvable for all $y \in \mathcal{H}$ even though A is not one-to-one, or $Ax = y$ may not be solvable for all $y \in \mathcal{H}$ even though A is one-to-one. We illustrate these possibilities with some examples.

Example 8.20 Consider the multiplication operator $M : L^2([0,1]) \to L^2([0,1])$ defined by

$$Mf(x) = xf(x).$$

Then $M^* = M$, and M is one-to-one, so every $g \in L^2([0,1])$ is orthogonal to $\ker M^*$; but the range of M is a proper dense subspace of $L^2([0,1])$, so $Mf = g$ is not solvable for every $g \in L^2([0,1])$ (see Example 9.5 for more details).

Example 8.21 The range of the right shift operator $S : \ell^2(\mathbb{N}) \to \ell^2(\mathbb{N})$, defined in Example 8.15, is closed since it consists of $y = (y_1, y_2, y_3, \ldots) \in \ell^2(\mathbb{N})$ such that $y_1 = 0$. The left shift operator $T = S^*$ is singular since its kernel is the one-dimensional space with basis $\{(1, 0, 0, \ldots)\}$. The equation $Sx = y$, or

$$(0, x_1, x_2, \ldots) = (y_1, y_2, y_3, \ldots),$$

is solvable if and only if $y_1 = 0$, or $y \perp \ker T$, which verifies Theorem 8.18 in this case. If a solution exists, then it is unique. On the other hand, the equation $Tx = y$ is solvable for every $y \in \ell^2(\mathbb{N})$, even though T is not one-to-one, and the solution is not unique.

These examples motivate the following definition.

Definition 8.22 A bounded linear operator A on a Hilbert space is a *Fredholm operator* if:

(a) $\operatorname{ran} A$ is closed;
(b) $\ker A$ and $\ker A^*$ are finite-dimensional.

The *index* of a Fredholm operator A, $\operatorname{ind} A$, is the integer

$$\operatorname{ind} A = \dim \ker A - \dim \ker A^*.$$

For example, a linear operator on a finite-dimensional Hilbert space and the identity operator on an infinite-dimensional Hilbert space are Fredholm operators with index zero. The right and left shift operators S and T in Example 8.21 are Fredholm, but their indices are nonzero. Since $\dim \ker S = 0$, $\dim \ker T = 1$, and

$S = T^*$, we have $\operatorname{ind} S = -1$ and $\operatorname{ind} T = 1$. The multiplication operator in Example 8.20 is not Fredholm because it does not have closed range.

It is possible to prove that if A is Fredholm and K is compact, then $A + K$ is Fredholm, and $\operatorname{ind}(A + K) = \operatorname{ind} A$. Thus the index of a Fredholm operator is unchanged by compact perturbations. In particular, compact perturbations of the identity are Fredholm operators with index zero, so they satisfy the Fredholm alternative in Definition 8.19. We will prove a special case of this result, for compact, self-adjoint perturbations of the identity, in Theorem 9.26.

8.4 Self-adjoint and unitary operators

Two of the most important classes of operators on a Hilbert space are the classes of self-adjoint and unitary operators. We begin by defining self-adjoint operators.

Definition 8.23 A bounded linear operator $A : \mathcal{H} \to \mathcal{H}$ on a Hilbert space $\mathcal{H}$ is *self-adjoint* if $A^* = A$.

Equivalently, a bounded linear operator A on $\mathcal{H}$ is self-adjoint if and only if

$$\langle x, Ay \rangle = \langle Ax, y \rangle \qquad \text{for all } x, y \in \mathcal{H}.$$

Example 8.24 From Example 8.14, a linear map on $\mathbb{R}^n$ with matrix A is self-adjoint if and only if A is *symmetric*, meaning that $A = A^T$, where A^T is the transpose of A. A linear map on $\mathbb{C}^n$ with matrix A is self-adjoint if and only if A is *Hermitian*, meaning that $A = A^*$.

Example 8.25 From Example 8.16, an integral operator $K : L^2([0,1]) \to L^2([0,1])$,

$$Kf(x) = \int_0^1 k(x,y) f(y) \, dy,$$

is self-adjoint if and only if $k(x,y) = \overline{k(y,x)}$.

Given a linear operator $A : \mathcal{H} \to \mathcal{H}$, we may define a sesquilinear form

$$a : \mathcal{H} \times \mathcal{H} \to \mathbb{C}$$

by $a(x,y) = \langle x, Ay \rangle$. If A is self-adjoint, then this form is *Hermitian symmetric*, or *symmetric*, meaning that

$$a(x,y) = \overline{a(y,x)}.$$

It follows that the associated quadratic form $q(x) = a(x,x)$, or

$$q(x) = \langle x, Ax \rangle, \tag{8.13}$$

is real-valued. We say that A is *nonnegative* if it is self-adjoint and $\langle x, Ax \rangle \geq 0$ for all $x \in \mathcal{H}$. We say that A is *positive*, or *positive definite*, if it is self-adjoint and $\langle x, Ax \rangle > 0$ for every nonzero $x \in \mathcal{H}$. If A is a positive, bounded operator, then

$$(x, y) = \langle x, Ay \rangle$$

defines an inner product on $\mathcal{H}$. If, in addition, there is a constant $c > 0$ such that

$$\langle x, Ax \rangle \geq c\|x\|^2 \qquad \text{for all } x \in \mathcal{H},$$

then we say that A is *bounded from below*, and the norm associated with $(\cdot, \cdot)$ is equivalent to the norm associated with $\langle \cdot, \cdot \rangle$.

The quadratic form associated with a self-adjoint operator determines the norm of the operator.

Lemma 8.26 If A is a bounded self-adjoint operator on a Hilbert space $\mathcal{H}$, then

$$\|A\| = \sup_{\|x\|=1} |\langle x, Ax \rangle|.$$

Proof. Let

$$\alpha = \sup_{\|x\|=1} |\langle x, Ax \rangle|.$$

The inequality $\alpha \leq \|A\|$ is immediate, since

$$|\langle x, Ax \rangle| \leq \|Ax\|\,\|x\| \leq \|A\|\,\|x\|^2.$$

To prove the reverse inequality, we use the definition of the norm,

$$\|A\| = \sup_{\|x\|=1} \|Ax\|.$$

For any $z \in \mathcal{H}$, we have

$$\|z\| = \sup_{\|y\|=1} |\langle y, z \rangle|.$$

It follows that

$$\|A\| = \sup \left\{ |\langle y, Ax \rangle| \mid \|x\| = 1, \|y\| = 1 \right\}. \tag{8.14}$$

The polarization formula (6.5) implies that

$$
\begin{aligned}
\langle y, Ax \rangle = \frac{1}{4} &\{ \langle x + y, A(x+y) \rangle - \langle x - y, A(x - y) \rangle \\
&- i\langle x + iy, A(x + iy) \rangle + i\langle x - iy, A(x - iy) \rangle \}.
\end{aligned}
$$

Since A is self-adjoint, the first two terms are real, and the last two are imaginary. We replace y by $e^{i\varphi}y$, where $\varphi \in \mathbb{R}$ is chosen so that $\langle e^{i\varphi}y, Ax \rangle$ is real. Then the imaginary terms vanish, and we find that

$$
\begin{aligned}
|\langle y, Ax \rangle|^2 &= \frac{1}{16} |\langle x + y, A(x + y) \rangle - \langle x - y, A(x - y) \rangle|^2 \\
&\leq \frac{1}{16} \alpha^2 (\|x + y\|^2 + \|x - y\|^2)^2 \\
&= \frac{1}{4} \alpha^2 (\|x\|^2 + \|y\|^2)^2,
\end{aligned}
$$

where we have used the definition of α and the parallelogram law. Using this result in (8.14), we conclude that $\|A\| \leq \alpha$. $\square$

As a corollary, we have the following result.

Corollary 8.27 If A is a bounded operator on a Hilbert space then $\|A^*A\| = \|A\|^2$. If A is self-adjoint, then $\|A^2\| = \|A\|^2$.

Proof. The definition of $\|A\|$, and an application Lemma 8.26 to the self-adjoint operator A^*A, imply that

$$
\|A\|^2 = \sup_{\|x\|=1} |\langle Ax, Ax \rangle| = \sup_{\|x\|=1} |\langle x, A^*Ax \rangle| = \|A^*A\|.
$$

Hence, if A is self-adjoint, then $\|A\|^2 = \|A^2\|$. $\square$

Next, we define orthogonal or unitary operators, on real or complex spaces, respectively.

Definition 8.28 A linear map $U : \mathcal{H}_1 \to \mathcal{H}_2$ between real or complex Hilbert spaces $\mathcal{H}_1$ and $\mathcal{H}_2$ is said to be *orthogonal* or *unitary*, respectively, if it is invertible and if

$$
\langle Ux, Uy \rangle_{\mathcal{H}_2} = \langle x, y \rangle_{\mathcal{H}_1} \qquad \text{for all } x, y \in \mathcal{H}_1.
$$

Two Hilbert spaces $\mathcal{H}_1$ and $\mathcal{H}_2$ are *isomorphic* as Hilbert spaces if there is a unitary linear map between them.

Thus, a unitary operator is one-to-one and onto, and preserves the inner product. A map $U : \mathcal{H} \to \mathcal{H}$ is unitary if and only if $U^*U = UU^* = I$.

Example 8.29 An $n \times n$ real matrix Q is orthogonal if $Q^T = Q^{-1}$. An $n \times n$ complex matrix U is unitary if $U^* = U^{-1}$.

Example 8.30 If A is a bounded self-adjoint operator, then

$$
e^{iA} = \sum_{n=0}^{\infty} \frac{1}{n!} (iA)^n
$$

is unitary, since

$$\left(e^{iA}\right)^* = e^{-iA} = \left(e^{iA}\right)^{-1}.$$

A bounded operator S is *skew-adjoint* if $S^* = -S$. Any skew-adjoint operator S on a complex Hilbert space may be written as $S = iA$ where A is a self-adjoint operator. The commutator $[A, B] = AB - BA$ is a Lie bracket on the space of bounded, skew-adjoint operators, and we say that this space is the Lie algebra of the Lie group of unitary operators.

Example 8.31 Let $\mathcal{H}$ be a finite dimensional Hilbert space. If $\{e_1, e_2, \ldots, e_n\}$ is an orthonormal basis of $\mathcal{H}$, then $U : \mathbb{C}^n \to \mathcal{H}$ defined by

$$U(z_1, z_2, \ldots, z_n) = z_1 e_1 + z_2 e_2 + \ldots + z_n e_n$$

is unitary. Thus, any n-dimensional, complex Hilbert space is isomorphic to $\mathbb{C}^n$.

Example 8.32 Suppose that $\mathcal{H}_1$ and $\mathcal{H}_2$ are two Hilbert spaces of the same, possibly infinite, dimension. Let $\{u_\alpha\}$ be an orthonormal basis of $\mathcal{H}_1$ and $\{v_\alpha\}$ an orthonormal basis of $\mathcal{H}_2$. Any $x \in \mathcal{H}_1$ can be written uniquely as

$$x = \sum_\alpha c_\alpha u_\alpha,$$

with coefficients $c_\alpha \in \mathbb{C}$. We define $U : \mathcal{H}_1 \to \mathcal{H}_2$ by

$$U\left(\sum_\alpha c_\alpha u_\alpha\right) = \sum_\alpha c_\alpha v_\alpha.$$

Then U is unitary. Thus, Hilbert spaces of the same dimension are isomorphic.

More generally, if $\lambda_\alpha = e^{i\varphi_\alpha}$ are complex numbers with $|\lambda_\alpha| = 1$, then $U : \mathcal{H}_1 \to \mathcal{H}_2$ defined by

$$Ux = \sum_\alpha \lambda_\alpha \langle u_\alpha, x \rangle v_\alpha$$

is unitary. For example, the periodic *Hilbert transform* $\mathbb{H} : L^2(\mathbb{T}) \to L^2(\mathbb{T})$ is defined by

$$\mathbb{H}\left(\sum_{n=-\infty}^{\infty} \hat{f}_n e^{inx}\right) = \sum_{n=-\infty}^{\infty} i\,(\operatorname{sgn} n)\, \hat{f}_n e^{inx},$$

where sgn is the sign function, defined in (5.8). The Hilbert transform is not a unitary mapping on $L^2(\mathbb{T})$ because $\mathbb{H}(1) = 0$; however, Parseval's theorem implies that it is a unitary mapping on the subspace $\mathcal{H}$ of square-integrable periodic functions with zero mean,

$$\mathcal{H} = \left\{f \in L^2(\mathbb{T}) \,\Big|\, \int_{\mathbb{T}} f(x)\,dx = 0\right\}.$$

Example 8.33 The operator $U : L^2(\mathbb{T}) \to \ell^2(\mathbb{Z})$ that maps a function to its Fourier coefficents is unitary. Explicitly, we have

$$Uf = (c_n)_{n \in \mathbb{Z}}, \qquad c_n = \frac{1}{\sqrt{2\pi}} \int_0^{2\pi} f(x) e^{-inx} \, dx.$$

Thus, the Hilbert space of square integrable functions on the circle is isomorphic to the Hilbert space of sequences on $\mathbb{Z}$. As this example illustrates, isomorphic Hilbert spaces may be given concretely in forms that, at first sight, do not appear to be the same.

Example 8.34 For $a \in \mathbb{T}$, we define the translation operator $T_a : L^2(\mathbb{T}) \to L^2(\mathbb{T})$ by

$$(T_a f)(x) = f(x - a).$$

Then T_a is unitary, and

$$T_{a+b} = T_a T_b.$$

We say that $\{T_a \mid a \in \mathbb{T}\}$ is a unitary representation of the additive group $\mathbb{R}/(2\pi\mathbb{Z})$ on the linear space $L^2(\mathbb{T})$.

An operator $T : \mathcal{H} \to \mathcal{H}$ is said to be *normal* if it commutes with its adjoint, meaning that $TT^* = T^*T$. Both self-adjoint and unitary operators are normal. An important feature of normal operators is that they have a nice spectral theory. We will discuss the spectral theory of compact, self-adjoint operators in detail in the next chapter.

8.5 The mean ergodic theorem

Ergodic theorems equate time averages with probabilistic averages, and they are important, for example, in understanding the statistical behavior of deterministic dynamical systems.

The proof of the following ergodic theorem, due to von Neumann, is a good example of Hilbert space methods.

Theorem 8.35 (von Neumann ergodic) Suppose that U is a unitary operator on a Hilbert space $\mathcal{H}$. Let $\mathcal{M} = \{x \in \mathcal{H} \mid Ux = x\}$ be the subspace of vectors that are invariant under U, and P the orthogonal projection onto $\mathcal{M}$. Then, for all $x \in \mathcal{H}$, we have

$$\lim_{N \to \infty} \frac{1}{N+1} \sum_{n=0}^{N} U^n x = Px. \tag{8.15}$$

That is, the averages of U^n converge strongly to P.

Proof.   It is sufficient to prove (8.15) for $x \in \ker P$ and $x \in \operatorname{ran} P$, because then the orthogonal decomposition $\mathcal{H} = \ker P \oplus \operatorname{ran} P$ implies that (8.15) holds for all $x \in \mathcal{H}$. Equation (8.15) is trivial when $x \in \operatorname{ran} P = \mathcal{M}$, since then $Ux = x$ and $Px = x$.

To complete the proof, we show that (8.15) holds when $x \in \ker P$. From the definition of P, we have $\operatorname{ran} P = \ker(I - U)$. If U is unitary, then $Ux = x$ if and only if $U^*x = x$. Hence, using Theorem 8.17, we find that

$$\ker P = \ker(I - U)^{\perp} = \ker(I - U^*)^{\perp} = \overline{\operatorname{ran}(I - U)}.$$

Therefore every $x \in \ker P$ may be approximated by vectors of the form $(I - U)y$. If $x = (I - U)y$, then

$$\frac{1}{N+1}\sum_{n=0}^{N} U^n x \;=\; \frac{1}{N+1}\sum_{n=0}^{N}\left(U^n - U^{n+1}\right)y$$

$$=\; \frac{1}{N+1}\left(y - U^{N+1}y\right)$$

$$\to\; 0 \qquad \text{as } N \to \infty.$$

If $x \in \ker P$, then there is a sequence of elements $x_k = (I - U)y_k$ with $x_k \to x$. Hence,

$$\lim_{N\to\infty}\left\|\frac{1}{N+1}\sum_{n=0}^{N}U^n x\right\| \;\leq\; \limsup_{N\to\infty}\left\|\frac{1}{N+1}\sum_{n=0}^{N}U^n(x - x_k)\right\|$$

$$+ \limsup_{N\to\infty}\left\|\frac{1}{N+1}\sum_{n=0}^{N}U^n x_k\right\|$$

$$\leq\; \|x - x_k\|.$$

Since k is arbitrary and $x_k \to x$, it follows that (8.15) holds for every $x \in \ker P$. $\square$

Next, we explain the implications of this result in probability theory. Suppose that P is a probability measure on a probability space Ω, as described in Section 6.4. A one-to-one, onto, measurable map $T : \Omega \to \Omega$ is said to be *measure preserving* if $P\left(T^{-1}(A)\right) = P(A)$ for all measurable subsets A of Ω. Here,

$$T^{-1}(A) = \{\omega \in \Omega \mid T(\omega) \in A\}.$$

The rotations of the unit circle, studied in Theorem 7.11, are a helpful example to keep in mind here. In that case, $\Omega = \mathbb{T}$, and P is the measure which assigns a probability of $\theta/2\pi$ to an interval on $\mathbb{T}$ of length θ. Any rotation of the circle is a measure preserving map.

If f is a random variable (that is, a measurable real- or complex-valued function on Ω) then the composition of T and f, defined by $f \circ T(\omega) = f(T(\omega))$, is also a random variable. Since T is measure preserving, we have $\mathbb{E}f = \mathbb{E}f \circ T$, or

$$\int_\Omega f\,dP = \int_\Omega f \circ T\,dP.$$

If $f = f \circ T$, then we say that f is invariant under T. This is always true if f is a constant function. If these are the only invariant functions, then we say that T is ergodic.

Definition 8.36 A one-to-one, onto, measure preserving map T on a probability space (Ω, P) is *ergodic* if the only functions $f \in L^2(\Omega, P)$ such that $f = f \circ T$ are the constant functions.

For example, rotations of the circle through an irrational multiple of 2π are ergodic, but rotations through a rational multiple of 2π are not. To make the connection between ergodic maps and Theorem 8.35 above, we define an operator

$$U : L^2(\Omega, P) \to L^2(\Omega, P)$$

on the Hilbert space $L^2(\Omega, P)$ of second-order random variables on Ω by

$$Uf = f \circ T. \tag{8.16}$$

Suppose that $f, g \in L^2(\Omega, P)$. Then, since T is measure preserving, we have

$$\langle Uf, Ug \rangle = \int_\Omega \overline{f(T(\omega))}g(T(\omega))\,dP(\omega) = \int_\Omega \overline{f(\omega)}g(\omega)\,dP(\omega) = \langle f, g \rangle,$$

so the map U is unitary. The subspace of functions invariant under U consists of the functions that are invariant under T. Thus, if T is ergodic, the invariant subspace of U consists of the constant functions, and the orthogonal projection onto the invariant subspace maps a random variable to its expected value. An application of the von Neumann ergodic theorem to the map U defined in (8.16) then gives the following result.

Theorem 8.37 A one-to-one, onto, measure preserving map $T : \Omega \to \Omega$ on a probability space (Ω, P) is ergodic if and only if for every $f \in L^2(\Omega, P)$

$$\lim_{N \to \infty} \frac{1}{N+1} \sum_{n=0}^{N} f \circ T^n = \int_\Omega f\,dP, \tag{8.17}$$

where the convergence is in the L^2-norm.

If we think of $T : \Omega \to \Omega$ as defining a discrete dynamical system $x_{n+1} = Tx_n$ on the state space Ω, as described in Section 3.2, then the left-hand side of (8.17) is the time average of f, while the right-hand side is the probabilistic (or "ensemble") average of f. Thus, the theorem states that time averages and probabilistic averages coincide for ergodic maps.

There is a second ergodic theorem, called the *Birkhoff ergodic theorem*, which states that the averages on the left-hand side of equation (8.17) converge almost surely to the constant on the right-hand side for every f in $L^1(\Omega, P)$.

8.6   Weak convergence in a Hilbert space

A sequence (x_n) in a Hilbert space $\mathcal{H}$ converges *weakly* to $x \in \mathcal{H}$ if

$$\lim_{n \to \infty} \langle x_n, y \rangle = \langle x, y \rangle \qquad \text{for all } y \in \mathcal{H}.$$

Weak convergence is usually written as

$$x_n \rightharpoonup x \qquad \text{as } n \to \infty,$$

to distinguish it from strong, or norm, convergence. From the Riesz representation theorem, this definition of weak convergence for sequences in a Hilbert space is a special case of Definition 5.59 of weak convergence in a Banach space. Strong convergence implies weak convergence, but the converse is not true on infinite-dimensional spaces.

Example 8.38 Suppose that $\mathcal{H} = \ell^2(\mathbb{N})$. Let

$$e_n = (0, 0, \ldots, 0, 1, 0, \ldots)$$

be the standard basis vector whose nth term is 1 and whose other terms are 0. If $y = (y_1, y_2, y_3, \ldots) \in \ell^2(\mathbb{N})$, then

$$\langle e_n, y \rangle = y_n \to 0 \qquad \text{as } n \to \infty,$$

since $\sum |y_n|^2$ converges. Hence $e_n \rightharpoonup 0$ as $n \to \infty$. On the other hand, $\|e_n - e_m\| = \sqrt{2}$ for all $n \neq m$, so the sequence (e_n) does not converge strongly.

It is a nontrivial fact that a weakly convergent sequence is bounded. This is a consequence of the uniform boundedness theorem, or Banach-Steinhaus theorem, which we prove next.

Theorem 8.39 (Uniform boundedness) Suppose that

$$\{\varphi_n : X \to \mathbb{C} \mid n \in \mathbb{N}\}$$

is a set of linear functionals on a Banach space X such that the set of complex numbers $\{\varphi_n(x) \mid n \in \mathbb{N}\}$ is bounded for each $x \in X$. Then $\{\|\varphi_n\| \mid n \in \mathbb{N}\}$ is bounded.

Proof.   First, we show that the functionals are uniformly bounded if they are uniformly bounded on any ball. Suppose that there is a ball

$$B(x_0, r) = \{x \in X \mid \|x - x_0\| < r\},$$

with $r > 0$, and a constant M such that

$$|\varphi_n(x)| \leq M \qquad \text{for all } x \in B(x_0, r) \text{ and all } n \in \mathbb{N}.$$

Then, for any $x \in X$ with $x \neq x_0$, the linearity of φ_n implies that

$$|\varphi_n(x)| \leq \frac{\|x - x_0\|}{r} \left| \varphi_n \left(r \frac{x - x_0}{\|x - x_0\|} \right) \right| + |\varphi_n(x_0)| \leq \frac{M}{r} \|x - x_0\| + |\varphi_n(x_0)|.$$

Hence, if $\|x\| \leq 1$, we have

$$|\varphi_n(x)| \leq \frac{M}{r}(1 + \|x_0\|) + |\varphi_n(x_0)|.$$

Thus, the set of norms $\{\|\varphi_n\| \mid n \in \mathbb{N}\}$ is bounded, because $\{|\varphi_n(x_0)| \mid n \in \mathbb{N}\}$ is bounded.

We now assume for contradiction that $\{\|\varphi_n\|\}$ is unbounded. It follows from what we have just shown that for every open ball $B(x_0, r)$ in X with $r > 0$, the set

$$\{|\varphi_n(x)| \mid x \in B(x_0, r) \text{ and } n \in \mathbb{N}\}$$

is unbounded. We may therefore pick $n_1 \in \mathbb{N}$ and $x_1 \in B(x_0, r)$ such that $|\varphi_{n_1}(x_1)| > 1$. By the continuity of φ_{n_1}, there is an $0 < r_1 < 1$ such that $|\varphi_{n_1}(x)| > 1$ for all $x \in B(x_1, r_1)$. Next, we pick $n_2 > n_1$ and $x_2 \in B(x_1, r_1)$ such that $|\varphi_{n_2}(x_2)| > 2$. We choose a sufficiently small $0 < r_2 < 1/2$ such that $B(x_2, r_2)$ is contained in $B(x_1, r_1)$ and $|\varphi_{n_2}(x)| > 2$ for all $x \in B(x_2, r_2)$. Continuing in this way, we obtain a subsequence (φ_{n_k}) of linear functionals, and a nested sequence of balls $B(x_k, r_k)$ such that $0 < r_k < 1/k$ and

$$|\varphi_{n_k}(x)| > k \qquad \text{for all } x \in B(x_k, r_k).$$

The sequence (x_k) is Cauchy, and hence $x_k \to \bar{x}$ since X is complete. But $\bar{x} \in B(x_k, r_k)$ for all $k \in \mathbb{N}$ so that $|\varphi_{n_k}(\bar{x})| \to \infty$ as $k \to \infty$, which contradicts the pointwise boundedness of $\{\varphi_n(\bar{x})\}$. $\qquad\square$

Thus, the boundedness of the pointwise values of a family of linear functional implies the boundedness of their norms. Next, we prove that a weakly convergent sequence is bounded, and give a useful necessary and sufficient condition for weak convergence.

Theorem 8.40 Suppose that (x_n) is a sequence in a Hilbert space $\mathcal{H}$ and D is a dense subset of $\mathcal{H}$. Then (x_n) converges weakly to x if and only if:

(a) $\|x_n\| \leq M$ for some constant M;
(b) $\langle x_n, y \rangle \to \langle x, y \rangle$ as $n \to \infty$ for all $y \in D$.

Proof. Suppose that (x_n) is a weakly convergent sequence. We define the bounded linear functionals φ_n by $\varphi_n(x) = \langle x_n, x \rangle$. Then $\|\varphi_n\| = \|x_n\|$. Since $(\varphi_n(x))$

converges for each $x \in \mathcal{H}$, it is a bounded sequence, and the uniform bounded-ness theorem implies that $\{\|\varphi_n\|\}$ is bounded. It follows that a weakly convergent sequence satisfies (a). Part (b) is trivial.

Conversely, suppose that (x_n) satisfies (a) and (b). If $z \in \mathcal{H}$, then for any $\epsilon > 0$ there is a $y \in D$ such that $\|z - y\| < \epsilon$, and there is an N such that $|\langle x_n - x, y \rangle| < \epsilon$ for $n \geq N$. Since $\|x_n\| \leq M$, it follows from the Cauchy-Schwarz inequality that for $n \geq N$

$$
\begin{aligned}
|\langle x_n - x, z \rangle| &\leq |\langle x_n - x, y \rangle| + |\langle x_n - x, z - y \rangle| \\
&\leq \epsilon + \|x_n - x\| \|z - y\| \\
&\leq (1 + M + \|x\|)\, \epsilon.
\end{aligned}
$$

Thus, $\langle x_n - x, z \rangle \to 0$ as $n \to \infty$ for every $z \in \mathcal{H}$, so $x_n \rightharpoonup x$. $\square$

Example 8.41 Suppose that $\{e_\alpha\}_{\alpha \in I}$ is an orthonormal basis of a Hilbert space. Then a sequence (x_n) converges weakly to x if and only if it is bounded and its coordinates converge, meaning that $\langle x_n, e_\alpha \rangle \to \langle x, e_\alpha \rangle$ for each $\alpha \in I$.

The boundedness of the sequence is essential to ensure weak convergence, as the following example shows.

Example 8.42 In Example 8.38, we saw that the bounded sequence (e_n) of stan-dard basis elements in $\ell^2(\mathbb{N})$ converges weakly to zero. The unbounded sequence $(n e_n)$, where

$$
n e_n = (0, 0, \dots, 0, n, 0, \dots),
$$

does not converge weakly, however, even though the coordinate sequences with respect to the basis (e_n) converge to zero. For example,

$$
x = \left(n^{-3/4} \right)_{n=1}^{\infty}
$$

belongs to $\ell^2(\mathbb{N})$, but $\langle n e_n, x \rangle = n^{1/4}$ does not converge as $n \to \infty$.

The next example illustrates *oscillation*, *concentration*, and *escape to infinity*, which are typical ways that a weakly convergent sequence of functions fails to con-verge strongly.

Example 8.43 The sequence $(\sin n\pi x)$ converges weakly to zero in $L^2([0, 1])$ be-cause

$$
\int_0^1 f(x) \sin n\pi x \, dx \to 0 \qquad \text{as } n \to \infty
$$

for all $f \in L^2([0, 1])$ (see Example 5.47). The sequence cannot converge strongly to zero since $\| \sin n\pi x \| = 1/\sqrt{2}$ is bounded away from 0. In this case, the functions oscillate more and more rapidly as $n \to \infty$. If a function

$$f(x) = \sum_{n=1}^{\infty} a_n \sin n\pi x$$

in $L^2([0,1])$ is represented by its sequence (a_n) of Fourier sine coefficients, then this example is exactly the same as Example 8.38.

The sequence (f_n) defined by

$$f_n(x) = \begin{cases} \sqrt{n} & \text{if } 0 \le x \le 1/n, \\ 0 & \text{if } 1/n \le x \le 1, \end{cases}$$

converges weakly to zero in $L^2([0,1])$. To prove this fact, we observe that, for any polynomial p,

$$\left| \int_0^1 p(x) f_n(x) \, dx \right| = \sqrt{n} \left| \int_0^{1/n} p(x) \, dx \right|$$

$$\le \frac{1}{\sqrt{n}} \left| n \int_0^{1/n} p(x) \, dx \right|$$

$$\to 0 \qquad \text{as } n \to \infty$$

since, by the continuity of p,

$$n \int_0^{1/n} p(x) \, dx = p(0) + n \int_0^{1/n} \{p(x) - p(0)\} \, dx \to p(0) \qquad \text{as } n \to \infty.$$

Thus, $\langle p, f_n \rangle \to 0$ as $n \to \infty$ for every polynomial p. Since the polynomials are dense in $L^2([0,1])$ and $\|f_n\| = 1$ for all n, Theorem 8.40 implies that $f_n \rightharpoonup 0$. The norms of the f_n are bounded away from 0, so they cannot converge strongly to zero. In this case the functions f_n have a singularity that concentrates at a point.

The sequence (f_n) defined by

$$f_n(x) = \begin{cases} 1 & \text{if } n < x < n+1, \\ 0 & \text{otherwise,} \end{cases}$$

converges weakly, but not strongly, to zero in $L^2(\mathbb{R})$. In this case, the functions f_n escape to infinity. The proof follows from the density of functions with compact support in $L^2(\mathbb{R})$.

As the above examples show, the norm of the limit of a weakly convergent sequence may be strictly less than the norms of the terms in the sequence, corresponding to a loss of "energy" in oscillations, at a singularity, or by escape to infinity in the weak limit. In each case, the expansion of f_n in any orthonormal basis contains coefficients that wander off to infinity. If the norms of a weakly convergent sequence converge to the norm of the weak limit, then the sequence converges strongly.

Proposition 8.44 If (x_n) converges weakly to x, then

$$\|x\| \leq \liminf_{n \to \infty} \|x_n\|. \tag{8.18}$$

If, in addition,

$$\lim_{n \to \infty} \|x_n\| = \|x\|,$$

then (x_n) converges strongly to x.

Proof. Using the weak convergence of (x_n) and the Cauchy-Schwarz inequality, we find that

$$\|x\|^2 = \langle x, x \rangle = \lim_{n \to \infty} \langle x, x_n \rangle \leq \|x\| \liminf_{n \to \infty} \|x_n\|,$$

which proves (8.18). Expansion of the inner product gives

$$\|x_n - x\|^2 = \|x_n\|^2 - \langle x_n, x \rangle - \langle x, x_n \rangle + \|x\|^2.$$

If $x_n \rightharpoonup x$, then $\langle x_n, x \rangle \to \langle x, x \rangle$. Hence, if we also have $\|x_n\| \to \|x\|$, then $\|x_n - x\|^2 \to 0$, meaning that $x_n \to x$ strongly. $\qquad\qquad\square$

One reason for the utility of weak convergence is that it is much easier for sets to be compact in the weak topology than in the strong topology; in fact, a set is weakly precompact if and only if it is bounded. This result provides a partial analog of the Heine-Borel theorem for infinite-dimensional spaces, and is illustrated by the orthonormal sequence of vectors in Example 8.38. The sequence converges weakly, but no subsequence converges strongly, so the terms of the sequence form a weakly precompact, but not a strongly precompact, set.

Theorem 8.45 (Banach-Alaoglu) The closed unit ball of a Hilbert space is weakly compact.

Proof. We will prove the result for a separable Hilbert space. The result remains true for nonseparable spaces, but the proof requires deeper topological arguments and we will not give it here. We will use a diagonal argument to show that any sequence in the unit ball of a separable, infinite-dimensional Hilbert space has a convergent subsequence. Sequential weak compactness implies weak compactness, although this fact is not obvious because the weak topology is not metrizable.

Suppose that (x_n) is a sequence in the unit ball of a Hilbert space $\mathcal{H}$. Let $D = \{y_n \mid n \in \mathbb{N}\}$ be a dense subset of $\mathcal{H}$. Then $(\langle x_n, y_1 \rangle)$ is a bounded sequence in $\mathbb{C}$, since

$$|\langle x_n, y_1 \rangle| \leq \|x_n\| \|y_1\| \leq \|y_1\|.$$

By the Heine-Borel theorem, there is a subsequence of (x_n), which we denote by $(x_{1,k})$, such that $(\langle x_{1,k}, y_1 \rangle)$ converges as $k \to \infty$. In a similar way, there is a subsequence $(x_{2,k})$ of $(x_{1,k})$ such that $(\langle x_{2,k}, y_2 \rangle)$ converges. Continuing in this way, we obtain successive subsequences $(x_{j,k})$ such that $(\langle x_{j,k}, y_i \rangle)$ converges as $k \to \infty$ for each $1 \leq i \leq j$. Taking the diagonal subsequence $(x_{k,k})$ of (x_n), we see that $(\langle x_{k,k}, y \rangle)$ converges as $k \to \infty$ for every $y \in D$. We define the linear functional $\varphi : D \subset \mathcal{H} \to \mathbb{C}$ by

$$\varphi(y) = \lim_{k \to \infty} \langle x_{k,k}, y \rangle.$$

Then $|\varphi(y)| \leq \|y\|$ since $\|x_{k,k}\| \leq 1$, so φ is bounded on D. It therefore has a unique extension to a bounded linear functional on $\mathcal{H}$, and the Riesz representation theorem implies that there is an $x \in \mathcal{H}$ such that $\varphi(y) = \langle x, y \rangle$. It follows from Theorem 8.40 that $x_{k,k} \rightharpoonup x$ as $k \to \infty$. Moreover, from Proposition 8.44,

$$\|x\| \leq \liminf_{k \to \infty} \|x_{k,k}\| \leq 1,$$

so x belongs to the closed unit ball of $\mathcal{H}$. Thus every sequence in the ball has a weakly convergent subsequence whose limit belongs to the ball, so the ball is weakly sequentially compact. □

An important application of Theorem 8.45 is to minimization problems. A function $f : K \to \mathbb{R}$ on a weakly closed set K is said to be weakly sequentially lower semicontinuous, or weakly lower semicontinuous for short, if

$$f(x) \leq \liminf_{n \to \infty} f(x_n)$$

for every sequence (x_n) in K such that $x_n \rightharpoonup x$. For example, from Proposition 8.44, the norm $\| \cdot \|$ is weakly lower semicontinuous.

Theorem 8.46 Suppose that $f : K \to \mathbb{R}$ is a weakly lower semicontinuous function on a weakly closed, bounded subset K of a Hilbert space. Then f is bounded from below and attains its infimum.

The proof of this theorem is exactly the same as the proof of Theorem 1.72. Weak precompactness is a less stringent condition than strong precompactness, but weak closure and weak lower semicontinuity are more stringent conditions than their strong counterparts because there are many more weakly convergent sequences than strongly convergent sequences in an infinite-dimensional space.

A useful sufficient condition that allows one to deduce weak lower semicontinuity, or closure, from strong lower semicontinuity, or closure, is convexity. Convex sets were defined in (1.3). Convex functions are defined as follows.

Definition 8.47 Let $f : C \to \mathbb{R}$ be a real-valued function on a convex subset C of a real or complex linear space. Then f is *convex* if

$$f(tx + (1-t)y) \leq tf(x) + (1-t)f(y)$$

for all $x, y \in C$ and $0 \leq t \leq 1$. If we have strict inequality in this equation whenever $x \neq y$ and $0 < t < 1$, then f is *strictly convex*.

The following result, called *Mazur's theorem*, explains the connection between convexity and weak convergence, and gives additional insight into weak convergence. We say that a vector y, in a real or complex linear space, is a *convex combination* of the vectors $\{x_1, x_2, \ldots, x_n\}$ if there are nonnegative real numbers $\{\lambda_1, \lambda_2, \ldots, \lambda_n\}$ such that

$$y = \sum_{k=1}^{n} \lambda_k x_k, \qquad \sum_{k=1}^{n} \lambda_k = 1.$$

Theorem 8.48 (Mazur) If (x_n) converges weakly to x in a Hilbert space, then there is a sequence (y_n) of finite convex combinations of $\{x_n\}$ such that (y_n) converges strongly to x.

Proof. Replacing x_n by $x_n - x$, we may assume that $x_n \rightharpoonup 0$. We will construct y_n as a mean of almost orthogonal terms of a subsequence of (x_n). We pick $n_1 = 1$, and choose $n_2 > n_1$ such that $\langle x_{n_1}, x_{n_2} \rangle \leq 1$. Given $n_1, \ldots, n_k$, we pick $n_{k+1} > n_k$ such that

$$\left| \langle x_{n_1}, x_{n_{k+1}} \rangle \right| \leq \frac{1}{k}, \ldots, \left| \langle x_{n_k}, x_{n_{k+1}} \rangle \right| \leq \frac{1}{k}. \tag{8.19}$$

This is possible because, by the weak convergence of (x_n), we have $\langle x_{n_i}, x_n \rangle \to 0$ as $n \to \infty$ for $1 \leq i \leq k$. Let

$$y_k = \frac{1}{k} \left(x_{n_1} + x_{n_2} + \ldots + x_{n_k} \right).$$

Then

$$\|y_k\|^2 = \frac{1}{k^2} \sum_{i=1}^{k} \|x_{n_k}\|^2 + \frac{2}{k^2} \operatorname{Re} \sum_{j=1}^{k} \sum_{i=1}^{j-1} \langle x_{n_i}, x_{n_j} \rangle.$$

Since (x_n) converges weakly, it is bounded, and there is a constant M such that $\|x_n\| \leq M$. Using (8.19), we obtain that

$$\|y_k\|^2 \leq \frac{M^2}{k} + \frac{2}{k^2} \sum_{j=1}^{k} \sum_{i=1}^{j-1} \frac{1}{j-1} \leq \frac{M^2 + 2}{k}.$$

Hence, $y_k \to 0$ as $k \to \infty$. $\square$

It follows immediately from this result that a strongly closed, convex set is weakly closed. This need not be true without convexity; for example, the closed unit ball $\{x \in \ell^2(\mathbb{N}) \mid \|x\| \leq 1\}$ is weakly closed, but the closed unit sphere $\{x \in \ell^2(\mathbb{N}) \mid \|x\| = 1\}$ is not. It also follows from Exercise 8.19 that a strongly lower semicontinuous, convex function is weakly lower semicontinuous. We therefore have the following basic result concerning the existence of a minimizer for a convex optimization problem.

Theorem 8.49 Suppose that $f : C \to \mathbb{R}$, is a strongly lower semicontinuous, convex function on a strongly closed, convex, bounded subset C of a Hilbert space. Then f is bounded from below and attains its infimum. If f is strictly convex, then the minimizer is unique.

For example, the norm on a Hilbert space is strictly convex, as well as weakly lower semicontinuous, so it follows that every convex subset of a Hilbert space has a unique point with minimum norm. The existence of a minimizer for a nonconvex variational problem is usually much harder to establish, if one exists at all (see Exercise 8.22).

As in the finite dimensional case (see Exercise 1.25), a similar result holds if $f : \mathcal{H} \to \mathbb{R}$ and f is *coercive*, meaning that

$$\lim_{\|x\| \to \infty} f(x) = \infty.$$

Theorem 8.50 Suppose that $f : \mathcal{H} \to \mathbb{R}$, is a coercive, strongly lower semicontinuous, convex function on a Hilbert space $\mathcal{H}$. Then f is bounded from below and attains its infimum.

Proof. Since f is coercive, there is an $R > 0$ such that

$$f(x) > \inf_{y \in \mathcal{H}} f(y) + 1 \qquad \text{for all } x \in \mathcal{H} \text{ with } \|x\| > R.$$

We may therefore restrict f to the closed, convex ball $\{x \in \mathcal{H} \mid \|x\| \leq R\}$, and apply Theorem 8.49. $\qquad\qquad\square$

The same theorems hold, with the same proofs, when C is a convex subset of a reflexive Banach space. We will use these abstract existence results to obtain a solution of Laplace's equation in Section 13.7.

8.7   References

For more about convex analysis, see Rockafellar [46]. For bounded linear operators in Hilbert spaces see, for example, Kato [26], Lusternik and Sobolev [33], Naylor and Sell [40], and Reed and Simon [45].

8.8   Exercises

Exercise 8.1 If M is a linear subspace of a linear space X, then the *quotient space* X/M is the set $\{x + M \mid x \in X\}$ of affine spaces

$$x + M = \{x + y \mid y \in M\}$$

parallel to M.

(a) Show that X/M is a linear space with respect to the operations

$$\lambda(x + M) = \lambda x + M, \qquad (x + M) + (y + M) = (x + y) + M.$$

(b) Suppose that $X = M \oplus N$. Show that N is linearly isomorphic to X/M.
(c) The *codimension* of M in X is the dimension of X/M. Is a subspace of a Banach space with finite codimension necessarily closed?

Exercise 8.2 If $\mathcal{H} = \mathcal{M} \oplus \mathcal{N}$ is an orthogonal direct sum, show that $\mathcal{M}^\perp = \mathcal{N}$ and $\mathcal{N}^\perp = \mathcal{M}$.

Exercise 8.3 Let $\mathcal{M}, \mathcal{N}$ be closed subspaces of a Hilbert space $\mathcal{H}$ and P, Q the orthogonal projections with $\operatorname{ran} P = \mathcal{M}$, $\operatorname{ran} Q = \mathcal{N}$. Prove that the following conditions are equivalent: (a) $\mathcal{M} \subset \mathcal{N}$; (b) $QP = P$; (c) $PQ = P$; (d) $\|Px\| \leq \|Qx\|$ for all $x \in \mathcal{H}$; (e) $\langle x, Px \rangle \leq \langle x, Qx \rangle$ for all $x \in \mathcal{H}$.

Exercise 8.4 Suppose that (P_n) is a sequence of orthogonal projections on a Hilbert space $\mathcal{H}$ such that

$$\operatorname{ran} P_{n+1} \supset \operatorname{ran} P_n, \qquad \bigcup_{n=1}^{\infty} \operatorname{ran} P_n = \mathcal{H}.$$

Prove that (P_n) converges strongly to the identity operator I as $n \to \infty$. Show that (P_n) does not converge to the identity operator with respect to the operator norm unless $P_n = I$ for all sufficiently large n.

Exercise 8.5 Let $\mathcal{H} = L^2(\mathbb{T}^3; \mathbb{R}^3)$ be the Hilbert space of 2π-periodic, square-integrable, vector-valued functions $\mathbf{u} : \mathbb{T}^3 \to \mathbb{R}^3$, with the inner product

$$\langle \mathbf{u}, \mathbf{v} \rangle = \int_{\mathbb{T}^3} \mathbf{u}(\mathbf{x}) \cdot \mathbf{v}(\mathbf{x}) \, d\mathbf{x}.$$

We define subspaces $\mathcal{V}$ and $\mathcal{W}$ of $\mathcal{H}$ by

$$\begin{aligned}
\mathcal{V} &= \left\{ \mathbf{v} \in C^\infty(\mathbb{T}^3; \mathbb{R}^3) \mid \nabla \cdot \mathbf{v} = 0 \right\}, \\
\mathcal{W} &= \left\{ \mathbf{w} \in C^\infty(\mathbb{T}^3; \mathbb{R}^3) \mid \mathbf{w} = \nabla \varphi \text{ for some } \varphi : \mathbb{T}^3 \to \mathbb{R} \right\}.
\end{aligned}$$

Show that $\mathcal{H} = \mathcal{M} \oplus \mathcal{N}$ is the orthogonal direct sum of $\mathcal{M} = \overline{\mathcal{V}}$ and $\mathcal{N} = \overline{\mathcal{W}}$.

Let P be the orthogonal projection onto $\mathcal{M}$. The velocity $\mathbf{v}(\mathbf{x}, t) \in \mathbb{R}^3$ and pressure $p(\mathbf{x}, t) \in \mathbb{R}$ of an incompressible, viscous fluid satisfy the *Navier-Stokes equations*

$$\begin{aligned}
\mathbf{v}_t + \mathbf{v} \cdot \nabla \mathbf{v} + \nabla p &= \nu \Delta \mathbf{v}, \\
\nabla \cdot \mathbf{v} &= 0.
\end{aligned}$$

Show that the velocity $\mathbf{v}$ satisfies the nonlocal equation

$$\mathbf{v}_t + P\left[\mathbf{v} \cdot \nabla \mathbf{v} \right] = \nu \Delta \mathbf{v}.$$

Exercise 8.6 Show that a linear operator $U : \mathcal{H}_1 \to \mathcal{H}_2$ is unitary if and only if it is an isometric isomorphism of normed linear spaces. Show that an invertible linear map is unitary if and only if its inverse is.

Exercise 8.7 If φ_y is the bounded linear functional defined in (8.5), prove that $\|\varphi_y\| = \|y\|$.

Exercise 8.8 Prove that $\mathcal{H}^*$ is a Hilbert space with the inner product defined by

$$\langle \varphi_x, \varphi_y \rangle_{\mathcal{H}^*} = \langle y, x \rangle_{\mathcal{H}}.$$

Exercise 8.9 Let $A \subset \mathcal{H}$ be such that

$$\mathcal{M} = \{x \in \mathcal{H} \mid x \text{ is a finite linear combination of elements in } A\}$$

is a dense linear subspace of $\mathcal{H}$. Prove that any bounded linear functional on $\mathcal{H}$ is uniquely determined by its values on A. If $\{u_\alpha\}$ is an orthonormal basis, find a necessary and sufficient condition on a family of complex numbers c_α for there to be a bounded linear functional φ such that $\varphi(u_\alpha) = c_\alpha$.

Exercise 8.10 Let $\{u_\alpha\}$ be an orthonormal basis of $\mathcal{H}$. Prove that $\{\varphi_{u_\alpha}\}$ is an orthonormal basis of $\mathcal{H}^*$.

Exercise 8.11 Prove that if $A : \mathcal{H} \to \mathcal{H}$ is a linear map and $\dim \mathcal{H} < \infty$, then

$$\dim \ker A + \dim \operatorname{ran} A = \dim \mathcal{H}.$$

Prove that, if $\dim \mathcal{H} < \infty$, then $\dim \ker A = \dim \ker A^*$. In particular, $\ker A = \{0\}$ if and only if $\ker A^* = \{0\}$.

Exercise 8.12 Suppose that $A : \mathcal{H} \to \mathcal{H}$ is a bounded, self-adjoint linear operator such that there is a constant $c > 0$ with

$$c\|x\| \leq \|Ax\| \qquad \text{for all } x \in \mathcal{H}.$$

Prove that there is a unique solution x of the equation $Ax = y$ for every $y \in \mathcal{H}$.

Exercise 8.13 Prove that an orthogonal set of vectors $\{u_\alpha \mid \alpha \in \mathcal{A}\}$ in a Hilbert space $\mathcal{H}$ is an orthonormal basis if and only if

$$\sum_{\alpha \in \mathcal{A}} u_\alpha \otimes u_\alpha = I.$$

Exercise 8.14 Suppose that $A, B \in \mathcal{B}(\mathcal{H})$ satisfy

$$\langle x, Ay \rangle = \langle x, By \rangle \qquad \text{for all } x, y \in \mathcal{H}.$$

Prove that $A = B$. Use the polarization equation to prove that if

$$\langle x, Ax \rangle = \langle x, Bx \rangle \qquad \text{for all } x \in \mathcal{H},$$

then $A = B$.

Exercise 8.15 Prove that for all $A, B \in \mathcal{B}(\mathcal{H})$, and $\lambda \in \mathbb{C}$, we have: (a) $A^{**} = A$; (b) $(AB)^* = B^* A^*$; (c) $(\lambda A)^* = \bar{\lambda} A^*$; (d) $(A + B)^* = A^* + B^*$; (e) $\|A^*\| = \|A\|$.

Exercise 8.16 Prove that the operator U defined in (8.16) is unitary.

Exercise 8.17 Prove that strong convergence implies weak convergence. Also prove that strong and weak convergence are equivalent in a finite-dimensional Hilbert space.

Exercise 8.18 Let (u_n) be a sequence of orthonormal vectors in a Hilbert space. Prove that $u_n \rightharpoonup 0$ weakly.

Exercise 8.19 Prove that a strongly lower-semicontinuous convex function $f : \mathcal{H} \to \mathbb{R}$ on a Hilbert space $\mathcal{H}$ is weakly lower-semicontinuous.

Exercise 8.20 Let $\mathcal{H}$ be a real Hilbert space, and $\varphi \in \mathcal{H}^*$. Define the quadratic functional $f : \mathcal{H} \to \mathbb{R}$ by

$$f(x) = \frac{1}{2}\|x\|^2 - \varphi(y).$$

Prove that there is a unique element $\bar{x} \in \mathcal{H}$ such that

$$f(\bar{x}) = \inf_{x \in \mathcal{H}} f(x).$$

Exercise 8.21 Show that a function is convex if and only if its epigraph, defined in Exercise 1.24, is a convex set.

Exercise 8.22 Consider the nonconvex functional

$$f : W^{1,4}([0, 1]) \to \mathbb{R},$$

defined by

$$f(u) = \int_0^1 \left\{ u^2 + \left[1 - (u')^2 \right]^2 \right\} dx,$$

where $W^{1,4}([0, 1])$ is the Sobolev space of functions that belong to $L^4([0, 1])$ and whose weak derivatives belong to $L^4([0, 1])$. Show that the infimum of f on $W^{1,4}([0, 1])$ is equal to zero, but that the infimum is not attained.

Chapter 9

The Spectrum of Bounded Linear Operators

In Chapter 7, we used Fourier series to solve various constant coefficient, linear partial differential equations, such as the heat equation. Consider, as an example, the following initial boundary value problem for a variable coefficient, linear equation

$$u_t = u_{xx} - q(x)u \qquad 0 < x < 1, \, t > 0,$$
$$u(0, t) = 0, \quad u(1, t) = 0 \qquad t \geq 0,$$
$$u(x, 0) = f(x) \qquad 0 \leq x \leq 1,$$

where q is a given coefficient function. This equation describes the temperature of a heat conducting bar with a nonuniform heat loss term given by $-q(x)u$. What would it take to express the solution for given initial data f as a series expansion similar to a Fourier series?

If we use separation of variables and look for a solution of the form

$$u(x, t) = \sum_{n=1}^{\infty} a_n(t)u_n(x),$$

where $\{u_n \mid n \in \mathbb{N}\}$ is a basis of $L^2([0, 1])$, then we find that the a_n satisfy

$$\frac{da_n}{dt} = -\lambda_n a_n$$

for some constants λ_n, and the u_n satisfy

$$-\frac{d^2 u_n}{dx^2} + q u_n = \lambda_n u_n.$$

Thus, the u_n should be eigenvectors of the linear operator A defined by

$$Au = -\frac{d^2 u}{dx^2} + qu,$$
$$u(0) = 0, \quad u(1) = 0.$$

215

We therefore want to find a complete set of eigenvectors of A, or, equivalently, to diagonalize A. The problem of diagonalizing a linear map on an infinite-dimensional space arises in many other ways, and is part of what is called spectral theory.

Spectral theory provides a powerful way to understand linear operators by decomposing the space on which they act into invariant subspaces on which their action is simple. In the finite-dimensional case, the spectrum of a linear operator consists of its eigenvalues. The action of the operator on the subspace of eigenvectors with a given eigenvalue is just multiplication by the eigenvalue. As we will see, the spectral theory of bounded linear operators on infinite-dimensional spaces is more involved. For example, an operator may have a *continuous spectrum* in addition to, or instead of, a *point spectrum* of eigenvalues. A particularly simple and important case is that of compact, self-adjoint operators. Compact operators may be approximated by finite-dimensional operators, and their spectral theory is close to that of finite-dimensional operators. We begin with a brief review of the finite-dimensional case.

9.1   Diagonalization of matrices

We consider an $n \times n$ matrix A with complex entries as a bounded linear map $A : \mathbb{C}^n \to \mathbb{C}^n$. A complex number λ is an *eigenvalue* of A if there is a nonzero vector $u \in \mathbb{C}^n$ such that

$$Au = \lambda u. \tag{9.1}$$

A vector $u \in \mathbb{C}^n$ such that (9.1) holds is called an *eigenvector* of A associated with the eigenvalue λ.

The matrix A is *diagonalizable* if there is a basis $\{u_1, \ldots, u_n\}$ of $\mathbb{C}^n$ consisting of eigenvectors of A, meaning that there are eigenvalues $\{\lambda_1, \ldots, \lambda_n\}$ in $\mathbb{C}$, which need not be distinct, such that

$$Au_k = \lambda_k u_k \qquad \text{for } k = 1, \ldots, n. \tag{9.2}$$

The set of eigenvalues of A is called the *spectrum* of A, and is denoted by $\sigma(A)$. The most useful bases of Hilbert spaces are orthonormal bases. A natural question is therefore: When does an $n \times n$ matrix have an orthonormal basis of eigenvectors?

If $\{u_1, \ldots, u_n\}$ is an orthonormal basis of $\mathbb{C}^n$, then the matrix $U = (u_1, \ldots, u_n)$, whose columns are the basis vectors, is a unitary matrix such that

$$Ue_k = u_k, \qquad U^*u_k = e_k,$$

where $\{e_1, \ldots, e_n\}$ is the standard basis of $\mathbb{C}^n$. If the basis vectors $\{u_1, \ldots, u_n\}$ are eigenvectors of A, as in (9.2), then

$$U^*AUe_k = \lambda_k e_k.$$

It follows that U^*AU is a diagonal matrix with the eigenvalues of A on the diagonal, so $A = UDU^*$ where

$$D = \begin{pmatrix} \lambda_1 & 0 & \cdots & 0 \\ 0 & \lambda_2 & \cdots & 0 \\ \vdots & \vdots & \ddots & \vdots \\ 0 & 0 & \cdots & \lambda_n \end{pmatrix}.$$

Conversely, if $A = UDU^*$ with U unitary and D diagonal, then the columns of U form an orthonormal basis of $\mathbb{C}^n$ consisting of eigenvectors of A. Thus, a matrix A can be diagonalized by a unitary matrix if and only if $\mathbb{C}^n$ has an orthonormal basis consisting of eigenvectors of A.

If $A = UDU^*$, then $A^* = U\overline{D}U^*$. Since any two diagonal matrices commute, it follows that A commutes with its Hermitian conjugate A^*:

$$A^*A = U\overline{D}U^*UDU^* = U\overline{D}DU^* = UD\overline{D}U^* = UDU^*U\overline{D}U^* = AA^*.$$

Matrices with this property are called *normal matrices*. For example, Hermitian matrices A, satisfying $A^* = A$, and unitary matrices U, satisfying $U^*U = I$, are normal. We have shown that any matrix with an orthonormal basis of eigenvectors is normal. A standard theorem in linear algebra, which we will not prove here, is that the converse also holds.

Theorem 9.1 An $n \times n$ complex matrix A is normal if and only if $\mathbb{C}^n$ has an orthonormal basis consisting of eigenvectors of A.

A normal matrix N can be written as the product of a unitary matrix V, and a nonnegative, Hermitian matrix A. This follows directly from the diagonal form of N. If $N = UDU^*$ has eigenvalues $\lambda_k = |\lambda_k|e^{i\varphi_k}$, we can write $D = \Phi|D|$, where Φ is a diagonal matrix with entries $e^{i\varphi_k}$ and $|D|$ a diagonal matrix with entries $|\lambda_k|$. Then

$$N = VA, \tag{9.3}$$

where $V = U\Phi U^*$ is unitary, and $A = U|D|U^*$ is nonnegative, meaning that,

$$u^*Au \geq 0 \qquad \text{for all } u \in \mathbb{C}^n.$$

It is straightforward to check that $VA = AV$. Equation (9.3) is called the *polar decomposition* of N. It is a matrix analog of the polar decomposition of a complex number $z = re^{i\theta}$ as the product of a nonnegative number r and a complex number $e^{i\theta}$ on the unit circle. The converse is also true: if $N = VA$, with V unitary, A Hermitian, and $VA = AV$, then N is normal.

The eigenvalues of a matrix A are the roots of the *characteristic polynomial* p_A of A, given by

$$p_A(\lambda) = \det(A - \lambda I).$$

If $p_A(\lambda) = 0$, then $A - \lambda I$ is singular, so $\ker(A - \lambda I) \neq \{0\}$ and there is an associated eigenvector. Since every polynomial has at least one root, it follows that every matrix has at least one eigenvalue, and each distinct eigenvalue has a nonzero subspace of eigenvectors.

It is not true that all matrices have a basis of eigenvectors, because the dimension of the space of eigenvectors associated with a multiple root of the characteristic polynomial may be strictly less than the algebraic multiplicity of the root. We call the dimension of the eigenspace associated with a given eigenvalue the *geometric multiplicity* of the eigenvalue, or the *multiplicity* for short.

Example 9.2 The 2×2 Jordan block

$$A = \begin{pmatrix} \lambda & 1 \\ 0 & \lambda \end{pmatrix}$$

has one eigenvalue λ. The eigenvectors associated with λ are scalar multiples of $u = (1, 0)^T$, so its multiplicity is one, and the eigenspace does not include a basis of $\mathbb{C}^2$. The matrix is not normal, since

$$[A, A^*] = \begin{pmatrix} 1 & 0 \\ 0 & -1 \end{pmatrix}.$$

9.2   The spectrum

A bounded linear operator on an infinite-dimensional Hilbert space need not have any eigenvalues at all, even if it is self-adjoint (see Example 9.5 below). Thus, we cannot hope to find, in general, an orthonormal basis of the space consisting entirely of eigenvectors. It is therefore necessary to define the spectrum of a linear operator on an infinite-dimensional space in a more general way than as the set of eigenvalues. We denote the space of bounded linear operators on a Hilbert space $\mathcal{H}$ by $\mathcal{B}(\mathcal{H})$.

Definition 9.3 The *resolvent set* of an operator $A \in \mathcal{B}(\mathcal{H})$, denoted by $\rho(A)$, is the set of complex numbers λ such that $(A - \lambda I) : \mathcal{H} \to \mathcal{H}$ is one-to-one and onto. The *spectrum* of A, denoted by $\sigma(A)$, is the complement of the resolvent set in $\mathbb{C}$, meaning that $\sigma(A) = \mathbb{C} \setminus \rho(A)$.

If $A - \lambda I$ is one-to-one and onto, then the open mapping theorem implies that $(A - \lambda I)^{-1}$ is bounded. Hence, when $\lambda \in \rho(A)$, both $A - \lambda I$ and $(A - \lambda I)^{-1}$ are one-to-one, onto, bounded linear operators.

As in the finite-dimensional case, a complex number λ is called an eigenvalue of A if there is a nonzero vector $u \in \mathcal{H}$ such that $Au = \lambda u$. In that case, $\ker(A - \lambda I) \neq \{0\}$, so $A - \lambda I$ is not one-to-one, and $\lambda \in \sigma(A)$. This is not the only way, however, that a complex number can belong to the spectrum. We subdivide the spectrum of a bounded linear operator as follows.

Definition 9.4 Suppose that A is a bounded linear operator on a Hilbert space $\mathcal{H}$.

(a) The *point spectrum* of A consists of all $\lambda \in \sigma(A)$ such that $A - \lambda I$ is not one-to-one. In this case λ is called an *eigenvalue* of A.

(b) The *continuous spectrum* of A consists of all $\lambda \in \sigma(A)$ such that $A - \lambda I$ is one-to-one but not onto, and $\operatorname{ran}(A - \lambda I)$ is dense in $\mathcal{H}$.

(c) The *residual spectrum* of A consists of all $\lambda \in \sigma(A)$ such that $A - \lambda I$ is one-to-one but not onto, and $\operatorname{ran}(A - \lambda I)$ is not dense in $\mathcal{H}$.

The following example gives a bounded, self-adjoint operator whose spectrum is purely continuous.

Example 9.5 Let $\mathcal{H} = L^2([0,1])$, and define the multiplication operator $M : \mathcal{H} \to \mathcal{H}$ by

$$Mf(x) = xf(x).$$

Then M is bounded with $\|M\| = 1$. If $Mf = \lambda f$, then $f(x) = 0$ a.e., so $f = 0$ in $L^2([0,1])$. Thus, M has no eigenvalues. If $\lambda \notin [0,1]$, then $(x - \lambda)^{-1}f(x) \in L^2([0,1])$ for any $f \in L^2([0,1])$ because $(x - \lambda)$ is bounded away from zero on $[0,1]$. Thus, $\mathbb{C} \setminus [0,1]$ is in the resolvent set of M. If $\lambda \in [0,1]$, then $M - \lambda I$ is not onto, because $c(x-\lambda)^{-1} \notin L^2([0,1])$ for $c \neq 0$, so the nonzero constant functions c do not belong to the range of $M - \lambda I$. The range of $M - \lambda I$ is dense, however. For any $f \in L^2([0,1])$, let

$$f_n(x) = \begin{cases} f(x) & \text{if } |x - \lambda| \geq 1/n, \\ 0 & \text{if } |x - \lambda| < 1/n. \end{cases}$$

Then $f_n \to f$ in $L^2([0,1])$, and $f_n \in \operatorname{ran}(M - \lambda I)$, since $(x-\lambda)^{-1}f_n(x) \in L^2([0,1])$. It follows that $\sigma(M) = [0,1]$, and that every $\lambda \in [0,1]$ belongs to the continuous spectrum of M. If M acts on the "delta function" supported at λ, which is a distribution (see Chapter 11) with the property that for every continuous function f,

$$f(x)\delta_\lambda(x) = f(\lambda)\delta_\lambda(x),$$

then $M\delta_\lambda = \lambda \delta_\lambda$. Thus, in some sense, there are eigenvectors associated with points in the continuous spectrum of M, but they lie outside the space $L^2([0,1])$ on which M acts.

If λ belongs to the resolvent set $\rho(A)$ of a linear operator A, then $A - \lambda I$ has an everywhere defined, bounded inverse. The operator

$$R_\lambda = (\lambda I - A)^{-1} \tag{9.4}$$

is called the *resolvent* of A at λ, or simply the resolvent of A. The resolvent of A is an operator-valued function defined on the subset $\rho(A)$ of $\mathbb{C}$.

An operator-valued function $F : \Omega \to \mathcal{B}(\mathcal{H})$, defined on an open subset Ω of the complex plane $\mathbb{C}$, is said to be *analytic* at $z_0 \in \Omega$ if there are operators $F_n \in \mathcal{B}(\mathcal{H})$ and a $\delta > 0$ such that

$$F(z) = \sum_{n=0}^{\infty} (z - z_0)^n F_n,$$

where the power series on the right-hand side converges with respect to the operator norm on $\mathcal{B}(\mathcal{H})$ in a disc $|z - z_0| < \delta$ for some $\delta > 0$. We say that F is *analytic* or *holomorphic* in Ω if it is analytic at every point in Ω. This definition is a straightforward generalization of the definition of an analytic complex-valued function $f : \Omega \subset \mathbb{C} \to \mathbb{C}$ as a function with a convergent power series expansion at each point of Ω. The fact that we are dealing with vector-valued, or operator-valued, functions instead of complex-valued functions makes very little difference.

Proposition 9.6 If A is a bounded linear operator on a Hilbert space, then the resolvent set $\rho(A)$ is an open subset of $\mathbb{C}$ that contains the exterior disc $\{\lambda \in \mathbb{C} \mid |\lambda| > \|A\|\}$. The resolvent R_λ is an operator-valued analytic function of λ defined on $\rho(A)$.

Proof. Suppose that $\lambda_0 \in \rho(A)$. Then we may write

$$\lambda I - A = (\lambda_0 I - A) \left[I - (\lambda_0 - \lambda)(\lambda_0 I - A)^{-1} \right].$$

If $|\lambda_0 - \lambda| < \|(\lambda_0 I - A)^{-1}\|^{-1}$, then we can invert the operator on the right-hand side by the Neumann series (see Exercise 5.17). Hence, there is an open disk in the complex plane with center λ_0 that is contained in $\rho(A)$. Moreover, the resolvent R_λ is given by an operator-norm convergent Taylor series in the disc, so it is analytic in $\rho(A)$. If $|\lambda| > \|A\|$, then the Neumann series also shows that $R_\lambda = \lambda (I - A/\lambda)$ is invertible, so $\lambda \in \rho(A)$. $\square$

Since the spectrum $\sigma(A)$ of A is the complement of the resolvent set, it follows that the spectrum is a closed subset of $\mathbb{C}$, and

$$\sigma(A) \subset \{z \in \mathbb{C} \mid |z| \leq \|A\|\}.$$

The *spectral radius* of A, denoted by $r(A)$, is the radius of the smallest disk centered at zero that contains $\sigma(A)$,

$$r(A) = \sup\{|\lambda| \mid \lambda \in \sigma(A)\}.$$

We can refine Proposition 9.6 as follows.

Proposition 9.7 If A is a bounded linear operator, then

$$r(A) = \lim_{n \to \infty} \|A^n\|^{1/n}. \tag{9.5}$$

If A is self-adjoint, then $r(A) = \|A\|$.

Proof. To prove (9.5), we first show that the limit on the right-hand side exists. Let

$$a_n = \log \|A^n\|.$$

We want to show that (a_n/n) converges. Since $\|A^{m+n}\| \le \|A^m\| \|A^n\|$, we have $a_n \le na_1$ and

$$a_{m+n} \le a_m + a_n.$$

We write $n = pm + q$ where $0 \le q < m$. It follows that

$$\frac{a_n}{n} \le \frac{p}{n} a_m + \frac{1}{n} a_q.$$

If $n \to \infty$ with m fixed, then $p/n \to 1/m$, so

$$\limsup_{n \to \infty} \frac{a_n}{n} \le \frac{a_m}{m}.$$

Taking the limit as $m \to \infty$, we obtain that

$$\limsup_{n \to \infty} \frac{a_n}{n} \le \liminf_{m \to \infty} \frac{a_m}{m},$$

which implies that (a_n/n) converges.

Equation (9.5) implies that the Neumann series

$$I + A + A^2 + \ldots + A^n + \ldots$$

converges if $r(A) < 1$ and diverges if $r(A) > 1$: if $r(A) < 1$, then there is an $r(A) < R < 1$ and an N such that $\|A^n\| \le R^n$ for all $n \ge N$; while if $r(A) > 1$, there is an $1 < R < r(A)$ and an N such that $\|A^n\| \ge R^n$ for all $n \ge N$. It follows that $\lambda I - A$ may be inverted by a Neumann series when $|\lambda| > r(A)$, so the spectrum of A is contained inside the disc $\{\lambda \in \mathbb{C} \mid |\lambda| \le r(A)\}$, and that the Neumann series must diverge, so $\lambda I - A$ is not invertible, for some $\lambda \in \mathbb{C}$ with $|\lambda| = r(A)$. For more details, see Reed and Simon [45], for example.

From Corollary 8.27, if A is self-adjoint, then $\|A^2\| = \|A\|^2$. The repeated use of this result implies that $\|A^{2^m}\| = \|A\|^{2^m}$, and hence (9.5), applied to the subsequence with $n = 2^m$, implies that $r(A) = \|A\|$. $\qquad\square$

Although the spectral radius of a self-adjoint operator is equal to its norm, the spectral radius does not provide a norm on the space of all bounded operators. In particular, $r(A) = 0$ does not imply that $A = 0$, as Exercise 5.13 illustrates. If $r(A) = 0$, then A is called a *nilpotent operator*.

Proposition 9.8 The spectrum of a bounded operator on a Hilbert space is nonempty.

Proof. Suppose that $A \in \mathcal{B}(\mathcal{H})$. Then the resolvent $R_\lambda = (\lambda I - A)^{-1}$ is an analytic function on $\rho(A)$. Therefore, for every $x, y \in \mathcal{H}$, the function $f : \rho(A) \to \mathbb{C}$ defined by

$$f(\lambda) = \langle x, R_\lambda y \rangle$$

is analytic in $\rho(A)$, and $\lim_{\lambda \to \infty} f(\lambda) = 0$. Suppose, for contradiction, that $\sigma(A)$ is empty. Then f is a bounded entire function, and Liouville's Theorem implies that $f : \mathbb{C} \to \mathbb{C}$ is a constant function, so $f = 0$. But if $f = 0$ for every $x, y \in \mathcal{H}$, then $R_\lambda = 0$ for all $\lambda \in \mathbb{C}$, which is impossible. Hence, $\sigma(A)$ is not empty. $\qquad \square$

The spectrum of a bounded operator may, however, consist of a single point (see Exercise 9.7 for an example).

9.3 The spectral theorem for compact, self-adjoint operators

In this section, we analyze the spectrum of a compact, self-adjoint operator. The spectrum consists entirely of eigenvalues, with the possible exception of zero, which may belong to the continuous spectrum. We begin by proving some basic properties of the spectrum of a bounded, self-adjoint operator.

Lemma 9.9 *The eigenvalues of a bounded, self-adjoint operator are real, and eigenvectors associated with different eigenvalues are orthogonal.*

Proof. If $A : \mathcal{H} \to \mathcal{H}$ is self-adjoint, and $Ax = \lambda x$ with $x \neq 0$, then

$$\lambda \langle x, x \rangle = \langle x, Ax \rangle = \langle Ax, x \rangle = \overline{\lambda} \langle x, x \rangle,$$

so $\lambda = \overline{\lambda}$, and $\lambda \in \mathbb{R}$. If $Ax = \lambda x$ and $Ay = \mu y$, where λ and μ are real, then

$$\lambda \langle x, y \rangle = \langle Ax, y \rangle = \langle x, Ay \rangle = \mu \langle x, y \rangle.$$

It follows that if $\lambda \neq \mu$, then $\langle x, y \rangle = 0$ and $x \perp y$. $\qquad \square$

As we will see in the next chapter, self-adjoint operators are a rich source of orthonormal bases.

A linear subspace $\mathcal{M}$ of $\mathcal{H}$ is called an *invariant subspace* of a linear operator A on $\mathcal{H}$ if $Ax \in \mathcal{M}$ for all $x \in \mathcal{M}$. In that case, the restriction $A|_{\mathcal{M}}$ of A to $\mathcal{M}$ is a linear operator on $\mathcal{M}$. Suppose that $\mathcal{H} = \mathcal{M} \oplus \mathcal{N}$ is a direct sum of invariant subspaces $\mathcal{M}$ and $\mathcal{N}$ of A. Then every $x \in \mathcal{H}$ may be written as $x = y + z$, with $y \in \mathcal{M}$ and $z \in \mathcal{N}$, and

$$Ax = A|_{\mathcal{M}} y + A|_{\mathcal{N}} z. \tag{9.6}$$

Thus, the action of A on $\mathcal{H}$ is determined by its actions on the invariant subspaces.

Example 9.10 Consider matrices acting on $\mathbb{C}^d = \mathbb{C}^m \oplus \mathbb{C}^n$ where $d = m + n$. A $d \times d$ matrix A leaves $\mathbb{C}^m$ invariant if it has the block form

$$A = \begin{pmatrix} B & D \\ 0 & C \end{pmatrix},$$

where B is an $m \times m$ matrix, D is $m \times n$, and C is $n \times n$. The matrix A leaves both $\mathbb{C}^m$ and the complementary space $\mathbb{C}^n$ invariant if $D = 0$.

An invariant subspace of a nondiagonalizable operator may have no complementary invariant subspace. However, the orthogonal complement of an invariant subspace of a self-adjoint operator is also invariant, as we prove in the following lemma. Thus, we can decompose the action of a self-adjoint operator on a linear space into actions on smaller orthogonal invariant subspaces.

Lemma 9.11 If A is a bounded, self-adjoint operator on a Hilbert space $\mathcal{H}$ and $\mathcal{M}$ is an invariant subspace of A, then $\mathcal{M}^\perp$ is an invariant subspace of A.

Proof. If $x \in \mathcal{M}^\perp$ and $y \in \mathcal{M}$, then

$$\langle y, Ax \rangle = \langle Ay, x \rangle = 0$$

because $A = A^*$ and $Ay \in \mathcal{M}$. Therefore, $Ax \in \mathcal{M}^\perp$. $\qquad\square$

Next, we show that the whole spectrum — not just the point spectrum — of a bounded, self-adjoint operator is real, and that the residual spectrum is empty. We begin with a preliminary proposition.

Proposition 9.12 If λ belongs to the residual spectrum of a bounded operator A on a Hilbert space, then $\overline{\lambda}$ is an eigenvalue of A^*.

Proof. If λ belongs to the residual spectrum of $A \in \mathcal{B}(\mathcal{H})$, then $\mathrm{ran}\,(A - \lambda I)$ is not dense in $\mathcal{H}$. By Theorem 6.13, there is a nonzero vector $x \in \mathcal{H}$ such that $x \perp \mathrm{ran}\,(A - \lambda I)$. Theorem 8.17 then implies that $x \in \ker\left(A^* - \overline{\lambda}I\right)$. $\qquad\square$

Lemma 9.13 If A is a bounded, self-adjoint operator on a Hilbert space, then the spectrum of A is real and is contained in the interval $[-\|A\|, \|A\|]$.

Proof. We have shown that $r(A) \leq \|A\|$, so we only have to prove that the spectrum is real. Suppose that $\lambda = a + ib \in \mathbb{C}$, where with $a, b \in \mathbb{R}$ and $b \neq 0$. For any $x \in \mathcal{H}$, we have

$$
\begin{aligned}
\|(A - \lambda I)x\|^2 &= \langle (A - \lambda I)x, (A - \lambda I)x \rangle \\
&= \langle (A - aI)x, (A - aI)x \rangle + \langle (-ib)x, (-ib)x \rangle \\
&\quad + \langle Ax, (-ib)x \rangle + \langle (-ib)x, Ax \rangle \\
&= \|(A - aI)x\|^2 + b^2 \|x\|^2 \\
&\geq b^2 \|x\|^2.
\end{aligned}
$$

It follows from this estimate and Proposition 5.30 that $A - \lambda I$ is one-to-one and has closed range. If $\operatorname{ran}(A - \lambda I) \neq \mathcal{H}$, then λ belongs to the residual spectrum of A, and, by Proposition 9.12, $\overline{\lambda} = a - ib$ is an eigenvalue of A. Thus A has an eigenvalue that does not belong to $\mathbb{R}$, which contradicts Lemma 9.9. It follows that $\lambda \in \rho(A)$ if λ is not real. $\square$

Corollary 9.14 The residual spectrum of a bounded, self-adjoint operator is empty.

Proof. From Lemma 9.13, the point spectrum and the residual spectrum are disjoint subsets of $\mathbb{R}$, so the result follows immediately from Proposition 9.12. $\square$

Bounded linear operators on an infinite-dimensional Hilbert space do not always behave like operators on a finite-dimensional space. We have seen in Example 9.5 that a bounded, self-adjoint operator may have no eigenvalues, while the identity operator on an infinite-dimensional Hilbert space has a nonzero eigenvalue of infinite multiplicity. The properties of compact operators are much closer to those of operators on finite-dimensional spaces, and we will study their spectral theory next.

Proposition 9.15 A nonzero eigenvalue of a compact, self-adjoint operator has finite multiplicity. A countably infinite set of nonzero eigenvalues has zero as an accumulation point, and no other accumulation points.

Proof. Suppose, for contradiction, that λ is a nonzero eigenvalue with infinite multiplicity. Then there is a sequence (e_n) of orthonormal eigenvectors. This sequence is bounded, but (Ae_n) does not have a convergent subsequence because $Ae_n = \lambda e_n$, which contradicts the compactness of A.

If A has a countably infinite set $\{\lambda_n\}$ of nonzero eigenvalues, then, since the eigenvalues are bounded by $\|A\|$, there is a convergent subsequence (λ_{n_k}). If $\lambda_{n_k} \to \lambda$ and $\lambda \neq 0$, then the orthogonal sequence of eigenvectors (f_{n_k}), where $f_{n_k} = \lambda_{n_k}^{-1} e_{n_k}$ and $\|e_{n_k}\| = 1$, would be bounded; but (Af_{n_k}) has no convergent subsequence since $Af_{n_k} = e_{n_k}$. $\square$

To motivate the statement of the spectral theorem for compact, self-adjoint operators, suppose that $x \in \mathcal{H}$ is given by

$$x = \sum_k c_k e_k + z, \tag{9.7}$$

where $\{e_k\}$ is an orthonormal set of eigenvectors of A with corresponding nonzero eigenvalues $\{\lambda_k\}$, $z \in \ker A$, and $c_k \in \mathbb{C}$. Then $Ax = \sum_k \lambda_k c_k e_k$. Let P_k denote the one-dimensional orthogonal projection onto the subspace spanned by e_k,

$$P_k x = \langle e_k, x \rangle e_k. \tag{9.8}$$

From Lemma 9.9, we have $z \perp e_k$, so $c_k = \langle e_k, x \rangle$ and

$$Ax = \sum_k \lambda_k P_k x. \qquad (9.9)$$

If λ_k has finite multiplicity $m_k > 1$, meaning that the dimension of the associated eigenspace $E_k \subset \mathcal{H}$ is greater than one, then we may combine the one-dimensional projections associated with the same eigenvalues. In doing so, we may represent A by a sum of the same form as (9.9) in which the λ_k are distinct, and the P_k are orthogonal projections onto the eigenspaces E_k.

The spectral theorem for compact, self-adjoint operators states that any $x \in \mathcal{H}$ can be expanded in the form (9.7), and that A can be expressed as a sum of orthogonal projections, as in (9.9).

Theorem 9.16 (Spectral theorem for compact, self-adjoint operators) Let $A : \mathcal{H} \to \mathcal{H}$ be a compact, self-adjoint operator on a Hilbert space $\mathcal{H}$. There is an orthonormal basis of $\mathcal{H}$ consisting of eigenvectors of A. The nonzero eigenvalues of A form a finite or countably infinite set $\{\lambda_k\}$ of real numbers, and

$$A = \sum_k \lambda_k P_k, \qquad (9.10)$$

where P_k is the orthogonal projection onto the finite-dimensional eigenspace of eigenvectors with eigenvalue λ_k. If the number of nonzero eigenvalues is countably infinite, then the series in (9.10) converges to A in the operator norm.

Proof. First we prove that if A is compact and self-adjoint, then $\lambda = \|A\|$ or $\lambda = -\|A\|$ (or both) is an eigenvalue of A. This is the crucial part of the proof. We use a variational argument to obtain an eigenvector.

There is nothing to prove if $A = 0$, so we suppose that $A \neq 0$. From Lemma 8.26, we have

$$\|A\| = \sup_{\|x\|=1} |\langle x, Ax \rangle|.$$

Hence, there is a sequence (x_n) in $\mathcal{H}$ with $\|x_n\| = 1$ such that

$$\|A\| = \lim_{n \to \infty} |\langle x_n, Ax_n \rangle|.$$

Since A is self-adjoint, $\langle x_n, Ax_n \rangle$ is real for all n, so there is a subsequence of (x_n), which we still denote by (x_n), such that

$$\lim_{n \to \infty} \langle x_n, Ax_n \rangle = \lambda, \qquad (9.11)$$

where $\lambda = \|A\|$ or $\lambda = -\|A\|$.

The sequence (x_n) consists of unit vectors, so it is bounded. The compactness of A implies that there is a subsequence, which we still denote by (x_n), such that (Ax_n) converges. We let $y = \lim_{n \to \infty} Ax_n$. We claim that y is an eigenvector of

A with eigenvalue λ. First, $y \neq 0$, since otherwise (9.11) would imply that $\lambda = 0$, which is not the case since $|\lambda| = \|A\|$ and $A \neq 0$. The fact that y is an eigenvector follows from the following computation:

$$
\begin{aligned}
\|(A - \lambda I)y\|^2 &= \lim_{n \to \infty} \|(A - \lambda I)Ax_n\|^2 \\
&\leq \|A\|^2 \lim_{n \to \infty} \|(A - \lambda I)x_n\|^2 \\
&= \|A\|^2 \lim_{n \to \infty} \left[\|Ax_n\|^2 + \lambda^2 \|x_n\|^2 - 2\lambda \langle x_n, Ax_n \rangle\right] \\
&\leq \|A\|^2 \lim_{n \to \infty} \left[\|A\|^2 \|x_n\|^2 + \lambda^2 \|x_n\|^2 - 2\lambda \langle x_n, Ax_n \rangle\right] \\
&= \|A\|^2 \left[\lambda^2 + \lambda^2 - 2\lambda^2\right] \\
&= 0.
\end{aligned}
$$

To complete the proof, we use an orthogonal decomposition of $\mathcal{H}$ into invariant subspaces to apply the result we have just proved to smaller and smaller subspaces of $\mathcal{H}$. We let $\mathcal{N}_1 = \mathcal{H}$ and $A_1 = A$. There is a normalized eigenvector of A_1, which we denote by e_1, with eigenvalue λ_1, where $|\lambda_1| = \|A_1\|$. Let $\mathcal{M}_2$ be the one-dimensional subspace of $\mathcal{N}_1$ generated by e_1. Then $\mathcal{M}_2$ is an invariant subspace of A_1. We decompose $\mathcal{N}_1 = \mathcal{M}_2 \oplus \mathcal{N}_2$, where $\mathcal{N}_2 = \mathcal{M}_2^\perp$. Lemma 9.11 implies that $\mathcal{N}_2$ is an invariant subspace of A_1. We denote the restriction of A_1 to $\mathcal{N}_2$ by A_2. Then A_2 is the difference of two compact operators, so it is compact by Proposition 5.43. We also have that $\|A_2\| \leq \|A_1\|$, since $\mathcal{N}_2 \subset \mathcal{N}_1$.

An application of the same argument to A_2 implies that A_2 has a normalized eigenvector e_2 with eigenvalue λ_2, where

$$|\lambda_2| = \|A_2\| \leq \|A_1\| = |\lambda_1|.$$

Moreover, $e_2 \perp e_1$. Repeating this procedure, we define A_n inductively to be the restriction of A to $\mathcal{N}_n = \mathcal{M}_n^\perp$, where $\mathcal{M}_n$ is the space spanned by $\{e_1, \dots e_{n-1}\}$, and e_n to be an eigenvector of A_n with eigenvalue λ_n. By construction, $|\lambda_n| = \|A_n\|$ and $(|\lambda_n|)$ is a nonincreasing sequence.

If $A_{n+1} = 0$ for some n, then A has only finitely many nonzero eigenvalues, and it is given by a finite sum of the form (9.10). If $\dim \mathcal{H} > n$, then the orthonormal set $\{e_k \mid k = 1, \dots n\}$ can be extended to an orthonormal basis of $\mathcal{H}$. All other basis vectors are eigenvectors of A with eigenvalue zero.

If $A_n \neq 0$ for every $n \in \mathbb{N}$, then we obtain an infinite sequence of nonzero eigenvalues and eigenvectors. From Proposition 9.15, the eigenvalues have finite multiplicities and

$$\lim_{n \to \infty} \lambda_n = 0. \tag{9.12}$$

For any $n \in \mathbb{N}$, we have

$$A = \sum_{k=1}^{n} \lambda_k P_k + A_{n+1},$$

where A_{n+1} is zero on the subspace spanned by $\{e_1, \ldots, e_n\}$, and $\|A_{n+1}\| = |\lambda_{n+1}|$. By (9.12), we have

$$\lim_{n \to \infty} \left\| A - \sum_{k=1}^{n} \lambda_k \langle e_k, \cdot \rangle e_k \right\| = \lim_{n \to \infty} |\lambda_{n+1}| = 0,$$

meaning that

$$A = \sum_{k=1}^{\infty} \lambda_k P_k,$$

where the sum converges in the operator norm.

If A has an infinite sequence of nonzero eigenvalues, then the range of A is

$$\operatorname{ran} A = \left\{ \sum_{k=1}^{\infty} c_k e_k \;\Big|\; \sum_{k=1}^{\infty} \frac{|c_k|^2}{|\lambda_k|^2} < \infty \right\}.$$

The range is not closed since $\lambda_n \to 0$ as $n \to \infty$. The closure of the range, $\mathcal{M} = \overline{\operatorname{ran} A}$, is the closed linear span of the set of eigenvectors $\{e_n \mid n \in \mathbb{N}\}$ with nonzero eigenvalues,

$$\mathcal{M} = \left\{ \sum_{k=1}^{\infty} c_k e_k \;\Big|\; \sum_{k=1}^{\infty} |c_k|^2 < \infty \right\}.$$

If $x \in \mathcal{M}^\perp$, then $Ax = A_n x$ for all $n \in \mathbb{N}$, so that

$$\|Ax\| \le \|A_n\| \, \|x\| \to 0 \qquad \text{as } n \to \infty.$$

Therefore, $\mathcal{M}^\perp = \ker A$ consists of eigenvectors of A with eigenvalue zero, and we can extend $\{e_n \mid n \in \mathbb{N}\}$ to an orthonormal basis of $\mathcal{H}$ consisting of eigenvectors by adding an orthonormal basis of $\ker A$. $\qquad\square$

A similar spectral theorem holds for compact, normal operators, which have orthogonal eigenvectors but possibly complex eigenvalues. A generalization also holds for bounded, self-adjoint or normal operators. In that case, however, the sum in (9.10) must be replaced by an integral of orthogonal projections with respect to an appropriate spectral measure that accounts for the possibility of a continuous spectrum. We will not discuss such generalizations in this book. Non-normal matrices on finite-dimensional linear spaces may be reduced to a Jordan canonical form, but the spectral theory of non-normal operators on infinite-dimensional spaces is more complicated.

We will discuss unbounded linear operators in the next chapter. The above theory may be used to study an unbounded operator whose inverse is compact or, more generally, an unbounded operator whose resolvent is compact, meaning that $(\lambda I - A)^{-1}$ is compact for some $\lambda \in \rho(A)$. For example, a regular Sturm-Liouville differential operator has a compact, self-adjoint resolvent, which explains why it has a complete orthonormal set of eigenvectors.

9.4   Compact operators

Before we can apply the spectral theorem for compact, self-adjoint operators, we
have to check that an operator is compact. In this section, we discuss some ways
to do this, and also give examples of compact operators.

The most direct way to prove that an operator A is compact is to verify the
definition by showing that if E is a bounded subset of $\mathcal{H}$, then the set $A(E) = \{Ax \mid x \in E\}$ is precompact. In many examples, this can be done by using an appropriate
condition for compactness, such as the Arzelà-Ascoli theorem or Rellich's theorem.
The following theorem characterizes precompact sets in a general, separable Hilbert
space.

Theorem 9.17 Let E be a subset of an infinite-dimensional, separable Hilbert
space $\mathcal{H}$.

(a) If E is precompact, then for every orthonormal set $\{e_n \mid n \in \mathbb{N}\}$ and every
$\epsilon > 0$, there is an N such that

$$\sum_{n=N+1}^{\infty} |\langle e_n, x \rangle|^2 < \epsilon \qquad \text{for all } x \in E. \tag{9.13}$$

(b) If E is bounded and there is an orthonormal basis $\{e_n\}$ of $\mathcal{H}$ with the
property that for every $\epsilon > 0$ there is an N such that (9.13) holds, then E
is precompact.

Proof.   First, we prove (a). A precompact set is bounded, so it is sufficient to
show that if E is bounded and (9.13) does not hold, then E is not precompact. If
(9.13) does not hold, then there is an $\epsilon > 0$ such that for each N there is an $x_N \in E$
with

$$\sum_{n=N+1}^{\infty} |\langle e_n, x_N \rangle|^2 \geq \epsilon. \tag{9.14}$$

We construct a subsequence of (x_N) as follows. Let $N_1 = 1$, and pick N_2 such that

$$\sum_{n=N_2+1}^{\infty} |\langle e_n, x_{N_1} \rangle|^2 \leq \frac{\epsilon}{4}.$$

Given N_k, pick N_{k+1} such that

$$\sum_{n=N_{k+1}+1}^{\infty} |\langle e_n, x_{N_k} \rangle|^2 \leq \frac{\epsilon}{4}. \tag{9.15}$$

We can always do this because the sum

$$\sum_{n=1}^{\infty} |\langle e_n, x_N \rangle|^2 = \|x_N\|^2$$

converges by Parseval's identity.

For any $N \geq 1$, we define the orthogonal projection P_N by

$$P_N x = \sum_{n=1}^{N} \langle e_n, x \rangle e_n.$$

For $k > l$, we have

$$
\begin{aligned}
\|x_{N_k} - x_{N_l}\|^2 &\geq \|(I - P_{N_k})(x_{N_k} - x_{N_l})\|^2 \\
&\geq \left[\|(I - P_{N_k}) x_{N_k}\| - \|(I - P_{N_k}) x_{N_l}\| \right]^2 \\
&= \left[\sqrt{\sum_{n=N_k+1}^{\infty} |\langle e_n, x_{N_k} \rangle|^2} - \sqrt{\sum_{n=N_k+1}^{\infty} |\langle e_n, x_{N_l} \rangle|^2} \right]^2 \\
&\geq \left[\epsilon^{1/2} - (\epsilon/4)^{1/2} \right]^2 \\
&\geq \frac{\epsilon}{4},
\end{aligned}
$$

where we have used (9.14), (9.15), and the fact that $N_k \geq N_{l+1}$. It follows that (x_{N_k}) does not have any convergent subsequences, contradicting the hypothesis that E is precompact.

To prove part (b), suppose that $\{e_n \mid n \in \mathbb{N}\}$ is an orthonormal basis with the stated property, and let (x_n) be any sequence in E. We will use a diagonalization argument to construct a convergent subsequence, thus proving that E is precompact. Without loss of generality we may assume that $\|x\| \leq 1$ for all $x \in E$. We choose $n_1 = 1$. Then

$$\|(I - P_{n_1}) x_n\| \leq 1 \qquad \text{for all } n \in \mathbb{N}.$$

Since $P_{n_1} x_n$ is in the finite-dimensional Hilbert subspace spanned by $e_1, \ldots, e_{n_1}$ for each $n \in \mathbb{N}$, there is subsequence $(x_{1,k})$ of (x_n) such that $P_{n_1} x_{1,k}$ converges. Therefore, we can pick the subsequence such that (see Exercise 1.18)

$$\|P_{n_1}(x_{1,k} - x_{1,l})\|^2 \leq \frac{1}{k} \qquad \text{for } k \leq l.$$

Next, we choose n_2 such that

$$\|(I - P_{n_2}) x_{1,k}\|^2 \leq \frac{1}{2} \qquad \text{for all } k \in \mathbb{N}.$$

This is possible because of (9.13). We then pick a subsequence $x_{2,k}$ of $x_{1,k}$ such that $(P_{n_2} x_{2,k})$ is Cauchy and

$$\|P_{n_2}(x_{2,k} - x_{2,l})\|^2 \leq \frac{1}{k} \qquad \text{for all } k \leq l.$$

Continuing in this way, we choose n_l such that

$$\|(I - P_{n_l})\, x_{l-1,k}\|^2 \leq \frac{1}{l} \qquad \text{for all } k \in \mathbb{N},$$

and then pick a subsequence $(x_{l,k})$ of $(x_{l-1,k})$ such that $(P_{n_l} x_{l,k})$ satisfies

$$\|P_{n_l}\, (x_{l,k} - x_{l,j})\|^2 \leq \frac{1}{k} \qquad \text{for all } k \leq j.$$

The diagonal sequence $(x_{k,k})$ is Cauchy, since

$$\|x_{m,m} - x_{n,n}\|^2 = \|P_k(x_{m,m} - x_{n,n})\|^2 + \|(I - P_k)\, x_{m,m} - x_{n,n}\|^2 \leq \frac{2}{k}$$

for all $m, n \geq k$.   □

Example 9.18 Let $\mathcal{H} = \ell^2(\mathbb{N})$. The *Hilbert cube*

$$C = \{(x_1, x_2, \ldots, x_n, \ldots) \mid |x_n| \leq 1/n\}$$

is closed and precompact. Hence C is a compact subset of $\mathcal{H}$.

Example 9.19 The diagonal operator $A : \ell^2(\mathbb{N}) \to \ell^2(\mathbb{N})$ defined by

$$A(x_1, x_2, x_3, \ldots, x_n, \ldots) = (\lambda_1 x_1, \lambda_2 x_2, \ldots, \lambda_n x_n, \ldots), \qquad (9.16)$$

where $\lambda_n \in \mathbb{C}$ is compact if and only if $\lambda_n \to 0$ as $n \to \infty$. Any compact, normal operator on a separable Hilbert space is unitarily equivalent to such a diagonal operator.

Proposition 5.43 implies that the uniform limit of compact operators is compact. An operator with finite rank is compact. Therefore, another way to prove that A is compact is to show that A is the limit of a uniformly convergent sequence of finite-rank operators. One such class of compact operators is the class of Hilbert-Schmidt operators.

Definition 9.20 A bounded linear operator A on a separable Hilbert space $\mathcal{H}$ is *Hilbert-Schmidt* if there is an orthonormal basis $\{e_n \mid n \in \mathbb{N}\}$ such that

$$\sum_{n=1}^{\infty} \|Ae_n\|^2 < \infty. \qquad (9.17)$$

If A is a Hilbert-Schmidt operator, then

$$\|A\|_{HS} = \sqrt{\sum_{n=1}^{\infty} \|Ae_n\|^2} \qquad (9.18)$$

is called the *Hilbert-Schmidt norm* of A.

One can show that the sum in (9.17) is finite in every orthonormal basis if it is finite in one orthonormal basis, and the norm (9.18) does not depend on the choice of basis.

Theorem 9.21 A Hilbert-Schmidt operator is compact.

Proof. Suppose that A is Hilbert-Schmidt and $\{e_n \mid n \in \mathbb{N}\}$ is an orthonormal basis. If P_N is the orthogonal projection onto the finite-dimensional space spanned by $\{e_1, \ldots e_N\}$, then $P_N A$ is a finite-rank operator, and one can check that $P_N A \to A$ uniformly as $N \to \infty$. □

Example 9.22 The diagonal operator $A : \ell^2(\mathbb{N}) \to \ell^2(\mathbb{N})$ defined in (9.16) is Hilbert-Schmidt if and only if

$$\sum_{n=1}^{\infty} |\lambda_n|^2 < \infty.$$

We say that A is a *trace class operator* if

$$\sum_{n=1}^{\infty} |\lambda_n| < \infty.$$

A trace class operator is Hilbert-Schmidt, and a Hilbert-Schmidt operator is compact.

Example 9.23 Let $\Omega \subset \mathbb{R}^n$. One can show that an integral operator K on $L^2(\Omega)$,

$$Kf(x) = \int_\Omega k(x,y)f(y)\,dy, \tag{9.19}$$

is Hilbert-Schmidt if and only if $k \in L^2(\Omega \times \Omega)$, meaning that

$$\int_{\Omega \times \Omega} |k(x,y)|^2 \, dxdy < \infty.$$

The Hilbert-Schmidt norm of K is

$$\|K\|_{HS} = \left(\int_{\Omega \times \Omega} |k(x,y)|^2 \, dxdy \right)^{1/2}.$$

If K is a self-adjoint, Hilbert-Schmidt operator then there is an orthonormal basis $\{\varphi_n \mid n \in \mathbb{N}\}$ of $L^2(\Omega)$ consisting of eigenvectors of K, such that

$$\int_\Omega k(x,y)\varphi_n(y)\,dy = \lambda_n \varphi_n(x).$$

Then

$$k(x,y) = \sum_{n=1}^{\infty} \lambda_n \varphi_n(x)\varphi_n(y),$$

where the series converges in $L^2(\Omega \times \Omega)$:

$$\lim_{N \to \infty} \int_{\Omega \times \Omega} \left| k(x,y) - \sum_{n=1}^{N} \lambda_n \varphi_n(x) \varphi_n(y) \right|^2 \, dx dy = 0.$$

For a proof, see, for example, Hochstadt [22].

Another way to characterize compact operators on a Hilbert space is in terms of weak convergence.

Theorem 9.24 A bounded linear operator on a Hilbert space is compact if and only if it maps weakly convergent sequences into strongly convergent sequences.

Proof. First, we show that a bounded operator $A : \mathcal{H} \to \mathcal{H}$ on a Hilbert space $\mathcal{H}$ maps weakly convergent sequences into weakly convergent sequences. If $x_n \rightharpoonup x$ as $n \to \infty$, then for every $z \in \mathcal{H}$ we have

$$\langle Ax_n - Ax, z \rangle = \langle x_n - x, A^*z \rangle \to 0 \qquad \text{as } n \to \infty.$$

Therefore, $Ax_n \rightharpoonup Ax$ as $n \to \infty$. Now suppose that A is compact, and $x_n \rightharpoonup x$. Since a weakly convergent sequence is bounded, the sequence (Ax_n) is contained in a compact subset of $\mathcal{H}$. Moreover, each strongly convergent subsequence is weakly convergent, so it converges to the same limit, namely Ax. It follows that the whole sequence converges strongly to Ax (see Exercise 1.27).

Conversely, suppose that A maps weakly convergent sequences into strongly convergent sequences, and E is a bounded set in $\mathcal{H}$. If (y_n) is a sequence in $A(E)$, then there is a sequence (x_n) in E such that $y_n = Ax_n$. By Theorem 8.45, the sequence (x_n) has a weakly convergent subsequence (x_{n_k}). The operator A maps this into a strongly convergent subsequence (y_{n_k}) of (y_n). Thus $A(E)$ is compact for any bounded set E, so A is compact. $\qquad \Box$

9.5   Functions of operators

The theory of functions of operators is called *functional calculus*. In this section, we describe some basic ideas of functional calculus in the special case of compact, self-adjoint operators.

If $q : \mathbb{C} \to \mathbb{C}$ is a polynomial function of degree d,

$$q(x) = \sum_{k=0}^{d} c_k x^k,$$

with coefficients $c_k \in \mathbb{C}$, then we define an operator-valued polynomial function

$q : \mathcal{B}(\mathcal{H}) \to \mathcal{B}(\mathcal{H})$ in the obvious way as

$$q(A) = \sum_{k=0}^{d} c_k A^k. \tag{9.20}$$

There are several ways to define more general functions of a linear operator than polynomials. We have already seen that if A is a bounded operator and the function $f : \mathbb{C} \to \mathbb{C}$ is analytic at zero, with a Taylor series whose radius of convergence is strictly greater than $\|A\|$, then we may define $f(A)$ by a norm-convergent power series. For example e^A is defined for any bounded operator A, and $(I - A)^{-1}$ is defined in this way for any operator A with $r(A) < 1$.

An alternative approach is to use spectral theory to define a continuous function of a self-adjoint operator. First suppose that

$$A = \sum_{n=1}^{N} \lambda_n P_n$$

is a finite linear combination of orthogonal projections P_n with orthogonal ranges, and q is the polynomial function defined in (9.20). Since $\{P_n\}$ is an orthogonal family of projections, we have

$$P_n^k = P_n, \quad P_n P_m = 0 \qquad \text{for } n \neq m, k \geq 1.$$

It follows that $A^k = \sum_{n=1}^{N} \lambda_n^k P_n$ and

$$q(A) = \sum_{n=1}^{N} q(\lambda_n) P_n.$$

If A is a compact, self-adjoint operator with the spectral representation

$$A = \sum_{n=1}^{\infty} \lambda_n P_n, \tag{9.21}$$

then one can check that (see Exercise 9.18)

$$q(A) = \sum_{n=0}^{\infty} q(\lambda_n) P_n. \tag{9.22}$$

If $f : \sigma(A) \to \mathbb{C}$ is a continuous function, then a natural generalization of the expression in (9.22) for $q(A)$ is

$$f(A) = \sum_{n=1}^{\infty} f(\lambda_n) P_n. \tag{9.23}$$

This series converges strongly for any continuous f, and uniformly if in addition $f(0) = 0$ (see Exercise 9.19). An equivalent way to define $f(A)$ is to choose a sequence (q_n) of polynomials that converges uniformly to f on $\sigma(A)$, and define

$f(A)$ as the uniform limit of $q_n(A)$ (see Exercise 9.18). If f is real-valued, then the operator $f(A)$ is self-adjoint, and if f is complex-valued, then $f(A)$ is normal.

As a consequence of the spectral representation of A in (9.21) and $f(A)$ in (9.23), we have the following result.

Theorem 9.25 (Spectral mapping) If A is a compact, self-adjoint operator on a Hilbert space and $f : \sigma(A) \to \mathbb{C}$ is continuous, then

$$\sigma\left(f(A)\right) = f\left(\sigma(A)\right).$$

Here, $\sigma\left(f(A)\right)$ is the spectrum of $f(A)$, and $f\left(\sigma(A)\right)$ is the image of the spectrum of A under f,

$$f\left(\sigma(A)\right) = \{\mu \in \mathbb{C} \mid \mu = f(\lambda) \text{ for some } \lambda \in \sigma(A)\}.$$

A result of this kind is called a spectral mapping theorem. A spectral mapping theorem holds for bounded operators on a Hilbert space, and many unbounded operators, but there exist nonnormal, unbounded operators for which it is false (see Exercise 10.19). Thus, in general, unlike the finite-dimensional case, a knowledge of the spectrum of an unbounded operator is not sufficient to determine the spectrum of a function of the operator, and some knowledge of the operator's structure is also required.

Consider a linear evolution equation that can be written in the form

$$x_t = Ax, \qquad x(0) = x_0, \tag{9.24}$$

where A is a compact, self-adjoint linear operator on a Hilbert space $\mathcal{H}$. The solution is

$$x(t) = e^{At}x_0.$$

If x_0 is an eigenvector of A with eigenvalue λ, then $e^{At}x_0 = e^{\lambda t}x_0$. The solution decays exponentially if $\operatorname{Re}\lambda < 0$, and grows exponentially if $\operatorname{Re}\lambda > 0$. From the spectral mapping theorem, if the spectrum $\sigma(A)$ is contained in a left-half plane $\{\lambda \in \mathbb{C} \mid \operatorname{Re}\lambda \leq \omega\}$, then the spectrum of e^{At} is contained in the disc $\{\lambda \in \mathbb{C} \mid |\lambda| \leq e^{\omega t}\}$.

If $\{e_n \mid n \in \mathbb{N}\}$ is an orthonormal basis of eigenvectors of A, then we may write the solution as

$$x(t) = \sum_{n=1}^{\infty} e^{\lambda_n t}\langle e_n, x_0\rangle e_n.$$

If $\lambda_n \leq \omega$ for all n, it follows that

$$\|x(t)\| = \sqrt{\sum_{n=1}^{\infty} |e^{\lambda_n t}\langle e_n, x_0\rangle|^2} \leq e^{\omega t}\sqrt{\sum_{n=1}^{\infty} |\langle e_n, x_0\rangle|^2} = e^{\omega t}\|x_0\|.$$

When $\omega < 0$, any solution decays exponentially to 0 as $t \to \infty$. In that case, we say that the equilibrium solution $x(t) = 0$ is *globally asymptotically stable*.

9.6 Perturbation of eigenvalues

Suppose that $A(\epsilon)$ is a family of operators on a Hilbert space that depends on a real or complex parameter ϵ. If we know the spectrum of $A(0)$, then we can use perturbation theory to obtain information about the spectrum of $A(\epsilon)$ for small ϵ. In this section, we consider the simplest case, when $A(\epsilon)$ is a compact, self-adjoint operator depending on a real parameter ϵ.

Before doing this, we prove a preliminary result of independent interest: the Fredholm alternative for a compact, self-adjoint perturbation of the identity.

Theorem 9.26 Suppose that A is a compact, self-adjoint operator on a Hilbert space and $\lambda \in \mathbb{C}$ is nonzero. Then the equation

$$(A - \lambda I) x = y \tag{9.25}$$

has a solution if and only if $y \perp z$ for every solution z of the homogeneous equation

$$(A - \lambda I) z = 0.$$

The solution space of the homogeneous equation is finite-dimensional.

Proof. If $A : \mathcal{H} \to \mathcal{H}$ is compact and self-adjoint, then there is an orthonormal basis $\{e_n \mid n \in \mathbb{N}\}$ of $\mathcal{H}$ consisting of eigenvectors of A, with $Ae_n = \lambda_n e_n$ for $\lambda_n \in \mathbb{R}$. We expand x and y with respect to this basis as

$$x = \sum_{n=1}^{\infty} x_n e_n, \qquad y = \sum_{n=1}^{\infty} y_n e_n,$$

where $x_n = \langle e_n, x \rangle$ and $y_n = \langle e_n, y \rangle$. With respect to this basis, equation (9.25) has the diagonalized form

$$(\lambda_n - \lambda) x_n = y_n \qquad \text{for } n \in \mathbb{N}. \tag{9.26}$$

If $\lambda_n \neq \lambda$ for all n, then $\lambda_n - \lambda$ is bounded away from zero, since $\lambda \neq 0$ and there are no nonzero accumulation points of the eigenvalues of a compact operator. Hence, equation (9.25) is uniquely solvable for every $y \in \mathcal{H}$, with the solution

$$x = \sum_{n=1}^{\infty} \frac{\langle e_n, y \rangle}{\lambda_n - \lambda} e_n.$$

If $\lambda = \lambda_n$ for some n, then there is a finite-dimensional subspace of eigenvectors with the nonzero eigenvalue λ. Suppose the corresponding eigenvectors are

$\{e_{n_1}, e_{n_2}, \ldots, e_{n_k}\}$. Then we can solve (9.26) if and only if

$$y_{n_1} = y_{n_2} = \ldots = y_{n_k} = 0,$$

meaning that y is orthogonal to the kernel of $(A - \lambda I)$.   □

Suppose that $A(\epsilon)$ is a compact, self-adjoint operator depending on a parameter $\epsilon \in \mathbb{R}$. We assume that A is a real-analytic function of ϵ at $\epsilon = 0$, meaning that it has a Taylor series expansion

$$A(\epsilon) = A_0 + \epsilon A_1 + \epsilon^2 A_2 + O(\epsilon^3)$$

that converges with respect to the operator norm in some interval $|\epsilon| < R$. The coefficient operators A_n are given by

$$A_0 = A(0), \quad A_1 = \dot{A}(\epsilon)\Big|_{\epsilon=0}, \quad A_2 = \frac{1}{2}\ddot{A}(\epsilon)\Big|_{\epsilon=0}, \ldots, \quad A_n = \frac{1}{n!}\frac{d^n}{d\epsilon^n}A(\epsilon)\Big|_{\epsilon=0}, \ldots$$

where the dot denotes a derivative with respect to ϵ.

We look for eigenvalues $\lambda(\epsilon)$ and eigenvectors $x(\epsilon)$ of $A(\epsilon)$ that satisfy

$$[A(\epsilon) - \lambda(\epsilon)]\, x(\epsilon) = 0. \tag{9.27}$$

It can be shown that the eigenvalues and eigenvectors of $A(\epsilon)$ have convergent Taylor series expansions

$$\lambda(\epsilon) = \lambda_0 + \epsilon\lambda_1 + \epsilon^2\lambda_2 + O(\epsilon^3), \tag{9.28}$$
$$x(\epsilon) = x_0 + \epsilon x_1 + \epsilon^2 x_2 + O(\epsilon^3), \tag{9.29}$$

where $\lambda_0 = \lambda(0)$, $\lambda_1 = \dot{\lambda}(0)$, and so on. We will not prove the convergence of these series here, but we will show how to compute the coefficients.

Setting $\epsilon = 0$ in (9.27), we obtain that

$$(A_0 - \lambda_0 I)\, x_0 = 0. \tag{9.30}$$

Thus, λ_0 is a nonzero eigenvalue of A_0 and x_0 is an eigenvector. For definiteness, we assume that λ_0 is a *simple eigenvalue* of A_0, meaning that it has multiplicity one, although eigenvalues of higher multiplicity can be treated in a similar way.

Differentiation of (9.27) with respect to ϵ implies that

$$(A - \lambda I)\,\dot{x} + \left(\dot{A} - \dot{\lambda}I\right)x = 0.$$

Setting ϵ equal to zero in this equation, we obtain that

$$(A_0 - \lambda_0 I)\, x_1 = \lambda_1 x_0 - A_1 x_0. \tag{9.31}$$

Since $\{x_0\}$ is a basis of $\ker(A_0 - \lambda_0 I)$, Theorem 9.26 implies that this equation is solvable for x_1 if and only if the right-hand side is orthogonal to x_0. It follows that

$$\lambda_1 = \frac{\langle x_0, A_1 x_0 \rangle}{\|x_0\|^2}.$$

Continuing in this way, we differentiate equation (9.27) n times with respect to ϵ and set ϵ equal to zero in the result, which gives an equation of the form

$$(A_0 - \lambda_0 I)\, x_n = \lambda_n x_0 + f_{n-1}\,(x_0,\ldots,x_{n-1},\lambda_0,\ldots,\lambda_{n-1})\,. \tag{9.32}$$

The Fredholm alternative implies that the right-hand side must be orthogonal to x_0. This condition determines λ_n, and we can then solve the equation for x_n. Thus, we can successively determine the coefficients in the expansions of $\lambda(\epsilon)$ and $x(\epsilon)$.

The solution of (9.32) for x_n includes an arbitrary multiple of x_0. This nonuniqueness is a consequence of the arbitrariness in the normalization of the eigenvector. If $c(\epsilon) = 1 + \epsilon c_1 + \epsilon^2 c_2 + O(\epsilon^3)$ is a scalar and $x(\epsilon) = x_0 + \epsilon x_1 + \epsilon^2 x_2 + O(\epsilon^3)$ is an eigenvector, then

$$c(\epsilon)x(\epsilon) = x_0 + \epsilon\,(x_1 + c_1 x_0) + \epsilon^2\,(x_2 + c_1 x_1 + c_2 x_0) + O(\epsilon^3)$$

is also an eigenvector. Each term in the expansion contains an arbitrary multiple of x_0.

An alternative way to derive the perturbation equations (9.30)–(9.32) is to use the Taylor series (9.28)–(9.29) in (9.27), expand, and equate coefficients of powers of ϵ in the result.

Example 9.27 Consider the eigenvalue problem

$$-u'' + V(x,\epsilon)u = \lambda u, \tag{9.33}$$

where $u \in L^2(\mathbb{R})$ and

$$V(x,\epsilon) = x^2 + \epsilon(x^4 - 4x^2)e^{-2x^2}. \tag{9.34}$$

In quantum mechanics, this problem corresponds to the determination of the energy levels of a slightly anharmonic oscillator. See Figure 9.1 for a graph of the potential for four different values of ϵ. For definiteness, we consider the energy of the ground state only, that is, the smallest eigenvalue, although the perturbation of other eigenvalues can be computed in exactly the same way.

The eigenvalue problem is of the form $Au = \lambda u$, where $A = A_0 + \epsilon A_1$ with

$$A = -\frac{d^2}{dx^2} + V, \quad A_0 = -\frac{d^2}{dx^2} + x^2, \quad A_1 = (x^4 - 4x^2)e^{-2x^2}.$$

From Exercise 6.14, the unperturbed ground state u_0 and the associated eigenvalue λ_0 of A_0 are given by

$$u_0(x) = e^{-x^2/2}, \quad \lambda_0 = 1.$$

The operator A is unbounded. We will assume that the perturbed operator has a ground state close to that of the unperturbed operator, and apply the above expansion without discussing the validity of the method in this case. See the book by Kato [26] for a comprehensive discussion.

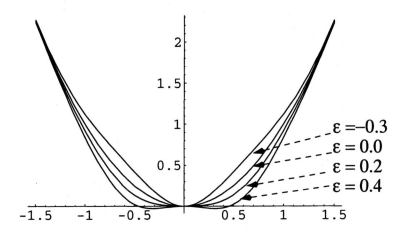

Fig. 9.1   The perturbed harmonic potential $V(x, \epsilon)$, defined in (9.34), for ϵ=-0.3, 0, 0.2, and 0.4.

The perturbed eigenvalue has the expansion $\lambda = 1 + \epsilon\lambda_1 + O(\epsilon^2)$, where

$$\lambda_1 = \frac{\langle u_0, (x^4 - 4x^2)e^{-2x^2}u_0\rangle}{\|u_0\|^2} = \frac{\int_{-\infty}^{\infty}(x^4 - 4x^2)e^{-3x^2}\,dx}{\int_{-\infty}^{\infty}e^{-x^2}\,dx}. \tag{9.35}$$

The expression in (9.35) for λ_1 may be evaluated in the following way. First, note that

$$\frac{\int_{-\infty}^{\infty}x^n e^{-3x^2}\,dx}{\int_{-\infty}^{\infty}e^{-x^2}\,dx} = \frac{1}{3^{(n+1)/2}}\frac{\int_{-\infty}^{\infty}x^n e^{-x^2}\,dx}{\int_{-\infty}^{\infty}e^{-x^2}\,dx},$$

for $n \geq 0$. We need to compute this ratio of integrals for $n = 2$ and $n = 4$. Let

$$J(a) = \int_{-\infty}^{\infty}e^{-x^2+2ax}\,dx. \tag{9.36}$$

Differentiating this expression n times with respect to a, we obtain that

$$J^{(n)}(a) = 2^n\int_{-\infty}^{\infty}x^n e^{-x^2+2ax}\,dx.$$

Hence, setting $a = 0$, we have

$$\frac{\int_{-\infty}^{\infty}x^n e^{-x^2}\,dx}{\int_{-\infty}^{\infty}e^{-x^2}\,dx} = \frac{1}{2^n}\frac{J^{(n)}(0)}{J(0)}.$$

To evaluate the right-hand side of this equation, we complete the square in the exponent of the integrand in (9.36) and change the integration variable from $x \mapsto x - a$. This gives

$$J(a) = e^{a^2}J(0).$$

It follows that

$$\frac{\int_{-\infty}^{\infty} x^n e^{-x^2}\, dx}{\int_{-\infty}^{\infty} e^{-x^2}\, dx} = \frac{1}{2^n} \left.\frac{d^n e^{a^2}}{da^n}\right|_{a=0}.$$

In particular, for $n = 2, 4$, we compute

$$\frac{d^2 e^{a^2}}{da^2} = \left(4a^2 + 2\right) e^{a^2}, \quad \frac{d^4 e^{a^2}}{da^4} = \left(16a^4 + 48a^2 + 12\right) e^{a^2}.$$

From these computations it follows that $\lambda_1 = -7/(24\sqrt{3})$, and

$$\lambda = 1 - \frac{7}{24\sqrt{3}}\epsilon + O(\epsilon^2).$$

If $\epsilon > 0$, corresponding to an oscillator that becomes "softer" for small amplitude oscillations, then the ground state energy decreases, while if $\epsilon < 0$, corresponding to an oscillator that becomes "stiffer" for small amplitude oscillations, then the ground state energy increases.

9.7 References

For additional discussion of the spectra of bounded and compact, normal operators, see Naylor and Sell [40]. The terminology of the classification of the spectrum is not entirely uniform (see Reed and Simon [45] for a further discussion). With the definitions we use here, the spectrum of a bounded operator is the disjoint union of its point, continuous, and residual spectrums. For an introduction to complex analysis and a proof of Liouville's theorem, see [36]. See Kato [26] for the perturbation theory of spectra.

9.8 Exercises

Exercise 9.1 Prove that $\rho(A^*) = \overline{\rho(A)}$, where $\overline{\rho(A)}$ is the set $\{\lambda \in \mathbb{C} \mid \overline{\lambda} \in \rho(A)\}$.

Exercise 9.2 If λ is an eigenvalue of A, then $\overline{\lambda}$ is in the spectrum of A^*. What can you say about the type of spectrum $\overline{\lambda}$ belongs to?

Exercise 9.3 Suppose that A is a bounded linear operator of a Hilbert space and $\lambda, \mu \in \rho(A)$. Prove that the resolvent R_λ of A satisfies the *resolvent equation*

$$R_\lambda - R_\mu = (\mu - \lambda)\, R_\lambda R_\mu.$$

Exercise 9.4 Prove that the spectrum of an orthogonal projection P is either $\{0\}$, in which case $P = 0$, or $\{1\}$, in which case $P = I$, or else $\{0, 1\}$.

Exercise 9.5 Let A be a bounded, nonnegative operator on a complex Hilbert space. Prove that $\sigma(A) \subset [0, \infty)$.

Exercise 9.6 Let G be a multiplication operator on $L^2(\mathbb{R})$ defined by

$$Gf(x) = g(x)f(x),$$

where g is continuous and bounded. Prove that G is a bounded linear operator on $L^2(\mathbb{R})$ and that its spectrum is given by

$$\sigma(G) = \overline{\{g(x) \mid x \in \mathbb{R}\}}.$$

Can an operator of this form have eigenvalues?

Exercise 9.7 Let $K : L^2([0, 1]) \to L^2([0, 1])$ be the integral operator defined by

$$Kf(x) = \int_0^x f(y)\,dy.$$

(a) Find the adjoint operator K^*.
(b) Show that $\|K\| = 2/\pi$.
(c) Show that the spectral radius of K is equal to zero.
(d) Show that 0 belongs to the continuous spectrum of K.

Exercise 9.8 Define the right shift operator S on $\ell^2(\mathbb{Z})$ by

$$S(x)_k = x_{k-1} \qquad \text{for all } k \in \mathbb{Z},$$

where $x = (x_k)_{k=-\infty}^{\infty}$ is in $\ell^2(\mathbb{Z})$. Prove the following facts.

(a) The point spectrum of S is empty.
(b) $\operatorname{ran}(\lambda I - S) = \ell^2(\mathbb{Z})$ for every $\lambda \in \mathbb{C}$ with $|\lambda| > 1$.
(c) $\operatorname{ran}(\lambda I - S) = \ell^2(\mathbb{Z})$ for every $\lambda \in \mathbb{C}$ with $|\lambda| < 1$.
(d) The spectrum of S consists of the unit circle $\{\lambda \in \mathbb{C} \mid |\lambda| = 1\}$ and is purely continuous.

Exercise 9.9 Define the *discrete Laplacian operator* Δ on $\ell^2(\mathbb{Z})$ by

$$(\Delta x)_k = x_{k-1} - 2x_k + x_{k+1}, \tag{9.37}$$

where $x = (x_k)_{k=-\infty}^{\infty}$. Show that $\Delta = S + S^* - 2I$. Prove that the spectrum of Δ is entirely continuous and consists of the interval $[-4, 0]$.

HINT: Consider $x_k = e^{ik\xi}$ where $-\pi \le \xi \le \pi$. Finite difference schemes for the numerical solution of differential equations may be written in terms of shift operators, and a study of their spectrum is useful in the stability analysis of finite difference schemes (see Strikwerder [53]).

Exercise 9.10 Define the right shift operator S on $\ell^2(\mathbb{N})$ by

$$S\left((x_1, x_2, x_3, \ldots)\right) = (0, x_1, x_2, \ldots), \tag{9.38}$$

and the left shift operator T on $\ell^2(\mathbb{N})$ by

$$T((x_1, x_2, x_3, \ldots)) = (x_2, x_3, x_4, \ldots). \tag{9.39}$$

Prove the following.

(a) The resolvent set of S is the exterior of the unit disc $\{\lambda \in \mathbb{C} \mid |\lambda| > 1\}$.
(b) Every $\lambda \in \mathbb{C}$ with $|\lambda| = 1$ belongs to the continuous spectrum of S.
(c) $\operatorname{ran}(\lambda I - S)$ is not dense in ℓ^2 for every $\lambda \in \mathbb{C}$ with $|\lambda| < 1$, meaning that the interior of the unit disc is contained in the residual spectrum of S.
(d) The resolvent set of T consists of all $\lambda \in \mathbb{C}$ such that $|\lambda| > 1$.
(e) The continuous spectrum of T is the unit circle.
(f) The point spectrum of T is the interior of the unit disc.
(g) The residual spectrum of T is empty.

Exercise 9.11 A complex number λ belongs to the *approximate spectrum* of a bounded linear operator $A : \mathcal{H} \to \mathcal{H}$ on a Hilbert space $\mathcal{H}$ if there is a sequence (x_n) of vectors in $\mathcal{H}$ such that $\|x_n\| = 1$ and $(A - \lambda I)x_n \to 0$ as $n \to \infty$. Prove that the approximate spectrum is contained in the spectrum, and contains the point and continuous spectrum. Give an example to show that a point in the residual spectrum need not belong to the approximate spectrum.

Exercise 9.12 Let $\mathcal{H}$ be a separable Hilbert space with an orthonormal basis $\{e_n\}$, and $A \in \mathcal{B}(\mathcal{H})$ such that

$$\sum_n \|Ae_n\|^2 < \infty.$$

(a) Prove that the Hilbert-Schmidt norm defined in (9.18) is independent of the basis. That is, show that for any other orthonormal basis $\{f_n\}$ one has

$$\sum_n \|Af_n\|^2 = \sum_n \|Ae_n\|^2.$$

(b) Prove that

$$\|A\|_{HS} = \|A^*\|_{HS}.$$

Exercise 9.13 Suppose that $L : \mathbb{R} \to \mathcal{B}(\mathcal{H})$ and $A : \mathbb{R} \to \mathcal{B}(\mathcal{H})$ are smooth, operator-valued functions of $t \in \mathbb{R}$, where $L(t)$ is self-adjoint and $A(t)$ is skew-adjoint. If $L(t)$ satisfies the ODE

$$\dot{L} = [L, A], \tag{9.40}$$

show that

$$L(t) = U^*(t)L(0)U(t),$$

where $\dot{U} = UA$ and $U(0) = I$. Show that $U(t)$ is unitary, and deduce that the eigenvalues of $L(t)$ are independent of t.

An equation that can be written in the form (9.40) for suitable operators $L(t)$ and $A(t)$ is said to be *completely integrable* because it possesses a large number of conserved quantities, namely, the eigenvalues of L. The pair of operators $\{L, A\}$ is called a *Lax pair* for the equation.

Exercise 9.14 Show that the $n \times n$, tridiagonal matrices

$$
L = \begin{pmatrix} b_1 & a_1 & 0 & \cdots & 0 \\ a_1 & b_2 & a_2 & \cdots & 0 \\ 0 & a_2 & b_3 & \cdots & 0 \\ \vdots & \vdots & \vdots & \ddots & \vdots \\ 0 & 0 & 0 & \cdots & b_n \end{pmatrix}, \quad A = \begin{pmatrix} 0 & a_1 & 0 & \cdots & 0 \\ -a_1 & 0 & a_2 & \cdots & 0 \\ 0 & -a_2 & 0 & \cdots & 0 \\ \vdots & \vdots & \vdots & \ddots & \vdots \\ 0 & 0 & 0 & \cdots & 0 \end{pmatrix}
$$

form a Lax pair for the *Toda lattice equations*

$$
\begin{aligned}
\dot{a}_k &= a_k \left(b_{k+1} - b_k \right) && \text{for } k = 1, \ldots, n-1, \\
\dot{b}_k &= 2 \left(a_k^2 - a_{k-1}^2 \right) && \text{for } k = 1, \ldots, n,
\end{aligned}
$$

where $a_0 = a_n = 0$ and $a_k > 0$. Write out the equations for $n = 2$, and determine explicitly their conserved quantities.

Exercise 9.15 Show that the KdV equation for $u(x, t)$,

$$
u_t = 6uu_x - u_{xxx},
$$

can be written in the form (9.40), where L and A are the following differential operators acting on smooth functions of x:

$$
L = -\partial_x^2 + u, \qquad A = -4\partial_x^3 + 3u\partial_x + 3\partial_x u.
$$

Exercise 9.16 Prove that (9.20) and (9.22) define the same operator $q(A)$.

Exercise 9.17 Let A be a compact, self-adjoint operator, on an infinite-dimensional separable Hilbert space, and $q : \sigma(A) \to \mathbb{C}$ a continuous function.

(a) Prove that the series in (9.22) converges in norm if and only if $q(0) = 0$.
(b) For an arbitrary value $q(0)$, prove that the series in (9.22) converges strongly.

Exercise 9.18 Suppose that A is a compact self-adjoint operator. Let $f \in C(\sigma(A))$, and consider $f(A)$ defined by (9.23). Prove that

$$
\|f(A)\| = \sup\{|f(\lambda_n)| \mid n \in \mathbb{N}\}.
$$

Let (q_N) be a sequence of polynomials of degree N, converging uniformly to f on $\sigma(A)$. The existence of such a sequence is a consequence of the Weierstrass approximation theorem. Prove that $(q_N(A))$ converges in norm, and that its limit equals $f(A)$ as defined in (9.23).

Exercise 9.19 Let A be a compact self-adjoint linear operator. Prove that the series in (9.23) is convergent in the strong operator topology for any $f \in C(\sigma(A))$, and that it converges uniformly if in addition $f(0) = 0$.

Exercise 9.20 Let A be a self-adjoint compact operator on a Hilbert space $\mathcal{H}$, and let $f : \sigma(A) \to \mathbb{C}$ be a continuous function. When is $f(A)$ compact?

Exercise 9.21 Consider the evolution equation $x_t = Ax$, where A is a bounded operator on a Hilbert space such that

$$\mathrm{Re}\,\langle x, Ax \rangle \leq 2\alpha \|x\|^2,$$

for some $\alpha \in \mathbb{R}$. By taking the inner product of the evolution equation with x, derive the energy estimate

$$\|x(t)\| \leq e^{\alpha t}\|x(0)\|.$$

Compare this result with that of the spectral method for self-adjoint and non-self-adjoint operators A.

Exercise 9.22 Suppose that A is a compact, nonnegative linear operator on a Hilbert space. Prove that there is a unique compact, nonnegative linear operator B such that $B^2 = A$. Thus, $B = A^{1/2}$ is the square root of A.

Exercise 9.23 Consider the eigenvalue problem

$$\int_{-\infty}^{\infty} e^{-|x|-|y|}u(y)\,dy + \epsilon x u(x) = \lambda u(x), \qquad -\infty < x < \infty,$$

where $\epsilon \in \mathbb{R}$. Show that if $\epsilon = 0$, then the spectrum consists purely of eigenvalues, and $\lambda = 1$ is a simple eigenvalue with eigenfunction $u(x) = e^{-|x|}$. Show that a formal perturbation expansion with respect to ϵ as $\epsilon \to 0$ gives

$$\lambda = 1 + \frac{1}{2}\epsilon^2 + \dots,$$

$$u(x) = e^{-|x|}\left(1 + \epsilon x + \epsilon^2\left(x^2 - \frac{3}{4}\right) + \dots\right).$$

Show, however, that there are no eigenfunctions $u \in L^2(\mathbb{R})$ when $\epsilon \neq 0$. (It is possible to show that then the spectrum is purely continuous.)

Chapter 10

Linear Differential Operators and Green's Functions

We have seen that linear differential operators on normed function spaces are not bounded. Differential operators are important for the study of differential equations and we would like to analyze them in spite of their lack of continuity. There are two main approaches to this problem. One is to use a weak topology, not derived from a norm, with respect to which differential operators are continuous. This is what is done in distribution theory, studied in Chapter 11. The other approach, which we follow in this chapter, is to consider special classes of unbounded operators that are defined on dense linear subspaces of a Hilbert, or Banach, space.

The inverse of a linear differential operator is an integral operator, whose kernel is called the *Green's function* of the differential operator. We may use the bounded inverse to study the properties of the unbounded differential operator. For example, if the inverse is a compact, self-adjoint operator, then the differential operator has a complete orthonormal set of eigenfunctions.

We begin by giving some general definitions for unbounded operators. We will consider unbounded linear operators acting in a Hilbert space, although similar ideas apply to unbounded operators acting in a Banach space.

10.1 Unbounded operators

One of the main new features of unbounded operators, in comparison with bounded operators, is that they are not defined on the whole space. For example, a general continuous function does not have a continuous derivative, so differential operators are defined on a subspace of differentiable functions. The definition of an unbounded linear operator

$$A : \mathcal{D}(A) \subset \mathcal{H} \to \mathcal{H}$$

acting in a Hilbert space $\mathcal{H}$ therefore includes the definition of its domain $\mathcal{D}(A)$. We will assume that the domain of A is a dense linear subspace of $\mathcal{H}$, unless we state explicitly otherwise. If the domain of A is not dense, then we may obtain a densely defined operator by setting A equal to zero on the orthogonal complement of its domain, so this assumption does not lead to any loss of generality.

An operator $\tilde{A}$ is an *extension* of A, or A is a *restriction* of $\tilde{A}$, if $\mathcal{D}(\tilde{A}) \supset \mathcal{D}(A)$ and $\tilde{A}x = Ax$ for all $x \in \mathcal{D}(A)$. We write this relationship as $\tilde{A} \supset A$, or $A \subset \tilde{A}$. From Theorem 5.19, if A is a bounded linear operator on a dense domain $\mathcal{D}(A)$ in $\mathcal{H}$, then A has a unique bounded extension to $\mathcal{H}$. Consequently, it is only useful to consider densely defined operators when the operator is unbounded.

The domain of a differential operator defines the somewhat technical property of the smoothness of the functions on which the operator acts. More importantly, it also encodes any boundary conditions associated with the operator. The following example, which we discuss in greater detail later on, illustrates differential operators and their domains.

Example 10.1 Let $A_k u = u''$ with $k = 1, 2, 3, 4$ be differential operators in $L^2([0,1])$ with domains

$$\mathcal{D}(A_1) = \left\{ u \in C^2([0,1]) \mid u(0) = u(1) = 0 \right\},$$
$$\mathcal{D}(A_2) = C^2([0,1]),$$
$$\mathcal{D}(A_3) = \left\{ u \in H^2((0,1)) \mid u(0) = u(1) = 0 \right\},$$
$$\mathcal{D}(A_4) = H^2((0,1)).$$

Here, $H^2((0,1))$ is the Sobolev space of functions whose weak derivatives of order less than or equal to two belong to $L^2([0,1])$. The Sobolev embedding theorem implies that $H^2((0,1)) \subset C^1([0,1])$, so it makes sense to use the pointwise values of u in defining $\mathcal{D}(A_3)$. Then $A_1 \subset A_2 \subset A_4$, and $A_1 \subset A_3 \subset A_4$.

The adjoint of an unbounded operator $A : \mathcal{D}(A) \subset \mathcal{H} \to \mathcal{H}$ is an operator

$$A^* : \mathcal{D}(A^*) \subset \mathcal{H} \to \mathcal{H}.$$

Generalizing the basic property in (8.9) of the adjoint of a bounded linear operator, we want

$$\langle Ax, y \rangle = \langle x, A^* y \rangle \qquad \text{for all } x \in \mathcal{D}(A) \text{ and all } y \in \mathcal{D}(A^*), \qquad (10.1)$$

where $\mathcal{D}(A^*)$ is the largest subspace of $\mathcal{H}$ for which (10.1) holds. In more detail, if $y \in \mathcal{H}$, then $\varphi_y(x) = \langle y, Ax \rangle$ defines a linear functional $\varphi_y : \mathcal{D}(A) \to \mathbb{C}$. We say that $y \in \mathcal{D}(A^*)$ if φ_y is bounded on $\mathcal{D}(A)$. In that case, since $\mathcal{D}(A)$ is dense in $\mathcal{H}$, the bounded linear transformation theorem in Theorem 5.19 implies that φ_y has a unique extension to a bounded linear functional on $\mathcal{H}$, and the Riesz representation theorem in Theorem 8.12 implies there is a unique vector $z \in \mathcal{H}$ such that $\varphi_y(x) = \langle z, x \rangle$. Then $\langle y, Ax \rangle = \langle z, x \rangle$ for all $x \in \mathcal{D}(A)$, and we define $A^* y = z$. Summarizing this procedure, we get the following definition.

Definition 10.2 Suppose that $A : \mathcal{D}(A) \subset \mathcal{H} \to \mathcal{H}$ is a densely defined unbounded linear operator on a Hilbert space $\mathcal{H}$. The *adjoint* $A^* : \mathcal{D}(A^*) \subset \mathcal{H} \to \mathcal{H}$ is the operator with domain

$$\mathcal{D}(A^*) = \{y \in \mathcal{H} \mid \text{there is a } z \in \mathcal{H} \text{ with } \langle Ax, y \rangle = \langle x, z \rangle \text{ for all } x \in \mathcal{D}(A)\}.$$

If $y \in \mathcal{D}(A^*)$, then we define $A^* y = z$, where z is the unique element such that $\langle Ax, y \rangle = \langle x, z \rangle$ for all $x \in \mathcal{D}(A)$.

It is possible that $\mathcal{D}(A^*)$ is not dense in $\mathcal{H}$, even if $\mathcal{D}(A)$ is dense, in which case we do not define A^{**} (see Exercise 10.15 for an example).

As we will see below, the adjoint of a differential operator is another differential operator, which we obtain by using integration by parts. The domain $\mathcal{D}(A)$ defines boundary conditions for A, and the domain $\mathcal{D}(A^*)$ defines adjoint boundary conditions for A^*. The boundary conditions ensure that the boundary terms arising in the integration by parts vanish.

A particularly important class of unbounded operators is the class of self-adjoint operators. Self-adjointness includes the equality of the domains of A and A^*. For differential operators, this equality of domains corresponds to the self-adjointness of the boundary conditions.

Definition 10.3 An unbounded operator A is *self-adjoint* if $A^* = A$, meaning that $\mathcal{D}(A^*) = \mathcal{D}(A)$ and $A^* x = Ax$ for all $x \in \mathcal{D}(A)$. An unbounded operator A is *symmetric* if A^* is an extension of A, meaning that $\mathcal{D}(A^*) \supset \mathcal{D}(A)$ and $A^* x = Ax$ for all $x \in \mathcal{D}(A)$.

It is usually straightforward to show that an operator is symmetric, but it may be more difficult to show that a symmetric operator is self-adjoint.

Example 10.4 For the differential operators defined in Example 10.1, we will see that $A_1^* = A_3$, so A_1 is symmetric but not self-adjoint, while $A_3^* = A_3$, so A_3 is self-adjoint. We will also see that $A_2^* = A_4^* = A_5$ where $A_5 u = u''$ with domain

$$\mathcal{D}(A_5) = \left\{u \in H^2([0,1]) \mid u(0) = u(1) = u'(0) = u'(1) = 0\right\}.$$

Since A_5 is not an extension of A_2 or A_4, neither A_2 nor A_4 is symmetric. We also have $A_5^* = A_4$, so A_5 is symmetric, but not self-adjoint.

Although differential operators are not continuous, they have a related property called *closedness*.

Definition 10.5 An operator $A : \mathcal{D}(A) \subset \mathcal{H} \to \mathcal{H}$ is *closed* if for every sequence (x_n) in $\mathcal{D}(A)$ such that $x_n \to x$ and $Ax_n \to y$, we have $x \in \mathcal{D}(A)$ and $Ax = y$.

Note carefully the difference between continuous and closed operators. For a continuous operator A, the convergence of the sequence (x_n) implies the convergence of (Ax_n), and

$$\lim_{n\to\infty} Ax_n = A\left(\lim_{n\to\infty} x_n\right). \tag{10.2}$$

For a closed operator A, the convergence of (x_n) does not imply the convergence of (Ax_n); but if both (x_n) and (Ax_n) converge, then (10.2) holds. The *graph* $\Gamma(A)$ of an operator $A : \mathcal{D}(A) \subset \mathcal{H} \to \mathcal{H}$ is the subset of $\mathcal{H} \times \mathcal{H}$ defined by

$$\Gamma(A) = \{(x,y) \mid x \in \mathcal{D}(A) \text{ and } y = Ax\}.$$

An operator is closed if and only if its graph is a closed subspace of $\mathcal{H} \times \mathcal{H}$.

An operator A is *closable* if it has the following property: for every sequence (x_n) of elements in $\mathcal{D}(A)$ such that $x_n \to 0$ and $Ax_n \to y$ for some $y \in \mathcal{H}$, we have $y = 0$. We define the *closure* $\overline{A}$ of a closable operator A to be the operator with domain

$$\mathcal{D}(\overline{A}) = \{x \in \mathcal{H} \mid \text{there is a sequence } (x_n) \text{ in } \mathcal{D}(A) \text{ and a } y \in \mathcal{H}$$
$$\text{such that } x_n \to x \text{ and } Ax_n \to y\}.$$

If $x_n \to x$ and $Ax_n \to y$, then we define $\overline{A}x = y$. Since A is closable, the value y does not depend on the sequence (x_n) in $\mathcal{D}(A)$ that is used to approximate x. The graph of $\overline{A}$ is the closure of the graph of A in $\mathcal{H} \times \mathcal{H}$, and $\overline{A}$ is the smallest closed extension of A. If A is not closable, then the closure of the graph of A is not the graph of an operator, and A has no closed extensions (see Exercise 10.8 for an example). Every symmetric operator is closable (see Exercise 10.2). We say that a symmetric operator A is *essentially self-adjoint* if its closure is self-adjoint.

Example 10.6 The operators A_1 and A_2 in Example 10.1 are not closed because we may choose a sequence of functions $u_n \in C^2([0,1])$ such that $u_n \to u$ and $u_n'' \to v$ in $L^2([0,1])$, where v is not continuous. Hence u is not C^2, and therefore does not belong to the domain of A_1 or A_2. The operators A_3 and A_4 are closed. Both A_1 and A_2 are closable, with $\overline{A_1} = A_3$, and $\overline{A_2} = A_4$. Thus, A_1 is essentially self-adjoint, but A_2 is not.

If $A : \mathcal{D}(A) \subset \mathcal{H} \to \mathcal{H}$ is one-to-one and onto, then we define the inverse operator $A^{-1} : \mathcal{H} \to \mathcal{H}$ by $A^{-1}y = x$ if and only if $Ax = y$. The range of A^{-1} is equal to the domain of A. If A is closed, then the closed graph theorem, which we do not prove here, implies that A^{-1} is bounded.

Proposition 10.7 If $A : \mathcal{D}(A) \subset \mathcal{H} \to \mathcal{H}$ is a densely defined linear operator on a Hilbert space $\mathcal{H}$ with a bounded inverse $A^{-1} : \mathcal{H} \to \mathcal{H}$, then $(A^*)^{-1} = (A^{-1})^*$.

Proof.   Since A^{-1} is bounded, it has a bounded adjoint. If $x \in \mathcal{D}(A^*)$ and $y \in \mathcal{H}$, then

$$\langle (A^{-1})^* A^* x, y \rangle = \langle A^* x, A^{-1} y \rangle = \langle x, A A^{-1} y \rangle = \langle x, y \rangle.$$

Therefore $(A^{-1})^* A^* x = x$ for $x \in \mathcal{D}(A^*)$. Moreover, if $x \in \mathcal{H}$ and $y \in \mathcal{D}(A)$, then

$$\langle A^* (A^{-1})^* x, y \rangle = \langle (A^{-1})^* x, A y \rangle = \langle x, A^{-1} A y \rangle = \langle x, y \rangle.$$

Since $\mathcal{D}(A)$ is dense in $\mathcal{H}$, it follows that $(A^{-1})^* x \in \mathcal{D}(A^*)$ and $A^* (A^{-1})^* x = x$. $\square$

The definitions of the resolvent set, spectrum, and resolvent operator for an unbounded operator $A : \mathcal{D}(A) \subset \mathcal{H} \to \mathcal{H}$ are analogous to those for a bounded operator. The resolvent set $\rho(A)$ of A consists of the complex numbers λ such that $A - \lambda I$ is a one-to-one, onto map from $\mathcal{D}(A)$ to $\mathcal{H}$, and $(A - \lambda I)^{-1}$ is bounded. The spectrum $\sigma(A)$ is the complement of the resolvent set in $\mathbb{C}$. If $\lambda \in \rho(A)$, then we define the resolvent operator $R_\lambda : \mathcal{H} \to \mathcal{H}$ by

$$R_\lambda = (\lambda I - A)^{-1}.$$

If A is closed, then the closed graph theorem implies that R_λ is bounded whenever $A - \lambda I$ is one-to-one and onto. Unlike bounded operators, unbounded operators may have an empty spectrum (see Exercise 10.13 for an example).

10.2   The adjoint of a differential operator

In this section, we consider differential operators acting on smooth functions, and explain how to determine their adjoints. We discuss the domain of the adjoint in more detail in Section 10.4.

A linear ordinary differential operator of order n is a linear map A that acts on an n-times continuously differentiable function u by

$$Au = \sum_{j=0}^{n} a_j u^{(j)},$$

where $u^{(j)}$ denotes the jth derivative of u, and the coefficients a_j are real or complex-valued functions. Our goal is to study BVPs (boundary value problems) for ODEs of the form

$$Au = f, \qquad Bu = 0, \tag{10.3}$$

where $Bu = 0$ denotes a set of linear boundary conditions.

For concreteness, we assume that all functions are defined on the interval $[0, 1]$, and we consider second-order ordinary differential operators A of the form

$$Au = au'' + bu' + cu, \tag{10.4}$$

where a, b, and c are sufficiently smooth functions on $[0, 1]$. The same ideas apply to linear ordinary differential operators of arbitrary order. We assume, unless stated otherwise, that $a(x) > 0$ for all $0 \leq x \leq 1$, so that A is second-order at every point.

For a second-order differential equation, we expect that we need to impose two boundary conditions to obtain a unique solution, although this is not always sufficient to guarantee uniqueness. Sometimes we may want to consider overdetermined or underdetermined boundary value problems with a larger or smaller number of boundary conditions. We always assume that the boundary condition $Bu = 0$ is a homogeneous system of linear equations that involves the values of u and u' at the endpoints $x = 0, 1$. Higher derivatives of u may be expressed in terms of u and u' by use of the differential equation.

Some common types of boundary conditions are:

$$u(0) = 0, \quad u(1) = 0 \qquad \text{Dirichlet;}$$
$$u'(0) = 0, \quad u'(1) = 0 \qquad \text{Neumann;}$$
$$u(0) = u(1), \quad u'(0) = u'(1) \qquad \text{periodic;}$$
$$\alpha_0 u(0) + \beta_0 u'(0) = 0, \quad \alpha_1 u(1) + \beta_1 u'(1) = 0 \qquad \text{mixed.}$$

In the mixed boundary condition, α_0, α_1, β_0, and β_1 are complex constants. Instead of imposing conditions that involve the solution at both endpoints, we can impose two conditions at one of the endpoints:

$$u(0) = 0, \quad u'(0) = 0 \qquad \text{initial;}$$
$$u(1) = 0, \quad u'(1) = 0 \qquad \text{final.}$$

For linear problems, nonhomogeneous boundary conditions may be reduced to homogeneous ones by subtraction of any function that satisfies the nonhomogeneous conditions: if $Au = f$, $Bu = b$, and $Bu_p = b$, then $v = u - u_p$ satisfies $Av = g$ and $Bv = 0$, where $g = f - Au_p$. In practice, it may be convenient to retain nonhomogeneous boundary conditions when using Green's formula below, but in developing the general theory it is simplest to assume that all boundary conditions have been reduced to homogeneous ones.

We begin by formulating the adjoint boundary value problem, using the following result.

Proposition 10.8 (Green's) Suppose that A is given by (10.4), where $a \in C^2([0, 1])$, $b \in C^1([0, 1])$, and $c \in C([0, 1])$. Let $\langle \cdot, \cdot \rangle$ denote the usual L^2-inner product,

$$\langle v, u \rangle = \int_0^1 \overline{v(x)} u(x)\, dx,$$

and define A^* by

$$A^* v = (\bar{a} v)'' - (\bar{b} v)' + \bar{c} v. \qquad (10.5)$$

Then, for every $u, v \in C^2([0, 1])$, we have

$$\langle v, Au \rangle - \langle A^*v, u \rangle = [a(\bar{v}u' - \bar{v}'u) + (b - a')\bar{v}u]_0^1. \qquad (10.6)$$

Proof. Integration by parts implies that

$$
\begin{aligned}
\langle v, Au \rangle &= \int_0^1 \bar{v}(au'' + bu' + cu)\, dx \\
&= [a\bar{v}u' + b\bar{v}u]_0^1 + \int_0^1 \{-(a\bar{v})'u' - (b\bar{v})'u + c\bar{v}u\}\, dx \\
&= [a\bar{v}u' - (a\bar{v})'u + b\bar{v}u]_0^1 + \int_0^1 \overline{\left((\bar{a}v)'' - (\bar{b}v)' + \bar{c}v\right)}u\, dx,
\end{aligned}
$$

which gives (10.6). $\square$

We call A^* in (10.5) the *formal adjoint* of A ("formal" because we have not specified its domain). The adjoint A^* depends on the inner product as well as on A (see Exercise 10.10). We will use the standard L^2-inner product, unless explicitly stated otherwise.

Example 10.9 Let D be the differentiation operator,

$$D = \frac{d}{dx}. \qquad (10.7)$$

Then $D^* = -D$, $(iD)^* = iD$, and $(D^2)^* = D^2$, so D is formally skew-adjoint, while iD and D^2 are formally self-adjoint.

Given boundary conditions B for A, we define adjoint boundary conditions B^* for A^* by the requirement that the boundary terms in (10.6) vanish. Thus, for $v \in C^2([0, 1])$, we say that $B^*v = 0$ if and only if

$$\langle v, Au \rangle = \langle A^*v, u \rangle \qquad \text{for all } u \in C^2([0, 1]) \text{ such that } Bu = 0.$$

For A given by (10.4), we have $B^*v = 0$ if and only if

$$[a(\bar{v}u' - \bar{v}'u) + (b - a')\bar{v}u]_0^1 = 0 \qquad \text{for all } u \text{ such that } Bu = 0.$$

We say that the BVP (10.3) is *self-adjoint* if $A = A^*$ and $B = B^*$.

Example 10.10 Suppose that $A = D^2$. Then Green's formula may be written as

$$\langle v, u'' \rangle - \langle v'', u \rangle = [\bar{v}u' - \bar{v}'u]_0^1.$$

If $Bu = 0$ is the Dirichlet conditions $u(0) = u(1) = 0$, then we have

$$[\bar{v}u' - \bar{v}'u]_0^1 = \overline{v(1)}u'(1) - \overline{v(0)}u'(0).$$

This vanishes for all values of $u'(0)$ and $u'(1)$ if and only if $v(0) = v(1) = 0$. Thus, the Dirichlet boundary value problem for D^2 is self-adjoint. Neumann, mixed, and periodic boundary conditions are also self-adjoint. For initial conditions $u(0) = u'(0) = 0$ the boundary terms reduce to

$$[\bar{v}u' - \bar{v}'u]_0^1 = \overline{v(1)}u'(1) - \overline{v'(1)}u(1).$$

These terms vanish if and only if $v(1) = v'(1) = 0$, so final conditions are the adjoint of initial conditions, and the initial or final value problem for D^2 is not self-adjoint.

If we impose no boundary conditions on u, then we must require that $v(0) = v'(0) = v(1) = v'(1) = 0$. The adjoint of an undetermined boundary value problem is therefore overdetermined, and conversely.

Let us find all the formally self-adjoint, second-order differential operators. Expanding the expression for A^* in (10.5) and equating it with the expression for A in (10.4), we find that

$$au'' + bu' + cu = \bar{a}u'' + (2\bar{a}' - \bar{b})u' + (\bar{a}'' - \bar{b}' + \bar{c})u$$

for every $u \in C^2([0,1])$. We must therefore have

$$a = \bar{a}, \quad b = 2\bar{a}' - \bar{b}, \quad c = \bar{a}'' - \bar{b}' + \bar{c}.$$

These relations are satisfied if and only if a is real, $\operatorname{Re} b = a'$, and $\operatorname{Im} c = -\operatorname{Im} b/2$, where $\operatorname{Re} z$ and $\operatorname{Im} z$ denote the real and imaginary parts of $z \in \mathbb{C}$, respectively. The coefficients of a self-adjoint, second-order ordinary differential operator A are therefore determined by three real functions: a, $\operatorname{Im} b$, and $\operatorname{Re} c$. For operators with real coefficients, there are only two independent real-valued coefficient functions, which we denote by p and q, where $a = -p$, $b = -p'$, and $c = q$. The resulting formally self-adjoint operator, called a *Sturm-Liouville* operator, is given by

$$Au = -(pu')' + qu, \tag{10.8}$$

or $A = -D(pD) + q$. For example, if $p = 1$ and $q = 0$, we get the second-derivative operator $A = -D^2$. By imposing self-adjoint boundary conditions on functions in the domain of a Sturm-Liouville operator, we obtain a self-adjoint operator.

For operators with imaginary coefficients, we find that $a = 0$, $b = 2ir$ and $c = ir'$, which gives

$$Au = i\left(2ru' + r'u\right),$$

or $A = i(rD + Dr)$, since $Dr = rD + r'$. Any real linear combination of these real and imaginary formally self-adjoint operators is formally self-adjoint.

10.3 Green's functions

For concreteness, we consider the Dirichlet boundary value problem for the second-order differential operator A defined in (10.4),

$$Au = f, \qquad u(0) = u(1) = 0, \tag{10.9}$$

where $f : [0,1] \to \mathbb{C}$ is a given continuous function.

We look for a solution of (10.9) in the form

$$u(x) = \int_0^1 g(x,y)f(y)\, dy, \tag{10.10}$$

where $g : [0,1] \times [0,1] \to \mathbb{C}$ is a suitable function, called the *Green's function* of (10.9). If we regard

$$A : \mathcal{D}(A) \subset C([0,1]) \to C([0,1])$$

as an operator in $C([0,1])$ with domain

$$\mathcal{D}(A) = \left\{ u \in C^2([0,1]) \mid u(0) = u(1) = 0 \right\},$$

then the integral operator $G : C([0,1]) \to \mathcal{D}(A)$ given by

$$Gf(x) = \int_0^1 g(x,y)f(y)dy \tag{10.11}$$

is the inverse of A.

We can write an equation for the Green's function g in terms of the *Dirac delta function* δ. We give a heuristic discussion here, and use it to motivate the classical definition of the Green's function in Definition 10.11 below. In Chapter 11, we will show that the delta function has a mathematically rigorous interpretation as a distribution.

We regard δ as a "function" on $\mathbb{R}$ that has unit integral concentrated at the origin, meaning that

$$\int_{-\infty}^{\infty} \delta(x)\, dx = 1, \qquad \delta(x) = 0 \quad \text{for } x \neq 0.$$

More generally, for any continuous function f, we have

$$\int_{-\infty}^{\infty} \delta(x-y)f(y)\, dy = f(x).$$

Formally, we also have

$$\int_{-\infty}^{x} \delta(y)\, dy = H(x),$$

where H is the *Heaviside step function*, defined by

$$H(x) = \begin{cases} 0 & \text{if } x < 0, \\ 1 & \text{if } x > 0. \end{cases}$$

The step function is constant on any interval that does not contain the origin and has a jump of one at zero. Conversely, the delta function,

$$\delta = H', \tag{10.12}$$

is the derivative of the step function. We will give a precise meaning to these results when we study distribution theory in Chapter 11.

The Green's function $g(x,y)$ associated with the boundary value problem in (10.9) is the solution of the following problem:

$$Ag(x,y) = \delta(x-y), \qquad g(0,y) = g(1,y) = 0. \tag{10.13}$$

Here, A is a differential operator with respect to x, and y plays the role of a parameter. If u is given by (10.10), then formally differentiating under the integral sign with respect to x, we find that for $0 < x < 1$

$$Au(x) = \int_0^1 Ag(x,y)f(y)\,dy = \int_0^1 \delta(x-y)f(y)\,dy = f(x).$$

Moreover, u satisfies the boundary conditions, since

$$u(0) = \int_0^1 g(0,y)f(y)\,dy = 0, \qquad u(1) = \int_0^1 g(1,y)f(y)\,dy = 0.$$

Thus, (10.10) provides an integral representation of the solution of (10.9).

We may reformulate (10.13) in classical, pointwise terms. From (10.4), (10.12), and (10.13), we want $g(x,y)$ to satisfy the homogeneous ODE (as a function of x) when $x \neq y$, and we want the jump in $a(x)g_x(x,y)$ across $x = y$ to equal one in order to obtain a delta function after taking a second x-derivative. We therefore make the following definition.

Definition 10.11 A function $g : [0,1] \times [0,1] \to \mathbb{C}$ is a *Green's function* for (10.9) if it satisfies the following conditions.

(a) The function $g(x,y)$ is continuous on the square $0 \leq x,y \leq 1$, and twice continuously differentiable with respect to x on the triangles $0 \leq x \leq y \leq 1$ and $0 \leq y \leq x \leq 1$, meaning that the partial derivatives exist in the interiors of the triangles and extend to continuous functions on the closures. The left and right limits of the partial derivatives on $x = y$ are not equal, however.

(b) The function $g(x,y)$ satisfies the ODE with respect to x and the boundary conditions:

$$Ag = 0 \qquad \text{in } 0 < x < y < 1 \text{ and } 0 < y < x < 1, \tag{10.14}$$
$$g(0,y) = g(1,y) = 0 \qquad \text{for } 0 \leq y \leq 1. \tag{10.15}$$

(c) The jump in g_x across the line $x = y$ is given by

$$g_x\left(y^+, y\right) - g_x\left(y^-, y\right) = \frac{1}{a(y)}, \qquad (10.16)$$

where the subscript x denotes a partial derivative with respect to the first variable in $g(x, y)$, and

$$g_x(y^+, y) = \lim_{x \to y^+} g_x(x, y), \qquad g_x(y^-, y) = \lim_{x \to y^-} g_x(x, y).$$

We will discuss the existence and construction of the Green's function below. First we show that if a function g satisfies the conditions in this definition, then the expression in (10.10) gives a solution of (10.9).

Proposition 10.12 Let A be given by (10.4), where $a, b, c \in C([0, 1])$ and $a(x) > 0$ for all $0 \le x \le 1$. If g satisfies (10.14)–(10.16) and $f \in C([0, 1])$, then Gf given by (10.11) is a solution of (10.9).

Proof. The proof is by direct computation. The only non-trivial part to check is that the function

$$u(x) = \int_0^1 g(x, y) f(y) \, dy$$

satisfies the ODE $Au = f$. We split the integration range into $0 \le y \le x$ and $x \le y \le 1$:

$$Au(x) = \left[a \frac{d^2}{dx^2} + b \frac{d}{dx} + c\right] \left[\int_0^x g(x, y) f(y) \, dy + \int_x^1 g(x, y) f(y) \, dy\right]. \qquad (10.17)$$

Leibnitz's formula for the differentiation of an integral with variable limits states that if $\alpha(x)$ and $\beta(x)$ are continuously differentiable functions of x, and $h(x, y)$ is a continuous function of (x, y) on $\alpha(x) \le y \le \beta(x)$ that has a continuous partial derivative $h_x(x, y)$ with respect to x on $\alpha(x) \le y \le \beta(x)$, then

$$\frac{d}{dx} \int_{\alpha(x)}^{\beta(x)} h(x, y) \, dy = \beta'(x) h\left(x, \beta(x)\right) - \alpha'(x) h\left(x, \alpha(x)\right) + \int_{\alpha(x)}^{\beta(x)} h_x(x, y) \, dy.$$

$$(10.18)$$

Using this formula to compute the derivatives in the expression on the right-hand side of (10.17), we find that

$$\frac{du}{dx} = \int_0^x g_x(x, y) f(y) \, dy + g(x, x) f(x) + \int_x^1 g_x(x, y) f(y) \, dy - g(x, x) f(x)$$

$$= \int_0^1 g_x(x, y) f(y) \, dy,$$

$$\frac{d^2u}{dx^2} = \int_0^1 g_{xx}(x, y) f(y) \, dy + \left[g_x(x, x^-) - g_x(x, x^+)\right] f(x).$$

Thus,

$$A\left(\int_0^1 g(x,y)f(y)\,dy\right) = \int_0^1 Ag(x,y)f(y)\,dy + a(x)\left[g_x(x,x^-) - g_x(x,x^+)\right]f(x).$$

Since $g(x,y)$ is smooth in $x \le y$ and $x \ge y$, we have $g_x(x,x^-) = g_x(x^+,x)$ and $g_x(x,x^+) = g_x(x^-,x)$. It follows from the properties of g that $Au = f$. $\qquad\square$

Thus, we can give an integral representation of the solution of (10.9) if we can construct the associated Green's function. We may write the Green's function in terms of the solutions of the homogeneous equations. When a, b, and c are continuous functions and $a(x) \ne 0$, the homogeneous ODE

$$au'' + bu' + cu = 0 \qquad (10.19)$$

has a two-dimensional space of solutions spanned by any linearly independent pair of solutions. For example, we may construct a basis $\{u_1, u_2\}$ of the solution space by solving (10.19) subject to the initial conditions $u(0) = 1$, $u'(0) = 0$ for $u = u_1$ and $u(0) = 0$, $u'(0) = 1$ for $u = u_2$. The solutions exist by the existence theorem for ODEs in Theorem 3.7. The uniqueness of solutions of the initial value problem implies that if u is a solution of (10.19), then $u = u(0)u_1 + u'(0)u_2$, so u is a linear combination of $\{u_1, u_2\}$.

In order to construct a function g satisfying the conditions of Definition 10.11, we choose nonzero solutions v_1 and v_2 of $Av = 0$ such that

$$v_1(0) = 0, \quad v_2(1) = 0. \qquad (10.20)$$

The pair $\{v_1, v_2\}$ is linearly independent if and only if the only solution of the homogeneous Dirichlet problem, $Au = 0$ with $u(0) = u(1) = 0$, is $u = 0$. The Green's function g then has the following form:

$$g(x,y) = \begin{cases} C(y)v_1(x)v_2(y) & \text{if } 0 \le x \le y, \\ C(y)v_1(y)v_2(x) & \text{if } y \le x \le 1. \end{cases} \qquad (10.21)$$

It is clear that g is continuous, satisfies $Ag = 0$ whenever $x \ne y$, and $g(0,y) = g(1,y) = 0$. The jump condition in (10.16) is satisfied if $C(y)$ is given by

$$C(y) = \frac{1}{a(y)W(y)}, \qquad (10.22)$$

where W is the *Wronskian* of v_1 and v_2:

$$W = \begin{vmatrix} v_1 & v_2 \\ v_1' & v_2' \end{vmatrix} = v_1v_2' - v_2v_1'. \qquad (10.23)$$

If a is nonzero at every point, then the Wronskian of two linearly independent solutions is nonzero at every point, so C in (10.22) is well-defined.

Thus, if the homogeneous Dirichlet problem has only the zero solution, then g defined by (10.21) has all the properties required in Proposition 10.12 and is unique.

Example 10.13 The stationary temperature distribution in a rod of unit length that has both ends kept at a constant zero temperature, with heat loss through its surface proportional to u, and that is subject to a given nonuniform heat source per unit length f, is the solution of

$$-u'' + u = f, \qquad u(0) = u(1) = 0. \tag{10.24}$$

To construct the Green's function we need two linearly independent solutions v_1, v_2 of the homogeneous version of (10.24) that satisfy $v_1(0) = 0$ and $v_2(1) = 0$. The general solution of the homogeneous equation is of the form

$$u(x) = c_1 e^x + c_2 e^{-x}.$$

For v_1 and v_2 we choose the solutions

$$v_1(x) = \sinh x, \qquad v_2(x) = \sinh(1 - x),$$

where

$$\sinh x = \frac{e^x - e^{-x}}{2}.$$

The Wronskian, $W = -\sinh 1$, of these solutions is a nonzero constant, so the solutions are linearly independent combinations of e^x and e^{-x}. The Green's function is given by

$$g(x,y) = \begin{cases} \sinh x \sinh(1 - y)/\sinh 1 & \text{if } 0 \le x \le y \le 1, \\ \sinh y \sinh(1 - x)/\sinh 1 & \text{if } 0 \le y \le x \le 1. \end{cases}$$

We may also write this equation as

$$g(x,y) = \frac{\sinh(x_<)\sinh(1 - x_>)}{\sinh 1},$$

where

$$x_< = \min(x,y), \qquad x_> = \max(x,y).$$

The Green's function is a symmetric function of (x,y), as is always the case for real, self-adjoint boundary value problems.

We can use the Green's function to study the relationship between the solvability of the direct and adjoint BVPs. The following argument shows that the Fredholm alternative in Definition 8.19 applies to linear BVPs for ODEs.

Suppose that the homogeneous BVP,

$$Au = 0, \qquad Bu = 0,$$

has only the zero solution and the coefficient a of the highest derivative never vanishes. Then we can construct its Green's function, and therefore the nonhomogeneous

BVP,

$$Au = f, \qquad Bu = 0,$$

has a unique solution $u \in C^2([0,1])$ for every $f \in C([0,1])$. If

$$A^* v = 0, \qquad B^* v = 0,$$

then for every $f \in C([0,1])$ we have

$$\langle f, v \rangle = \langle Au, v \rangle = \langle u, A^* v \rangle = 0.$$

Hence, $v = 0$, and the homogeneous adjoint BVP has only the zero solution. We can then construct the adjoint Green's function $g^*(x,y)$, and the adjoint BVP $A^* v = h$, $B^* v = 0$ has a unique solution $v \in C^2([0,1])$ for every $h \in C([0,1])$.

Since $(A^*)^{-1} = (A^{-1})^*$, the direct and adjoint Green's functions are related by

$$g^*(x,y) = \overline{g(y,x)}. \tag{10.25}$$

If A is self-adjoint, then g is Hermitian symmetric.

If A is singular, then A^* is also singular. In that case, it is possible to define a generalized inverse of A, whose kernel is called the *modified Green's function* of A, and show that the direct BVP is solvable if and only if the right-hand side is orthogonal to the kernel of the adjoint (see Exercise 10.9 for an example, and Stakgold [52] for further discussion).

Finally, we describe the Green's function representation of the solution of a BVP with nonhomogeneous boundary conditions. We begin by giving a formal derivation of the representation. For definiteness, we consider the Dirichlet problem for a real, second-order ODE,

$$Au = f(x), \qquad u(0) = \alpha_0, \qquad u(1) = \alpha_1, \tag{10.26}$$

where A is defined in (10.4). A similar derivation applies to other types of boundary conditions. The adjoint Green's function $g^*(x,y)$ satisfies

$$A^* g^* = \delta(x - y), \qquad g^*(0,y) = g^*(1,y) = 0,$$

where A^* is a differential operator in x, and y plays the role of a parameter. Using Green's identity (10.6), we find that

$$\int_0^1 \{ g^*(x,y) Au(x) - A^* g^*(x,y) u(x) \} \, dx$$
$$= [a(x) g^*(x,y) u_x(x) - a(x) g_x^*(x,y) u(x) + \{ b(x) - a_x(x) \} g^*(x,y) u(x)]_{x=0}^1 ,$$

where the x-subscript denotes a derivative with respect to x. We have formally that

$$\int_0^1 A^* g^*(x,y) u(x) \, dx = \int_0^1 \delta(x - y) u(x) \, dx = u(y).$$

Hence, using the equations satisfied by g^* and u, and rearranging the result, we get

$$u(y) = \int_0^1 g^*(x,y)f(x)\,dx + [a(x)g_x^*(x,y)u(x)]_{x=0}^1.$$

Exchanging x and y in this equation, and using (10.25) to replace g^* by g, we obtain the following Green's function representation of the solution of (10.26):

$$u(x) = \int_0^1 g(x,y)f(y)\,dy + a(1)g_y(x,1)\alpha_1 - a(0)g_y(x,0)\alpha_0.$$

The above derivation of this representation does not constitute a proof. We can, however, verify the correctness of the result directly. From Proposition 10.12, the function

$$u_p(x) = \int_0^1 g(x,y)f(y)\,dy$$

is the solution of the nonhomogeneous equation $Au_p = f$ that satisfies the homogeneous boundary conditions $u_p(0) = u_p(1) = 0$. On the other hand, it follows from (10.20)–(10.23) that

$$u_h(x) = a(1)g_y(x,1)\alpha_1 - a(0)g_y(x,0)\alpha_0$$

is the solution of the homogeneous equation $Au_h = 0$ that satisfies the nonhomogeneous boundary conditions $u_h(0) = \alpha_0$, $u_h(1) = \alpha_1$.

10.4 Weak derivatives

In the previous sections, we considered "classical" differential operators that act on continuously differentiable functions. The resulting differential operators lack a number of desirable properties; for example, they are not closed or self-adjoint. To obtain such operators, we need to extend the domains of the classical differentiation operators to include functions whose weak derivatives belong to L^2. In this section, we define the notion of a weak L^2-derivative in terms of integration against test functions. We show that weakly differentiable functions can be approximated by smooth functions, and we use this fact to study some of their basic properties. We also define the Sobolev spaces H^k of functions with k square-integrable derivatives, and use them to give a precise description of the domains of some simple self-adjoint ordinary differential operators.

We begin by considering functions defined on $\mathbb{R}$. We say that $\varphi : \mathbb{R} \to \mathbb{C}$ is a *test function* if it has compact support and continuous derivatives of all orders. We denote the space of test functions by $C_c^\infty(\mathbb{R})$. The following example shows that there are many test functions.

Example 10.14 The function

$$\varphi(x) = \begin{cases} \exp\left[-1/(1-x^2)\right] & \text{if } |x| < 1, \\ 0 & \text{if } |x| \geq 1, \end{cases}$$

belongs to $C_c^\infty(\mathbb{R})$. All its derivatives exist and are equal to zero at $x = \pm 1$. This function is not analytic at $x = \pm 1$, however, since its Taylor series at these points converge to zero, rather than to the function itself. Rescaling this function,

$$\psi(x) = c\varphi\left(\frac{x - x_0}{\delta}\right),$$

we obtain a test function ψ supported on the interval $|x - x_0| \leq \delta$ whose integral has any desired value.

Before defining weak derivatives, we show that $C_c^\infty(\mathbb{R})$ is dense in $L^2(\mathbb{R})$. To do this, we approximate an L^2-function by its convolution with a smooth approximate identity, a technique called *mollification*.

Let $\varphi \in C_c^\infty(\mathbb{R})$ be a nonnegative test function with support $[-1, 1]$ and

$$\int_{\mathbb{R}} \varphi(x)\, dx = 1.$$

For $\epsilon > 0$, we let

$$\varphi_\epsilon(x) = \frac{1}{\epsilon}\varphi\left(\frac{x}{\epsilon}\right).$$

We call such a function φ_ϵ a *mollifier* or *averaging kernel*. The family $\{\varphi_\epsilon \mid \epsilon > 0\}$ is an approximate identity as $\epsilon \to 0^+$ since the support of φ_ϵ shrinks to the origin and each φ_ϵ has unit integral. If $u \in L^2(\mathbb{R})$, we define the *mollification* $u_\epsilon = \varphi_\epsilon * u$ of u, meaning that

$$u_\epsilon(x) = \int_{\mathbb{R}} \varphi_\epsilon(x - y)u(y)\, dy. \tag{10.27}$$

The function u_ϵ belongs to $C^\infty(\mathbb{R})$ because

$$u_\epsilon^{(k)}(x) = \int_{\mathbb{R}} \varphi_\epsilon^{(k)}(x - y)u(y)\, dy. \tag{10.28}$$

The differentiation under the integral sign is justified by the dominated convergence theorem.

Lemma 10.15 If $u \in L^2(\mathbb{R})$ and u_ϵ is defined by (10.27), where φ_ϵ is a mollifier, then $\|u_\epsilon\| \leq \|u\|$, where $\|\cdot\|$ denotes the L^2-norm.

Proof. Using the fact that φ_ϵ is nonnegative and has unit integral over $\mathbb{R}$, we find from the Cauchy-Schwarz inequality that

$$|u_\epsilon(x)| = \left| \int_{\mathbb{R}} \varphi_\epsilon^{1/2}(x - y)\varphi_\epsilon^{1/2}(x - y)u(y)\,dy \right|$$

$$\leq \left(\int_{\mathbb{R}} \varphi_\epsilon(x - y)\,dy \right)^{1/2} \left(\int_{\mathbb{R}} \varphi_\epsilon(x - y)|u(y)|^2\,dy \right)^{1/2}$$

$$= \left(\int_{\mathbb{R}} \varphi_\epsilon(x - y)|u(y)|^2\,dy \right)^{1/2}.$$

Squaring this equation and integrating the result with respect to x, we obtain that

$$\int_{\mathbb{R}} |u_\epsilon(x)|^2\,dx \leq \int_{\mathbb{R}} \left(\int_{\mathbb{R}} \varphi_\epsilon(x - y)|u(y)|^2\,dy \right)\,dx.$$

Exchanging the order of integration, which is justified by Fubini's theorem, we find that

$$\|u_\epsilon\|^2 \leq \int_{\mathbb{R}} \left(\int_{\mathbb{R}} \varphi_\epsilon(x - y)\,dx \right) |u(y)|^2\,dy = \|u\|^2.$$

$\square$

Using this lemma, we prove that the mollifications u_ϵ converge to u in L^2.

Theorem 10.16 The space $C_c^\infty(\mathbb{R})$ is dense in $L^2(\mathbb{R})$. If $u \in L^2(\mathbb{R})$, then $u_\epsilon \to u$ strongly in $L^2(\mathbb{R})$ as $\epsilon \to 0^+$.

Proof. Suppose that $u \in L^2(\mathbb{R})$. Let $\eta > 0$ be arbitrary. The space $C_c(\mathbb{R})$ of continuous functions with compact support is dense in $L^2(\mathbb{R})$ (see Theorem 12.50), so there is a $v \in C_c(\mathbb{R})$ such that $\|u - v\| < \eta/3$. We define $v_\epsilon = \varphi_\epsilon * v \in C_c^\infty(\mathbb{R})$. Then, from Lemma 10.15, we have

$$\|u_\epsilon - v_\epsilon\| \leq \|u - v\| < \eta/3.$$

The supports of v and v_ϵ are contained in a compact set. The argument in the proof of Theorem 7.2 implies that $v_\epsilon \to v$ uniformly as $\epsilon \to 0^+$, and therefore $v_\epsilon \to v$ in $L^2(\mathbb{R})$. There is a $\delta > 0$ such that $\|v - v_\epsilon\| < \eta/3$ for $0 < \epsilon < \delta$, and then

$$\|u - u_\epsilon\| \leq \|u - v\| + \|v - v_\epsilon\| + \|v_\epsilon - u_\epsilon\| < \eta.$$

It follows that $u_\epsilon \to u$ in $L^2(\mathbb{R})$.

$\square$

To motivate the definition of a weak L^2-derivative, we first consider $u \in C^1(\mathbb{R})$ with a "classical" pointwise, continuous derivative

$$v(x) = u'(x). \tag{10.29}$$

The use of this formula, followed by an integration by parts, implies that

$$\int_{\mathbb{R}} v\varphi\,dx = -\int_{\mathbb{R}} u\varphi'\,dx \qquad \text{for all } \varphi \in C_c^\infty(\mathbb{R}). \tag{10.30}$$

The boundary terms are zero because φ vanishes outside a compact set. Conversely, if $u \in C^1(\mathbb{R})$ satisfies (10.30) for some $v \in L^2(\mathbb{R})$, then another integration by parts implies that

$$\int_{\mathbb{R}} v\varphi \, dx = \int_{\mathbb{R}} u'\varphi \, dx \qquad \text{for all } \varphi \in C_c^\infty(\mathbb{R}).$$

Hence, $v = u'$ because $C_c^\infty(\mathbb{R})$ is dense in $L^2(\mathbb{R})$. Thus, (10.29) and (10.30) are equivalent when u is continuously differentiable. Equation (10.30) makes sense, however, if u and v are only square-integrable, because the derivative acts on the test function. Rewriting the integrals as inner products, we obtain the following definition of a weak derivative.

Definition 10.17 A function $u \in L^2(\mathbb{R})$ has a weak derivative $v = u' \in L^2(\mathbb{R})$ if

$$\langle v, \varphi \rangle = -\langle u, \varphi' \rangle \qquad \text{for all } \varphi \in C_c^\infty(\mathbb{R}).$$

The Sobolev space $H^k(\mathbb{R})$ consists of the functions with k square-integrable weak derivatives,

$$H^k(\mathbb{R}) = \left\{ u \in L^2(\mathbb{R}) \mid u, u', \ldots, u^{(k)} \in L^2(\mathbb{R}) \right\},$$

equipped with the following norm and inner product:

$$\|u\|_{H^k} = \left(\int_{\mathbb{R}} \left\{ |u|^2 + |u'|^2 + \ldots + |u^{(k)}|^2 \right\} dx \right)^{1/2},$$

$$\langle u, v \rangle_{H^k} = \int_{\mathbb{R}} \left\{ \overline{u}v + \overline{u}'v' + \ldots + \overline{u}^{(k)}v^{(k)} \right\} dx.$$

Proposition 10.18 The differentiation operator $D : H^1(\mathbb{R}) \subset L^2(\mathbb{R}) \to L^2(\mathbb{R})$ defined by $Du = u'$ is closed.

Proof. Suppose that $u_n \to u$ and $Du_n \to v$ in $L^2(\mathbb{R})$. It follows from this convergence and the definition of the weak derivative that for every test function φ,

$$\langle v, \varphi \rangle = \lim_{n \to \infty} \langle u'_n, \varphi \rangle = - \lim_{n \to \infty} \langle u_n, \varphi' \rangle = -\langle u, \varphi' \rangle.$$

Hence $u \in H^1(\mathbb{R})$, and $Du = v$, so D is closed. $\qquad\square$

The closedness of D implies that $H^k(\mathbb{R})$ is complete and therefore a Hilbert space. If a sequence (u_n) is Cauchy in H^k, then $(u_n^{(j)})$ is Cauchy in L^2 for each $j \leq k$. Since L^2 is complete, there are functions $v, v_j \in L^2$ such that $u_n \to v$ and $u_n^{(j)} \to v_j$ as $n \to \infty$. Since D is closed, it follows that $v_j = v^{(j)}$ for each $j \leq k$, so $u_n \to v$ in H^k.

An alternative, but equivalent, way to define L^2-derivatives is as the L^2-limit of smooth derivatives. Thus, we say that $u' = v$ if there is a sequence of smooth functions u_n such that $u_n \to u$ and $u'_n \to v$ in L^2. The equivalence of these definitions follows from the following theorem, which shows that any H^k-function can be approximated in the H^k-norm by a test function.

Theorem 10.19 The space $C_c^\infty(\mathbb{R})$ is dense in $H^k(\mathbb{R})$. If $u \in H^k(\mathbb{R})$ and $u_\epsilon = \varphi_\epsilon * u$, where φ_ϵ is a mollifier, then $u_\epsilon \to u$ strongly in $H^k(\mathbb{R})$ as $\epsilon \to 0^+$.

Proof. Suppose that $u \in H^k(\mathbb{R})$, and $u_\epsilon = \varphi_\epsilon * u$, where φ_ϵ is a mollifier. The function $\varphi_{\epsilon,x} : \mathbb{R} \to \mathbb{R}$ defined by

$$\varphi_{\epsilon,x}(y) = \varphi_\epsilon(x - y)$$

is a test function in $C_c^\infty(\mathbb{R})$. It therefore follows from (10.28) and the definition of the weak derivative that

$$
\begin{aligned}
u_\epsilon^{(j)}(x) &= \int_\mathbb{R} \frac{\partial^j}{\partial x^j} \left[\varphi_\epsilon(x - y)\right] u(y)\, dy \\
&= (-1)^j \int_\mathbb{R} \frac{\partial^j}{\partial y^j} \left[\varphi_\epsilon(x - y)\right] u(y)\, dy \\
&= \int_\mathbb{R} \varphi_\epsilon(x - y) u^{(j)}(y)\, dy
\end{aligned}
$$

for every $j \le k$ and $x \in \mathbb{R}$. Theorem 10.16 implies that $u_\epsilon^{(j)} \to u^{(j)}$ in $L^2(\mathbb{R})$, so $u_\epsilon \to u$ in $H^k(\mathbb{R})$.

If u does not have compact support, then u_ϵ does not have compact support either. To show that $C_c^\infty(\mathbb{R})$ is dense in $H^k(\mathbb{R})$, we truncate u before mollification. We choose $\psi \in C_c^\infty(\mathbb{R})$ such that

$$
\psi(x) = \begin{cases} 1 & \text{if } |x| \le 1, \\ 0 & \text{if } |x| \ge 2, \end{cases}
$$

and define $\psi_n(x) = \psi(x/n)$. Then $u_n = \psi_n u$ has compact support, and $u_n \in H^k(\mathbb{R})$ when $u \in H^k(\mathbb{R})$. One can show that $u_n \to u$ in $H^k(\mathbb{R})$ as $n \to \infty$, and we have just proved that $\varphi_\epsilon * u_n \to u_n$ as $\epsilon \to 0^+$. Since $\varphi_\epsilon * u_n \in C_c^\infty(\mathbb{R})$, the density follows. $\qquad \square$

As an illustration of the use of mollification, we show that integration by parts holds for H^1-functions.

Proposition 10.20 Suppose that $u, v \in H^1(\mathbb{R})$, then

$$\int_\mathbb{R} uv'\, dx = -\int_\mathbb{R} u'v\, dx. \tag{10.31}$$

Proof. From Theorem 10.19, there are sequences (u_n) and (v_n) in $C_c^\infty(\mathbb{R})$ such that $u_n \to u$ and $v_n \to v$ in $H^1(\mathbb{R})$. Since u_n and v_n vanish outside a compact set, we have

$$\int_\mathbb{R} u_n v_n'\, dx = -\int_\mathbb{R} u_n' v_n\, dx.$$

Taking the limit of this equation as $n \to \infty$, and using the continuity of the L^2-inner product with respect to L^2-convergence, we obtain (10.31). $\square$

Example 10.21 Let $A : H^1(\mathbb{R}) \subset L^2(\mathbb{R}) \to L^2(\mathbb{R})$ be the operator $A = iD$, meaning that $Au = iu'$. We claim that A is self adjoint. For every $u, v \in H^1(\mathbb{R})$, we have

$$\langle Au, v \rangle = -i \int_{\mathbb{R}} \overline{u}'v \, dx = i \int_{\mathbb{R}} \overline{u}v' \, dx = \langle u, Av \rangle.$$

Hence, A is symmetric and $\mathcal{D}(A^*) \supset H^1(\mathbb{R})$. If $v \in \mathcal{D}(A^*)$, then there is a $w \in L^2(\mathbb{R})$ such that

$$\langle iu', v \rangle = \langle u, w \rangle \qquad \text{for all } u \in H^1(\mathbb{R}).$$

Since $H^1(\mathbb{R})$ contains $C_c^{\infty}(\mathbb{R})$, it follows that

$$\langle \varphi', v \rangle = \langle \varphi, iw \rangle \qquad \text{for all } \varphi \in C_c^{\infty}(\mathbb{R}),$$

which means that $v \in H^1(\mathbb{R})$ and $w = iv'$. Thus, $\mathcal{D}(A^*) \subset H^1(\mathbb{R})$, so $\mathcal{D}(A^*) = H^1(\mathbb{R})$, and A is self-adjoint.

We now consider functions defined on a bounded open interval $(0, 1)$. The space of test functions $C_c^{\infty}((0, 1))$ consists of smooth functions that vanish outside a closed interval contained strictly inside $(0, 1)$. A function $v \in L^2([0, 1])$ is the weak L^2-derivative of $u \in L^2([0, 1])$ if

$$\int_0^1 v\varphi \, dx = - \int_0^1 u\varphi' \, dx \qquad \text{for all } \varphi \in C_c^{\infty}((0, 1)).$$

The Sobolev space $H^k((0, 1))$ consists of the functions in $L^2([0, 1])$ with k weak derivatives in $L^2([0, 1])$.

Theorem 10.22 *The space $C^{\infty}([0, 1])$ is dense in $H^k((0, 1))$.*

Proof.   We would like to obtain a smooth approximation of $u \in H^k((0, 1))$ by extending u to a function

$$\widetilde{u}(x) = \begin{cases} u(x) & \text{if } x \in (0, 1), \\ 0 & \text{if } x \notin (0, 1), \end{cases}$$

in $L^2(\mathbb{R})$, and mollifying the extension $\widetilde{u}$. However, $\widetilde{u}$ need not belong to $H^k(\mathbb{R})$ because it may be discontinuous at the endpoints of $(0, 1)$, so we cannot conclude immediately that the restriction of $\varphi_\epsilon * \widetilde{u}$ to $(0, 1)$ converges to u in $H^k((0, 1))$. The proof therefore requires a more complicated argument, which we outline without giving all the details. For $\delta > 0$, we define the stretching map $L_\delta : (0, 1) \to (-\delta, 1+\delta)$ by

$$L_\delta(x) = (1 + 2\delta)\left(x - \frac{1}{2}\right) + \frac{1}{2}.$$

We define $u_\delta \in H^k((-\delta, 1+\delta))$ by $u_\delta = u \circ L_\delta^{-1}$. Then one can show that the restriction of u_δ to $(0,1)$ converges to u in $H^k((0,1))$ as $\delta \to 0^+$. We extend u_δ by zero to obtain $\tilde{u}_\delta \in L^2(\mathbb{R})$. Let φ_ϵ be a mollifier, and

$$\varphi_{\epsilon,x}(y) = \varphi_\epsilon(x - y).$$

For $x \in (0,1)$ and $\epsilon < \delta$, we have $\varphi_{\epsilon,x} \in C_c^\infty((-\delta, 1+\delta))$. The restriction of $\varphi_\epsilon * \tilde{u}_\delta$ to $(0,1)$ is therefore a C^∞ function on $[0,1]$ that converges to the restriction of u_δ to $(0,1)$ in $H^k((0,1))$ as $\epsilon \to 0^+$. The result then follows. $\qquad\square$

Although $C_c^\infty(\mathbb{R})$ is dense in $H^k(\mathbb{R})$ and $C_c^\infty((0,1))$ is dense in $L^2([0,1])$, it is not true that $C_c^\infty((0,1))$ is dense in $H^k((0,1))$ for $k \geq 1$.

Definition 10.23 The Sobolev space

$$H_0^k((0,1)) = \overline{C_c^\infty((0,1))} \subset H^k((0,1))$$

is the closure of $C_c^\infty((0,1))$ in $H^k((0,1))$.

It follows from the Sobolev embedding theorem below that $H_0^k((0,1))$ consists of the functions in $H^k((0,1))$ whose derivatives of order less than or equal to $k-1$ vanish at the endpoints of $(0,1)$.

In Section 7.2, we proved the Sobolev embedding theorem for periodic functions by using Fourier series. Here we give a different proof, for which we need the following lemma.

Lemma 10.24 Suppose that $h : [0,1] \to \mathbb{R}$ is a continuous function such that

$$\int_0^1 h(x)\, dx = 1.$$

Define $k : [0,1] \times [0,1] \to \mathbb{R}$ by

$$k(x,y) = \begin{cases} \int_0^y h(t)\, dt & \text{if } 0 \leq y \leq x, \\ -\int_y^1 h(t)\, dt & \text{if } x < y \leq 1. \end{cases}$$

If $u \in C^1([0,1])$, then

$$u(x) = \int_0^1 u(y)h(y)\, dy + \int_0^1 k(x,y)u'(y)\, dy \qquad \text{for all } 0 \leq x \leq 1.$$

Proof. If $u \in C^1([0,1])$, the fundamental theorem of calculus implies that, for every $x, y \in [0,1]$,

$$u(x) = u(y) + \int_y^x u'(t)\, dt.$$

Multiplying this equation by $h(y)$ and integrating the result, we obtain that

$$u(x) = \int_0^1 u(y)h(y)\, dy + \int_0^1 \left(\int_y^x u'(t)\, dt \right) h(y)\, dy.$$

Exchanging the order of integration, we find that

$$\int_0^1 \left(\int_y^x u'(t)\, dt \right) h(y)\, dy = \int_0^x \left(\int_a^t h(y)\, dy \right) u'(t)\, dt$$

$$- \int_x^1 \left(\int_t^1 h(y)\, dy \right) u'(t)\, dt$$

$$= \int_0^1 k(x,t)u'(t)\, dt,$$

and the result follows. $\square$

Theorem 10.25 (Sobolev embedding) The space $H^1((0,1))$ is a subset of $C([0,1])$. There is a constant $C > 0$ such that

$$\|u\|_\infty \le C\|u\|_{H^1} \qquad \text{for all } u \in H^1((0,1)). \tag{10.32}$$

Proof.   First, suppose that $u \in C^\infty([0,1])$. Then, from Lemma 10.24 and the Cauchy-Schwarz inequality, we find that

$$
\begin{aligned}
|u(x)| &\le \left| \int_0^1 u(y)h(y)\, dy \right| + \left| \int_0^1 k(x,y)u'(y)\, dy \right| \\
&\le \|h\|_{L^2}\|u\|_{L^2} + \|k(x,\cdot)\|_{L^2}\|u'\|_{L^2} \\
&\le C\|u\|_{H^1},
\end{aligned}
$$

since $\|k(x,\cdot)\|_{L^2}$ is bounded uniformly in x for a continuous function h. Taking the supremum of this inequality with respect to x, we obtain that $\|u\|_\infty \le C\|u\|_{H^1}$ for every $u \in C^\infty([0,1])$. Since C^∞ is dense in H^1, it follows that this inequality holds for every $u \in H^1$. Furthermore, every $u \in H^1$ is the uniform limit of a sequence of C^∞-functions, and is therefore continuous. $\square$

Strictly speaking, an element of H^1 is an equivalence class of square-integrable functions that are equal almost everywhere, and the embedding theorem states that each such equivalence class contains a continuous function. An alternative way to state this result is that there is a continuous map, or embedding,

$$J : H^1((0,1)) \to C([0,1])$$

that identifies a function u, regarded as an element of $H^1((0,1))$, with the same function u, regarded as an element of $C([0,1])$. The following theorem shows that this embedding is compact.

Theorem 10.26 (Rellich) A bounded subset of $H^1((0,1))$ is a precompact subset of $C([0,1])$.

Proof. Since $C^\infty([0,1])$ is dense in $H^1((0,1))$, it is sufficient to show that a subset of $C^\infty([0,1])$ that is bounded in $H^1((0,1))$ is precompact in $C([0,1])$. Suppose that $\mathcal{F}$ is a subset of $C^\infty([0,1])$ such that there is a constant M with

$$\|u\|_{H^1} \le M \qquad \text{for all } u \in \mathcal{F}.$$

From (10.32), the set $\mathcal{F}$ is bounded in $C([0,1])$. Moreover, by the fundamental theorem of calculus and the Cauchy-Schwarz inequality, we have for every $u \in \mathcal{F}$ and $x, y \in [0,1]$ that

$$
\begin{aligned}
|u(x) - u(y)| &= \left| \int_x^y u'(t)\,dt \right| \\
&= \left| \int_0^1 \chi_{[x,y]}(t) u'(t)\,dt \right| \\
&\le |x - y|^{1/2} \left(\int_0^1 |u'(t)|^2\,dt \right)^{1/2} \\
&\le M |x - y|^{1/2}.
\end{aligned}
$$

Here, $\chi_{[x,y]}$ is the characteristic function of the interval $[x,y]$. Thus $\mathcal{F}$ is equicontinuous, and therefore the Arzelà-Ascoli theorem implies that it is precompact in $C([0,1])$. $\qquad\square$

A function $u \in C([0,1])$ that satisfies

$$|u(x) - u(y)| \le M|x - y|^r \qquad \text{for all } x, y \in [0,1]$$

for constants $M > 0$ and $0 < r \le 1$ is said to be *Hölder continuous* with exponent r. Thus, the proof of Theorem 10.26 shows that every $u \in H^1((0,1))$ is Hölder continuous with exponent $1/2$. For a generalization of this result, see Theorem 12.73.

Proposition 10.27 If A is the second-order ordinary differential operator defined in (10.4), where a, b, c are smooth coefficient functions, then Green's formula (10.6) holds for all $u, v \in H^2((0,1))$.

Proof. If $u, v \in H^2((0,1))$, then there are sequences (u_n), (v_n) in $C^\infty([0,1])$ such that $u_n \to u$ and $v_n \to v$ in $H^2((0,1))$. From Green's formula, we have

$$\langle Au_n, v_n \rangle - \langle u_n, A^* v_n \rangle = \left[a\left(\overline{u_n}' v_n - \overline{u_n} v_n' \right) + (c - a') \overline{u_n} v_n \right]_0^1.$$

Letting $n \to \infty$, we obtain Green's formula for u and v, because $Au_n \to Au$, $Av_n \to Av$ in L^2, and, from the Sobolev embedding theorem, the boundary terms converge pointwise. $\qquad\square$

Example 10.28 Let us prove that the second derivative operator $A = -D^2$ with domain

$$\mathcal{D}(A) = \left\{ u \in H^2((0,1)) \mid u(0) = u(1) = 0 \right\}$$

is self-adjoint. If $v \in \mathcal{D}(A^*)$, then there is a $w \in L^2([0,1])$ such that

$$\langle -u'', v \rangle = \langle u, w \rangle \qquad \text{for all } u \in \mathcal{D}(A).$$

Since $\mathcal{D}(A) \supset C_c^\infty((0,1))$, it follows from the definition of the weak derivative that $v \in H^2((0,1))$ and $w = -v''$. Hence, $\mathcal{D}(A^*) \subset H^2((0,1))$, and $A^* = -D^2$ on its domain. If $u \in \mathcal{D}(A)$ and $v \in H^2((0,1))$, then an integration by parts implies that

$$\langle -u'', v \rangle = \langle u, -v'' \rangle + [-\overline{u}'v]_0^1.$$

Thus, v belongs to $\mathcal{D}(A^*)$ if and only if $v(0) = v(1) = 0$, so $\mathcal{D}(A) = \mathcal{D}(A^*)$ and $A = A^*$.

As the previous example illustrates, the direct verification of self-adjointness may be nontrivial even for the simplest unbounded operators. The following result, which we state without proof, gives a basic criterion for self-adjointness.

Theorem 10.29 Let A be a closed, symmetric operator on a Hilbert space $\mathcal{H}$. Then the following statements are equivalent:

(a) A is self-adjoint;
(b) $\ker(A^* \pm iI) = \{0\}$;
(c) $\operatorname{ran}(A \pm iI) = \mathcal{H}$.

If $m = \dim \ker(A^* - iI)$ and $n = \dim \ker(A^* + iI)$, then the pair (m,n) is called the *deficiency index* of A. Thus, a closed, symmetric operator is self-adjoint if and only if its deficiency index is $(0,0)$.

10.5   The Sturm-Liouville eigenvalue problem

In this section, we study the *Sturm-Liouville eigenvalue problem*

$$-(pu')' + qu = \lambda u, \qquad (10.33)$$
$$u(0) = u(1) = 0,$$

where the coefficients p, q are given real-valued functions, and $\lambda \in \mathbb{R}$. For definiteness, we consider the Dirichlet problem, but other self-adjoint boundary conditions can be analyzed in a similar way. Equation (10.33) is the spectral problem for the self-adjoint *Sturm-Liouville operator* $A : \mathcal{D}(A) \subset L^2([0,1]) \to L^2([0,1])$ defined by

$$Au = -(pu')' + qu, \qquad (10.34)$$
$$\mathcal{D}(A) = \{u \in H^2((0,1)) \mid u(0) = u(1) = 0\}. \qquad (10.35)$$

Theorem 10.30 Suppose that $p \in C^1([0,1])$, $q \in C([0,1])$ are real-valued functions and $p(x) > 0$ for all $x \in [0,1]$. There is an orthonormal basis of $L^2([0,1])$ that consists of eigenfunctions of the Sturm-Liouville eigenvalue problem (10.33). The eigenvalues $\lambda_1 < \lambda_2 < \ldots$ are real and simple, and $\lambda_n \to \infty$ as $n \to \infty$.

Proof.   We begin by showing that if λ is real and sufficiently negative, then the only solution of (10.33) is $u = 0$, so λ is not an eigenvalue of A. We take the inner product of (10.33) with u, and integrate the result by parts. This gives:

$$\int_0^1 \left\{ p\,|u'|^2 + q\,|u|^2 \right\} \, dx = \lambda \int_0^1 |u|^2 \, dx. \tag{10.36}$$

We let

$$\alpha = \min_{0 \le x \le 1} p(x), \quad \beta = \min_{0 \le x \le 1} q(x). \tag{10.37}$$

Since $p > 0$, we have $\alpha > 0$; if $q > 0$, then $\beta > 0$ also, but we may have $\beta \le 0$. Using (10.37) in (10.36), and rearranging the result, we find that

$$\alpha \int_0^1 |u'|^2 \, dx + (\beta - \lambda) \int_0^1 |u|^2 \, dx \le 0.$$

It follows that if $\lambda < \beta$, then

$$\int_0^1 |u'|^2 \, dx = \int_0^1 |u|^2 \, dx = 0,$$

so $u = 0$.

This result shows that the kernel of $A - \lambda I$ is zero when $\lambda < \beta$. From what we have shown previously, the Green's function g_λ of $A - \lambda I$ exists. Therefore, λ is in the resolvent set of A, and the self-adjoint resolvent operator R_λ is given by

$$R_\lambda = (\lambda I - A)^{-1} : L^2([0,1]) \to L^2([0,1]),$$
$$R_\lambda f(x) = -\int_0^1 g_\lambda(x,y) f(y) \, dy.$$

Since g_λ is continuous, we certainly have

$$\int_0^1 \int_0^1 [g_\lambda(x,y)]^2 \, dx dy < \infty,$$

so R_λ is Hilbert-Schmidt and hence compact. The spectral theorem for compact, self-adjoint operators implies that there is an orthonormal basis of $L^2([0,1])$ consisting of eigenvectors $\{u_n \mid n \in \mathbb{N}\}$ of R_λ with eigenvalues $\{\mu_n \mid n \in \mathbb{N}\}$ such that $\mu_n \to 0$ as $n \to \infty$. Since $(\lambda I - A)R_\lambda = I$, we have $u_n \in \mathcal{D}(A)$ and $Au_n = \lambda_n u_n$, where

$$\beta \le \lambda_n = \lambda - \frac{1}{\mu_n},$$

so $\lambda_n \to \infty$ as $n \to \infty$. The Sturm-Liouville operator A therefore has a complete orthonormal set of eigenvectors that forms a basis of $L^2([0,1])$.

If an eigenvalue λ is not simple, then (10.33) has a pair of linearly independent solutions. It follows that every solution of the Sturm-Liouville equation

$$-(pu')' + qu = \lambda u$$

vanishes at $x = 0, 1$ since it is a linear combination of eigenvectors. This contradicts the existence of a solution of the initial value problem with nonzero initial data for $u(0)$. $\qquad\qquad\square$

The compactness of the resolvent may also be obtained as a consequence of Rellich's theorem, in Theorem 10.26. We define a symmetric, sesquilinear form a on $H_0^1\left((0,1)\right)$ by

$$a(u,v) = \int_0^1 \{p\bar{u}'v' + q\bar{u}v\}\, dx. \qquad (10.38)$$

We call a the *Dirichlet form* of A. For $u, v \in \mathcal{D}(A)$, we have

$$a(u,v) = \langle Au, v\rangle = \langle u, Av\rangle.$$

The set $\mathcal{D}(A) \times \mathcal{D}(A)$ is dense in $H_0^1\left((0,1)\right) \times H_0^1\left((0,1)\right)$, and the form extends continuously to the larger space. The associated quadratic form $a(u,u)$ on $H_0^1((0,1))$ is given by

$$a(u,u) = \int_0^1 \left\{p\,|u'|^2 + q\,|u|^2\right\}\, dx.$$

As we saw above, we have the estimate

$$a(u,u) \geq \alpha \int_0^1 |u'|^2\, dx + \beta \int_0^1 |u|^2\, dx.$$

It follows that if $u \in \mathcal{D}(A)$, then

$$\langle (A - \lambda I)u, u\rangle \geq \alpha \int_0^1 |u'|^2\, dx + (\beta - \lambda) \int_0^1 |u|^2\, dx.$$

If $\lambda < \beta$, this estimate implies that $(\lambda I - A)^{-1}$ maps bounded sets in L^2 to bounded sets in H_0^1, which are precompact in L^2 by Rellich's theorem. Hence, A has a compact resolvent.

The operator A is diagonal in a basis of eigenvectors. We may therefore solve the Sturm-Liouville BVP

$$-(pu')' + qu = \lambda u + f,$$
$$u(0) = u(1) = 0,$$

by expanding u and f with respect to the orthonormal basis of eigenvectors. Assuming that λ is not an eigenvalue of A, the solution is

$$u(x) = \sum_{n=1}^{\infty} \frac{\langle f_n, u \rangle}{\lambda_n - \lambda} u_n(x),$$

where the series converges in $L^2([0,1])$. We may write the operator A as

$$A = \sum_{n=1}^{\infty} \lambda_n u_n \otimes u_n, \quad \mathcal{D}(A) = \left\{ \sum_{n=1}^{\infty} c_n u_n \in \mathcal{H} \,\middle|\, \sum_{n=1}^{\infty} (1 + \lambda_n^2) |c_n|^2 < \infty \right\},$$

where the sum converges strongly on the domain of A. The resolvent operator of A is

$$R_\lambda = \sum_{n=1}^{\infty} \frac{u_n \otimes u_n}{\lambda - \lambda_n},$$

where the sum converges uniformly for $\lambda \in \rho(A)$, and the Green's function g_λ of $A - \lambda I$ is given by

$$g_\lambda(x,y) = \sum_{n=1}^{\infty} \frac{u_n(x)u_n(y)}{\lambda_n - \lambda},$$

where the sum converges in $L^2([0,1] \times [0,1])$. The resolvent operator and the Green's function, regarded as functions of λ, have poles at the eigenvalues of A.

Example 10.31 The simplest example of a Sturm-Liouville eigenvalue problem is

$$-u'' = \lambda u, \qquad u(0) = u(1) = 0.$$

The eigenfunctions u_n and eigenvalues λ_n, where $n = 1, 2, 3, \ldots$, are given by

$$u_n(x) = \sqrt{2} \sin(n\pi x), \qquad \lambda_n = n^2 \pi^2.$$

The associated eigenfunction expansion is a Fourier sine expansion. Neumann boundary conditions lead to a Fourier cosine expansion. Thus, Theorem 10.30 provides another proof of the completeness of the Fourier basis. In this example, the nth eigenfunction has $n - 1$ zeros inside the interval $(0, 1)$. This property holds for all regular Sturm-Liouville eigenvalue problems (see Coddington and Levinson [6]).

A Sturm-Liouville problem is said to be *regular* if it is posed on a bounded interval $[a, b]$ and $p(x) \neq 0$ for every $a \leq x \leq b$; otherwise, it is said to be *singular*. We have just proved that a regular Sturm-Liouville eigenvalue problem has a complete orthonormal set of eigenvectors. The resolvent operator of a singular Sturm-Liouville operator may or may not be compact and, if it is not compact, then the corresponding Sturm-Liouville eigenvalue problem may have a continuous spectrum as well as, or instead of, a point spectrum.

Example 10.32 The function $p(x) = 1 - x^2$ vanishes at the boundaries of the interval $[-1, 1]$. The corresponding singular Sturm-Liouville eigenvalue problem on $[-1, 1]$, with $q = 0$, is

$$- \left[(1 - x^2)u'\right]' = \lambda u \qquad \text{for } -1 < x < 1,$$

$$\int_{-1}^{1} \left\{ (1 - x^2) |u'|^2 + |u|^2 \right\} dx < \infty. \qquad (10.39)$$

This eigenvalue problem has a complete orthogonal set of eigenvectors, the Legendre polynomials, with eigenvalues $\lambda_n = n(n + 1)$ (see Exercise 6.12). Since $\lambda_n = n(n + 1) \to \infty$ as $n \to \infty$, the resolvent operator $(-I - A)^{-1}$ is compact. No boundary conditions are required at the singular endpoints. The condition in (10.39) rules out singular solutions which are unbounded at $x = \pm 1$.

More generally, if $m \in \mathbb{N}$ is a positive integer, then the singular Sturm-Liouville problem,

$$- \left[(1 - x^2)u'\right]' + \frac{m^2}{1 - x^2}u = \lambda u \qquad \text{for } -1 < x < 1,$$

$$\int_{-1}^{1} \left\{ (1 - x^2) |u'|^2 + \frac{|u|^2}{1 - x^2} \right\} dx < \infty,$$

has eigenvalues $\lambda_n = n(n + 1)$, where $n = m, m + 1, \ldots$. The corresponding eigenfunctions are the *Legendre functions* $u = P_n^m$. They may be expressed in terms of the Legendre polynomial $P_n = P_n^0$ as

$$
\begin{aligned}
P_n^m(x) &= (-1)^m \left(1 - x^2\right)^{m/2} \frac{d^m}{dx^m} P_n(x) \\
&= (-1)^m \left(1 - x^2\right)^{m/2} \frac{1}{2^n n!} \frac{d^{m+n}}{dx^{m+n}} \left(x^2 - 1\right)^n.
\end{aligned}
$$

Example 10.33 An example of a Sturm-Liouville operator on the whole of $\mathbb{R}$ with a compact inverse is the quantum harmonic oscillator,

$$Au = -u'' + x^2 u, \qquad \mathcal{D}(A) = \left\{ u \in H^2(\mathbb{R}) \mid x^2 u \in L^2(\mathbb{R}) \right\}.$$

Its eigenvectors are the Hermite functions (see Exercise 6.14), which form a complete orthonormal set in $L^2(\mathbb{R})$.

Example 10.34 An example of a Sturm-Liouville operator on $L^2(\mathbb{R})$ with a non-compact inverse is given by

$$Au = -u'' + u, \qquad \mathcal{D}(A) = H^2(\mathbb{R}).$$

The inverse of A is given by

$$A^{-1}u(x) = \frac{1}{2} \int_{\mathbb{R}} e^{-|x-y|} u(y)\, dy.$$

The spectrum of A is $[1, \infty)$ and is continuous. For $\lambda \in \mathbb{C} \setminus [1, \infty)$, the resolvent operator $R_\lambda = (\lambda I - A)^{-1}$ is given by

$$R_\lambda u(x) = -\frac{1}{2\sqrt{1-\lambda}} \int_{\mathbb{R}} e^{-\sqrt{1-\lambda}|x-y|} u(y) \, dy,$$

where we use the branch of the square-root with $\mathrm{Re}\,\sqrt{z} > 0$ in order to ensure that the kernel decays at infinity. The resolvent operator has a branch cut along the continuous spectrum of A.

10.6 Laplace's equation

Adjoint operators and Green's functions can be defined for partial differential equations as well as ordinary differential equations. If the partial differential operator has a compact, self-adjoint inverse (or resolvent), then it has a complete orthonormal set of eigenvectors. In this section, we discuss Laplace's equation, which is one of the most important linear PDEs. We will consider classical solutions in this section. Weak solutions are discussed further in Section 12.11.

Let Ω be a bounded, open, connected set in $\mathbb{R}^n$, with a sufficiently regular boundary $\partial\Omega$. We will not make the required regularity assumptions precise here (see Gilbarg and Trudinger [15] for a detailed discussion). We denote by $C^k(\Omega)$ the space of functions that are k-times continuously differentiable in Ω, and by $C^k(\overline{\Omega})$ the space of functions whose partial derivatives of order less than or equal to k exist in Ω and extend to a continuous function on the closure $\overline{\Omega}$.

If $\mathbf{F} : \overline{\Omega} \to \mathbb{R}^n$ is a continuously differentiable vector field on $\overline{\Omega}$, then the divergence theorem states that

$$\int_\Omega \nabla \cdot \mathbf{F} \, dx = \int_{\partial\Omega} \mathbf{F} \cdot \mathbf{n} \, dS, \tag{10.40}$$

where $\mathbf{n}$ is the unit outward normal to $\partial\Omega$, and dS is an element of $(n-1)$-dimensional surface area on $\partial\Omega$.

The Laplacian operator $-\Delta$ acting on a function $u(x)$ where $x = (x_1, \dots, x_n) \in \mathbb{R}^n$ is given by

$$-\Delta u = -\sum_{i=1}^{n} \frac{\partial^2 u}{\partial x_i^2}.$$

It is convenient to introduce a minus sign in the definition of the Laplacian operator because Δ is a negative operator. The Dirichlet problem for the Laplacian on Ω is

$$-\Delta u = f \quad \text{in } \Omega, \tag{10.41}$$

$$u = h \quad \text{on } \partial\Omega.$$

Other types of boundary conditions, such as Neumann conditions

$$\frac{\partial u}{\partial n} = h \qquad \text{on } \partial\Omega,$$

where $\partial u/\partial n = \nabla u \cdot \mathbf{n}$ is the outward normal derivative of u, can be treated in a similar way. First, we show that the Laplacian is formally self-adjoint.

Theorem 10.35 (Green's) If $u, v \in C^2(\overline{\Omega})$, then

$$\int_\Omega (u\Delta v - v\Delta u)\, dx = \int_{\partial\Omega} \left(u\frac{\partial v}{\partial n} - v\frac{\partial u}{\partial n} \right) dS.$$

Proof. The result follows from an integration of the vector identity

$$u\Delta v - v\Delta u = \nabla \cdot (u\nabla v - v\nabla u)$$

over Ω and an application of the divergence theorem. ∎

If $Bu = 0$ is a boundary condition for the Laplacian, then we define the adjoint boundary condition $B^*v = 0$ by the requirement that the boundary terms in Green's formula vanish. If

$$\langle u, v \rangle = \int_\Omega \overline{u}v\, dx$$

is the L^2-inner product on Ω, then we have that

$$\langle \Delta u, v \rangle = \langle u, \Delta v \rangle \qquad \text{for all } u, v \in C^2(\overline{\Omega}) \text{ such that } Bu = B^*v = 0.$$

For example, the adjoint boundary condition to $u = 0$ is $v = 0$, so the Dirichlet problem for Laplace's equation is self-adjoint.

The n-dimensional δ-function has the formal properties

$$\delta(x) = 0 \quad \text{for } x \neq y, \qquad \int_{\mathbb{R}^n} \delta(x)\, dx = 1, \qquad \int_{\mathbb{R}^n} \delta(x-y)f(y)\, dy = f(y)$$

for any continuous function $f : \mathbb{R}^n \to \mathbb{C}$. The Green's function $g(x, y)$ of the Dirichlet problem for the Laplacian is the distributional solution of

$$-\Delta g = \delta(x - y) \qquad \text{for } x \in \Omega, \tag{10.42}$$

$$g(x, y) = 0 \qquad \text{for } x \in \partial\Omega.$$

The self-adjointness of the boundary value problem implies that g is symmetric, meaning that $g(x, y) = g(y, x)$.

The Green's function representation of the solution of (10.41) follows formally from Green's formula:

$$\int_\Omega \{g\Delta u - u\Delta g\}\, dx = \int_{\partial\Omega} \left\{ g\frac{\partial u}{\partial n} - u\frac{\partial g}{\partial n} \right\} dS.$$

Using (10.41) and (10.42) in this equation, and indicating the integration variable explicitly, we obtain that

$$\int_\Omega \{-g(x,y)f(x) + u(x)\delta(x-y)\}\, dx = -\int_{\partial\Omega} h(x)\frac{\partial g(x,y)}{\partial n(x)}\, dS(x).$$

Evaluating the integral involving the delta function, exchanging x and y, and using the symmetry of the Green's function, we find that

$$u(x) = \int_\Omega g(x,y)f(y)\, dy - \int_{\partial\Omega} h(y)\frac{\partial g(x,y)}{\partial n(y)}\, dS(y).$$

Thus, we can represent the solution of (10.41) for general data $f : \Omega \to \mathbb{R}$ and $h : \partial\Omega \to \mathbb{R}$ in terms of the Green's function.

To give a nondistributional characterization of the Green's function, we integrate (10.42) over a small ball $B_\epsilon(y)$ of radius ϵ centered at y, use the divergence theorem, and let $\epsilon \to 0^+$. This gives

$$\lim_{\epsilon \to 0^+} \int_{\partial B_\epsilon(y)} \frac{\partial g}{\partial n}(x,y)\, dx = -1 \qquad \text{for every } y \in \Omega, \tag{10.43}$$

where $\partial/\partial n$ is the unit outward normal derivative to the ball. We also have

$$-\Delta g = 0 \qquad \text{for } x,y \in \Omega \text{ and } x \neq y,$$
$$g(x,y) = 0 \qquad \text{for } x \in \partial\Omega \text{ and } y \in \Omega.$$

For most domains Ω, it is not possible to obtain an explicit analytical expression for g. A simple solvable case is that of the free-space Green's function g_f defined on $\mathbb{R}^n$. In view of the rotational invariance of the Laplacian, we look for a solution $g_f = g_f(r)$ that depends only on $r = |x - y|$, where $|\cdot|$ denotes the Euclidean norm on $\mathbb{R}^n$. The polar form of Laplace's equation implies that

$$\frac{1}{r^{n-1}}\frac{d}{dr}\left(r^{n-1}\frac{dg_f}{dr}\right) = 0 \qquad \text{for } r > 0. \tag{10.44}$$

The solution of (10.44) is

$$g_f = c_2 \log\left(\frac{1}{r}\right) \qquad \text{when } n = 2,$$

$$g_f = \frac{c_n}{r^{n-2}} \qquad \text{when } n \geq 3, \tag{10.45}$$

where c_n is a constant, and we omit an arbitrary additive constant.

The radial derivative of g is constant on any sphere centered at y, so the singularity condition in (10.43) implies that

$$\lim_{r \to 0^+} r^{n-1}\frac{dg_f}{dr} = -\frac{1}{\omega_n}, \tag{10.46}$$

where ω_n is the area of the unit sphere in $\mathbb{R}^n$. This area is given by

$$\omega_n = \frac{2\pi^{n/2}}{\Gamma(n/2)},$$

where the *Gamma function* Γ is defined, for $x > 0$, by

$$\Gamma(x) = \int_0^\infty e^{-t} t^{x-1}\, dt.$$

One can show that

$$\Gamma(x+1) = x\Gamma(x), \qquad \Gamma(1) = 1, \qquad \Gamma\left(\frac{1}{2}\right) = \sqrt{\pi}.$$

Hence, for each $n \in \mathbb{N}$,

$$\Gamma(n) = (n-1)!, \qquad \Gamma\left(n+\frac{1}{2}\right) = \left(k - \frac{1}{2}\right)\left(k - \frac{3}{2}\right) \cdots \left(\frac{1}{2}\right)\sqrt{\pi},$$

which gives $\omega_2 = 2\pi$ (the length of the unit circle), and $\omega_3 = 4\pi$ (the area of the unit sphere).

Using (10.45) in (10.46), we find that

$$c_2 = \frac{1}{\omega_2}, \qquad c_n = \frac{1}{(n-2)\omega_n} \quad \text{when } n \geq 3.$$

Thus, the free-space Green's function g_f of Laplace's equation in two and three space dimensions is given by

$$g_f(x,y) = \frac{1}{2\pi} \log\left(\frac{1}{|x-y|}\right) \qquad \text{when } n = 2,$$

$$g_f(x,y) = \frac{1}{4\pi|x-y|} \qquad \text{when } n = 3.$$

In contrast to the one-dimensional case, the Green's function is unbounded at $x = y$.

Returning to the Green's function for Laplace's equation on a bounded domain, we may write the solution of (10.42) in the form

$$g(x,y) = g_f(x-y) + \varphi(x,y),$$

where $\varphi(x,y)$ satisfies

$$-\Delta\varphi = 0 \qquad x \in \Omega,$$
$$\varphi(x,y) = -g_f(x-y) \qquad x \in \partial\Omega.$$

If $y \in \Omega$, then the boundary data $-g_f(x-y)$ is smooth for $x \in \partial\Omega$. The solution of an elliptic PDE, like Laplace's equation, on a smooth domain with smooth boundary data is smooth, and therefore $\varphi(x,y)$ is a C^∞ function on $\Omega \times \Omega$. We have used the free-space Green's function to "subtract off" the singularity in the Green's function on a bounded domain.

The eigenvalue problem for the Laplacian is

$$-\Delta u = \lambda u \qquad \text{in } \Omega,$$

$$u = 0 \qquad \text{on } \partial\Omega.$$

(10.47)

We again assume Dirichlet boundary conditions for definiteness, when the eigenvalues are strictly positive. If Ω is a bounded domain with a sufficiently regular boundary, then one can show that the Green's operator is a compact, self-adjoint operator on $L^2(\Omega)$. Consequently, it has a complete orthonormal set of eigenfunctions.

Using the divergence theorem, we find that

$$\int_\Omega u\,(-\Delta - \lambda I)\,u\,dx \;=\; \int_\Omega \left\{ -\nabla \cdot (u\nabla u) + |\nabla u|^2 - \lambda u^2 \right\} dx$$

$$= \int_\Omega \left\{ |\nabla u|^2 - \lambda u^2 \right\} dx.$$

The boundary terms vanish because $u = 0$ on $\partial\Omega$. Hence, if λ is an eigenvalue of the Dirichlet problem for $-\Delta$ with eigenfunction u, then

$$\int_\Omega \left\{ |\nabla u|^2 - \lambda u^2 \right\} dx = 0.$$

Since $u \neq 0$, it follows that $\lambda \geq 0$. If $\lambda = 0$, then $\nabla u = 0$ in Ω, so $u = \text{constant}$. The boundary condition implies that $u = 0$, so $\lambda = 0$ is not an eigenvalue of the Dirichlet problem. A similar argument applies to the Neumann problem for the Laplacian, with boundary condition $\partial u/\partial n = 0$ on $\partial\Omega$, except that $\lambda = 0$ is an eigenvalue with constant eigenfunction $u = 1$.

It is not possible to compute the eigenvalues and eigenfunctions of the Laplacian explicitly unless the domain Ω has a special shape. For example, if the boundary of Ω is made up of coordinate surfaces of a coordinate system in which the Laplacian separates, then we may use the method of separation of variables illustrated in the next two examples.

Example 10.36 The eigenvalue problem for the Laplacian with Dirichlet boundary conditions on the rectangle $\Omega = [0, a] \times [0, b] \subset \mathbb{R}^2$ is

$$-(u_{xx} + u_{yy}) = \lambda u, \qquad 0 < x < a, \quad 0 < y < b,$$

$$u(0, y) = u(a, y) = 0,$$

$$u(x, 0) = u(x, b) = 0.$$

The eigenfunctions $u = u_{m,n}$ and eigenvalues $\lambda = \lambda_{m,n}$, where $m, n = 1, 2, 3, \ldots,$ are given by

$$u_{m,n}(x, y) = \frac{2}{\sqrt{ab}} \sin\left(\frac{m\pi x}{a}\right) \sin\left(\frac{n\pi y}{b}\right),$$

$$\lambda_{m,n} = \pi^2 \left(\frac{m^2}{a^2} + \frac{n^2}{b^2} \right).$$

The corresponding eigenfunction expansion is a Fourier sine expansion. The lowest eigenvalue is simple, but higher eigenvalues need not be. For example, in the case of a square, $a = b$, the eigenvalue $\lambda = 50\pi^2/a^2$ has multiplicity 3 corresponding to $(m, n) = (5, 5), (1, 7), (7, 1)$. The Green's function $g(x, y; \xi, \eta)$ satisfies

$$- (g_{xx} + g_{yy}) = \delta(x - \xi)\delta(y - \eta),$$
$$g(0, y; \xi, \eta) = g(a, y; \xi, \eta) = 0,$$
$$g(x, 0; \xi, \eta) = g(x, b; \xi, \eta) = 0.$$

The eigenfunction expansion of the Green's function is

$$g(x, y; \xi, \eta) = \frac{4}{ab} \sum_{m=1}^{\infty} \sum_{n=1}^{\infty} \frac{\sin{(m\pi x/a)} \sin{(n\pi y/b)} \sin{(m\pi\xi/a)} \sin{(n\pi\eta/b)}}{\pi^2 \left(m^2/a^2 + n^2/b^2 \right)},$$

where the series converges in $L^2(\Omega \times \Omega)$.

Example 10.37 The Dirichlet eigenvalue problem for the Laplacian in the three-dimensional unit ball

$$\Omega = \left\{ x \in \mathbb{R}^3 \mid |x| < 1 \right\}$$

may be solved using spherical polar coordinates (r, θ, φ), where

$$x = r \sin\theta \cos\varphi, \quad y = r \sin\theta \sin\varphi, \quad z = r \cos\theta,$$

and $0 \le r$, $0 \le \theta \le \pi$, $0 \le \varphi \le 2\pi$. The eigenvalue problem (10.47) for Laplace's equation is

$$- \left[\frac{1}{r^2} \frac{\partial}{\partial r} \left(r^2 \frac{\partial u}{\partial r} \right) + \frac{1}{r^2 \sin\theta} \frac{\partial}{\partial \theta} \left(\sin\theta \frac{\partial u}{\partial \theta} \right) + \frac{1}{r^2 \sin^2\theta} \frac{\partial^2 u}{\partial \varphi^2} \right] = k^2 u, \quad \text{for } r < 1,$$
$$u = 0 \quad \text{for } r = 1.$$

Here, we write $\lambda = k^2$, since $\lambda > 0$. First, we separate the radial and angular dependence, and look for solutions of the form $u(r, \theta, \varphi) = R(r)Y(\theta, \varphi)$. This gives

$$\frac{1}{r^2} \left(r^2 R' \right)' + \left(k^2 - \frac{\mu}{r^2} \right) R = 0, \quad r < 1, \quad R(1) = 0, \qquad (10.48)$$

$$- \left[\frac{1}{\sin\theta} \frac{\partial}{\partial \theta} \left(\sin\theta \frac{\partial Y}{\partial \theta} \right) + \frac{1}{\sin^2\theta} \frac{\partial^2 Y}{\partial \varphi^2} \right] = \mu Y, \qquad (10.49)$$

where μ is a constant. Equation (10.49) is the eigenvalue problem for the *Laplace-Beltrami equation* on the unit sphere. The nonzero, square-integrable solutions that are 2π-periodic in φ are parametrized by two integers (l, m), where $l \ge 0$ and

$m = -l, -l+1, \ldots, l-1, l$. The eigenvalues are $\mu = l(l+1)$ and the eigenfunctions are the *spherical harmonics* $Y = Y_l^m$, given by

$$Y_l^m(\theta, \varphi) = (-1)^m \left[\frac{2l+1}{4\pi} \frac{(l-m)!}{(l+m)!} \right]^{1/2} P_l^m(\cos\theta) e^{im\varphi}.$$

Here, $P_l^m(x)$ is the Legendre function defined in Example 10.32. The set

$$\{Y_l^m \mid l \geq 0 \text{ and } |m| \leq l\}$$

forms a complete orthonormal basis of $L^2(\partial\Omega)$, where $\partial\Omega = \{x \in \mathbb{R}^3 \mid |x| = 1\}$ is the two-dimensional unit sphere in $\mathbb{R}^3$.

Up to an arbitrary multiplicative constant, the solution of the radial equation (10.48) that is bounded at $r = 0$ is given by

$$R(r) = j_l(kr),$$

where $j_l(x)$ is the lth order *spherical Bessel function* that satisfies

$$u'' + \frac{2}{x} u' + \left[1 - \frac{l(l+1)}{x^2} \right] u = 0.$$

The boundary condition $R(1) = 0$ implies that $j_l(k) = 0$, so that $k = z_{l,n}$ where $x = z_{l,n}$ with $n = 1, 2, \ldots$ is the nth positive zero of $j_l(x)$. The corresponding eigenvalues are therefore $\lambda = \lambda_{l,n}$, where $\lambda_{l,n} = z_{l,n}^2$, and $\lambda_{l,n}$ has a multiplicity of $(2l+1)$ corresponding to the different possible choices of $-l \leq m \leq l$. The eigenfunctions,

$$u_{l,m,n}(r, \theta, \varphi) = j_l\left(\sqrt{\lambda_{l,n}} r\right) P_l^m(\cos\theta) e^{im\varphi}, \qquad l \geq 0, |m| \leq l, n \geq 1,$$

form a complete orthogonal basis of $L^2(\Omega)$.

Finally, we consider two examples of partial differential operators that are not formally self-adjoint.

Example 10.38 The *advection-diffusion operator* is

$$A = \mathbf{a} \cdot \nabla + \Delta,$$

where $\mathbf{a}$ is a smooth vector field. The equation $Au = 0$, or

$$\Delta u + \mathbf{a} \cdot \nabla u = 0$$

describes the steady state of a quantity u, such as temperature or the density of a pollutant, subject to diffusion and advection by a velocity field $\mathbf{a}$. We consider A as acting on $C^2(\overline{\Omega})$, where Ω is a smooth, bounded domain in $\mathbb{R}^n$. For simplicity, we suppose all functions are real-valued. Then, using the vector identity

$$\nabla \cdot (u\mathbf{a}) = \nabla u \cdot \mathbf{a} + u \nabla \cdot \mathbf{a},$$

and the divergence theorem, we find that

$$\int_\Omega (\mathbf{a} \cdot \nabla u + \Delta u)\, v\, dx \;=\; \int_\Omega u\,(-\nabla \cdot (\mathbf{a} v) + \Delta v)\, dx$$

$$+ \int_{\partial\Omega} \left(uv\mathbf{a} \cdot \mathbf{n} + v\frac{\partial u}{\partial n} - u\frac{\partial v}{\partial n} \right) dS.$$

Thus, the formal adjoint of A is

$$A^* = -\nabla \cdot \mathbf{a} + \Delta.$$

Example 10.39 The heat operator is

$$A = -\partial_t + \Delta.$$

We consider A as acting on real-valued functions $u(x,t)$ in $C^2(\overline{\Omega} \times [0,T])$, where Ω is a smooth, bounded domain in $\mathbb{R}^n$, and $T > 0$. Then

$$\int_0^T \int_\Omega (-u_t + \Delta u)\, v\, dx dt \;=\; \int_0^T \int_\Omega u\,(v_t + \Delta v)\, dx dt$$

$$- \int_\Omega [uv]_0^T\, dx + \int_0^T \int_{\partial\Omega} \left(v\frac{\partial u}{\partial n} - u\frac{\partial v}{\partial n} \right) dS dt.$$

Thus, for example, the adjoint problem to the initial value problem for the heat equation,

$$u_t = \Delta u + f \qquad \text{in } \Omega \times [0,T],$$
$$u(x,t) = 0 \qquad \text{for } x \in \partial\Omega,$$
$$u(x,0) = u_0(x),$$

is the final value problem for the backward heat equation,

$$-v_t = \Delta v + g \qquad \text{in } \Omega \times [0,T],$$
$$v(x,t) = 0 \qquad \text{for } x \in \partial\Omega,$$
$$v(x,T) = v_T(x).$$

10.7   References

For more on unbounded operators and a proof of the closed graph theorem, see Kato [26] or Reed and Simon [45]. For Sturm-Liouville problems, see Coddington and Levinson [6]. For an introduction to Green's functions for PDEs, see Zauderer [57]. An extensive collection of Green's functions for various boundary value problems for PDEs is given in Morse and Feshbach [39]. Mikhlin [38] gives a detailed and careful analytical discussion of Green's functions for Laplace's equation. Further analysis of spectral problems for ODEs and PDEs is given in Vol. 3 of Dautry and Lions [7]. For the definition and properties of special functions, such as the Gamma function,

Bessel functions, and spherical harmonics, see Hochstadt [23] or Lebedev [31]. For a summary of formulae and integrals, including ones that involve special functions, see Gradshteyn and Ryzhik [16].

10.8 Exercises

Exercise 10.1 Prove that if A^{**} exists, then it is an extension of A.

Exercise 10.2 Prove that a symmetric operator is closable.

Exercise 10.3 Show that the operator A on $\mathcal{H} = L^2(\mathbb{T})$, with domain

$$\mathcal{D}(A) = \left\{ f(x) = \sum_{n \in \mathbb{Z}} a_n e^{inx} \,\Big|\, \sum_{n \in \mathbb{Z}} n^4 |a_n|^2 < \infty \right\},$$

defined by

$$A \left(\sum_{n \in \mathbb{Z}} a_n e^{inx} \right) = \sum_{n \in \mathbb{Z}} n^2 a_n e^{inx}$$

is a self-adjoint extension of the classical differentiation operator $-d^2/dx^2$ with domain $C^2(\mathbb{T})$.

Exercise 10.4 Let $M : \mathcal{D}(M) \subset L^2(\mathbb{R}) \to L^2(\mathbb{R})$ be the multiplication operator $Mf = xf$ with

$$\mathcal{D}(M) = \left\{ f \in L^2(\mathbb{R}) \mid xf \in L^2(\mathbb{R}) \right\}.$$

Show that M is self-adjoint.

Exercise 10.5 Suppose that $\{e_n \mid n \in \mathbb{N}\}$ is an orthonormal basis of a separable Hilbert space $\mathcal{H}$, and $\lambda_n \in \mathbb{R}$. For $x \in \mathcal{H}$, let $x_n = \langle e_n, x \rangle \in \mathbb{C}$, so

$$x = \sum_{n=1}^{\infty} x_n e_n.$$

Define an operator $A : \mathcal{D}(A) \subset \mathcal{H} \to \mathcal{H}$ by

$$\mathcal{D}(A) = \left\{ x \in H \,\Big|\, \sum_{n=1}^{\infty} (1 + \lambda_n^2) |x_n|^2 < \infty \right\},$$

$$A \left(\sum_{n=1}^{\infty} x_n e_n \right) = \sum_{n=1}^{\infty} \lambda_n x_n e_n.$$

Prove that A is self-adjoint.

Exercise 10.6 Let A and B be two linear operators on a Hilbert space $\mathcal{H}$ with domains $\mathcal{D}(A)$ and $\mathcal{D}(B)$, respectively, and assume $\mathcal{D}(A) \cap \mathcal{D}(B)$ is dense. Define an operator C by $\mathcal{D}(C) = \mathcal{D}(A) \cap \mathcal{D}(B)$ and $Cx = Ax + Bx$ for all $x \in \mathcal{D}(C)$. Prove that C^* is an extension of $A^* + B^*$. Define $\mathcal{D}(AB)$ and $\mathcal{D}(B^*A^*)$ by

$$
\begin{aligned}
\mathcal{D}(AB) &= \{x \in \mathcal{D}(B) \mid Bx \in \mathcal{D}(A)\} \\
\mathcal{D}(B^*A^*) &= \{x \in \mathcal{D}(A^*) \mid A^*x \in \mathcal{D}(B^*)\}
\end{aligned}
$$

and assume that $\mathcal{D}(AB)$ and $\mathcal{D}(B^*A^*)$ are dense. Define operators AB and B^*A^* on their respective domains in the obvious way. Prove that $(AB)^*$ is an extension of B^*A^*.

Exercise 10.7 Prove that the adjoint of a densely defined, unbounded operator in a Hilbert space is closed.

Exercise 10.8 Let $\{x_n \mid n \in \mathbb{N}\}$ be an orthonormal basis of a separable Hilbert space $\mathcal{H}$, and let y an element of $\mathcal{H}$ that is not a linear combination of a finite number of basis elements x_n. Define a linear operator A in $\mathcal{H}$, whose domain $\mathcal{D}(A)$ consists of finite linear combinations of the x_n and y, as follows:

$$
A\left(\sum_{n=1}^N a_n x_n + by\right) = by, \quad \mathcal{D}(A) = \left\{\sum_{n=1}^N a_n x_n + by \,\middle|\, a_n, b \in \mathbb{C}\right\}.
$$

Show that A is not closable.

Exercise 10.9 Consider a singular self-adjoint BVP,

$$
\begin{aligned}
-(pu')' + qu &= f, \\
u(0) = u(1) &= 0.
\end{aligned}
$$

Suppose that the null space of the homogeneous problem is one-dimensional with orthonormal basis $\{\varphi\}$. Define the modified Green's operator $G : L^2([0,1]) \to L^2([0,1])$ where $u = Gf$ if and only if u satisfies the problem

$$
\begin{aligned}
-(pu')' + qu &= f - \langle \varphi, f \rangle \varphi, \\
u(0) = u(1) = 0, \qquad \langle \varphi, u \rangle &= 0.
\end{aligned}
$$

Prove that G is well defined, and show that G is an integral operator of the form

$$
Gf(x) = \int_0^1 g(x,y) f(y) \, dy.
$$

Compute the modified Green's function g in terms of φ.

Exercise 10.10 Let $r : [0,1] \to \mathbb{R}$ be a smooth, nonnegative function. Let $\mathcal{H}$ be the Hilbert space of (equivalence classes) of Lebesgue measurable functions $u : [0,1] \to \mathbb{C}$ such that

$$\int_0^1 r(x)|u(x)|^2 \, dx < \infty,$$

with the inner product

$$\langle u, v \rangle = \int_0^1 r(x)\overline{u(x)}v(x) \, dx.$$

Determine the formally self-adjoint, second-order differential operators on $\mathcal{H}$.

Exercise 10.11 Prove that the Wronskian $W(x)$ of the Sturm-Liouville operator (10.8) satisfies $p(x)W(x) = $ constant. Verify directly that the Green's function is symmetric.

Exercise 10.12 The following linearized BBM (Benjamin-Bona-Mahoney) equation for $u(x,t)$, where $x, t \in \mathbb{R}$, arises in the analysis of water waves:

$$-u_{xxt} + u_t = u_x,$$
$$u(x,0) = u_0(x).$$

Use a Green's function to reformulate this equation as an evolution equation

$$u_t = Ku,$$

for a suitable integral operator $K : L^2(\mathbb{R}) \to L^2(\mathbb{R})$, and deduce that there is a global in time solution with $u(\cdot, t) \in L^2(\mathbb{R})$ for any initial data $u_0 \in L^2(\mathbb{R})$. Show that the L^2-norm of u is conserved.

Exercise 10.13 For $k = 1, 2, 3$, let $A_k : \mathcal{D}(A_k) \subset L^2([0,1]) \to L^2([0,1])$ be the first-order differential operators $A_k u = iu'$ with domains

$$
\begin{aligned}
\mathcal{D}(A_1) &= H^1((0,1)), \\
\mathcal{D}(A_2) &= \left\{ u \in H^1((0,1)) \mid u(0) = u(1) \right\}, \\
\mathcal{D}(A_3) &= \left\{ u \in H^1((0,1)) \mid u(0) = 0 \right\}.
\end{aligned}
$$

Show that the spectrum of A_1 is $\mathbb{C}$, the spectrum of A_2 is the set $\{2n\pi \mid n \in \mathbb{Z}\}$, and the spectrum of A_3 is empty.

Exercise 10.14 Consider the operators A_1, A_2 defined by $A_k u = iu'$, with

$$
\begin{aligned}
\mathcal{D}(A_1) &= \left\{ u \in H^1((0,1)) \mid u(0) = u(1) \right\}, \\
\mathcal{D}(A_2) &= \left\{ u \in H^1((0,1)) \mid u(0) = u(1) = 0 \right\}.
\end{aligned}
$$

Show that both operators are closed and symmetric. Compute $\text{ran}\,(A_k \pm i)$ and $\ker(A_k^* \pm i)$. Use Theorem 10.29 to determine whether or not these operators are self-adjoint.

Exercise 10.15 Let φ be a nonzero function in $L^2(\mathbb{R})$ and define an operator $A : \mathcal{D}(A) \subset L^2(\mathbb{R}) \to L^2(\mathbb{R})$ by

$$Au = \left(\int_{\mathbb{R}} u(x)dx \right) \varphi, \qquad \mathcal{D}(A) = L^1(\mathbb{R}) \cap L^2(\mathbb{R}).$$

Show that A is a densely defined unbounded linear operator in $L^2(\mathbb{R})$ that is *not* closed. Show that

$$\mathcal{D}(A^*) = \{\varphi\}^\perp$$

and $A^* = 0$ on $\mathcal{D}(A^*)$, so the domain of A^* is not dense in $L^2(\mathbb{R})$.

Exercise 10.16 If $u \in H^1((0, \infty))$ and $u(0) = 0$, prove *Hardy's inequality*:

$$\int_0^\infty \frac{|u|^2}{x^2}\,dx \le 4 \int_0^\infty |u'|^2\,dx.$$

Exercise 10.17 Suppose that u_1 and u_2 are two solutions of the Dirichlet problem for Laplace's equation

$$-\Delta u = f \quad x \in \Omega,$$
$$u = h \quad x \in \partial\Omega,$$

where Ω is a smooth, bounded domain in $\mathbb{R}^n$ and $f : \Omega \to \mathbb{R}$ and $h : \partial\Omega \to \mathbb{R}$ are given functions. Show that if $v = u_1 - u_2$ then

$$\int_\Omega |\nabla v|^2\,dx = 0,$$

and deduce that the solution is unique. What can you say about solutions of the Neumann problem, with boundary condition

$$\frac{\partial u}{\partial n} = h \quad x \in \partial\Omega?$$

Exercise 10.18 According to Maxwell's equations, the magnetic field $\mathbf{B}$ generated in three-dimensional space by a steady current distribution $\mathbf{J}$ satisfies

$$\text{curl}\,\mathbf{B} = \mathbf{J}, \qquad \text{div}\,\mathbf{B} = 0.$$

A mathematically identical problem arises in fluid mechanics in reconstructing an incompressible velocity field $\mathbf{u}$, with $\text{div}\,\mathbf{u} = 0$, from the vorticity $\omega = \text{curl}\,\mathbf{u}$. Derive the *Biot-Savart law*,

$$\mathbf{B}(\mathbf{x}) = \int \frac{\mathbf{J}(\mathbf{y}) \times (\mathbf{x} - \mathbf{y})}{4\pi\,|\mathbf{x} - \mathbf{y}|}\,d\mathbf{y}.$$

HINT. Write $\mathbf{B} = \text{curl }\mathbf{A}$ and derive a Laplace equation for $\mathbf{A}$.

Exercise 10.19 Let N_n be the $n \times n$ Jordan block

$$N_n = \begin{pmatrix} 0 & 1 & 0 & \cdots & 0 \\ 0 & 0 & 1 & \cdots & 0 \\ \vdots & \vdots & \vdots & \ddots & \vdots \\ 0 & 0 & 0 & \cdots & 1 \\ 0 & 0 & 0 & \cdots & 0 \end{pmatrix},$$

and let $c_n = n^{-1/2}(1, 1, \ldots, 1)^T \in \mathbb{C}^n$. Show that for each $n \in \mathbb{N}$ and $t \geq 0$:

$$\left\| e^{tN_n} \right\| \leq e^t; \quad \|N_n c_n - c_n\| \leq n^{-1/2}; \quad \left\| e^{tN_n} c_n - e^t c_n \right\| \leq n^{-1/2} t e^t.$$

Let $\mathcal{H} = \bigoplus_{n=1}^{\infty} \mathbb{C}^n$, meaning that $x \in \mathcal{H}$ is of the form

$$x = (x_1, x_2, \ldots, x_n, \ldots), \quad x_n \in \mathbb{C}^n, \quad \sum_{n=1}^{\infty} |x_n|^2 < \infty.$$

Here, $|\cdot|$ denotes the Euclidean norm on $\mathbb{C}^n$. Let $A_n = N_n + in I_n$, where I_n is the $n \times n$ identity, and define $A : \mathcal{D}(A) \subset \mathcal{H} \to \mathcal{H}$ by

$$A = \bigoplus_{n=1}^{\infty} A_n, \quad A(x_1, x_2, \ldots, x_n, \ldots) = (A_1 x_1, A_2 x_2, \ldots, A_n x_n, \ldots),$$

where $\mathcal{D}(A) = \{x \in \mathcal{H} \mid Ax \in \mathcal{H}\}$. We define the associated C_0-semigroup $T(t) = e^{tA}$ for $t \geq 0$, where $T(t) : \mathcal{H} \to \mathcal{H}$, by

$$T(t)(x_1, x_2, \ldots, x_n, \ldots) = \left(e^{tA_1} x_1, e^{tA_2} x_2, \ldots, e^{tA_n} x_n, \ldots\right).$$

Show that the spectrum of A is $\{in \in \mathbb{C} \mid n \in \mathbb{N}\}$, and consists entirely of eigenvalues, so it is contained in the left-half plane $\{\lambda \in \mathbb{C} \mid \text{Re }\lambda \leq 0\}$. Show, however, that the spectral radius of $T(t)$ is greater than or equal to e^t, so the spectral mapping theorem does not hold for A.

HINT. Consider the action of $T(t)$ on the vectors $(0, 0, \ldots, 0, c_n, 0, \ldots) \in X$. This example of an operator with arbitrarily large Jordan blocks illustrates some of the pathologies that can arise for unbounded, nonnormal operators on a Hilbert space.

Exercise 10.20 Consider heat flow in a rod with rapidly varying thermal conductivity $a_n(x) = a(nx)$, where $n \in \mathbb{N}$ and $a(y)$ is a strictly positive periodic function with period one, assumed continuous for simplicity. If the ends of the rod are held at an equal fixed temperature, and there is a given heat source $f(x)$ per unit length, the temperature $u_n(x)$ satisfies the boundary value problem

$$-\frac{d}{dx}\left(a_n(x)\frac{d}{dx}u_n\right) = f(x), \quad 0 < x < 1, \quad u_n(0) = u_n(1) = 0.$$

Linear Differential Operators and Green's Functions

Integrate this ODE and solve for $u_n(x)$. Let $H_0^1([0,1])$ be the Sobolev space

$$H_0^1([0,1]) = \left\{ u : [0,1] \to \mathbb{R} \mid u, u' \in L^2([0,1]), \ u(0) = u(1) = 0 \right\},$$

with the inner product

$$\langle u, v \rangle = \int_0^1 u'v' \, dx.$$

Show that $u_n \rightharpoonup u$ weakly in $H_0^1([0,1])$, where u is the solution of the *homogenized equation*

$$-\frac{d}{dx}\left(a^h \frac{d}{dx} u \right) = f(x), \quad 0 < x < 1, \qquad u(0) = u(1) = 0,$$

and the effective conductivity a^h is the harmonic mean of the original conductivity,

$$\frac{1}{a^h} = \int_0^1 \frac{1}{a(y)} \, dy.$$

Chapter 11

Distributions and the Fourier Transform

A distribution is a continuous linear functional on a space of test functions. Distributions provide a simple and elegant extension of functions that clarifies many aspects of analysis. For example, the delta function may be interpreted as a distribution. An advantage of distributions is that every distribution is differentiable, and differentiation is a continuous operation on spaces of distributions. Moreover, every tempered distribution has a Fourier transform, and a function whose Fourier transform is not defined as a function may nevertheless have a distributional transform. One limitation on the use of distributions is that there is no product of distributions that preserves the usual properties of the pointwise product of functions. Therefore, when studying nonlinear problems involving distributions, one must make sure that any products of distributions that appear are well defined.

11.1 The Schwartz space

In this section, we define a space of test functions on $\mathbb{R}^n$ called the *Schwartz space* that consists of smooth, rapidly decreasing functions.

We begin by introducing a concise notation for partial derivatives. Let

$$\mathbb{Z}_+ = \{n \in \mathbb{Z} \mid n \geq 0\}$$

denote the nonnegative integers. A *multi-index*

$$\alpha = (\alpha_1, \ldots, \alpha_n) \in \mathbb{Z}_+^n$$

is an n-tuple of nonnegative integers $\alpha_i \geq 0$. For multi-indices $\alpha = (\alpha_1, \ldots, \alpha_n)$ and $\beta = (\beta_1, \ldots, \beta_n)$, we define

$$|\alpha| = \sum_{i=1}^n \alpha_i, \qquad \alpha! = \prod_{i=1}^n \alpha_i!,$$
$$\alpha + \beta = (\alpha_1 + \beta_1, \ldots, \alpha_n + \beta_n),$$
$$\alpha \geq \beta \quad \text{if and only if } \alpha_i \geq \beta_i \text{ for } i = 1, \ldots, n.$$

If $x = (x_1, \ldots, x_n) \in \mathbb{R}^n$ and $\alpha = (\alpha_1, \ldots, \alpha_n) \in \mathbb{Z}_+^n$, then we define

$$\partial^\alpha = \left(\frac{\partial}{\partial x_1}\right)^{\alpha_1} \cdots \left(\frac{\partial}{\partial x_n}\right)^{\alpha_n},$$

$$x^\alpha = \prod_{i=1}^{n} x_i^{\alpha_i},$$

$$|x| = \sqrt{x_1^2 + \ldots + x_n^2}.$$

We use the notation $x^\alpha f$ to denote the function whose value at x is $x^\alpha f(x)$.

The Taylor remainder theorem for $f \in C^k(\mathbb{R}^n)$ may be written as

$$f(x) = \sum_{|\alpha| \leq k} \frac{1}{\alpha!} \partial^\alpha f(x_0)(x - x_0)^\alpha + r_k(x), \qquad (11.1)$$

where the remainder term r_k satisfies

$$\lim_{x \to x_0} \frac{r_k(x)}{|x - x_0|^k} = 0.$$

The Leibnitz rule for the derivative of the product of $f, g \in C^k(\mathbb{R}^n)$ may be written as

$$\partial^\alpha(fg) = \sum_{\beta + \gamma = \alpha} \frac{\alpha!}{\beta!\gamma!} \left(\partial^\beta f\right)\left(\partial^\gamma g\right). \qquad (11.2)$$

For multi-indices $\alpha, \beta \in \mathbb{Z}_+^n$, and $\varphi \in C^\infty(\mathbb{R}^n)$, we define

$$p_{\alpha,\beta}(\varphi) = \sup_{x \in \mathbb{R}^n} \left|x^\alpha \partial^\beta \varphi(x)\right|. \qquad (11.3)$$

We also write $p_{\alpha,\beta}(\varphi)$ as $\|\varphi\|_{\alpha,\beta}$.

Definition 11.1 (Schwartz space) The *Schwartz space* $\mathcal{S}(\mathbb{R}^n)$, or $\mathcal{S}$ for short, consists of all functions $\varphi \in C^\infty(\mathbb{R}^n)$ such that $p_{\alpha,\beta}(\varphi)$ in (11.3) is finite for every pair of multi-indices $\alpha, \beta \in \mathbb{Z}_+^n$.

If $\varphi \in \mathcal{S}$, then for every $d \in \mathbb{N}$ and $\alpha \in \mathbb{Z}_+^n$ there is a constant $C_{d,\alpha}$ such that

$$|\partial^\alpha \varphi(x)| \leq \frac{C_{d,\alpha}}{(1 + |x|^2)^{d/2}} \qquad \text{for all } x \in \mathbb{R}^n.$$

Thus, an element of $\mathcal{S}$ is a smooth function such that the function and all of its derivatives decay faster than any polynomial as $|x| \to \infty$. Elements of $\mathcal{S}$ are called *Schwartz functions*, or *test functions*. There are many functions in $\mathcal{S}$. For example, every function of the form

$$q(x)e^{-c|x - x_0|^2},$$

where $c > 0$, $x_0 \in \mathbb{R}^n$, and

$$q(x) = \sum_{|\alpha| \le d} c_\alpha x^\alpha$$

is a polynomial function on $\mathbb{R}^n$, is a Schwartz function.

In order to define a notion of the convergence of test functions, we want to put a topology on $\mathcal{S}$. As we will see, the appropriate topology is not derived from a norm, but instead from the countable family $\{p_{\alpha,\beta}\}$ of seminorms. We therefore first discuss topologies defined by seminorms in more generality.

Definition 11.2 Suppose that X is a real or complex linear space. A function $p : X \to \mathbb{R}$ is a *seminorm* on X if it has the following properties:

(a) $p(x) \ge 0$ for all $x \in X$;
(b) $p(x + y) \le p(x) + p(y)$ for all $x, y \in X$;
(c) $p(\lambda x) = |\lambda| p(x)$ for every $x \in X$ and $\lambda \in \mathbb{C}$.

A seminorm p has the same properties as a norm, except that $p(x) = 0$ need not imply $x = 0$. Suppose that $\{p_\alpha\}_{\alpha \in \mathcal{A}}$ is a countable or uncountable family of seminorms, indexed by a set $\mathcal{A}$, defined on a linear space X. Then X is a topological linear space with the following base $\mathcal{N}$ of open neighborhoods:

$$\mathcal{N} = \{N_{x;\alpha_1,\ldots,\alpha_n;\epsilon} \mid x \in X, \alpha_1 \ldots, \alpha_n \in \mathcal{A}, \text{ and } \epsilon > 0\},$$
$$N_{x;\alpha_1,\ldots,\alpha_n;\epsilon} = \{y \in X \mid p_{\alpha_i}(x - y) < \epsilon \text{ for } i = 1, \ldots, n\}.$$

A sequence (x_n) converges to $x \in X$ in this topology if and only if $p_\alpha(x - x_n) \to 0$ as $n \to \infty$ for each $\alpha \in \mathcal{A}$.

We say that a family $\{p_\alpha\}_{\alpha \in \mathcal{A}}$ of seminorms *separates points* if $p_\alpha(x) = 0$ for every $\alpha \in \mathcal{A}$ implies that $x = 0$. In that case, the associated topology is Hausdorff. A topological linear space whose topology may be derived from a family of seminorms that separates points is called a *locally convex space*.

If the family of seminorms $\{p_1, \ldots, p_n\}$ is finite and separates points, then

$$\|x\| = p_1(x) + \ldots + p_n(x)$$

defines a norm on X. Thus, there is no additional generality in using a finite family of seminorms instead of a norm. The main case of interest to us here is that of a locally convex space X whose topology is generated by a countably infinite family of seminorms $\{p_n \mid n \in \mathbb{N}\}$. In that case, the topology is metrizable because

$$d(x, y) = \sum_{n \in \mathbb{N}} \frac{1}{2^n} \frac{p_n(x - y)}{1 + p_n(x - y)} \tag{11.4}$$

defines a metric on X with the same collection of open sets as those generated by the family of seminorms (see Exercise 11.2). A metrizable, locally convex space that is complete as a metric space is called a *Fréchet space*.

The function $p_{\alpha,\beta}$ in (11.3) is a seminorm on $\mathcal{S}$. We equip $\mathcal{S}$ with the topology generated by the countable family of seminorms

$$\{p_{\alpha,\beta} \mid \alpha, \beta \in \mathbb{Z}_+^n\}. \tag{11.5}$$

This family separates points, since $p_{0,0}$ is just the sup-norm. The following proposition shows that $\mathcal{S}$ is a Fréchet space.

Proposition 11.3 The Schwartz space $\mathcal{S}$ with the metrizable topology generated by the countable family of seminorms (11.5), where $p_{\alpha,\beta}$ is given by (11.3), is complete.

Proof. Let (φ_n) be a Cauchy sequence in $\mathcal{S}$. We have to prove that (φ_n) converges in the topology of $\mathcal{S}$ to a function $\varphi \in \mathcal{S}$. The sequence (φ_n) is Cauchy with respect to the sup-norm $p_{0,0}$. Since the space of bounded continuous functions on $\mathbb{R}$ with the supremum norm is complete, there is a bounded continuous function φ such that $\varphi_n \to \varphi$ uniformly. For each multi-index α, the sequence $\partial^\alpha \varphi_n$ is Cauchy with respect to the sup-norm, and hence converges uniformly to a bounded continuous function ψ_α. We claim that

$$\psi_\alpha = \partial^\alpha \varphi \qquad \text{for every multi-index } \alpha. \tag{11.6}$$

We prove (11.6) by induction on $|\alpha|$. The equation holds for $|\alpha| = 0$. Suppose we have proved (11.6) for every α with $|\alpha| \leq m$. Then, if $|\beta| = m + 1$, there exists an $\alpha \in \mathbb{Z}_+^n$ such that $|\alpha| = m$ and $\beta = \alpha + e_j$ for some j, where e_j is the jth standard basis vector of $\mathbb{Z}^n$. The fundamental theorem of calculus implies that

$$\partial^\alpha \varphi_n(x + te_j) - \partial^\alpha \varphi_n(x) = \int_0^t \partial^{e_j} \partial^\alpha \varphi_n(x + se_j)\, ds.$$

Clearly, $\partial^{e_j} \partial^\alpha = \partial^\beta$. Letting $n \to \infty$, we obtain that

$$\partial^\alpha \varphi(x + te_j) - \partial^\alpha \varphi(x) = \int_0^t \psi_\beta(x + se_j)\, ds.$$

We divide this expression by t and take the limit of the resulting expression as $t \to 0^+$. Using the definition of derivative and the continuity of ψ_β, we find that

$$\partial^\beta \varphi(x) = \psi_\beta(x).$$

Finally, for every pair of multi-indices (α, β), the sequence $(x^\alpha \partial^\beta \varphi_n)$ is Cauchy with respect to the uniform norm, so it converges uniformly. The uniform limit is equal to the pointwise limit $x^\alpha \partial^\beta \varphi$, so $p_{\alpha,\beta}(\varphi_n - \varphi) \to 0$ for all multi-indices, and therefore (φ_n) converges in $\mathcal{S}$. $\qquad \square$

One main motivation for the use of this topology on $\mathcal{S}$ is that differentiation is a continuous operation.

Proposition 11.4 For each $\alpha \in \mathbb{Z}_+^n$, the partial differentiation operator $\partial^\alpha : S \to S$ is a continuous linear operator on S.

Proof. The fact that ∂^α is a linear map of S into S is obvious. To prove the continuity, suppose that $\varphi_n \to \varphi$ in S. Then $p_{\beta,\gamma}(\varphi_n - \varphi) \to 0$ as $n \to \infty$ for all $\beta, \gamma \in \mathbb{Z}_+^n$. Therefore,

$$p_{\beta,\gamma}(\partial^\alpha \varphi_n - \partial^\alpha \varphi) = p_{\beta,\alpha+\gamma}(\varphi_n - \varphi) \to 0$$

as $n \to \infty$ for all $\beta, \gamma \in \mathbb{Z}_+^n$, so $\partial^\alpha \varphi_n \to \partial^\alpha \varphi$ in S. $\qquad\square$

The Schwartz space is not the only possible space of test functions. Another useful choice is the smaller space $\mathcal{D} = C_c^\infty(\mathbb{R}^n)$ of smooth functions with compact support. The appropriate topology on $\mathcal{D}$ is, however, harder to describe than the topology on S because it is not metrizable.

11.2 Tempered distributions

The topological dual space of S, denoted by S^* or S', is the space of continuous linear functionals $T : S \to \mathbb{C}$. Elements of S^* are called *tempered distributions*. The space S^* is a linear space under the pointwise addition and scalar multiplication of functionals.

Since S is a metric space, a functional $T : S \to \mathbb{C}$ is continuous if and only if for every convergent sequence $\varphi_n \to \varphi$ in S, we have

$$\lim_{n \to \infty} T(\varphi_n) = T(\varphi).$$

The continuity of a linear functional T is implied by an estimate of the form

$$|T(\varphi)| \leq \sum_{|\alpha|,|\beta| \leq d} c_{\alpha,\beta} \|\varphi\|_{\alpha,\beta}$$

for some $d \in \mathbb{Z}_+$ and constants $c_{\alpha,\beta} \geq 0$. Conversely, one can show that if T is continuous, then such an estimate holds for some d and $c_{\alpha,\beta}$.

Example 11.5 The fundamental example of a distribution is the *delta function*. The name "delta function" is a misnomer because it is not a function on $\mathbb{R}^n$, but a functional on S. We define $\delta : S \to \mathbb{C}$ by evaluation at 0:

$$\delta(\varphi) = \varphi(0).$$

The linearity of δ is trivial. To show the continuity, suppose that $\varphi_n \to \varphi$ in S. Then $\varphi_n \to \varphi$ uniformly, and therefore $\varphi_n(0) \to \varphi(0)$. Hence, $\delta(\varphi_n) \to \delta(\varphi)$, so $\delta \in S^*$ is a continuous linear functional. Similarly, for each $x_0 \in \mathbb{R}^n$, we define the delta function supported at x_0 by evaluation at x_0:

$$\delta_{x_0}(\varphi) = \varphi(x_0).$$

Example 11.6 Suppose that f is a continuous, or Lebesgue measurable, function on $\mathbb{R}^n$ such that

$$|f(x)| \le g(x) \left(1 + |x|^2\right)^{d/2}$$

a.e. in $\mathbb{R}^n$ for a nonnegative integer $d \ge 0$ and a nonnegative, integrable function $g : \mathbb{R}^n \to \mathbb{R}$. Then

$$T_f(\varphi) = \int_{\mathbb{R}^n} f(x)\varphi(x)\, dx \tag{11.7}$$

defines a tempered distribution, as follows from the estimate:

$$
\begin{aligned}
|T_f(\varphi)| &\le \int_{\mathbb{R}^n} g(x) \left(1 + |x|^2\right)^{d/2} |\varphi(x)|\, dx \\
&\le \left[\int_{\mathbb{R}^n} g(x)\, dx\right] \sup_{x \in \mathbb{R}^n} \left[\left(1 + |x|^2\right)^{d/2} |\varphi(x)|\right].
\end{aligned}
$$

Moreover, the function f is uniquely determined, up to pointwise-a.e. equivalence, by the distribution T_f. To see this, let $\{\varphi_\epsilon \mid \epsilon > 0\}$ be an approximate identity in $\mathcal{S}(\mathbb{R}^n)$, for example the Gaussian approximate identity,

$$\varphi_\epsilon(x) = \frac{1}{(2\pi\epsilon)^{n/2}} \exp\left(-\frac{|x|^2}{2\epsilon}\right).$$

Then for each $\epsilon > 0$ and $x \in \mathbb{R}^n$, the function $\varphi_{\epsilon,x}$ defined by

$$\varphi_{\epsilon,x}(y) = \varphi_\epsilon(x - y)$$

is an element of $\mathcal{S}(\mathbb{R}^n)$, and

$$T_f\left(\varphi_{\epsilon,x}\right) = (\varphi_\epsilon * f)(x).$$

Since we can recover f pointwise-a.e. from its convolutions with an approximate identity, we see that f is determined by T_f.

Distributions of the form (11.7) that are given by the integration of a test function φ against a function f are called *regular distributions*, and distributions, such as the delta function, that are not of this form are called *singular distributions*. Thus, we may regard tempered distributions as a generalization of locally integrable functions with polynomial growth.

A function that has a nonintegrable singularity, or a function that grows faster than a polynomial (such as $e^{c|x|^2}$ where $c > 0$), does not define a regular tempered distribution since its integral against a Schwartz function need not be finite.

Example 11.7 The function $(1/x) : \mathbb{R} \setminus 0 \to \mathbb{R}$ has a nonintegrable singularity at $x = 0$, so it does not define a regular distribution. We can, however, use a limiting procedure to define a singular distribution called a *principal value distribution*, denoted by p.v.$(1/x)$. We define its action on a test function $\varphi \in \mathcal{S}(\mathbb{R})$ by

$$\text{p.v.}\frac{1}{x}(\varphi) = \lim_{\epsilon \to 0+} \int_{|x|>\epsilon} \frac{\varphi(x)}{x}\, dx.$$

The limit is finite because of a cancellation between the nonintegrable contributions of $1/x$ for $x < 0$ and $x > 0$:

$$\text{p.v.}\frac{1}{x}(\varphi) = \lim_{\epsilon \to 0+} \int_{\epsilon}^{\infty} \frac{\varphi(x) - \varphi(-x)}{x}\, dx = \int_{0}^{\infty} \frac{\varphi(x) - \varphi(-x)}{x}\, dx.$$

The integrand is bounded at $x = 0$ since φ is smooth. For $x > 0$, we have

$$\left| \frac{\varphi(x) - \varphi(-x)}{x} \right| \le \frac{1}{x} \int_{-x}^{x} |\varphi'(t)|\, dt \le 2\|\varphi'\|_\infty,$$

so the continuity of p.v.$(1/x)$ on $\mathcal{S}$ follows from the estimate

$$\begin{aligned}
\left| \text{p.v.}\frac{1}{x}(\varphi) \right| &\le \int_{0}^{1} \left| \frac{\varphi(x) - \varphi(-x)}{x} \right| dx + \int_{1}^{\infty} \left| \frac{x\,[\varphi(x) - \varphi(-x)]}{x^2} \right| dx \\
&\le 2\|\varphi'\|_\infty + 2\|x\varphi\|_\infty \\
&= 2\left(\|\varphi\|_{0,1} + \|\varphi\|_{1,0} \right).
\end{aligned}$$

Example 11.8 The function $1/|x|^2 : \mathbb{R}^n \setminus \{0\} \to \mathbb{R}$ has an integrable singularity at the origin when $n \ge 3$ since the radial integral

$$\int_{0}^{1} r^{-2}\, r^{n-1}\, dr$$

is finite. In that case

$$\frac{1}{|x|^2}(\varphi) = \int_{\mathbb{R}^n} \frac{\varphi(x)}{|x|^2}\, dx$$

defines a regular distribution in $\mathcal{S}^*(\mathbb{R}^n)$. If $n = 2$, the function is not integrable, but we can define an associated singular distribution, called a *finite part distribution*, denoted by f.p.$(1/|x|^2)$:

$$\text{f.p.}\frac{1}{|x|^2}(\varphi) = \int_{|x|<1} \frac{\varphi(x) - \varphi(0)}{|x|^2}\, dx + \int_{|x|>1} \frac{\varphi(x)}{|x|^2}\, dx.$$

The action of the elements of the dual space $\mathcal{S}^*$ on $\mathcal{S}$ may be represented by a *duality pairing*, which resembles an inner product:

$$\langle \cdot, \cdot \rangle : \mathcal{S}^* \times \mathcal{S} \to \mathbb{C}.$$

We write the action of a distribution T on a test function φ as

$$T(\varphi) = \langle T, \varphi \rangle.$$

If $\mathcal{H}$ is a Hilbert space, then the duality pairing on $\mathcal{H}^* \times \mathcal{H}$ can be identified with the inner product on $\mathcal{H}$ by the Riesz representation theorem. Note, however, that in the case of an inner product on a Hilbert space, the duality pairing is antilinear in one of the variables, whereas the duality pairing on $\mathcal{S}^* \times \mathcal{S}$ is linear in both variables.

Another notation for the action of $T \in \mathcal{S}^*$ on $\varphi \in \mathcal{S}$ is

$$T(\varphi) = \int T(x)\varphi(x)\, dx,$$

as if $\mathcal{S}^*$ were a function space. If T_f is the regular distribution defined in (11.7), then this notation amounts to the identification of T_f with f. The action of the distribution δ_{x_0} is then written as

$$\delta_{x_0}(\varphi) = \int \delta(x - x_0)\varphi(x)\, dx = \varphi(x_0).$$

Since the pairing on $\mathcal{S}^* \times \mathcal{S}$ shares a number of properties with inner products defined through an integral, this notation is often convenient in computations, provided one remembers that it is just a way to write continuous linear functionals.

The tempered distributions are a subspace of the space $\mathcal{D}^*$ of *distributions* that are continuous linear functionals on the space $\mathcal{D}$ of smooth, compactly supported test functions. Unlike tempered distributions, distributions in $\mathcal{D}^*$ can grow faster than any polynomial at infinity. The Fourier transform of a distribution in $\mathcal{D}^*$ need not belong to $\mathcal{D}^*$, however, whereas we will see that every distribution in $\mathcal{S}^*$ has a Fourier transform that is also in $\mathcal{S}^*$. To be specific, we therefore restrict our discussion to tempered distributions, although similar ideas apply to distributions defined on other spaces of test functions.

11.3   Operations on distributions

We say that a continuous function $f : \mathbb{R}^n \to \mathbb{C}$ is of *polynomial growth* if there is an integer d and a constant C such that

$$|f(x)| \leq C \left(1 + |x|^2\right)^{d/2} \qquad \text{for all } x \in \mathbb{R}^n.$$

If $T \in \mathcal{S}^*$ and $f \in C^\infty(\mathbb{R}^n)$ is such that f and $\partial^\alpha f$ have polynomial growth for every $\alpha \in \mathbb{Z}_+^n$, then we define the product $fT \in \mathcal{S}^*$ by

$$\langle fT, \varphi \rangle = \langle T, f\varphi \rangle \qquad \text{for all } \varphi \in \mathcal{S}.$$

This definition makes sense because $f\varphi \in \mathcal{S}$ when $\varphi \in \mathcal{S}$. It is straightforward to check that fT is a continuous linear map on $\mathcal{S}$ if T is.

Example 11.9 If $T = \delta$ is the delta function, then

$$\langle f\delta, \varphi \rangle = \langle \delta, f\varphi \rangle = f(0)\varphi(0) = \langle f(0)\delta, \varphi \rangle.$$

Hence, $f\delta = f(0)\delta$.

The definition of products may be extended further; for example, the product $f\delta = f(0)\delta$ makes sense for any continuous function f. It is not possible, however, to define a product $ST \in S^*$ for general distributions $S, T \in S^*$ with the same algebraic properties as the pointwise product of functions (see Exercise 11.7).

Next, we define the derivative of a distribution. To motivate the definition, we first consider the regular distribution T_f associated with a Schwartz function f. Integrating by parts, we find that the action of the regular distribution $T_{\partial^\alpha f}$, associated with the αth partial derivative of f, on a test function φ is given by

$$\langle T_{\partial^\alpha f}, \varphi \rangle = \int (\partial^\alpha f)\,\varphi\,dx = (-1)^{|\alpha|} \int f\,(\partial^\alpha \varphi)\,dx = (-1)^{|\alpha|}\langle T_f, \partial^\alpha \varphi \rangle.$$

The following definition extends the differentiation of functions to the differentiation of distributions.

Definition 11.10 Suppose that T is a tempered distribution and α is a multi-index. The αth *distributional derivative* of T is the tempered distribution $\partial^\alpha T$ defined by

$$\langle \partial^\alpha T, \varphi \rangle = (-1)^{|\alpha|}\langle T, \partial^\alpha \varphi \rangle \qquad \text{for all } \varphi \in S. \tag{11.8}$$

Equation (11.8) does define a distribution. The linearity of the map $\partial^\alpha T : S \to \mathbb{C}$ is obvious. The continuity of $\partial^\alpha T$ follows from the continuity of T and ∂^α on S. If $\varphi_n \to \varphi$ in S, then $\partial^\alpha \varphi_n \to \partial^\alpha \varphi$ in S, so

$$\langle \partial^\alpha T, \varphi_n \rangle = (-1)^{|\alpha|}\langle T, \partial^\alpha \varphi_n \rangle \to (-1)^{|\alpha|}\langle T, \partial^\alpha \varphi \rangle = \langle \partial^\alpha T, \varphi \rangle.$$

Thus, every tempered distribution is differentiable. The space of distributions is therefore an extension of the space of functions that is closed under differentiation. The following structure theorem, whose proof we omit, shows that S is a minimal extension of the space of functions of polynomial growth that is closed under differentiation.

Theorem 11.11 For every $T \in S^*$ there is a continuous function $f : \mathbb{R}^n \to \mathbb{C}$ of polynomial growth and a multi-index $\alpha \in \mathbb{Z}_+^n$ such that $T = \partial^\alpha f$.

If T_f is a regular distribution whose distributional derivative is also a regular distribution T_g, then

$$\int_{\mathbb{R}^n} g\varphi\,dx = (-1)^{|\alpha|}\int_{\mathbb{R}^n} f\partial^\alpha \varphi\,dx \qquad \text{for all } \varphi \in S.$$

In this case, we say that the function g is the *weak* or *distributional* derivative of the function f, and we write $g = \partial^\alpha f$. Thus, the weak L^2-derivatives considered in Section 10.4 were a special case of the distributional derivative. If f does not have a weak derivative g, then the distributional derivative of T_f still exists, but it is a singular distribution not associated with a function.

Example 11.12 Let $f : \mathbb{R} \to \mathbb{R}$ be the function

$$f(x) = \begin{cases} 0 & \text{if } x < 0, \\ x & \text{if } x \geq 0. \end{cases}$$

Then f is Lipschitz continuous on $\mathbb{R}$, but it is not differentiable pointwise at $x = 0$, where its graph has a corner. Integrating by parts, and using the rapid decrease of a test function, we find that the action of the distributional derivative of f on a test function φ is given by

$$\langle f', \varphi \rangle = -\langle f, \varphi' \rangle = -\int_0^\infty x \varphi' \, dx = \int_0^\infty \varphi \, dx = \langle H, \varphi \rangle,$$

where H is the step function,

$$H(x) = \begin{cases} 0 & \text{if } x < 0, \\ 1 & \text{if } x > 0. \end{cases}$$

Thus, f is weakly differentiable, and its weak derivative is the step function H.

Example 11.13 The distributional derivative of the step function is given by

$$\langle H', \varphi \rangle = -\langle H, \varphi' \rangle = -\int_0^\infty \varphi'(x) \, dx = \varphi(0) = \langle \delta, \varphi \rangle.$$

Hence, the step function is not weakly differentiable. Its distributional derivative is the delta function, as stated in (10.12).

Example 11.14 The derivative of the one-dimensional delta function δ is given by

$$\langle \delta', \varphi \rangle = -\langle \delta, \varphi' \rangle = -\varphi'(0).$$

More generally, the kth distributional derivative of δ is given by

$$\langle \delta^{(k)}, \varphi \rangle = (-1)^k \varphi^{(k)}(0).$$

Example 11.15 The pointwise derivative of the Cantor function F, defined in Exercise 1.19, exists a.e. and is equal to zero except on the Cantor set. The function is not constant, however, and its distributional derivative is not zero. One can show that the distributional derivative of F is the Lebesgue-Stieltjes measure μ_F associated with the Cantor function, described in Example 12.15, meaning that

$$\langle F', \varphi \rangle = \int_{-\infty}^\infty \varphi(x) \, d\mu_F(x).$$

The use of duality to extend differentiation from test functions to distributions may be applied to other operations. Suppose that $K, K' : \mathcal{S} \to \mathcal{S}$ are continuous linear transformations on $\mathcal{S}$ such that

$$\int_{\mathbb{R}^n} (Kf)\varphi \, dx = \int_{\mathbb{R}^n} f(K'\varphi) \, dx \qquad \text{for all } f, \varphi \in \mathcal{S}. \tag{11.9}$$

We call K' the transpose of K. The transpose K' differs from the L^2-Hilbert space adjoint K^* of K because, unlike the L^2-inner product, we do not use a complex-conjugate in the duality pairing. If T is a tempered distribution, then we define the tempered distribution KT by

$$\langle KT, \varphi \rangle = \langle T, K'\varphi \rangle \qquad \text{for all } \varphi \in \mathcal{S}.$$

If T_f is the regular distribution associated with a test function $f \in \mathcal{S}$, then we have $KT_f = T_{Kf}$, so this definition is consistent with the definition for test functions.

Example 11.16 For each $h \in \mathbb{R}^n$, we define the *translation operator* $\tau_h : \mathcal{S} \to \mathcal{S}$ by

$$\tau_h f(x) = f(x - h).$$

We then have that

$$\int_{\mathbb{R}^n} (\tau_h f)\varphi \, dx = \int_{\mathbb{R}^n} f(\tau_{-h}\varphi) \, dx \qquad \text{for all } f, \varphi \in \mathcal{S}.$$

We therefore define the translation $\tau_h T$ of a distribution T by

$$\langle \tau_h T, \varphi \rangle = \langle T, \tau_{-h}\varphi \rangle \qquad \text{for all } \varphi \in \mathcal{S}.$$

For instance, we have $\delta_{x_0} = \tau_{x_0}\delta$.

Example 11.17 Let $R : \mathcal{S} \to \mathcal{S}$ be the *reflection operator*,

$$Rf(x) = f(-x).$$

Then

$$\int_{\mathbb{R}^n} (Rf)\varphi \, dx = \int_{\mathbb{R}^n} f(R\varphi) \, dx \qquad \text{for all } f, \varphi \in \mathcal{S}.$$

Thus, for $T \in \mathcal{S}^*$, we define the reflection $RT \in \mathcal{S}^*$ by

$$\langle RT, \varphi \rangle = \langle T, R\varphi \rangle \qquad \text{for all } \varphi \in \mathcal{S}.$$

We end this section by defining the convolution of a test function and a distribution. The convolution $\varphi * \psi$ of two test functions $\varphi, \psi \in \mathcal{S}$ is defined by

$$(\varphi * \psi)(x) = \int_{\mathbb{R}^n} \varphi(x - y)\psi(y) \, dy. \tag{11.10}$$

The following properties of the convolution are straightforward to prove.

Proposition 11.18 For any $\varphi, \psi, \omega \in \mathcal{S}$, we have:

(a) $\varphi * \psi = \psi * \varphi$;

(b) $(\varphi * \psi) * \omega = \varphi * (\psi * \omega)$,

(c) $\tau_h(\varphi * \psi) = (\tau_h\varphi) * \psi = \varphi * (\tau_h\psi)$ for every $h \in \mathbb{R}^n$.

It is clear from (11.10) that the definition of convolution can be extended from test functions to more general functions provided that the integral converges. For example, the convolution of a continuous function with compact support and an arbitrary continuous function exists, and the convolution of two L^1-functions exists and belongs to L^1. On the other hand, the convolution of two functions neither of which decays at infinity need not be well defined.

Using the translation and reflection operators defined in Example 11.16 and Example 11.17, we may write the convolution in (11.10) as

$$(\varphi * \psi)(x) = \int_{\mathbb{R}^n} (R\tau_x \varphi)(y)\psi(y)\,dy.$$

We therefore define the convolution $\varphi * T : \mathbb{R}^n \to \mathbb{C}$ of a test function $\varphi \in \mathcal{S}$ and a tempered distribution $T \in \mathcal{S}^*$ by

$$(\varphi * T)(x) = \langle T, R\tau_x \varphi \rangle.$$

One can prove that $\varphi * T \in C^\infty(\mathbb{R}^n)$, and is of at most polynomial growth.

Example 11.19 The convolution of a test function with the delta function is given by

$$(\varphi * \delta)(x) = \langle \delta, R\tau_x \varphi \rangle = (R\tau_x \varphi)(0) = (R\varphi)(-x) = \varphi(x),$$

meaning that $\varphi * \delta = \varphi$. This fact provides a distributional interpretation of the formula

$$\int \delta(x-y)\varphi(y)\,dy = \varphi(x).$$

Similarly, the convolution with a derivative of the delta function is

$$(\varphi * \partial^\alpha \delta)(x) = (-1)^{|\alpha|} \langle \delta, \partial^\alpha R\tau_x \varphi \rangle = \partial^\alpha \varphi(x).$$

More general convolutions of distributions may also be defined (for example, $\partial^\alpha \delta * T = \partial^\alpha T$ for any $T \in \mathcal{S}^*$), but we will not give a detailed description here.

11.4 The convergence of distributions

Let (T_n) be a sequence in $\mathcal{S}^*$. We say that (T_n) converges to T in $\mathcal{S}^*$ if and only if

$$\lim_{n \to \infty} \langle T_n, \varphi \rangle = \langle T, \varphi \rangle \qquad \text{for every } \varphi \in \mathcal{S}. \qquad (11.11)$$

We denote convergence in the space of distributions by

$$T_n \rightharpoonup T \qquad \text{as } n \to \infty.$$

Example 11.20 Let T_n be the distribution in $\mathcal{S}(\mathbb{R})$ defined by

$$\langle T_n, \varphi \rangle = n^3 \int_{\mathbb{R}} e^{inx} \varphi(x) \, dx.$$

Integrating by parts four times, and using the rapid decrease of $\varphi \in \mathcal{S}$, we find that

$$|\langle T_n, \varphi \rangle| = \left| \int_{\mathbb{R}} \frac{e^{inx}}{n} \varphi^{(4)}(x) \, dx \right| \to 0 \qquad \text{as } n \to \infty.$$

Thus, we have $T_n \rightharpoonup 0$ in $\mathcal{S}^*(\mathbb{R})$. The cancellation of oscillations for large n in the integration of $n^3 e^{inx}$ against a smooth test function outweighs the polynomial growth in n.

For each $\varphi \in \mathcal{S}$, the map

$$T \mapsto \langle T, \varphi \rangle \tag{11.12}$$

is a linear functional on $\mathcal{S}^*$. The convergence of distributions defined in (11.11) corresponds to convergence with respect to the weakest topology such that every functional of the form (11.12) is continuous. This topology, called the *weak-∗ topology* of $\mathcal{S}^*$, is the locally convex topology generated by the uncountable family of seminorms $\{p_\varphi \mid \varphi \in \mathcal{S}\}$, where

$$p_\varphi(T) = |\langle T, \varphi \rangle| \qquad \text{for } T \in \mathcal{S}^*. \tag{11.13}$$

Sequences of distributions that converge to the delta function are particularly important. Such sequences are called *delta sequences*. We have already encountered several examples of delta sequences, without thinking of them in terms of distributions.

Example 11.21 A simple delta sequence in $\mathcal{S}(\mathbb{R})$ is given by

$$T_n(\varphi) = \frac{n}{2} \int_{-1/n}^{1/n} \varphi(x) \, dx.$$

For any continuous function φ, we have

$$T_n(\varphi) \to \varphi(0) = \delta(\varphi) \qquad \text{as } n \to \infty,$$

so $T_n \rightharpoonup \delta$. Any approximate identity gives a delta sequence; for example, the Gaussian approximate identity

$$\varphi_n(x) = \sqrt{\frac{n}{2\pi}} e^{-nx^2/2} \tag{11.14}$$

is a delta sequence that consists of elements of $\mathcal{S}(\mathbb{R})$.

The following proposition gives a useful delta sequence of oscillatory functions. We define the *sinc function* by

$$\text{sinc}\, x = \begin{cases} \sin x / x & \text{if } x \neq 0, \\ 1 & \text{if } x = 0. \end{cases}$$

The integral of the absolute value of the sinc function does not converge, since it decays like $1/x$ as $|x| \to \infty$, but a contour integral argument gives the following improper Riemann integral

$$\lim_{R \to \infty} \int_{-R}^{R} \text{sinc}\, x \, dx = \pi. \tag{11.15}$$

Proposition 11.22 For $n \in \mathbb{N}$, let

$$\sigma_n(x) = \frac{\sin nx}{\pi x}. \tag{11.16}$$

Then $\sigma_n \rightharpoonup \delta$ in $\mathcal{S}^*(\mathbb{R})$ as $n \to \infty$.

Proof. From (11.15), we see that

$$\sigma_n(x) = \frac{n}{\pi} \text{sinc}\, nx$$

has unit integral for every $n \in \mathbb{N}$. To avoid difficulties caused by the lack of absolute convergence of the integral of σ_n at infinity, we split the integral of σ_n against a test function $\varphi \in \mathcal{S}$ into two terms:

$$\int_{-\infty}^{\infty} \frac{\sin nx}{\pi x} \varphi(x) \, dx = \int_{|x| \geq 1} \frac{\sin nx}{\pi x} \varphi(x) \, dx + \int_{|x| \leq 1} \frac{\sin nx}{\pi x} \varphi(x) \, dx. \tag{11.17}$$

An integration by parts implies that the first integral on the right-hand side tends to zero as $n \to \infty$, since

$$\int_{|x| \geq 1} \frac{\sin nx}{\pi x} \phi(x) dx = \frac{1}{n} \left[\cos nx \frac{\phi(x)}{x} \right]_{-1}^{1} + \frac{1}{n} \int_{|x| \geq 1} \cos nx \left(\frac{\phi(x)}{x} \right)' dx.$$

We write the second term on the right-hand side of (11.17) as

$$\int_{|x| \leq 1} \frac{\sin nx}{\pi x} \varphi(x) \, dx = \int_{|x| \leq 1} \frac{\sin nx}{\pi x} [\varphi(x) - \varphi(0)] \, dx + \varphi(0) \int_{|x| \leq 1} \frac{\sin nx}{\pi x} \, dx.$$

We may write $\varphi(x) = \varphi(0) + x\psi(x)$ where $\psi \in C^\infty$. The first integral on the right-hand side is therefore given by

$$\frac{1}{\pi} \int_{|x| \leq 1} \sin nx \, \psi(x) \, dx,$$

and an integration by parts shows this approaches zero as $n \to \infty$. Making the change of variable $nx \mapsto x$ and using (11.15), we see that the second term approaches

$\varphi(0)$ as $n \to \infty$, which proves the result. Note that the proof shows that $\sigma_n * \varphi \to \varphi$ uniformly for every $\varphi \in S$. $\qquad\qquad\qquad\qquad\qquad\qquad\qquad\qquad$ $\square$

The identification $\varphi \mapsto T_\varphi$ continuously embeds the Schwartz space S into the space S^* of tempered distributions. This embedding is clearly not onto, but the next result, whose proof we only outline, states that S is dense in S^*.

Theorem 11.23 The Schwartz space is dense in the space of tempered distributions.

Proof. Let (φ_n) be an approximate identity in S. Then $(\varphi_n * T)$ is a sequence of C^∞-functions of polynomial growth that converges to T in S^*. The Schwartz functions $(\varphi_n * T)e^{-\epsilon|x|^2}$ therefore converge to T in S^* as $n \to \infty$ and $\epsilon \to 0^+$. $\quad$ $\square$

A similar proof shows that S is dense in C_0, with respect to uniform convergence, and in L^p for $1 \leq p < \infty$.

11.5 The Fourier transform of test functions

In this section, we define the Fourier transform of a Schwartz function, and show that the Fourier transform is a continuous, one-to-one map from S onto S. In the next section, we will extend the transform by duality to a continuous, one-to-one map from S^* onto S^*.

Definition 11.24 If $\varphi \in S(\mathbb{R}^n)$, then the *Fourier transform* $\hat{\varphi} : \mathbb{R}^n \to \mathbb{C}$ is the function defined by

$$\hat{\varphi}(k) = \frac{1}{(2\pi)^{n/2}} \int_{\mathbb{R}^n} \varphi(x)e^{-ik \cdot x} \, dx \qquad \text{for } k \in \mathbb{R}^n. \tag{11.18}$$

There are many different conventions for where to place the factors of 2π and the signs in the Fourier transform. In the next proposition, we show that the transform of a rapidly decaying function is smooth, and the transform of a smooth function is rapidly decaying. As a result, the Fourier transform maps the Schwartz space into itself. We define the Fourier transform operator $\mathcal{F} : S \to S$ by $\mathcal{F}\varphi = \hat{\varphi}$.

Proposition 11.25 If $\varphi \in S(\mathbb{R}^n)$, then:

(a) $\hat{\varphi} \in C^\infty(\mathbb{R}^n)$, and

$$\partial^\alpha \hat{\varphi} = \mathcal{F}[(-ix)^\alpha \varphi]; \tag{11.19}$$

(b) $k^\alpha \hat{\varphi}$ is bounded for every multi-index $\alpha \in \mathbb{Z}^n_+$, and

$$(ik)^\alpha \hat{\varphi} = \mathcal{F}[\partial^\alpha \varphi]. \tag{11.20}$$

The Fourier transform $\mathcal{F} : S(\mathbb{R}^n) \to S(\mathbb{R}^n)$ is a continuous linear map on $S(\mathbb{R}^n)$.

Proof.   Equation (11.19) follows by differentiation under the integral sign in (11.18). This differentiation is justified by the dominated convergence theorem and the integrability of $x^\alpha \varphi$ for every $\alpha \in \mathbb{Z}^n_+$. Equation (11.20) follows from an integration by parts in the formula

$$
\begin{aligned}
\widehat{\partial^\alpha \varphi}(k) &= \frac{1}{(2\pi)^{n/2}} \int e^{-ik\cdot x} \partial^\alpha \varphi(x)\, dx \\
&= \frac{1}{(2\pi)^{n/2}} \int (ik)^\alpha e^{-ik\cdot x} \varphi(x)\, dx \\
&= (ik)^\alpha \hat{\varphi}(k).
\end{aligned}
$$

Thus, for every $\alpha, \beta \in \mathbb{Z}^n_+$, we have

$$
(ik)^\alpha \partial^\beta \hat{\varphi} = \mathcal{F}\left[\partial^\alpha (-ix)^\beta \varphi\right]. \tag{11.21}
$$

If $\varphi \in \mathcal{S}$, then

$$
\begin{aligned}
|\hat{\varphi}(k)| &= \frac{1}{(2\pi)^{n/2}} \left| \int e^{-ik\cdot x} \varphi(x)\, dx \right| \\
&\leq \frac{1}{(2\pi)^{n/2}} \int \frac{\left(1+|x|^2\right)^{n/2+1} |\varphi(x)|}{\left(1+|x|^2\right)^{n/2+1}}\, dx \\
&\leq C \sup_{x \in \mathbb{R}^n} \left[\left(1+|x|^2\right)^{n/2+1} |\varphi(x)|\right],
\end{aligned}
$$

where the constant C is given by

$$
C = \frac{1}{(2\pi)^{n/2}} \int_{\mathbb{R}^n} \frac{1}{\left(1+|x|^2\right)^{n/2+1}}\, dx < \infty.
$$

Taking the supremum of (11.21) with respect to k, using the Leibnitz rule to expand the function on the right-hand side, and estimating the result, we find for the seminorms in (11.3) that

$$
\|\hat{\varphi}\|_{\alpha,\beta} \leq \sum_{\alpha',\beta'} C_{\alpha',\beta'} \|\varphi\|_{\beta',\alpha'}
$$

for suitable constants $C_{\alpha',\beta'}$, where $|\alpha'| \leq |\alpha|$ and $|\beta'| \leq |\beta| + n + 2$. Hence, the Fourier transform is a continuous linear map on $\mathcal{S}$.   □

An important example of the Fourier transform of a Schwartz function is the transform of a Gaussian, which is another Gaussian.

Proposition 11.26 Suppose that A is an $n \times n$ symmetric, positive definite matrix. The Fourier transform of the n-dimensional Gaussian

$$
\varphi(x) = \exp\left(-\frac{1}{2}x \cdot Ax\right) \tag{11.22}
$$

is given by

$$\hat{\varphi}(k) = \frac{1}{\sqrt{\det A}} \exp\left(-\frac{1}{2}k \cdot A^{-1}k\right). \tag{11.23}$$

Proof. First, we consider the one-dimensional Gaussian

$$\varphi(x) = \exp\left(-\frac{ax^2}{2}\right),$$

where $a > 0$. We claim that

$$\hat{\varphi}(k) = \frac{1}{\sqrt{a}} \exp\left(-\frac{k^2}{2a}\right). \tag{11.24}$$

To prove this result, it suffices to consider the case $a = 1$. The formula for $a > 0$ then follows from the change of variables $x \mapsto \sqrt{a}x$. Thus, we just need to show that

$$\frac{1}{\sqrt{2\pi}} \int e^{-x^2/2} e^{-ikx} \, dx = e^{-k^2/2}.$$

The left-hand side of this equation may be written as

$$\frac{1}{\sqrt{2\pi}} e^{-k^2/2} \int e^{-(x+ik)^2/2} \, dx.$$

So we want to show that

$$\frac{1}{\sqrt{2\pi}} \int e^{-(x+ik)^2/2} \, dx = 1. \tag{11.25}$$

This integral is independent of k, since

$$
\begin{aligned}
\frac{d}{dk}\left(\frac{1}{\sqrt{2\pi}} \int e^{-(x+ik)^2/2} \, dx\right) &= -i\frac{1}{\sqrt{2\pi}} \int (x+ik)e^{-(x+ik)^2/2} \, dx \\
&= i\frac{1}{\sqrt{2\pi}} \int \frac{d}{dx} e^{-(x+ik)^2/2} \, dx \\
&= i\frac{1}{\sqrt{2\pi}} e^{-(x+ik)^2/2}\Big|_{-\infty}^{\infty} \\
&= 0,
\end{aligned}
$$

so (11.25) follows from the standard Gaussian integral,

$$\int_{-\infty}^{\infty} e^{-x^2/2} \, dx = \sqrt{2\pi}.$$

Now we consider the n-dimensional case. The Fourier tranform of the Gaussian in (11.22) is given by

$$\hat{\varphi}(k) = \frac{1}{(2\pi)^{n/2}} \int e^{-x \cdot Ax/2} e^{-ik \cdot x} \, dx. \tag{11.26}$$

Since A is positive definite, there is an orthogonal matrix Q such that $Q^T A Q = \Lambda$, where $\Lambda = \text{diag}(\lambda_1, \ldots, \lambda_n)$ is a diagonal matrix with the eigenvalues $\lambda_i > 0$ of A on the main diagonal. We make the change of variables $x = Q\overline{x}$ and $k = Q\overline{k}$ in (11.26). The Jacobian of the transformation $x \mapsto \overline{x}$ is $\det Q = 1$. The resulting expression factors into a product of one-dimensional Fourier integrals, which we may evaluate using the one-dimensional computation:

$$
\begin{aligned}
\hat{\varphi}(k) &= \frac{1}{(2\pi)^{n/2}} \int e^{-\overline{x}\cdot\Lambda\overline{x}/2} e^{-i\overline{k}\cdot\overline{x}} \, d\overline{x} \\
&= \prod_{j=1}^{n} \frac{1}{\sqrt{2\pi}} \int e^{-\lambda_j \overline{x}_j^2/2} e^{-i\overline{k}_j \overline{x}_j} \, d\overline{x}_j \\
&= \prod_{j=1}^{n} \frac{1}{\sqrt{\lambda_j}} e^{-\overline{k}_j^2/(2\lambda_j)}.
\end{aligned}
$$

Rewriting this result in terms of k, and using the facts that $\det A = (\lambda_1 \lambda_2 \ldots \lambda_n)$ and $A^{-1} = Q^T \Lambda^{-1} Q$, we obtain (11.23). $\qquad\square$

The covariance matrix A of the transform of a Gaussian is the inverse of the covariance matrix of the Gaussian. Thus, the transform of a Gaussian that is localized near the origin is delocalized, and conversely. The intuitive explanation of this behavior is that more high-frequency Fourier components are required to represent a rapidly varying, localized function than a slowly varying, delocalized function. For example, the Fourier transform of the Gaussian approximate identity

$$
\varphi_\epsilon(x) = \frac{1}{(2\pi\epsilon)^{n/2}} \exp\left(-\frac{|x|^2}{2\epsilon}\right)
$$

is given by

$$
\hat{\varphi}_\epsilon(k) = \frac{1}{(2\pi)^{n/2}} \exp\left(-\frac{\epsilon|x|^2}{2}\right).
$$

As $\epsilon \to 0^+$, we have $\varphi_\epsilon \rightharpoonup \delta$ and $\hat{\varphi}_\epsilon \rightharpoonup (2\pi)^{-n/2}$ in S^*. The spectrum of the Gaussian becomes flatter as it concentrates at the origin. These limits are consistent with the result below that $\hat{\delta} = (2\pi)^{-n/2}$.

The following proposition, whose proof we leave to Exercise 11.13, gives the formulae for the Fourier transform of translates and convolutions. An important result is the fact that the Fourier transform maps the convolution product of two functions to their pointwise product. We will see in Section 11.9 that this is related to the translational invariance of the convolution.

Proposition 11.27 If $\varphi, \psi \in S$ and $h \in \mathbb{R}^n$, then:

$$
\widehat{\tau_h \varphi} = e^{-ik\cdot h} \hat{\varphi}, \tag{11.27}
$$

$$
\widehat{e^{ix\cdot h} \varphi} = \tau_h \hat{\varphi}, \tag{11.28}
$$

$$\widehat{\varphi * \psi} = (2\pi)^{n/2}\hat{\varphi}\hat{\psi}. \tag{11.29}$$

Finally, we prove that $\mathcal{F}$ is invertible on $\mathcal{S}$ with a continuous inverse. First, we give a formula for the inverse.

Definition 11.28 If $\varphi \in \mathcal{S}$, then the *inverse Fourier transform* $\check{\varphi}$ is given by

$$\check{\varphi}(x) = \frac{1}{(2\pi)^{n/2}} \int_{\mathbb{R}^n} e^{ik\cdot x}\varphi(k)\, dk.$$

We define $\mathcal{F}^* : \mathcal{S} \to \mathcal{S}$ by $\mathcal{F}^*\varphi = \check{\varphi}$.

We will prove that $\mathcal{F}^* = \mathcal{F}^{-1}$, meaning that

$$\check{\hat{\varphi}} = \varphi = \hat{\check{\varphi}} \qquad \text{for every } \varphi \in \mathcal{S}. \tag{11.30}$$

To motivate the proof of the inversion formula, we first give a formal calculation, based on the completeness formula in (11.33) below. Writing out $\mathcal{F}^*\hat{\varphi}$, and exchanging the order of integration, we find that

$$
\begin{aligned}
\mathcal{F}^*\hat{\varphi}(x) &= \frac{1}{(2\pi)^n} \int e^{ik\cdot x} \left[\int e^{-ik\cdot y}\varphi(y)\, dy \right] dk \\
&= \int_{-\infty}^{\infty} \frac{1}{(2\pi)^n} \left[\int_{-\infty}^{\infty} e^{ik(x-y)}\, dk \right] \varphi(y)\, dy \\
&= \int \delta(x-y)\varphi(y)\, dy \\
&= \varphi(x).
\end{aligned}
$$

The exchange of integration in this calculation is not justified by Fubini's theorem because the integral is not absolutely convergent. To make the argument rigorous, we introduce an "ultraviolet cut-off" in the integral before exchanging the order of integration.

Proposition 11.29 The map $\mathcal{F}^*$ is a continuous linear transformation on $\mathcal{S}$, and $\mathcal{F}^* = \mathcal{F}^{-1}$.

Proof. We have $\mathcal{F}^* = R \circ \mathcal{F}$, where R is the reflection defined by $R\varphi(x) = \varphi(-x)$, so the continuity of $\mathcal{F}^*$ on $\mathcal{S}$ follows from the continuity of R and $\mathcal{F}$.

The n-dimensional Fourier transform is the composition of one-dimensional Fourier transforms in each of the components x_i of $x \in \mathbb{R}^n$, $i = 1, \dots, n$, so it suffices to prove the result for $n = 1$. Introducing a cut-off in the k-integral, and using Fubini's theorem to exchange the order of integration, we find that

$$
\begin{aligned}
\mathcal{F}^*\hat{\varphi}(x) &= \frac{1}{2\pi} \int_{-\infty}^{\infty} e^{ikx} \left[\int_{-\infty}^{\infty} e^{-iky}\varphi(y)\, dy \right] dk \\
&= \frac{1}{2\pi} \lim_{R\to\infty} \int_{-R}^{R} \left[\int_{-\infty}^{\infty} e^{ik(x-y)}\varphi(y)\, dy \right] dk
\end{aligned}
$$

$$= \frac{1}{2\pi} \lim_{R \to \infty} \int_{-\infty}^{\infty} \left[\int_{-R}^{R} e^{ik(x-y)} \, dk \right] \varphi(y) \, dy$$

$$= \lim_{R \to \infty} \int_{-\infty}^{\infty} \left[\frac{\sin R(x-y)}{\pi(x-y)} \, dx \right] \varphi(y) \, dy.$$

From Proposition 11.22, the sequence $(\sin Rx)/\pi x$ is a delta sequence as $R \to \infty$, so $\mathcal{F}^*\hat{\varphi} = \varphi$. An identical argument shows that $\mathcal{F}\check{\varphi} = \varphi$. Therefore $\mathcal{F}, \mathcal{F}^* : S \to S$ are one-to-one, onto continuous maps, and $\mathcal{F}^* = \mathcal{F}^{-1}$. □

We could have instead introduced a Gaussian regularization,

$$\mathcal{F}^*\hat{\varphi}(x) = \frac{1}{2\pi} \lim_{\epsilon \to 0+} \int_{-\infty}^{\infty} e^{ikx - \epsilon k^2/2} \left[\int_{\infty}^{\infty} e^{iky}\varphi(y) \, dy \right] dk,$$

exchanged the order of integration, and passed to the limit in the resulting Gaussian approximate identity.

11.6   The Fourier transform of tempered distributions

In this section, we define the Fourier transform of a tempered distribution. First, suppose that $f, \varphi \in S$. Using the definition of the transform and exchanging the order of integration, which is justified by Fubini's theorem, we find that the action of the Fourier transform $\hat{f}$ on a test function φ is given by

$$\int \hat{f}(k)\varphi(k) \, dk = \int \frac{1}{(2\pi)^{n/2}} \left(\int f(x)e^{-ik \cdot x} \, dx \right) \varphi(k) \, dk$$

$$= \int f(x) \frac{1}{(2\pi)^{n/2}} \left(\int \varphi(k)e^{-ik \cdot x} \, dk \right) dx$$

$$= \int f(x)\hat{\varphi}(x) \, dx. \tag{11.31}$$

In the notation of (11.9), this result means that $\mathcal{F}' = \mathcal{F}$. We therefore define the Fourier transform of a tempered distribution as follows.

Definition 11.30 The Fourier transform of a tempered distribution T is the tempered distribution $\hat{T} = \mathcal{F}T$ defined by

$$\langle \hat{T}, \varphi \rangle = \langle T, \hat{\varphi} \rangle \qquad \text{for all } \varphi \in S. \tag{11.32}$$

The inverse Fourier transform $\check{T} = \mathcal{F}^{-1}T$ on S^* is defined by

$$\langle \check{T}, \varphi \rangle = \langle T, \check{\varphi} \rangle \qquad \text{for all } \varphi \in S.$$

The linearity and continuity of the Fourier transform on S implies that $\hat{T}$ is a tempered distribution. The map $\mathcal{F} : S^* \to S^*$ is a continuous, one-to-one transformation of S^* onto itself. The fact that $\mathcal{F}^{-1}$ is the inverse of $\mathcal{F}$ on S^* follows immediately from the corresponding result on $\mathcal{S}$, since

$$\langle \check{\hat{T}}, \varphi \rangle = \langle T, \check{\hat{\varphi}} \rangle = \langle T, \varphi \rangle \qquad \text{for all } \varphi \in \mathcal{S}.$$

The formulae for the Fourier transform of derivatives, translates, and convolutions carry over directly to distributions (see Exercise 11.13). For example,

$$\widehat{\partial^\alpha T} = (ik)^\alpha \hat{T}.$$

We also write the Fourier transform using the integral notation,

$$\hat{T}(k) = \frac{1}{(2\pi)^{n/2}} \int T(x) e^{-ik \cdot x}\, dx,$$

as if T were a function, with an analogous expression for the inverse. This notation should be interpreted simply as a short-hand for the definition in (11.32).

Example 11.31 Let us compute the Fourier transform of the delta function. From (11.32), we have

$$\langle \hat{\delta}, \varphi \rangle = \langle \delta, \hat{\varphi} \rangle = \hat{\varphi}(0).$$

From the formula for the Fourier transform on $\mathcal{S}$, we have

$$\hat{\varphi}(0) = \frac{1}{(2\pi)^{n/2}} \int \varphi(x)\, dx = \frac{1}{(2\pi)^{n/2}} \langle 1, \varphi \rangle.$$

Hence, the Fourier transform of the delta function is a constant,

$$\hat{\delta} = \frac{1}{(2\pi)^{n/2}}.$$

Using the integral notation, we get from the inversion formula the following Fourier representation of the delta function:

$$\delta(x) = \frac{1}{(2\pi)^n} \int_{\mathbb{R}^n} e^{ik \cdot x}\, dk. \tag{11.33}$$

The formula for the transform of the derivative implies that the transform of the αth derivative of the delta function is a monomial,

$$\widehat{\partial^\alpha \delta} = \frac{1}{(2\pi)^{n/2}} (ik)^\alpha.$$

11.7   The Fourier transform on L^1

The Fourier integral

$$\hat{f}(k) = \frac{1}{(2\pi)^{n/2}} \int_{\mathbb{R}^n} f(x) e^{-ik \cdot x}\, dx \tag{11.34}$$

converges if and only if $f \in L^1(\mathbb{R}^n)$, meaning that

$$\int_{\mathbb{R}^n} |f(x)| \, dx < \infty.$$

We define the Fourier transform $\hat{f}$ of an L^1-function f by (11.34). This definition is consistent with the distributional definition, since Fubini's theorem justifies the exchange in the order of integration in (11.31) when $f \in L^1(\mathbb{R}^n)$.

Example 11.32 Let $f = \chi_{[-R,R]}$ be the characteristic function of the interval $[-R, R]$, sometimes called a "box" function. Then

$$\hat{f}(k) = \sqrt{2\pi}\,\frac{\sin Rk}{\pi k}.$$

Thus, the Fourier transform of a box function is a sinc function. The slow rate of decay of the Fourier transform as $k \to \infty$, of the order k^{-1}, is caused by the discontinuities in f. Although f belongs to L^1, the transform $\hat{f}$ does not. Thus, we cannot recover f from $\hat{f}$ by use of the inverse Fourier integral, but we can use the distributional definition of the inverse Fourier transform.

Example 11.33 For $a > 0$, let $f(x) = \exp(-a|x|)$. Then

$$\hat{f}(k) = \sqrt{\frac{2}{\pi}}\,\frac{a}{a^2 + k^2}.$$

The following result, called the Riemann-Lebesgue lemma, gives a basic property of the Fourier transform of L^1-functions. We denote by $C_0(\mathbb{R}^n)$ the space of continuous functions f that approach zero at infinity, meaning that for every $\epsilon > 0$ there is an R such that $|f(x)| < \epsilon$ when $|x| > R$. This space is the completion of $C_c(\mathbb{R}^n)$ with respect to the supremum norm, and is a Banach space.

Theorem 11.34 (Riemann-Lebesgue) If $f \in L^1(\mathbb{R}^n)$, then $\hat{f} \in C_0(\mathbb{R}^n)$, and

$$(2\pi)^{n/2}\|\hat{f}\|_\infty \le \|f\|_1.$$

Proof. To prove the claim, we first observe that if $\varphi \in \mathcal{S}$, then

$$(2\pi)^{n/2}|\hat{\varphi}(k)| = \left|\int e^{-ik \cdot x}\varphi(x)\,dx\right|$$

$$\le \int |\varphi(x)|\,dx.$$

Taking the supremum of this inequality over k, we find that

$$(2\pi)^{n/2}\,\|\hat{\varphi}\|_\infty \le \|\varphi\|_1\,.$$

The Schwartz space $\mathcal{S}$ is dense in L^1. Hence, if $f \in L^1$, there is sequence (φ_m) in $\mathcal{S}$ that converges to f with respect to the L^1-norm. Then $(\hat{\varphi}_m)$ is Cauchy in the supremum norm, since

$$(2\pi)^{n/2} \|\hat{\varphi}_m - \hat{\varphi}_\ell\|_\infty \leq \|\varphi_m - \varphi_\ell\|_1 \,.$$

Since $\mathcal{S}$ is contained in C_0, and C_0 is complete, there is a function $\hat{g} \in C_0$ such that $\hat{\varphi}_m \to \hat{g}$ uniformly. Moreover, $\hat{g} = \hat{f}$ because

$$
\begin{aligned}
(2\pi)^{n/2} \left| \hat{g}(k) - \hat{f}(k) \right| &= (2\pi)^{n/2} \lim_{m \to \infty} \left| \hat{\varphi}_m(k) - \hat{f}(k) \right| \\
&= \lim_{m \to \infty} \left| \int [\varphi_m(x) - f(x)] \, e^{-ik \cdot x} \, dx \right| \\
&\leq \liminf_{m \to \infty} \|\varphi_m - f\|_1 = 0.
\end{aligned}
$$

$\square$

The Fourier transform is therefore a bounded linear map from L^1 into C_0. We may make L^1 into an algebra with the convolution product, and C_0 into an algebra with the pointwise product. The following proposition shows that the Fourier transform maps the convolution product into the pointwise product, up to a factor of $(2\pi)^{n/2}$, which depends on the normalization of the Fourier transform. Thus the Fourier transform is an algebra isomorphism of L^1 and its image $\mathcal{F}(L^1) \subset C_0$. The image $\mathcal{F}(L^1)$ is strictly smaller than C_0, but a precise description of it is difficult.

Theorem 11.35 (Convolution) If $f, g \in L^1(\mathbb{R}^n)$, then $f * g \in L^1(\mathbb{R}^n)$ and

$$\widehat{f * g} = (2\pi)^{n/2} \hat{f} \hat{g}.$$

Proof. Fubini's theorem implies that

$$
\begin{aligned}
\int |f * g(x)| \, dx &= \int \left| \int f(x - y) g(y) \, dy \right| dx \\
&\leq \int \left[\int |f(x - y)| \, dx \right] |g(y)| \, dy \\
&= \left(\int |f(z)| \, dz \right) \left(\int |g(y)| \, dy \right),
\end{aligned}
$$

which shows that $f * g \in L^1(\mathbb{R}^n)$. Moreover, the absolute convergence of this integral implies that we can exchange the order of integration in the integral for the Fourier transform of $f * g$:

$$
\begin{aligned}
\widehat{f * g}(k) &= \frac{1}{(2\pi)^{n/2}} \int e^{-ik \cdot x} \left[\int f(x - y) g(y) \, dy \right] dx \\
&= \frac{1}{(2\pi)^{n/2}} \int e^{-ik \cdot y} \left[\int e^{-ik \cdot (x - y)} f(x - y) \, dx \right] g(y) \, dy \\
&= \frac{1}{(2\pi)^{n/2}} \left(\int e^{-ik \cdot z} f(z) \, dz \right) \left(\int e^{-ik \cdot y} g(y) \, dy \right)
\end{aligned}
$$

$$= (2\pi)^{n/2} \hat{f}\hat{g}.$$

$\square$

One use of the convolution theorem is the computation of the inverse Fourier transform of the product of two functions whose inverse Fourier transform we know.

Example 11.36 For $a > 0$, we have that

$$\mathcal{F}\left[\frac{1}{\pi}\frac{a}{a^2 + x^2}\right] = \frac{1}{\sqrt{2\pi}}e^{-a|k|}.$$

Taking the inverse Fourier transform of the equation

$$\frac{1}{\sqrt{2\pi}}e^{-(a+b)|k|} = \sqrt{2\pi}\,\frac{1}{\sqrt{2\pi}}e^{-a|k|}\,\frac{1}{\sqrt{2\pi}}e^{-b|k|},$$

where $a, b > 0$, and using the convolution theorem, we obtain the semigroup relation

$$\frac{1}{\pi}\frac{a+b}{(a+b)^2 + x^2} = \left(\frac{1}{\pi}\frac{a}{a^2 + x^2}\right) * \left(\frac{1}{\pi}\frac{b}{b^2 + x^2}\right).$$

Finally, we make a few comments about the extension of the Fourier transform to a function of a complex variable, called the *Fourier-Laplace transform*. If $f : \mathbb{R}^n \to \mathbb{C}$ is an integrable function with compact support, then (11.34) defines an entire function $\hat{f} : \mathbb{C}^n \to \mathbb{C}$ (meaning that $\hat{f}(k)$ is a differentiable, or analytic, function of the complex variable k for all $k \in \mathbb{C}^n$), since the integral obtained by differentiation under the integral sign converges for every $k \in \mathbb{C}^n$. The *Paley-Wiener theorem*, which we do not state here, gives a precise characterization of the Fourier transforms of compactly supported functions. Similarly, considering the case of one variable for simplicity, if f is integrable and the support of $f(x)$ is contained in the half-line $x \geq 0$, then the Fourier transform $\hat{f}(k)$ is an analytic function of k in the lower-half plane $\text{Im}\,k < 0$, because in that case the exponential e^{-ikx} decays as $x \to +\infty$. Setting $k = -iz$, and omitting the normalization factor of $\sqrt{2\pi}$, we obtain the *Laplace transform* of f,

$$\tilde{f}(z) = \int_0^\infty f(x)e^{-xz}\,dx,$$

which is analytic in the right-half plane $\text{Re}\,z > 0$. More generally, if $\text{supp}\,f \subset [0, \infty)$ and $f(x)e^{-ax}$ is integrable for some $a \in \mathbb{R}$, then $\tilde{f}(z)$ is analytic in the right-half plane $\text{Re}\,z > a$. Methods from complex analysis, such as contour integration, may be used to study and invert the Fourier-Laplace transform.

The space of Fourier transforms of test functions in $\mathcal{D} = C_c^\infty$ is a space $\mathcal{L}$ of entire functions. Continuous linear functionals on $\mathcal{L}$, equipped with an appropriate topology, are called *ultradistributions*. The space $\mathcal{L}^*$ of ultradistributions contains the space $\mathcal{S}^*$ of tempered distributions, and the Fourier transform of an arbitrary distribution in $\mathcal{D}^*$ may be defined as an ultradistribution, even if it has exponential growth at infinity. For example, the Fourier transform

$$\widehat{e^{x^2}} = \mathcal{F}\left(\sum_{n=0}^{\infty} \frac{x^{2n}}{n!}\right) = \sqrt{2\pi} \sum_{n=0}^{\infty} \frac{\delta^{(2n)}}{n!}$$

is well-defined as an ultradistribution. The series on the right-hand side does not converge in $\mathcal{S}^*$, however, since Schwartz functions need not have convergent Taylor series expansions.

11.8   The Fourier transform on L^2

We have seen that the Fourier transform is an isomorphism on both the Schwartz space and on the space of tempered distributions equipped with their appropriate topologies. In this section, we will see that the Fourier transform is also an isomorphism on the Hilbert space $L^2(\mathbb{R}^n)$ of square-integrable functions. To avoid confusion with our notation for the duality pairing on $\mathcal{S}^* \times \mathcal{S}$, we denote the L^2-inner product by

$$(f, g) = \int_{\mathbb{R}^n} \overline{f(x)} g(x) \, dx.$$

The duality pairing and inner-product of $f \in L^2$ and $\varphi \in \mathcal{S}$ are related by

$$\langle T_{\overline{f}}, \varphi \rangle = (f, \varphi).$$

Not every square-integrable function on $\mathbb{R}^n$ is integrable; for example, the function $(1 + x^2)^{-1/2}$ belongs to $L^2(\mathbb{R})$ but not $L^1(\mathbb{R})$. Thus, we cannot define the Fourier transform of a general L^2-function directly by means of its Fourier integral. Instead, we will use the L^2-boundedness of the Fourier transform to extend it from $\mathcal{S}(\mathbb{R}^n)$ to $L^2(\mathbb{R}^n)$.

If $\varphi \in \mathcal{S}$, then $\overline{\hat{\varphi}} = \check{\overline{\varphi}}$, since

$$\frac{1}{(2\pi)^{n/2}} \overline{\int \varphi(x) e^{-ik \cdot x} \, dx} = \frac{1}{(2\pi)^{n/2}} \int \overline{\varphi(x)} e^{ik \cdot x} \, dx.$$

Using (11.30) and (11.31), we see that for every $\varphi, \psi \in \mathcal{S}$

$$(\hat{\varphi}, \hat{\psi}) = \int_{\mathbb{R}^n} \overline{\hat{\varphi}(x)} \hat{\psi}(x) \, dx = \int_{\mathbb{R}^n} \check{\overline{\varphi}}(x) \hat{\psi}(x) \, dx = \int_{\mathbb{R}^n} \overline{\varphi(x)} \check{\hat{\psi}}(x) \, dx = (\varphi, \psi).$$

Thus, the Fourier transform is an isometric map

$$\mathcal{F} : \mathcal{S} \subset L^2 \to \mathcal{S} \subset L^2.$$

The Schwartz space $\mathcal{S}$ is dense in L^2, so the bounded linear transformation theorem implies that there is a unique isometric extension $\mathcal{F} : L^2 \to L^2$. Moreover, $\mathcal{F}^{-1} =$

$\mathcal{F}^*$, where $\mathcal{F}^*$ is the Hilbert space adjoint of $\mathcal{F}$. Consequently, we have the following theorem.

Theorem 11.37 (Plancherel) The Fourier transform $\mathcal{F} : L^2(\mathbb{R}^n) \to L^2(\mathbb{R}^n)$ is a unitary map. For every $f, g \in L^2(\mathbb{R}^n)$, we have

$$(\hat{f}, \hat{g}) = (f, g), \tag{11.35}$$

where $\hat{f} = \mathcal{F}f$. In particular,

$$\int_{\mathbb{R}^n} |f(x)|^2 \, dx = \int_{\mathbb{R}^n} \left|\hat{f}(k)\right|^2 \, dk. \tag{11.36}$$

To compute the Fourier transform of a general function $f \in L^2$, we choose any sequence (φ_n) in $\mathcal{S}$ (or, more generally, in $L^1 \cap L^2$) that converges to f in L^2. Then $\hat{f}$ is the L^2-limit of $(\hat{\varphi}_n)$. For example,

$$
\begin{aligned}
\hat{f}(k) &= \lim_{R \to \infty} \int_{|x| \le R} f(x) e^{-ik \cdot x} \, dx \\
&= \lim_{\epsilon \to 0^+} \int_{\mathbb{R}^n} f(x) e^{-ik \cdot x - \epsilon |x|^2} \, dx, \tag{11.37}
\end{aligned}
$$

where the limits are understood in the L^2-sense. The inverse transform may be computed in a similar way.

The Fourier transform is a unitary operator on $L^2(\mathbb{R}^n)$, so its spectrum lies on the unit circle in $\mathbb{C}$. The spectrum turns out to consist entirely of eigenvalues. We will describe it, without proof, in the one-dimensional case. Multi-dimensional eigenfunctions may be constructed from products of one-dimensional eigenfunctions in each of the coordinates.

Since $R\mathcal{F}^{-1} = \mathcal{F}$, where R is the reflection operator on L^2, we have $\mathcal{F}^2 = R$, and $\mathcal{F}^4 = I$. It follows that if $\lambda \in \mathbb{C}$ is an eigenvalue of $\mathcal{F}$, then $\lambda^4 = 1$, so $\lambda \in \{1, i, -1, -i\}$. Each of these values is an eigenvalue of infinite multiplicity. A complete orthonormal set of eigenfunctions is given by the Hermite functions,

$$\varphi_n(x) = \frac{1}{\sqrt{2^n n! \sqrt{\pi}}} e^{x^2/2} \frac{d^n}{dx^n} e^{-x^2}, \tag{11.38}$$

where $n = 0, 1, 2, \ldots$. One can prove that

$$\mathcal{F}\varphi_n = (-i)^n \varphi_n.$$

From Exercise 6.14, the Hermite functions are eigenfunctions of the differential operator

$$Au = -u'' + x^2 u.$$

Taking the Fourier transform of this expression, we find that the terms involving derivatives and multiplication by powers exchange places, so

$$\mathcal{F}Au = k^2\hat{u} - \hat{u}'' = A\mathcal{F}u.$$

Thus, A and $\mathcal{F}$ commute, which explains why they share a common set of eigenfunctions.

Once we know that the Hermite functions form an orthonormal basis of $L^2(\mathbb{R})$, we can give an alternative definition of the L^2-Fourier transform as

$$\mathcal{F}\left(\sum_{n=0}^{\infty} c_n\varphi_n\right) = \sum_{n=0}^{\infty}(-i)^n c_n\varphi_n.$$

The unitarity of the Fourier transform on L^2 can be seen immediately from this formula.

Just as we used Fourier series to define Sobolev spaces of periodic functions, we can use the Fourier transform to define Sobolev spaces of functions with square-integrable derivatives on $\mathbb{R}^n$. Since

$$\widehat{\partial^\alpha f} = (ik)^\alpha \hat{f},$$

the partial derivatives of f of order less than or equal to s are square-integrable if and only if $(ik)^\alpha \hat{f}$ is a square-integrable function for $|k| \leq s$. This is the case if the function

$$\left(1 + |k|^2\right)^{s/2} \hat{f}$$

is square-integrable. More generally, we can define Sobolev spaces of distributions with fractional, or even negative, order L^2-derivatives.

Definition 11.38 Let $s \in \mathbb{R}$. The Sobolev space $H^s(\mathbb{R}^n)$ consists of all distributions $f \in S^*$ whose Fourier transform $\hat{f} : \mathbb{R}^n \to \mathbb{C}$ is a regular distribution and

$$\int_{\mathbb{R}^n} \left(1 + |k|^2\right)^s \left|\hat{f}(k)\right|^2 dk < \infty.$$

A similar proof to the proof of the Sobolev embedding theorem for periodic functions, in Theorem 7.9, shows that if $f \in H^s(\mathbb{R}^n)$ for $s > n/2$, then $f \in C_0(\mathbb{R}^n)$ (see Exercise 11.12).

11.9 Translation invariant operators

There is a close connection between the Fourier transform and the group of translation operators τ_h defined in Example 11.16. Since

$$\tau_h e^{-ik\cdot x} = e^{ik\cdot h}e^{-ik\cdot x},$$

the exponential functions $e^{-ik \cdot x}$, with $k \in \mathbb{R}^n$, are eigenvectors of τ_h in $\mathcal{S}^*$ with eigenvalues $e^{ik \cdot h}$. The Fourier transform is therefore an expansion of a function or distribution with respect to the eigenvectors of τ_h. If $A : \mathcal{S}^* \to \mathcal{S}^*$ is a linear translation invariant operator, meaning that $A\tau_h = \tau_h A$, then we expect that there is a basis of common eigenvectors of τ_h and A, so that A can be diagonalized by use of the Fourier transform. In that case, the action of A on a distribution is to multiply the Fourier transform of the distribution by a C^∞-function $\hat{a}$ of polynomial growth,

$$\widehat{AT} = \hat{a}\hat{T}.$$

The function $\hat{a}$ is called the *symbol* of the operator A. Inverting the transform, we find from the convolution theorem that

$$AT = \frac{1}{(2\pi)^{n/2}} a * T,$$

with a suitable definition of the convolution $a * T$. Since $\tau_h(a * T) = a * (\tau_h T)$, the convolution is translation invariant.

Example 11.39 A constant coefficient linear differential operator $P : \mathcal{S}^* \to \mathcal{S}^*$ is translation invariant, and is given by

$$PT = \sum_{|\alpha| \leq d} c_\alpha \partial^\alpha T$$

for constants c_α. The Fourier representation is $\widehat{PT} = \hat{p}\hat{T}$, where

$$\hat{p}(k) = \sum_{|\alpha| \leq d} c_\alpha (ik)^\alpha.$$

Thus, the symbol of a differential operator is a polynomial. The convolution form of the operator is

$$PT = \left(\sum_{|\alpha| \leq d} c_\alpha \partial^\alpha \delta \right) * T.$$

It can be much simpler to define an operator in terms of its symbol than by an explicit formula for its action on a function.

Example 11.40 The symbol of the differential operator $(-\Delta + I)$ is the quadratic polynomial $(|k|^2 + 1)$. The square-root $(-\Delta + I)^{1/2}$ is the nonlocal operator with symbol $(|k|^2 + 1)^{1/2}$. Its action on a distribution T is given by

$$(-\Delta + I)^{1/2}T = \mathcal{F}^{-1}\left[(|k|^2 + 1)^{1/2}\hat{T} \right].$$

The inverse operator $(-\Delta + I)^{-1}$ has symbol $(|k|^2 + 1)^{-1}$, so

$$(-\Delta + I)^{-1}T = g * T,$$

where g is the Green's function of $(-\Delta + I)$, given by

$$g = \frac{1}{(2\pi)^{n/2}} \mathcal{F}^{-1} \left[\frac{1}{|k|^2 + 1} \right].$$

For $n = 3$, where n is the number of space dimensions, computation of the inverse Fourier transform gives

$$g(x) = \frac{1}{4\pi} \frac{e^{-|x|}}{|x|}.$$

For $n = 2$, the Green's function may be expressed in terms of Bessel functions. We will study some other examples of Green's functions in the next section.

We may also consider translation invariant operators defined on a subspace of S^*. For example, any bounded function $\hat{a} \in L^\infty(\mathbb{R}^n)$ is the symbol of a translation invariant operator $A : L^2(\mathbb{R}^n) \to L^2(\mathbb{R}^n)$ defined by $\widehat{Af} = \hat{a}\hat{f}$.

Example 11.41 For $g \in L^1(\mathbb{R}^n)$, we define the convolution integral operator $G : L^2(\mathbb{R}^n) \to L^2(\mathbb{R}^n)$ by

$$Gf(x) = \frac{1}{(2\pi)^{n/2}} \int_{\mathbb{R}^n} g(x - y) f(y) \, dy. \tag{11.39}$$

The symbol of G is $\hat{g}$. Since $g \in L^1$, the Riemann-Lebesgue lemma (Theorem 11.34) implies that $\hat{g} \in C_0$. Thus, the Fourier transform $\mathcal{F}$ diagonalizes G, and $G = \mathcal{F}^* \hat{g} \mathcal{F}$ is unitarily equivalent to multiplication by $\hat{g}$. Unless $\hat{g} = $ constant on a set of nonzero measure, the multiplication operator has a continuous spectrum, given by the closure of the range of $\hat{g}$, so this is also the spectrum of G.

More generally, the map G is well defined on L^2 whenever $\hat{g} \in L^\infty$ is bounded. For example, suppose that $\hat{f}_R$ is the function obtained by truncating the Fourier transform of $f \in L^2(\mathbb{R})$ at wavenumbers k with $|k| \leq R$:

$$\hat{f}_R(x) = \begin{cases} \hat{f}(k) & \text{if } |k| \leq R, \\ 0 & \text{if } |k| > R. \end{cases}$$

Then $\hat{f}_R = \chi_{[-R,R]}\hat{f}$. Since

$$\mathcal{F}^{-1}\left(\chi_{[-R,R]}\right) = \sqrt{\frac{2}{\pi}} R \, \text{sinc}(Rx),$$

the function $f_R = \mathcal{F}^{-1}\left[\hat{f}_R\right]$ is given by

$$f_R = \frac{R}{\pi} \text{sinc}(Rx) * f.$$

Example 11.42 The symbol of the translation operator τ_h itself is $e^{-ik \cdot h}$. The translation operators $\{\tau_h \mid h \in \mathbb{R}^n\}$ form a unitary group acting on $L^2(\mathbb{R}^n)$. If $h \neq 0$, then the spectrum of τ_h is the unit circle in $\mathbb{C}$ and is purely continuous.

Example 11.43 The operator $\mathbb{H} : L^2(\mathbb{R}) \to L^2(\mathbb{R})$ with symbol

$$\hat{h}(k) = i \operatorname{sgn} k$$

is called the *Hilbert transform*. Since the modulus of the symbol is equal to one, Plancherel's theorem implies that $\mathbb{H}$ is a unitary map of $L^2(\mathbb{R})$ onto itself. Since $\hat{h}^2 = -1$, we have $\mathbb{H}^2 = -I$. From Exercise 11.22 and the convolution theorem,

$$\mathbb{H}f = -\frac{1}{\pi}\left(\text{p.v.}\frac{1}{x}\right) * f.$$

The Hilbert transform is one of the simplest examples of a *singular integral operator*. Its properties are much more transparent from the Fourier representation than the convolution form.

Example 11.44 The operator $R_{pq} : L^2(\mathbb{R}^n) \to L^2(\mathbb{R}^n)$ with symbol

$$\hat{r}_{pq} = \frac{k_p k_q}{|k|^2}$$

is called the *Riesz transform*. Since $|\hat{r}_{pq}| \leq 1$, R_{pq} is a bounded linear map on $L^2(\mathbb{R}^n)$. The Riesz transform recovers the second derivatives of a differentiable function from its Laplacian:

$$\frac{\partial^2 f}{\partial x_p \partial x_q} = R_{pq}\Delta f.$$

One can also define *pseudodifferential operators*, whose symbol $\hat{a}(x, k)$ is a function belonging to a suitable class that is allowed to depend on both x and k, so that

$$\begin{aligned} Af(x) &= \frac{1}{(2\pi)^{n/2}} \int \hat{a}(x, k)\hat{f}(k)e^{ik\cdot x}\, dk \\ &= \frac{1}{(2\pi)^n} \int \hat{a}(x, k)e^{ik\cdot(x-y)} f(y)\, dy dk. \end{aligned}$$

These operators are not translation invariant, and they allow the use of Fourier methods in the analysis of variable coefficient, linear partial differential equations.

11.10   Green's functions

Constant coefficient, linear partial differential equations on $\mathbb{R}^n$ may be solved by use of the Fourier transform. In particular, we can use the distributional Fourier transform to compute their Green's functions.

The Green's function g of the Laplacian on $\mathbb{R}^n$ is a distributional solution of the equation

$$-\Delta g = \delta. \tag{11.40}$$

The delta function has the physical interpretation of the density of a point source located at the origin, and the Green's function g is the potential of the point source. Taking the Fourier transform of (11.40), we find that

$$|k|^2 \hat{g} = \frac{1}{(2\pi)^{n/2}}.$$

A complication in solving this equation for g is that the symbol $|k|^2$ of the Laplacian vanishes at $k = 0$. We therefore need to interpret division by $|k|^2$ in an appropriate sense. From Example 11.8, if $n \geq 3$, then a solution for $\hat{g}$ is the regular distribution

$$\hat{g}(k) = \frac{1}{(2\pi)^{n/2}} \frac{1}{|k|^2},$$

and the Green's function is

$$g(x) = \frac{1}{(2\pi)^{n/2}} \mathcal{F}^{-1}\left(\frac{1}{|k|^2}\right).$$

The solution is not unique. We may add an arbitrary linear combination of δ and first-order partial derivatives of δ to $\hat{g}$. The inverse transform of this distribution is a linear polynomial in x, which is a solution of the homogeneous Laplace equation. We omit this function of integration for simplicity.

We will compute the inverse transform of $\hat{g}$ explicitly when $n = 3$; the computation for $n \geq 4$ is similar. Since $\hat{g}(k)$ decays too slowly as $|k| \to \infty$ to be integrable, we introduce a cut-off, as in (11.37). Using spherical polar coordinates (r, θ, φ) in k-space, with the x-direction corresponding to $\theta = 0$, we find from the inversion formula and the sinc integral in (11.15), that

$$
\begin{aligned}
g(x) &= \frac{1}{(2\pi)^3} \lim_{R \to \infty} \int_{|x| < R} \frac{e^{ik \cdot x}}{|x|^2} \, dk \\
&= \frac{1}{(2\pi)^3} \lim_{R \to \infty} \int_0^R \int_0^\pi \int_0^{2\pi} \frac{e^{ir|x|\cos\theta}}{r^2} r^2 \sin\theta \, d\varphi \, d\theta \, dr \\
&= \frac{1}{(2\pi)^2} \lim_{R \to \infty} \int_0^R \frac{2 \sin r|x|}{r|x|} \, dr \\
&= \frac{1}{(2\pi)^2} \frac{\pi}{|x|}.
\end{aligned}
$$

It follows that the three-dimensional, free-space Green's function for Laplace's equation is

$$g(x) = \frac{1}{4\pi|x|},$$

as we found in Section 10.6 by a different method. For $n = 2$, a solution for $\hat{g}$ is

$$\hat{g}(k) = \frac{1}{2\pi} \text{f.p.} \frac{1}{|k|^2},$$

where the finite part distribution is defined in Example 11.8. One can show that the inverse Fourier transform of this distribution is of the form

$$g(x) = \frac{1}{2\pi} \log\left(\frac{1}{|x|}\right) + C$$

for a suitable constant C, also in agreement with our previous result.

Next, we consider the initial value problem for the heat or diffusion equation. The Green's function $g(x,t)$ is the solution of the following initial value problem:

$$g_t = \frac{1}{2}\Delta g \qquad \text{for } x \in \mathbb{R}^n,\, t > 0,$$
$$g(\cdot, t) \in S^*(\mathbb{R}^n) \qquad \text{for } t > 0,$$
$$g(x, 0) = \delta(x) \qquad \text{for } x \in \mathbb{R}^n.$$

Taking the Fourier transform $\mathcal{F}_x$ with respect to x of this equation, we find that $\hat{g}(k,t) = \mathcal{F}_x g(x,t)$ satisfies the ODE

$$\hat{g}_t = -\frac{1}{2}|k|^2 \hat{g}, \qquad \hat{g}(k, 0) = \frac{1}{(2\pi)^{n/2}}.$$

The solution is given by

$$\hat{g}(k, t) = \frac{1}{(2\pi)^{n/2}} e^{-t|k|^2/2}.$$

Using Proposition 11.26 to invert the transform, we obtain that

$$g(x, t) = \frac{1}{(2\pi t)^{n/2}} e^{-|x|^2/(2t)}.$$

The solution $u(x,t)$ of the heat equation with initial condition

$$u(x, 0) = f(x),$$

is given by a convolution with the Green's function:

$$u(x, t) = \frac{1}{(2\pi t)^{n/2}} \int_{\mathbb{R}^n} e^{-|x-y|^2/(2t)} f(y)\, dy.$$

Since the Green's function is a Schwartz function, this expression makes sense as a convolution for any initial data $f \in S^*$. The solution is C^∞ in both x and t when $t > 0$. This is the *smoothing property* of the heat equation. It can be shown that the solution of the initial value problem for the heat equation is not unique (see Exercise 11.24). There is, however, a unique solution of polynomial growth, and this is the one obtained by use of the Fourier transform.

11.11   The Poisson summation formula

The *Poisson summation formula* states that a large class of functions $f : \mathbb{R} \to \mathbb{C}$ satisfy the following identity:

$$\sum_{n=-\infty}^{\infty} f(2\pi n) = \frac{1}{\sqrt{2\pi}} \sum_{n=-\infty}^{\infty} \hat{f}(n). \tag{11.41}$$

The presence or absence of factors 2π in this equation depends on the normalization of the Fourier transform. This formula may be used to derive identities between infinite series, or even to sum a series explicitly. It can also be used to connect the Fourier series of a periodic function with the Fourier transform.

Theorem 11.45 Suppose that $f \in C^1(\mathbb{R})$, and there exist constants $C > 0$, $\epsilon > 0$ such that

$$\left| \left(1 + x^2\right)^{1/2+\epsilon} f(x) \right| \leq C, \quad \left| \left(1 + x^2\right)^{1/2+\epsilon} f'(x) \right| \leq C \tag{11.42}$$

for all $x \in \mathbb{R}$. Then we have the identity

$$\sum_{n=-\infty}^{\infty} f(x + 2\pi n) = \frac{1}{\sqrt{2\pi}} \sum_{n=-\infty}^{\infty} e^{inx} \hat{f}(n). \tag{11.43}$$

Proof.   The condition in (11.42) implies that the sum

$$g(x) = \sum_{n=-\infty}^{\infty} f(x + 2\pi n) \tag{11.44}$$

converges uniformly, and g is a continuously differentiable 2π-periodic function. Therefore, from Lemma 7.8, the Fourier series of g converges uniformly, and

$$g(x) = \frac{1}{2\pi} \sum_{n=-\infty}^{\infty} \left(\int_0^{2\pi} e^{-iny} g(y) \, dy \right) e^{inx}.$$

Since g is related to f by (11.44), we can rewrite this as (11.43). $\square$

Evaluation of (11.43) at $x = 0$ gives the Poisson summation formula (11.41).

Example 11.46 The *Jacobi theta function* θ is defined for $t > 0$ by

$$\theta(t) = \sum_{n=-\infty}^{\infty} e^{-\pi n^2 t}.$$

The Poisson summation formula implies that the theta function has the following symmetry property:

$$\theta(t) = \frac{1}{\sqrt{t}} \theta(1/t). \tag{11.45}$$

Theta functions have important connections with Riemann surfaces and the theory of integrable systems. They also arise in the solution of the heat equation on the circle, as in (7.21).

The Poisson summation formula holds, in particular, for Schwartz functions. The convergence of the series on the left-hand side of (11.41) for every $f \in \mathcal{S}$ implies that the series $\sum_{n=-\infty}^{\infty} \delta_{2\pi n}$ converges in $\mathcal{S}^*$. The series on the right-hand side of (11.41) may be written as:

$$\frac{1}{\sqrt{2\pi}} \sum_{n=-\infty}^{\infty} \hat{f}(n) = \frac{1}{2\pi} \sum_{n=-\infty}^{\infty} \int e^{-inx} f(x)\, dx.$$

Hence, the series $\sum_{n=-\infty}^{\infty} e^{-inx}$ also converges in $\mathcal{S}^*$. Changing n to $-n$ in this sum, we obtain the following identity of tempered distributions:

$$\sum_{n=-\infty}^{\infty} \delta_{2\pi n}(x) = \frac{1}{2\pi} \sum_{n=-\infty}^{\infty} e^{inx}. \tag{11.46}$$

This equation may be interpreted as the Fourier series expansion of a periodic array of delta functions (sometimes called the "delta comb"). Its Fourier coefficients are constants, independent of n.

More generally, we say that a distribution $T \in \mathcal{S}(\mathbb{R})$ is periodic with period 2π if $\tau_{2\pi} T = T$. In that case, one can show that

$$\hat{T} = \frac{1}{\sqrt{2\pi}} \sum_{n=-\infty}^{\infty} \hat{T}_n \delta_{2\pi n}$$

for suitable Fourier coefficients $\hat{T}_n \in \mathbb{C}$. The Fourier coefficients have polynomial growth in n, meaning that there are constants $C > 0$ and $d \in \mathbb{N}$ such that

$$\left| \hat{T}_n \right| \leq C \left(1 + n^2 \right)^{d/2}.$$

Thus, the Fourier transform of a periodic function or distribution is an $\mathcal{S}^*$-convergent linear combination of delta functions supported at $2\pi n$. The strengths of the delta functions give the Fourier coefficients of the periodic function. The distribution T is given by the $\mathcal{S}^*$-convergent Fourier series

$$T(x) = \frac{1}{\sqrt{2\pi}} \sum_{n=-\infty}^{\infty} \hat{T}_n e^{inx}.$$

11.12   The central limit theorem

A *random variable* describes the observation, or measurement, of a number associated with a random event. We say that a real-valued random variable X is *absolutely continuous* if its distribution may be described by a *probability density* $p \in L^1(\mathbb{R})$, meaning that for any $a \leq b$, the probability that X has a value between a and b is given by

$$\Pr\left(a \leq X \leq b\right) = \int_a^b p(x)\, dx. \tag{11.47}$$

Since a probability is a number between zero and one, the density function p is nonnegative and

$$\int_{-\infty}^{\infty} p(x)\, dx = 1. \tag{11.48}$$

We call any function p with these properties, a probability density. If X is not absolutely continuous (for example, because it takes integer values with probability one), then its distribution is described by a probability measure on $\mathbb{R}$ that does not have a probability density function. We consider absolutely continuous random variables for simplicity, but the central limit theorem does not depend on this restriction.

The expected value of a function $f(X)$ of X is given in terms of the density p by

$$\mathbb{E}[f(X)] = \int_{-\infty}^{\infty} f(x)p(x)\, dx,$$

provided that this integral converges. The *mean* μ and the *variance* σ^2 of X are given by

$$\mu = \mathbb{E}[X], \qquad \sigma^2 = \mathbb{E}\left[(X - \mathbb{E}[X])^2\right].$$

The expected deviation of X from its mean is therefore of the order of the *standard deviation* σ. If the mean and variance of X are finite, then the random variable Y defined by $X = \mu + \sigma Y$ has mean zero and variance one, so we can normalize the mean of X to zero and the variance of X to one by an affine transformation. In that case,

$$\int_{-\infty}^{\infty} xp(x)\, dx = 0, \qquad \int_{-\infty}^{\infty} x^2 p(x)\, dx = 1. \tag{11.49}$$

Example 11.47 We say that a real random variable X is a *Gaussian*, or *normal*, random variable with mean μ and variance σ^2 if its probability density p is given by

$$p(x) = \frac{1}{\sqrt{2\pi}\sigma} e^{-(x-\mu)^2/(2\sigma^2)}.$$

If $\mu = 0$ and $\sigma^2 = 1$, then we say that X is a standard Gaussian.

We say that N random variables $\{X_1, X_2, \ldots, X_N\}$ are *independent* if

$$\Pr\left(a_1 \le X_1 \le b_1, a_2 \le X_2 \le b_2, \ldots, a_N \le X_N \le b_N\right)$$
$$= \Pr\left(a_1 \le X_1 \le b_1\right)\Pr\left(a_2 \le X_2 \le b_2\right)\ldots\Pr\left(a_N \le X_N \le b_N\right).$$

In that case,

$$\mathbb{E}[f_1(X_1)f_2(X_2)\ldots f_N(X_N)] = \mathbb{E}[f_1(X_1)]\,\mathbb{E}[f_2(X_2)]\ldots\mathbb{E}[f_N(X_N)].$$

Suppose that $\{X_1, X_2, \ldots, X_N\}$ have a *joint probability density* $p(x_1, x_2, \ldots, x_N)$, meaning that

$$\Pr\left(a_1 \le X_1 \le b_1, a_2 \le X_2 \le b_2, \ldots, a_N \le X_N \le b_N\right)$$
$$= \int_{a_N}^{b_N} \ldots \int_{a_2}^{b_2} \int_{a_1}^{b_1} p(x_1, x_2, \ldots, x_N)\, dx_1 dx_2 \ldots dx_N.$$

Then the random variables are independent if and only if p has the form

$$p(x_1, x_2, \ldots, x_N) = p_1(x_1)p_2(x_2)\ldots p_N(x_N).$$

Intuitively, independence means that the value taken by one of the random variables has no influence on the values taken by the others.

In many applications it is important to consider the sum of a large number of independent, identically distributed random variables. For example, a standard way to reduce nonsystematic errors in the experimental measurement of a given quantity is to measure the quantity many times and take the average. The central limit theorem explains how this error reduction works, and also gives an estimate of the expected difference between the measured value of the quantity and its true value. As we will see, if the experimental measurements are independent and randomly distributed with mean equal to the true value and with finite variance σ^2, then for sufficiently large N the distribution of the average value of N measurements is approximately Gaussian with mean equal to the true value and variance σ^2/N. Thus, one needs to take four times as many measurements in order to double the accuracy. This example is the original application that led Gauss to introduce the Gaussian distribution.

A second example is the discrete-time *random walk*. Considering the case of one space dimension for simplicity, we suppose that a particle starts at the origin at time zero and moves a random distance $X_n \in \mathbb{R}$ at time $n \in \mathbb{N}$, where X_m and X_n are independent, identically distributed random variables for $m \ne n$. The particle then takes random steps up and down the real line. The total distance moved by the particle after N steps is

$$S_N = \sum_{n=1}^{N} X_n. \tag{11.50}$$

A natural question is: What is the probability distribution of the position S_N of the particle after N steps, given the probability distribution of each individual step? The central limit theorem describes the limiting behavior of S_N as $N \to \infty$. For instance, if each individual step has mean zero and variance one, then the distribution of S_N approaches a Gaussian distribution with mean zero and variance N. The corresponding $\sqrt{N}$-growth of S_N is characteristic of sums of N independent random variables: the sums do not remain bounded as $N \to \infty$, but there is a large amount of cancellation, so the sums grow at a slower rate than the number of their terms.

The Gaussian distribution is universal, in the sense that the distribution of any sum of a large number of independent, identically distributed random variables with finite mean and variance is approximately Gaussian, whatever the details of the probability distribution of the individual random variables. The central limit theorem remains true for sums of non-identical, independent random variables, under a suitable, mild condition (such as the *Lindeberg condition*) that ensures the distribution of the sum is not dominated by the distribution of a small number of the individual random variables. Moreover, some weak dependence between the variables may also be permitted.

Suppose that X and Y are independent random variables with probability density functions p_X and p_Y, respectively. Then

$$\Pr\left(a \leq X + Y \leq b\right) = \int\int_{a \leq x+y \leq b} p_X(x) p_Y(y)\, dx dy$$
$$= \int_a^b \left(\int_{-\infty}^{\infty} p_X(z-y) p_Y(y)\, dy \right) dz.$$

Thus, the probability density of $X + Y$ is the convolution of the probability densities of X and Y. Hence, the convolution theorem implies that the Fourier transforms of the densities multiply:

$$\hat{p}_{X+Y} = \sqrt{2\pi}\, \hat{p}_X \hat{p}_Y.$$

We can obtain the same result by an equivalent probabilistic argument. The *characteristic function* φ_X of a random variable X is defined by

$$\varphi_X(k) = \mathbb{E}\left[e^{ikX}\right].$$

What we have called the characteristic function χ_A of a set A is then referred to as the *indicator function* of the set. If X is absolutely continuous with probability density p, then

$$\varphi_X(k) = \int_{-\infty}^{\infty} e^{ikx} p(x)\, dx = \sqrt{2\pi}\, \hat{p}(-k).$$

Thus, up to normalization conventions, the characteristic function is the Fourier transform of the probability density. If X and Y are independent, then we have

$$\varphi_{X+Y}(k) = \mathbb{E}\left[e^{ik(X+Y)}\right] = \mathbb{E}\left[e^{ikX}\right]\mathbb{E}\left[e^{ikY}\right] = \varphi_X(k)\varphi_Y(k),$$

which agrees with the previous result. Because the Fourier transform maps convolutions to products, in studying sums of independent random variables it is much simpler to consider the characteristic functions rather than the densities themselves, and we shall use this observation to prove the central limit theorem.

Example 11.48 If X is a Gaussian random variable with mean μ and variance σ^2, then the formula for the Fourier transform of a Gaussian implies that

$$\mathbb{E}\left[e^{ikX}\right] = e^{i\mu k}e^{-\sigma^2 k^2/2}.$$

The product of such characteristic functions is another function of the same form, in which the means and variances add together. Consequently, the sum of independent Gaussian random variables is a Gaussian whose mean and variance is the sum of the individual means and variances, and problems involving Gaussian random variables are stochastically linear. The product of two independent Gaussian random variables is not Gaussian, however.

Suppose that $\{X_1, X_2, \ldots, X_N\}$ is a sequence of independent, identically distributed, random variables with finite mean and variance, and probability density p. By making an affine transformation, we may assume that the mean is zero and the variance is one without loss of generality. The probability density p_N of the sum $S_N = X_1 + X_2 + \ldots + X_N$ is given by

$$p_N = \underbrace{p * p * \cdots * p}_{N \text{ times}}.$$

We denote the probability density of $S_N/\sqrt{N}$ by q_N. Then

$$q_N(x) = \sqrt{N}p_N\left(\sqrt{N}x\right). \tag{11.51}$$

We will prove that $S_N/\sqrt{N}$ converges to a standard Gaussian as $N \to \infty$ in the following sense.

Definition 11.49 A sequence (X_n) of random variables *converges in distribution* to a random variable X if

$$\lim_{n\to\infty} \mathbb{E}[f(X_n)] = \mathbb{E}[f(X)] \qquad \text{for every } f \in C_b(\mathbb{R}),$$

where $C_b(\mathbb{R})$ is the space of bounded, continuous functions $f : \mathbb{R} \to \mathbb{C}$. A sequence (p_n) of probability densities *converges weakly* to a probability density p if

$$\lim_{n\to\infty} \int f(x)p_n(x)\,dx = \int f(x)p(x)\,dx \qquad \text{for every } f \in C_b(\mathbb{R}). \tag{11.52}$$

If X_n and X are absolutely continuous random variables with probability densities p_n and p, respectively, then X_n converges in distribution to X if and only if p_n converges weakly to p. Approximating the characteristic function of the interval $[a, b]$ by continuous functions, one can then also show that

$$\lim_{n \to \infty} \Pr(a \le X_n \le b) = \Pr(a \le X \le b).$$

The following theorem, called Lévy's continuity theorem, provides a useful necessary and sufficient condition for weak convergence.

Theorem 11.50 (Continuity) A sequence (p_n) of probability densities converges weakly to a probability density p if and only if $(\hat{p}_n)$ converges pointwise to $\hat{p}$.

Proof. If p_n converges weakly to p, then (11.52) with $f(x) = e^{-ikx}$ implies that $\hat{p}_n(k)$ converges to $\hat{p}(k)$.

We prove the converse statement in several steps. First, we show that if $\hat{p}_n$ converges pointwise to $\hat{p}$, then (11.52) holds for every Schwartz function $f \in S$. Since the Fourier transform maps S onto S, an equivalent statement is that

$$\lim_{n \to \infty} \int \hat{f}(x) p_n(x) \, dx = \int \hat{f}(x) p(x) \, dx \qquad \text{for every } f \in S.$$

From Fubini's theorem, as in (11.31), this statement is equivalent to

$$\lim_{n \to \infty} \int f(k) \hat{p}_n(k) \, dk = \int f(k) \hat{p}(k) \, dk \qquad \text{for every } f \in S. \qquad (11.53)$$

Since p_n is a probability density, the Riemann-Lebesgue lemma, Theorem 11.34, implies that $\hat{p}_n$ is a continuous function with $|\hat{p}_n(k)| \le 1/\sqrt{2\pi}$ for every $k \in \mathbb{R}$. Hence (11.53) follows from the pointwise convergence of $\hat{p}_n$, the integrability of f, and the Lebesgue dominated convergence theorem.

If $f \in C_0(\mathbb{R})$ is a continuous function that vanishes at infinity, then there is a sequence (φ_m) of Schwartz functions that converges uniformly to f. The estimate

$$\left| \int f(x) \left[p_n(x) - p(x) \right] dx \right| \le \left| \int \left[f(x) - \varphi_m(x) \right] p_n(x) \, dx \right|$$

$$+ \left| \int \varphi_m(x) \left[p_n(x) - p(x) \right] dx \right|$$

$$+ \left| \int \left[\varphi_m(x) - f(x) \right] p(x) \, dx \right|$$

$$\le 2\|f - \varphi_m\|_\infty + \left| \int \varphi_m(x) \left[p_n(x) - p(x) \right] dx \right|,$$

and (11.52) for $f = \varphi_m \in S$, implies that (11.52) holds for $f \in C_0(\mathbb{R})$.

In order to show (11.52) for $f \in C_b(\mathbb{R})$, we first prove that

$$\lim_{R \to \infty} \limsup_{n \to \infty} \int_{|x| \ge R} p_n(x) \, dx = 0. \qquad (11.54)$$

This condition means that probability cannot escape to infinity as $n \to \infty$. The family of probability measures associated with the densities p_n is then said to be *tight*. For each $R > 0$, we choose $\varphi_R \in C_0(\mathbb{R})$ such that $0 \leq \varphi_R(x) \leq 1$ for all $x \in \mathbb{R}$ and $\varphi_R(x) = 1$ when $|x| \leq R$. Then, by the dominated convergence theorem, and the fact that $\int p(x)\, dx = 1$, we have

$$\lim_{R \to \infty} \int \varphi_R(x) p(x)\, dx = 1. \tag{11.55}$$

Also, since $\varphi_R \in C_0(\mathbb{R})$,

$$\lim_{n \to \infty} \int \varphi_R(x) p_n(x)\, dx = \int \varphi_R(x) p(x)\, dx.$$

Using the fact that $\int p_n(x)\, dx = 1$, we therefore have

$$\limsup_{n \to \infty} \int_{|x| \geq R} p_n(x)\, dx \;\geq\; \limsup_{n \to \infty} \int [1 - \varphi_R(x)]\, p_n(x)\, dx$$

$$\geq\; 1 - \int \varphi_R(x) p(x)\, dx,$$

and (11.54) follows from (11.55).

For $f \in C_b(\mathbb{R})$ and $R > 0$, we define $f_R = \varphi_R f$. Since $f(x) = f_R(x)$ for $|x| \leq R$, we have the following estimate:

$$\left| \int f(x)\, [p_n(x) - p(x)]\, dx \right| \;\leq\; \left| \int [f(x) - f_R(x)]\, p_n(x)\, dx \right|$$

$$+ \left| \int f_R(x)\, [p_n(x) - p(x)]\, dx \right|$$

$$+ \left| \int [f_R(x) - f(x)]\, p(x)\, dx \right|$$

$$\leq\; 2\|f\|_\infty \left[\int_{|x| \geq R} [p_n(x) + p(x)]\, dx \right]$$

$$+ \left| \int f_R(x)\, [p_n(x) - p(x)]\, dx \right|$$

It then follows from (11.52) for $f = f_R \in C_0(\mathbb{R})$ and (11.54) that we can make the right hand side of this equation arbitrarily small for all sufficiently large n. Hence, equation (11.52) holds for all $f \in C_b(\mathbb{R})$. $\qquad\square$

We can now prove the following central limit theorem.

Theorem 11.51 (Central limit) Let S_N be the sum of N independent, identically distributed, absolutely continuous, real random variables with mean zero and variance one. Then $S_N/\sqrt{N}$ converges in distribution as $N \to \infty$ to a Gaussian random variable with mean zero and variance one.

Proof. From Theorem 11.50, we just have to show that the Fourier transform of the density q_N of $S_N/\sqrt{N}$ converges pointwise to the Fourier transform of the standard Gaussian density. Taking the Fourier transform of (11.51), we find that

$$\hat{q}_N(k) = \hat{p}_N\left(\frac{k}{\sqrt{N}}\right). \tag{11.56}$$

Since p_N is the N-fold convolution of p, the convolution theorem, Theorem 11.35, implies that

$$\hat{p}_N\left(\frac{k}{\sqrt{N}}\right) = (2\pi)^{(N-1)/2}\left[\hat{p}\left(\frac{k}{\sqrt{N}}\right)\right]^N. \tag{11.57}$$

We Taylor expand e^{-iz} as

$$e^{-iz} = 1 - iz - \frac{1}{2}z^2\left[1 + r(z)\right],$$

where $r(z)$ is a continuous function that vanishes at $z = 0$ and is uniformly bounded on the real line. Using the conditions in (11.48) and (11.49), we find that

$$
\begin{aligned}
\hat{p}\left(\frac{k}{\sqrt{N}}\right) &= \frac{1}{\sqrt{2\pi}}\int e^{-ikx/\sqrt{N}}p(x)\,dx \\
&= \frac{1}{\sqrt{2\pi}}\int\left\{1 - \frac{ikx}{\sqrt{N}} - \frac{k^2x^2}{2N}\left[1 + r\left(\frac{kx}{\sqrt{N}}\right)\right]\right\}p(x)\,dx \\
&= \frac{1}{\sqrt{2\pi}}\left[1 - \frac{k^2}{2N}(1 + R_N)\right],
\end{aligned}
$$

where

$$R_N = \int x^2 r\left(\frac{kx}{\sqrt{N}}\right)p(x)\,dx.$$

The integrand converges pointwise to zero as $N \to \infty$, and the dominated convergence theorem implies that $\lim_{N\to\infty} R_N = 0$. Computing the Nth power of $\hat{p}(k/\sqrt{N})$, and using (11.56)–(11.57), we obtain that

$$\lim_{N\to\infty}\hat{q}_N(k) = \lim_{N\to\infty}\frac{1}{\sqrt{2\pi}}\left[1 - \frac{k^2}{2N}(1 + R_N)\right]^N = \frac{1}{\sqrt{2\pi}}e^{-k^2/2}.$$

$\square$

We can rescale the discrete random walk (11.50) to obtain a continuous-time stochastic process $W(t)$, called *Brownian motion*, or the *Wiener process*, that satisfies

$$W(t) = \lim_{N\to\infty}\frac{S_{Nt}}{\sqrt{N}}.$$

Here, we extend S_N to a function S_t of a continuous time variable t by supposing, for example, that the particle moves at a constant velocity from its position at time

N to its position at time $N + 1$. This limit has to be interpreted in an appropriate probabilistic sense, which we will not make precise here.

As the central limit theorem suggests, Brownian motion $W(t)$ is a Gaussian process of mean 0 and variance t. Its sample paths are continuous, nowhere differentiable functions of time with probability one. The probability density $p(x, t)$ of finding the particle at position $W(t) = x$ at time t, assuming that $W(0) = 0$, is given by

$$p(x,t) = \frac{1}{\sqrt{2\pi t}} e^{-x^2/(2t)}.$$

The density p is the Green's function of the heat equation

$$p_t = \frac{1}{2} p_{xx}, \qquad p(x,0) = \delta(x).$$

Brownian motion is the simplest, and most fundamental, example of a *diffusion process*. These processes may also be described by *stochastic differential equations*, and they have widespread applications, from statistical physics to the modeling of financial markets.

11.13   References

See Hochstadt [22] for proofs and further discussion of the eigenfunctions of the Fourier transform. Distributions are discussed in Reed and Simon [45]. For an introduction to the theory of stochastic differential equations, see Øksendal [41].

11.14   Exercises

Exercise 11.1 Let X be a locally convex space. Prove the following.

    (a) The addition of vectors in X and the multiplication by a scalar are continuous.

    (b) A topology defined by a family of seminorms has a base of convex open neighborhoods. Such a topological space is called *locally convex.*

    (c) If for all $x \in X$ there exists $\alpha \in \mathcal{A}$ such that $p_\alpha(x) > 0$, then the topology defined by $\{p_\alpha \mid \alpha \in \mathcal{A}\}$ is Hausdorff.

Exercise 11.2 Suppose that $\{p_1, p_2, p_3, \ldots\}$ is a countable family of seminorms on a linear space X. Prove that (11.4) defines a metric on X, and prove that metric topology defined by d coincides with the one defined by the family of seminorms $\{p_1, p_2, p_3, \ldots\}$.

Exercise 11.3 Let (x_n) be a sequence in a locally convex space whose topology is defined by a countably infinite set of seminorms. Prove that (x_n) is a Cauchy sequence for the metric d defined in (11.4) if and only if for every $\alpha \in \mathcal{A}$ and $\epsilon > 0$, there is an N such that $p_\alpha(x_n - x_m) < \epsilon$ for all $n, m \geq N$.

Exercise 11.4 If $\varphi \in \mathcal{S}(\mathbb{R})$, prove that

$$\varphi \delta' = \varphi(0)\delta' - \varphi'(0)\delta.$$

Exercise 11.5 Prove that

$$\lim_{\epsilon \to 0^+} \frac{1}{x + i\epsilon} = \text{p.v.} \frac{1}{x} - i\pi\delta(x) \qquad \text{in } \mathcal{S}^*(\mathbb{R}).$$

Exercise 11.6 Show that the distributional derivative of $\log|x| : \mathbb{R} \to \mathbb{R}$ is p.v.$1/x$.

Exercise 11.7 Show that there is no product $\cdot : \mathcal{S}^* \times \mathcal{S}^* \to \mathcal{S}^*$ on the space of tempered distributions that is commutative, associative, and agrees with the usual product of a tempered distribution and a smooth function of polynomial growth. HINT. Compute the product $x \cdot \delta(x) \cdot \text{p.v.}(1/x)$ in two different ways.

Exercise 11.8 Suppose that $\omega \in \mathcal{S}(\mathbb{R})$ is a test function such that

$$\int_{\mathbb{R}} \omega(x)\,dx = 1.$$

Show that every test function $\varphi \in \mathcal{S}(\mathbb{R})$ may be written as

$$\varphi(x) = \omega(x)\left(\int_{\mathbb{R}} \varphi(y)\,dy\right) + \psi'(x)$$

for some test function $\psi \in \mathcal{S}(\mathbb{R})$. Deduce that if T is a tempered distribution such that $T' = 0$, then T is constant.

Exercise 11.9 Let $k \in \mathcal{S}$ and define the convolution operator

$$Kf(x) = \int k(x - y)f(y)\,dy \qquad \text{for all } f \in \mathcal{S}.$$

Prove that $K : \mathcal{S} \to \mathcal{S}$ is a continuous linear operator for the topology of $\mathcal{S}$.

Exercise 11.10 For every $h \in \mathbb{R}^n$ define a linear transformation $\tau_h : \mathcal{S} \to \mathcal{S}$ by $\tau_h(f)(x) = f(x - h)$.

    (a) Prove that for all $h \in \mathbb{R}^n$, τ_h is continuous in the topology of $\mathcal{S}$.

    (b) Prove that for all $f \in \mathcal{S}$, the map $h \mapsto \tau_h f$ is continuous from $\mathbb{R}^n$ to $\mathcal{S}$.

HINT. For (b), prove that for $f \in C(\mathbb{R}^n)$ one has $\lim_{h \to 0} \|\tau_h f - f\|_\infty = 0$ if and only if f is uniformly continuous. Also, note that it is sufficient to prove continuity at $h = 0$, due to the group property of τ_h.

Exercise 11.11 The density ρ of an array of N point masses of mass $m_j > 0$ located at $x_j \in \mathbb{R}^n$ is a sum of δ functions

$$\rho(x) = \sum_{j=1}^{N} m_j \delta(x - x_j).$$

Compute the Fourier transform $\hat{\rho}$ of ρ. Show that for any $\varphi \in \mathcal{S}$, and for any $k_1, \ldots, k_N \in \mathbb{R}^n$, $z_1, \ldots, z_N \in \mathbb{C}$ we have

$$\int \overline{\varphi(k)} \hat{\rho}(k - \ell) \varphi(\ell) \, dk d\ell \geq 0, \qquad \sum_{p,q=1}^{N} \bar{z}_p \hat{\rho}(k_p - k_q) z_q \geq 0.$$

The Fourier transform $\hat{\rho}$ is said to be of *positive type*.

Exercise 11.12 Prove that if $s > n/2$, then $H^s(\mathbb{R}^n) \subset C_0(\mathbb{R}^n)$, and there is a constant C such that

$$\|f\|_\infty \leq C \|f\|_{H^s}, \qquad \text{for all } f \in H^s(\mathbb{R}^n).$$

Exercise 11.13 Prove equations (11.27)–(11.29) for the Fourier transform of translates and convolutions. Prove the corresponding results for derivatives and translates of tempered distributions and for the convolution of a test function with a tempered distribution.

Exercise 11.14 The *Airy equation* is the ODE

$$u'' - xu = 0.$$

The solutions, called *Airy functions*, are the simplest functions that make a transition from oscillatory behavior (for $x < 0$) to exponential behavior (for $x > 0$). Take the Fourier transform, and deduce that

$$u(x) = c \int e^{ikx + ik^3/3} \, dk,$$

where c is an arbitrary constant. This nonconvergent integral is a simple example of an *oscillatory integral*. Here, it may be interpreted distributionally as an inverse Fourier transform. Why do you find only one linearly independent solution?

Exercise 11.15 Let $f_n : \mathbb{R} \to \mathbb{R}$ be the function

$$f_n(x) = \begin{cases} n^2 & \text{if } -1/n < x < 0, \\ -n^2 & \text{if } 0 < x < 1/n, \\ 0 & \text{otherwise.} \end{cases}$$

Show that the sequence (f_n) converges in $\mathcal{S}^*(\mathbb{R})$ as $n \to \infty$, and determine its distributional limit.

Exercise 11.16 Let $f \in L^1(\mathbb{R}^3)$ be a rotationally invariant function in the sense that there is a function $g : \mathbb{R}^+ \to \mathbb{C}$ such that

$$f(x) = g(|x|).$$

Prove that the Fourier transform of f is a continuous function $\hat{f}$ that is also rotation invariant, and $\hat{f}(k) = h(|k|)$, where

$$h(k) = \frac{1}{k}\sqrt{\frac{2}{\pi}} \int_0^\infty r \sin(kr) g(r) \, dr.$$

Exercise 11.17 Show that

$$A^{-1} = \frac{1}{(2\pi)^{n/2}} \int_{\mathbb{R}^n} x \otimes x \exp\left(-\frac{1}{2} x \cdot Ax\right) dx.$$

We therefore call A^{-1} the *covariance matrix* of the n-dimensional Gaussian probability distribution with density $(2\pi)^{-n/2} \exp\left(-x \cdot Ax/2\right)$.

Exercise 11.18 Prove that if $g \in L^2$ satisfies $g(-x) = \overline{g(x)}$, then $\hat{g}$ is real-valued.

Exercise 11.19 Give a counterexample to show that the Riemann-Lebesgue lemma does not hold for all functions in L^2. That is, find a function $f \in L^2(\mathbb{R})$ such that $\hat{f}$ is not continuous.

Exercise 11.20 Show that $\delta \in H^s(\mathbb{R}^n)$ if and only if $s < -n/2$.

Exercise 11.21 Show that the integral equation

$$u(x) + \int_{-\infty}^{\infty} e^{-(x-y)^2/2} u(y) \, dy = f(x)$$

has a unique solution $u \in L^2(\mathbb{R})$ for every $f \in L^2(\mathbb{R})$, and give an expression for u in terms of f.

Exercise 11.22 Show that

$$\mathcal{F}^{-1}\left[i\operatorname{sgn} k\right] = -\frac{2}{\sqrt{\pi}}\left(\text{p.v.}\frac{1}{x}\right).$$

Exercise 11.23 Show that the solution of the heat equation on a one-dimensional semi-infinite rod,

$$u_t = \frac{1}{2} u_{xx} \qquad 0 < x < \infty, \, t > 0,$$
$$u(0,t) = 0 \qquad t > 0,$$
$$u(x,0) = f(x) \qquad 0 < x < \infty,$$

is given by

$$u(x,t) = \int_0^\infty \frac{e^{-(x-y)^2/(2t)} - e^{-(x+y)^2/(2t)}}{\sqrt{2\pi t}} u_0(y) \, dy.$$

This solution illustrates the *method of images.*

Exercise 11.24 Let

$$f(t) = \begin{cases} \exp(-1/t^2) & \text{if } t > 0, \\ 0 & \text{if } t \le 0. \end{cases}$$

Show that

$$u(x,t) = \sum_{n=0}^{\infty} \frac{f^{(n)}(t)}{(2n)!} x^{2n}$$

is a nonzero solution of the one-dimensional heat equation $u_t = u_{xx}$ with zero initial data $u(x,0) = 0$.

Exercise 11.25 Find the Green's function $g(x,t)$ of the one-dimensional wave equation,

$$g_{tt} - g_{xx} = 0,$$
$$g(x,0) = 0, \quad g_t(x,0) = \delta(x).$$

Exercise 11.26 Consider the wave equation

$$u_{tt} = \Delta u$$

for $u(x,t)$, where $t > 0$ and $x \in \mathbb{R}^n$, with initial data

$$u(x,0) = 0, \quad u_t(x,0) = v_0(x).$$

For simplicity, assume that $v_0 \in \mathcal{S}(\mathbb{R}^n)$. Find the equation satsified by the Fourier transform of u,

$$\hat{u}(k,t) = \frac{1}{(2\pi)^{n/2}} \int e^{-ik \cdot x} u(x,t)\, dx,$$

and show that

$$u(x,t) = \frac{1}{(2\pi)^{n/2}} \int \frac{\sin(|k|t)}{|k|} e^{ik \cdot x} \hat{v}_0(k)\, dk,$$

where $\hat{v}_0$ is the Fourier transform of v_0 and $|\cdot|$ denotes the Euclidean norm.

For $n = 3$, let $d\Omega_t$ denote the surface integration measure on the sphere of radius t, so $\int_{|x|=t} d\Omega_t = 4\pi t^2$. Prove that the solution can be written as

$$u(x,t) = \frac{1}{4\pi t} \int_{|y|=t} v_0(x + y)\, d\Omega_t.$$

For $n = 2$, prove that the solution is

$$u(x,t) = \frac{1}{2\pi} \int_{|y|\le t} \frac{v_0(x + y)}{\sqrt{t^2 - |y|^2}}\, dy.$$

Interpret these formulae physically.

Exercise 11.27 Prove (11.45).

Exercise 11.28 Prove the following identity for all $a > 0$:

$$\sum_{n=-\infty}^{\infty} \frac{1}{n^2 + a^2} = \frac{\pi}{a} \frac{1 + e^{-2\pi a}}{1 - e^{-2\pi a}}.$$

By consideration of the limit $a \to 0^+$, show that

$$\sum_{n=1}^{\infty} \frac{1}{n^2} = \frac{\pi^2}{6}.$$

Exercise 11.29 We define the *Wigner distribution* $W(x, k)$ of a Schwartz function $\varphi(x)$, where $x, k \in \mathbb{R}^n$, by

$$W(x, k) = \frac{1}{(2\pi)^n} \int_{\mathbb{R}^n} \varphi\left(x - \frac{y}{2}\right) \overline{\varphi\left(x + \frac{y}{2}\right)} e^{ik \cdot y} \, dy.$$

Compute the Wigner distribution of a Gaussian $\exp(-x \cdot Ax)$, where A is a positive definite matrix. Show that W is real-valued, and

$$W(x, k) = \frac{1}{(2\pi)^n} \int_{\mathbb{R}^n} \hat{\varphi}\left(k - \frac{\ell}{2}\right) \overline{\hat{\varphi}\left(k + \frac{\ell}{2}\right)} e^{-i\ell \cdot x} \, d\ell,$$

$$\int_{\mathbb{R}^n} W(x, k) \, dk = |\varphi(x)|^2, \qquad \int_{\mathbb{R}^n} W(x, k) \, dx = |\hat{\varphi}(k)|^2.$$

Thus, the Wigner distribution W has some properties of a phase space (that is, an (x, k)-space) density of φ. Show, however, that W is not necessarily nonnegative.

Exercise 11.30 Let $\varphi : \mathbb{R} \to \mathbb{C}$ be any Schwartz function such that

$$\int_{-\infty}^{\infty} |\varphi(x)|^2 \, dx = 1,$$

and define

$$E_x = \int_{-\infty}^{\infty} x^2 |\varphi(x)|^2 \, dx, \qquad E_k = \int_{-\infty}^{\infty} k^2 |\hat{\varphi}(k)|^2 \, dk.$$

Prove the *Heisenberg uncertainty principle*:

$$E_x E_k \geq \frac{1}{4}.$$

Show that equality is attained when φ is a suitable Gaussian.

Chapter 12

Measure Theory and Function Spaces

In this chapter, we describe the basic ideas of measure theory and L^p spaces. We also define Sobolev spaces and summarize some of their main properties. Many results will be stated without proof, and we will not construct the most important example of a measure, namely, Lebesgue measure. Nevertheless, we hope that this discussion will allow the reader to use the concepts and results of measure theory as they are required in various applications.

12.1 Measures

The notion of measure generalizes the notion of volume. A *measure* μ on a set X associates to a subset A of X a nonnegative number $\mu(A)$, called the measure of A. It is convenient to allow for the possibility that the measure of a set may be infinite. It is too restrictive, in general, to require that the measure of every subset of X is well defined. Some sets may be too wild to define their measures in a consistent way. Sets that do have a well-defined measure are called *measurable* sets. Thus, a measure μ is a nonnegative, extended real-valued function defined on a collection of measurable subsets of X. We require that the measurable sets form a σ-algebra, meaning that complements, countable unions, and countable intersections of measurable sets are measurable. Moreover, as suggested by the properties of volumes, we require that the measure be *countably additive*, meaning that the measure of a countable union of disjoint sets is the sum of the measures of the individual sets. First, we give the formal definition of a σ-algebra.

Definition 12.1 A σ-*algebra* on a set X is a collection $\mathcal{A}$ of subsets of X such that:

(a) $\emptyset \in \mathcal{A}$;
(b) if $A \in \mathcal{A}$, then $A^c = X \setminus A \in \mathcal{A}$;
(c) if $\{A_i \mid i \in \mathbb{N}\}$ is a countable family of sets in $\mathcal{A}$, then $\bigcup_{i=1}^{\infty} A_i \in \mathcal{A}$.

A *measurable space* $(X, \mathcal{A})$ is a set X and a σ-algebra $\mathcal{A}$ on X. The elements of $\mathcal{A}$

are called *measurable sets*.

It follows from the definition that $X \in \mathcal{A}$, and $\mathcal{A}$ is closed under countable intersections, since

$$\bigcap_{i=1}^{\infty} A_i = \left(\bigcup_{i=1}^{\infty} A_i^c\right)^c.$$

Example 12.2 The smallest σ-algebra on an arbitrary set X is $\{\emptyset, X\}$. The largest σ-algebra is the power set $\mathcal{P}(X)$, that is, the collection of all subsets of X.

If $\mathcal{F}$ is an arbitrary collection of subsets of a set X, then the σ-algebra $\mathcal{A}(\mathcal{F})$ *generated* by $\mathcal{F}$ is the smallest σ-algebra on X that contains $\mathcal{F}$. This σ-algebra is the intersection of all σ-algebras on X that contain $\mathcal{F}$.

Example 12.3 Suppose that $(X, \mathcal{T})$ is a topological space, where $\mathcal{T}$ is the collection of open sets in X. The σ-algebra on X generated by $\mathcal{T}$ is called the *Borel σ-algebra* of X. We denote it by $\mathcal{R}(X)$. Since a σ-algebra is closed under complements, the Borel σ-algebra contains all closed sets, and is also generated by the collection of closed sets in X. Elements of the Borel σ-algebra are called *Borel sets*.

Example 12.4 The Borel σ-algebra of $\mathbb{R}$, with its usual topology, is generated by the collection of all open intervals in $\mathbb{R}$, since every open set is a countable union of open intervals. The collection of half-open intervals $\{(a, b] \mid a < b\}$ also generates $\mathcal{R}(\mathbb{R})$ (see Exercise 12.1). More generally, the Borel σ-algebra of $\mathbb{R}^n$ is generated by the collection of all cubes C of the form

$$C = (a_1, b_1) \times (a_2, b_2) \times \ldots \times (a_n, b_n), \tag{12.1}$$

where $a_i < b_i$. It is tempting to try and construct the Borel sets by forming the collection of countable unions of closed sets, or the collection of countable intersections of open sets. The union of these collections, however, is not a σ-algebra. It can be shown that an uncountably infinite iteration of the formation of countable intersections and unions is required to obtain the Borel σ-algebra on $\mathbb{R}^n$, starting from the open sets. Thus, the structure of a general Borel set is complicated. This fact explains the nonexplicit definition of the σ-algebra generated by a collection of sets.

Next, we define measures and introduce some convenient terminology.

Definition 12.5 A *measure* μ on a set X is a map $\mu : \mathcal{A} \to [0, \infty]$ on a σ-algebra $\mathcal{A}$ of X, such that:

(a) $\mu(\emptyset) = 0$;

(b) if $\{A_i \mid i \in \mathbb{N}\}$ is a countable family of mutually disjoint sets in $\mathcal{A}$, meaning that $A_i \cap A_j = \emptyset$ for $i \neq j$, then

$$\mu\left(\bigcup_{i=1}^{\infty} A_i\right) = \sum_{i=1}^{\infty} \mu(A_i). \tag{12.2}$$

The measure is *finite* if $\mu(X) < \infty$, and *σ-finite* if there is a countable family $\{A_i \in \mathcal{A} \mid i = 1, 2, \ldots\}$ of measurable subsets of X such that $\mu(A_i) < \infty$ and

$$X = \bigcup_{i=1}^{\infty} A_i.$$

A *measure space* is a triple $(X, \mathcal{A}, \mu)$ consisting of a set X, a σ-algebra $\mathcal{A}$ on X, and a measure $\mu : \mathcal{A} \to [0, \infty]$.

In the countable additivity condition (12.2), we make the natural convention that the sum of a divergent series of nonnegative terms is ∞. This countable additivity condition on μ makes sense because $\mathcal{A}$ is closed under countable unions. We will often write a measure space as (X, μ), or X, when the σ-algebra $\mathcal{A}$, or the measure μ, is clear from the context. It is also useful to consider *signed measures*, which take positive or negative values, *complex measures*, which take complex values, and *vector-valued measures*, which take values in a linear space, but we will not do so here.

Example 12.6 Let X be an arbitrary set and $\mathcal{A}$ the σ-algebra consisting of all subsets of X. The *counting measure* ν on X is defined by

$$\nu(A) = \text{the number of elements of } A,$$

with the convention that if A is an infinite set, then $\nu(A) = \infty$. The counting measure is finite if X is a finite set, and σ-finite if X is countable.

Example 12.7 We define the delta measure δ_{x_0} supported at $x_0 \in \mathbb{R}^n$ on the Borel σ-algebra $\mathcal{R}(\mathbb{R}^n)$ of $\mathbb{R}^n$ by

$$\delta_{x_0}(A) = \begin{cases} 1 & \text{if } x_0 \in A, \\ 0 & \text{if } x_0 \notin A. \end{cases}$$

This measure describes a "mass" distribution on $\mathbb{R}^n$ corresponding to a unit mass located at x_0. The formal density of this distribution is the delta function supported at x_0.

If a σ-algebra $\mathcal{A}$ is generated by a collection of sets $\mathcal{F}$, then we would like to define a measure on $\mathcal{A}$ by specifying its values on $\mathcal{F}$. The following theorem gives a useful sufficient condition to do this. A separate question, which we do not consider here, is when a function $\mu : \mathcal{F} \to [0, \infty]$ may be extended to a measure on the σ-algebra $\mathcal{A}(\mathcal{F})$ generated by $\mathcal{F}$.

Theorem 12.8 Suppose that $\mathcal{A}$ is the σ-algebra on X generated by the collection of sets $\mathcal{F}$. Let μ and ν be two measures on $\mathcal{A}$ such that

$$\mu(A) = \nu(A) \qquad \text{for every } A \in \mathcal{F}.$$

If there is a countable family of sets $\{A_i\} \subset \mathcal{F}$ such that $\bigcup_i A_i = X$ and $\mu(A_i) < \infty$, then $\mu = \nu$.

The following example of Lebesgue measure is fundamental.

Example 12.9 The Borel σ-algebra of $\mathbb{R}^n$, defined in Example 12.3, is generated by the collection of cubes C in (12.1). *Lebesgue measure* is the measure λ on the Borel σ-algebra $\mathcal{R}(\mathbb{R}^n)$ such that $\lambda(C) = \text{Vol}(C)$, meaning that

$$\lambda\left((a_1, b_1) \times (a_2, b_2) \times \ldots \times (a_n, b_n)\right) = (b_1 - a_1)(b_2 - a_2) \cdots (b_n - a_n).$$

Lebesgue measure is σ-finite, since

$$\mathbb{R}^n = \bigcup_{i=1}^{\infty} (-i, i)^n.$$

As we discuss in Example 12.14 below, Lebesgue measure may be extended to the larger σ-algebra $\mathcal{L}(\mathbb{R}^n)$ of Lebesgue measurable sets, which is the completion of the Borel σ-algebra with respect to Lebesgue measure.

We have not explained why Lebesgue measure should exist at all, but Theorem 12.8 implies that it is unique if it exists. One can prove the existence of Lebesgue measure by construction, although the proof is not easy. The construction shows the following result.

Theorem 12.10 A subset A of $\mathbb{R}^n$ is Lebesgue measurable if and only if for every $\epsilon > 0$, there is a closed set F and an open set G such that $F \subset A \subset G$ and $\lambda(G \setminus F) < \epsilon$. Moreover,

$$\begin{aligned} \lambda(A) &= \inf \{\lambda(U) \mid U \text{ is open and } U \supset A\} \\ &= \sup \{\lambda(K) \mid K \text{ is compact and } K \subset A\}. \end{aligned}$$

Thus, a Lebesgue measurable set may be approximated from the outside by open sets, and from the inside by compact sets.

Lebesgue measure has several natural geometrical properties. It is translationally invariant, meaning that for every $A \in \mathcal{L}(\mathbb{R}^n)$ and $h \in \mathbb{R}^n$, we have $\lambda(\tau_h A) = \lambda(A)$, where

$$\tau_h A = \{y \in \mathbb{R}^n \mid y = x + h \text{ for some } x \in A\}.$$

The space $\mathbb{R}^n$ is a commutative group with respect to addition. An invariant measure on a locally compact group, such as Lebesgue measure, is called a *Haar measure*. If $T : \mathbb{R}^n \to \mathbb{R}^n$ is a linear map and

$$TA = \{y \in \mathbb{R}^n \mid y = Tx \text{ for some } x \in A\},$$

then $\lambda(TA) = |\det T|\lambda(A)$. Thus, Lebesgue measure is rotationally invariant, and it has the scaling property that $\lambda(tA) = t^n \lambda(A)$ for $t > 0$.

If $(X, \mathcal{A}, \mu)$ is a measure space, a subset A of X is said to have *measure zero* if it is measurable and $\mu(A) = 0$. Sets of measure zero play a particularly important role in measure theory and integration.

Example 12.11 A subset A of $\mathbb{R}^n$ is of measure zero with respect to the delta-measure δ_{x_0} defined in Example 12.7 if and only if A is a Borel set and $x_0 \notin A$.

Example 12.12 For each $x \in \mathbb{R}$, the Lebesgue measure of the set $\{x\}$ is equal to zero, since

$$\lambda(\{x\}) = \lim_{\epsilon \to 0^+} \lambda(\{y \mid |x - y| < \epsilon\}) = \lim_{\epsilon \to 0^+} 2\epsilon = 0.$$

Hence every countable subset $A = \{x_i \mid i \in \mathbb{N}\}$ of $\mathbb{R}$ has measure zero, since the countable additivity of Lebesgue measure implies that

$$\lambda(A) = \sum_{i=1}^{\infty} \lambda(\{x_i\}) = 0.$$

One can show that a subset A of $\mathbb{R}^n$ has Lebesgue measure zero if and only if for every $\epsilon > 0$, A is contained in a not necessarily disjoint union of open cubes, the sum of whose volumes is less than ϵ.

It follows from the additivity and nonegativity of a measure that any measurable subset of a set of measure zero has zero measure. We may extend a measure in a unique fashion to every subset of a set of measure zero by defining it to be zero.

Definition 12.13 A measure space is *complete* if every subset of a set of measure zero is measurable. If $(X, \mathcal{A}, \mu)$ is a measure space, the *completion* $\overline{\mathcal{A}}$ of the σ-algebra $\mathcal{A}$ with respect to a measure μ on $\mathcal{A}$ consists of all subsets A of X such that there exists sets E and F in $\mathcal{A}$ with

$$E \subset A \subset F, \quad \text{and} \quad \mu(F \setminus E) = 0.$$

The completion $\overline{\mu}$ of μ is defined on $\overline{\mathcal{A}}$ by

$$\overline{\mu}(A) = \mu(E) = \mu(F).$$

The complete measure space $(X, \overline{\mathcal{A}}, \overline{\mu})$ is called the *completion* of $(X, \mathcal{A}, \mu)$.

Example 12.14 The Borel σ-algebra $\mathcal{R}(\mathbb{R}^n)$ is not complete. Its completion with respect to Lebesgue measure is the σ-algebra $\mathcal{L}(\mathbb{R}^n)$ of Lebesgue measurable sets. We will use the same notation λ to denote Lebesgue measure on the Borel sets and the Lebesgue measurable sets.

A property that holds except on a set of measure zero is said to hold *almost everywhere*, or *a.e.* for short. When we want to make explicit the measure μ with respect to which a set has measure zero, we write μ-a.e. We define the *essential supremum* of a set of real numbers $A \subset \mathbb{R}$ by

$$\operatorname{ess\,sup} A = \inf\{C \mid x \leq C \text{ for all } x \in A \setminus N, \text{ where } \mu(N) = 0\}.$$

A *Borel measure* is a measure defined on the Borel σ-algebra of a topological space. Thus, the delta-measure and Lebesgue measure defined on $\mathcal{R}(\mathbb{R}^n)$ are examples of Borel measures. The following example gives a useful class of Borel measures on $\mathbb{R}$.

Example 12.15 Let $F : \mathbb{R} \to \mathbb{R}$ be an increasing, right-continuous function, meaning that $F(x) \leq F(y)$ for $x \leq y$, and

$$F(x) = \lim_{y \to x^+} F(y).$$

There is a unique measure μ_F on the Borel σ-algebra of $\mathbb{R}$ such that

$$\mu_F((a, b]) = F(b) - F(a).$$

From Exercise 12.3, if $b_n \to b^+$ is a decreasing sequence and $a < b$, then

$$\mu_F((a, b_n]) \to \mu_F((a, b])^+,$$

which explains why F must be right-continuous. This measure is called a *Lebesgue-Stieltjes measure*, and F is called the *distribution function* of the measure. For example, if $F(x) = x$, then we obtain Lebesgue measure. If F is the right-continuous step function,

$$F(x) = \begin{cases} 1 & \text{if } x \geq 0, \\ 0 & \text{if } x < 0, \end{cases}$$

then we obtain the delta measure supported at the origin. If F is the Cantor function, defined in Exercise 1.19, then we obtain a measure such that the Cantor set C has measure one and $\mathbb{R} \setminus C$ has measure zero. Despite the fact that μ_F is supported on a set of Lebesgue measure zero, we have $\mu_F(\{x\}) = 0$ for every $x \in \mathbb{R}$.

Kolmogorov observed in the 1920s that measure theory provides the mathematical foundation of probability theory.

Definition 12.16 A *probability space* $(\Omega, \mathcal{A}, \mu)$ is a measure space such that $\mu(\Omega) = 1$. The measure μ is called a *probability measure*.

In modeling a random trial or experiment, we form a *sample space* Ω that consists of all possible outcomes of the trial, including outcomes that may occur with probability zero. An *event* is a measurable subset of Ω, and the collection of events forms a σ-algebra $\mathcal{A}$ on Ω. The *probability* $0 \leq \mu(A) \leq 1$ of an event $A \in \mathcal{A}$ is given by an appropriate probability measure defined on $\mathcal{A}$. The σ-algebra of a probability space has a natural interpretation as the collection of events about which information is available.

Example 12.17 Let $\Omega = \{n \in \mathbb{Z} \mid n \geq 0\}$ be the nonnegative integers and $\mathcal{A}$ the set of all subsets of Ω. Let μ be the measure on $\mathcal{A}$ such that

$$\mu(\{n\}) = \frac{\lambda^n}{n!} e^{-\lambda}, \qquad \mu(A) = \sum_{n \in A} \frac{\lambda^n}{n!} e^{-\lambda}.$$

This measure is called the *Poisson distribution*.

Example 12.18 Suppose that $\Omega = \mathbb{R}^n$ and $\mathcal{A}$ is the Lebesgue σ-algebra. The standard Gaussian probability measure on $\mathbb{R}^n$ is given by

$$\mu(A) = \frac{1}{(2\pi)^{n/2}} \int_A e^{-|x|^2/2} \, dx.$$

Here, the integral is the Lebesgue integral, defined below.

12.2 Measurable functions

Measurable functions are the natural mappings between measurable spaces. They play an analogous role to continuous functions between topological spaces.

Definition 12.19 Let $(X, \mathcal{A})$ and $(Y, \mathcal{B})$ be measurable spaces. A *measurable function* is a mapping $f : X \to Y$ such that

$$f^{-1}(B) \in \mathcal{A} \qquad \text{for every } B \in \mathcal{B}.$$

The measurability of $f : X \to Y$ depends only on the σ-algebras on X and Y, and not on what measure, if any, is defined on X or Y. When a measure μ is defined on X, we say that two measurable functions $f : X \to Y$ and $g : X \to Y$ are *equal a.e.* if

$$\mu\left(\{x \in X \mid f(x) \neq g(x)\}\right) = 0.$$

Two functions on a measure space that are equal a.e. will often be regarded as equivalent.

Example 12.20 A measurable map $T : X \to X$ on a measure space $(X, \mathcal{A}, \mu)$ is said to be *measure preserving* if

$$\mu\left(T^{-1}(A)\right) = \mu(A)$$

for all measurable sets A. Measure preserving maps arise naturally in physics and other applications. Ergodic theory studies the properties of various kinds of measure preserving maps (see Theorem 7.11 and Theorem 8.35, for example).

Example 12.21 A measurable map $X : \Omega \to \mathbb{R}$ on a probability space Ω is called a *random variable*.

If $\mathcal{B}$ is the σ-algebra generated by $\mathcal{F}$, then the condition that

$$f^{-1}(F) \in \mathcal{A} \qquad \text{for all } F \in \mathcal{F}$$

is sufficient to ensure that f is measurable. This follows from the fact that $\{f^{-1}(B) \mid B \in \mathcal{B}\}$ is the σ-algebra generated by $\{f^{-1}(F) \mid F \in \mathcal{F}\}$, and is therefore contained in $\mathcal{A}$.

Example 12.22 Every continuous function between topological spaces is Borel measurable. A continuous function $f : \mathbb{R}^n \to \mathbb{R}$ is measurable with respect to the Lebesgue σ-algebra on the domain $\mathbb{R}^n$ and the Borel σ-algebra on the range $\mathbb{R}$.

From now on, we will restrict our attention to real-valued functions defined on a measure space $(X, \mathcal{A}, \mu)$. Complex-valued functions may be treated by splitting them into their real and imaginary parts. The fact that the real numbers are totally ordered makes it particularly easy to develop the integral in this case. The theory applies to real-valued functions defined on a general measure space X, but it is helpful to keep in mind the case when X is $\mathbb{R}^n$ equipped with Lebesgue measure.

It is often convenient to allow measures and functions to take on the values $-\infty$ or ∞. We therefore introduce the *extended real numbers* $\overline{\mathbb{R}} = [-\infty, \infty]$. We make the following definitions of algebraic operations involving $x \in \mathbb{R}$ and ∞:

$$x + \infty = \infty, \quad x - \infty = -\infty,$$
$$x \cdot \infty = \infty, \quad x \cdot (-\infty) = -\infty \qquad \text{if } x > 0,$$
$$x \cdot \infty = -\infty, \quad x \cdot (-\infty) = \infty \qquad \text{if } x < 0,$$
$$|\infty| = |-\infty| = \infty.$$

We also define

$$0 \cdot \infty = 0 \cdot (-\infty) = 0.$$

For example, we will define the integral of a function that is infinite on a set of measure zero, or the integral of a function that is zero on a set of infinite measure, to be zero. We do not define $\infty - \infty$, and any expression of this form is meaningless.

We use the natural ordering and topology on $\overline{\mathbb{R}}$; as far as its ordering and topology are concerned, $\overline{\mathbb{R}}$ is isomorphic to the closed interval $[-1, 1]$. Any monotone sequence $\{x_n\}$ of points in $\overline{\mathbb{R}}$ has a limit. The limit of a monotone increasing sequence is $\sup\{x_n\}$ if the sequence is bounded, and ∞ if it is unbounded. The limit of a monotone decreasing sequence is $\inf\{x_n\}$ if the sequence is bounded, and $-\infty$ if it is unbounded. We equip $\overline{\mathbb{R}}$ with the Borel σ-algebra $\mathcal{R}(\overline{\mathbb{R}})$. This σ-algebra is generated by the semi-infinite intervals of the form $\{[-\infty, c) : c \in \mathbb{R}\}$, so we have the following criterion for the measurability of a function $f : X \to \overline{\mathbb{R}}$.

Proposition 12.23 Let $(X, \mathcal{A})$ be a measurable space. A function $f : X \to \overline{\mathbb{R}}$ is measurable if and only if the set $\{x \in X \mid f(x) < c\}$ belongs to $\mathcal{A}$ for every $c \in \mathbb{R}$.

In this proposition, the sets $\{f(x) \leq c\}$, $\{f(x) > c\}$, or $\{f(x) \geq c\}$ could be used equally well. A complex-valued function $f : X \to \mathbb{C}$ is measurable if and only if $f = g + ih$ where $g, h : X \to \mathbb{R}$ are measurable.

We say that a sequence of functions (f_n) from a measure space $(X, \mathcal{A}, \mu)$ to $\overline{\mathbb{R}}$ *converges pointwise* to a function $f : X \to \overline{\mathbb{R}}$ if

$$\lim_{n \to \infty} f_n(x) = f(x) \qquad \text{for every } x \in X.$$

The sequence *converges pointwise-a.e.* to f if it converges pointwise to f on $X \setminus N$, where N is a set of measure zero. The following result explains why all functions encountered in analysis are measurable.

Theorem 12.24 If (f_n) is a sequence of measurable functions that converges pointwise to a function f, then f is measurable. If $(X, \mathcal{A}, \mu)$ is a complete measure space and (f_n) converges pointwise-a.e. to f, then f is measurable.

A measurable function that takes on finitely many, finite values is called a *simple function*. Any measurable function may be approximated by simple functions.

Definition 12.25 A function $\varphi : X \to \mathbb{R}$ on a measurable space $(X, \mathcal{A})$ is a *simple function* if there are measurable sets $A_1, A_2, \ldots, A_n$ and real numbers $c_1, c_2, \ldots, c_n$ such that

$$\varphi = \sum_{i=1}^{n} c_i \chi_{A_i}. \qquad (12.3)$$

Here, χ_A is the characteristic function of the set A, meaning that

$$\chi_A(x) = \begin{cases} 1 & \text{if } x \in A, \\ 0 & \text{if } x \notin A. \end{cases}$$

The representation of a simple function as a sum of characteristic functions is not unique. A standard representation uses disjoint sets A_i and distinct values c_i. In that case, we have $\varphi(x) = c_i$ if and only if $x \in A_i$. Since the sets A_i are required to be measurable, a simple function is measurable.

Theorem 12.26 Let $f : X \to [0, \infty]$ be a nonnegative, measurable function. There is a monotone increasing sequence $\{\varphi_n\}$ of simple functions that converges pointwise to f.

Proof.   For each $n \in \mathbb{N}$, we subdivide the range of f into $2^{2n} + 1$ intervals

$$I_{n,k} = \left[\frac{k-1}{2^n}, \frac{k}{2^n}\right) \quad \text{for } k = 1, 2, \ldots, 2^{2n}, \quad I_{n,2^{2n}+1} = [2^n, \infty]$$

of length 2^{-n}. We define the measurable sets

$$A_{n,k} = f^{-1}(I_{n,k}) \qquad \text{for } k = 1, 2, \ldots, 2^{2n} + 1.$$

Then the increasing sequence of simple functions

$$\varphi_n = \sum_{k=1}^{2^{2n}+1} \left(\frac{k-1}{2^n}\right) \chi_{A_{n,k}}$$

converges pointwise to f as $n \to \infty$.   □

An arbitrary measurable function $f : X \to \overline{\mathbb{R}}$ may be written in a canonical way as the difference of two nonnegative measurable functions,

$$f = f_+ - f_-, \qquad f_+ = \max\{f, 0\}, \quad f_- = \max\{-f, 0\}. \tag{12.4}$$

We call f_+ the *positive part* of f and f_- the *negative part*. We may then approximate each part by simple functions.

12.3   Integration

The Lebesgue integral provides an extension of the Riemann integral which applies to highly discontinuous and unbounded functions, and which behaves very well with respect to limiting operations. To construct the Lebesgue integral, we first define the integral of a simple function. We then define the integral of a general measurable function using approximations by simple functions.

Suppose that

$$\varphi = \sum_{i=1}^{n} c_i \chi_{A_i}$$

is a simple function on a measure space $(X, \mathcal{A}, \mu)$. We define the integral of φ with respect to the measure μ by

$$\int \varphi \, d\mu = \sum_{i=1}^{n} c_i \mu(A_i).$$

The value of the sum on the right-hand side is independent of how φ is represented as a sum of characteristic functions.

This definition is already well outside the scope of the Riemann integral. For instance, the characteristic function of the rationals is not Riemann integrable, but its Lebesgue integral is zero. The Riemann integral of a function $f : [a, b] \to \mathbb{R}$

is based upon the approximation of the function by simple step functions φ, in which the sets A_i are intervals. It may not be possible to approximate a highly discontinuous function by step functions, and then the Riemann definition of the integral fails. The Lebesgue integral uses approximations of the function by simple functions in which the sets A_i are general measurable sets. Because of the way the approximating simple functions are constructed in Theorem 12.26, the Lebesgue approach to integration is sometimes contrasted with the Riemann approach by saying that it subdivides the range of the function instead of the domain.

Definition 12.27 Let $f : X \to [0, \infty]$ be a nonnegative measurable function on a measure space $(X, \mathcal{A}, \mu)$. We define

$$\int f \, d\mu = \sup \left\{ \int \varphi \, d\mu \, \bigg| \, \varphi \text{ is simple and } \varphi \le f \right\}.$$

If $f : X \to \overline{\mathbb{R}}$ and $f = f_+ - f_-$, where f_+ and f_- are the positive and negative parts of f defined in (12.4), then we define

$$\int f \, d\mu = \int f_+ \, d\mu - \int f_- \, d\mu,$$

provided that at least one of the integrals on the right hand side is finite. If

$$\int |f| \, d\mu = \int f_+ \, d\mu + \int f_- \, d\mu < \infty,$$

then we say that f is *integrable* or *summable*. A complex-valued function $f : X \to \mathbb{C}$ is *integrable* if $f = g + ih$ where $g, h : X \to \mathbb{R}$ are integrable, and then

$$\int f \, d\mu = \int g \, d\mu + i \int h \, d\mu.$$

If A is a measurable subset of X, we define

$$\int_A f \, d\mu = \int f \chi_A \, d\mu.$$

The Lebesgue integral does not assign a value to the integral of a highly oscillatory function f for which both $\int f_+ \, d\mu$ and $\int f_- \, d\mu$ are infinite.

Example 12.28 The function

$$f(x) = \frac{d}{dx} \left[x^2 \sin \left(\frac{1}{x^2} \right) \right] = -\frac{2}{x} \cos \left(\frac{1}{x^2} \right) + 2x \sin \left(\frac{1}{x^2} \right),$$

is not Lebesgue integrable on $[0, 1]$. The function is not Riemann integrable on $[0, 1]$ either, since it is unbounded. Nevertheless, the improper Riemann integral

$$\lim_{\epsilon \to 0+} \int_\epsilon^1 f(x) \, dx = \sin 1$$

exists, because of the cancellation between the large positive and negative oscillations in the integrand.

Depending on the context, we may write the integral in any of the following ways

$$\int f, \quad \int f \, d\mu, \quad \int f(x) \, d\mu(x), \quad \int f(x) \, \mu(dx).$$

We will also write the integral $\int f \, d\lambda$ of a function defined on $\mathbb{R}^n$ with respect to Lebesgue measure λ as

$$\int f \, dx, \quad \int f(x) \, dx.$$

Example 12.29 If δ_{x_0} is the delta-measure, and $f : \mathbb{R}^n \to \mathbb{R}$ is a Borel measurable function, then

$$\int f \, d\delta_{x_0} = f(x_0).$$

We have $f = g$ a.e. with respect to δ_{x_0} if and only if $f(x_0) = g(x_0)$.

Example 12.30 Let ν be the counting measure on the set $\mathbb{N}$ of natural numbers defined in Example 12.6. If $f : \mathbb{N} \to \mathbb{R}$, then

$$\int f \, d\nu = \sum_{n=1}^{\infty} f_n,$$

where $f_n = f(n)$. This integral is well defined if f is nonnegative, or if the sum on the right converges absolutely, in which case f is integrable with respect to ν. Thus, nonnegative and absolutely convergent series are a special case of the general Lebesgue integral.

Example 12.31 We denote the integral with respect to the Lebesgue-Stieltjes measure μ_F on $\mathbb{R}$ defined in Example 12.15 by

$$\int f \, d\mu_F = \int f \, dF.$$

If F is a piecewise smooth, monotone increasing function with a countable number of jump discontinuities, then the Lebesgue-Stieltjes integral includes a continuous integral from the smooth parts of F, and a discrete sum from the jumps.

12.4 Convergence theorems

Suppose that a sequence of functions (f_n) converges pointwise to a limiting function f. When can we assert that $\int f_n \, d\mu$ converges to $\int f \, d\mu$? The following example shows that some condition is required to ensure the convergence of the integrals.

Example 12.32 Define $f_n : [0,1] \to \mathbb{R}$ by

$$f_n(x) = \begin{cases} n & \text{if } 0 < x < 1/n, \\ 0 & \text{otherwise.} \end{cases}$$

Then we have $f_n(x) \to 0$ as $n \to \infty$ for every $x \in [0,1]$, but

$$\int_0^1 f_n(x)\, dx = 1$$

for every n, so $\int_0^1 f_n(x)\, dx$ does not tend to 0 as $n \to \infty$.

Two simple conditions that ensure the convergence of the integrals are the monotone convergence of the sequence, and a uniform bound on the sequence by an integrable function. The corresponding theorems, called the monotone convergence theorem and the Lebesgue dominated convergence theorem, are among the most important and frequently used results in integration theory.

A sequence of functions (f_n), where $f_n : X \to \overline{\mathbb{R}}$, is *monotone increasing* if

$$f_1(x) \le \ldots \le f_{n-1}(x) \le f_n(x) \le f_{n+1}(x) \le \ldots \quad \text{for every } x \in X.$$

Theorem 12.33 (Monotone convergence) Suppose that (f_n) is a monotone increasing sequence of nonnegative, measurable functions $f_n : X \to [0,\infty]$ on a measure space $(X, \mathcal{A}, \mu)$. Let $f : X \to [0,\infty]$ be the pointwise limit,

$$f(x) = \lim_{n \to \infty} f_n(x).$$

Then

$$\lim_{n \to \infty} \int f_n\, d\mu = \int f\, d\mu.$$

The convergence of the sequence in Example 12.32 is not monotone. A generalization of this result, called Fatou's lemma, is the following.

Theorem 12.34 (Fatou) If (f_n) is any sequence of nonnegative measurable functions $f_n : X \to [0,\infty]$ on a measure space $(X, \mathcal{A}, \mu)$, then

$$\int \left(\liminf_{n \to \infty} f_n \right) d\mu \le \liminf_{n \to \infty} \int f_n\, d\mu. \tag{12.5}$$

Example 12.32 shows that we may have strict inequality in (12.5). Intuitively, "mass" may "leak out to infinity" as $n \to \infty$, so the integral of the lim inf may be less than or equal to the lim inf of the integrals.

The crucial hypothesis in the next theorem is that every function in the sequence (f_n) is bounded independently of n by the same integrable function g. This theorem is the one of the most useful for applications.

Theorem 12.35 (Lebesgue dominated convergence) Suppose that (f_n) is a sequence of integrable functions, $f_n : X \to \overline{\mathbb{R}}$, on a measure space $(X, \mathcal{A}, \mu)$ that converges pointwise to a limiting function $f : X \to \overline{\mathbb{R}}$. If there is an integrable function $g : X \to [0, \infty]$ such that

$$|f_n(x)| \leq g(x) \qquad \text{for all } x \in X \text{ and } n \in \mathbb{N},$$

then f is integrable and

$$\lim_{n \to \infty} \int f_n \, d\mu = \int f \, d\mu.$$

The sequence in Example 12.32 is bounded uniformly in n by the function

$$g(x) = \begin{cases} 1/x & \text{if } 0 < x \leq 1, \\ 0 & \text{if } x = 0, \end{cases}$$

but this function is not integrable on $[0, 1]$. The same result applies if X is a complete measure space, $f_n \to f$ pointwise-a.e., and $|f_n(x)| \leq g(x)$ pointwise-a.e.

A corollary of the dominated convergence theorem is the following result for differentiation under an integral, which is proved by approximation of the derivative by difference quotients.

Corollary 12.36 (Differentiation under an integral) Suppose that $(X, \mathcal{A}, \mu)$ is a complete measure space, $I \subset \mathbb{R}$ is an open interval, and $f : X \times I \to \overline{\mathbb{R}}$ is a measurable function such that:

(a) $f(\cdot, t)$ is integrable on X for each $t \in I$;
(b) $f(x, \cdot)$ is differentiable in I for each $x \in X \setminus N$ where $\mu(N) = 0$;
(c) there is an integrable function $g : X \to [0, \infty]$ such that

$$\left| \frac{\partial f}{\partial t}(x, t) \right| \leq g(x) \qquad \text{a.e. in } X \text{ for every } t \in I.$$

Then

$$\varphi(t) = \int_X f(x, t) \, d\mu(x)$$

is a differentiable function of t in I, and

$$\frac{d\varphi}{dt}(t) = \int_X \frac{\partial f}{\partial t}(x, t) \, d\mu(x).$$

12.5 Product measures and Fubini's theorem

One of the most elementary geometrical facts is that the area of a rectangle in the plane is the product of the lengths of its sides. From the point of view of measure theory, this means that Lebesgue measure on $\mathbb{R}^2$ is the product of Lebesgue measures on $\mathbb{R}$. We will describe a general construction of product measures here. A closely related result from elementary calculus is that the double integral of a continuous function over a smooth region in $\mathbb{R}^2$ can be computed as two iterated one-dimensional integrals. Fubini's theorem provides a generalization of this result, which states that an integral of a function on a product space can be computed as iterated integrals over the individual components of the product space. Fubini's theorem is another of the most useful results in the theory of Lebesgue integration.

The key hypothesis in Fubini's theorem is that the function is integrable on the product space. The following example shows that the equality of double and iterated integrals is not true, in general, without an integrability condition.

Example 12.37 Define $f : [0,1] \times [0,1] \to \mathbb{R}$ by

$$f(x,y) = \frac{x^2 - y^2}{(x^2 + y^2)^2}.$$

Then a straightforward computation shows that

$$\int_0^1 \left(\int_0^1 f(x,y)\, dy \right) dx = \int_0^1 \frac{1}{1+x^2}\, dx = \frac{\pi}{4},$$

$$\int_0^1 \left(\int_0^1 f(x,y)\, dx \right) dy = -\int_0^1 \frac{1}{1+y^2}\, dx = -\frac{\pi}{4}.$$

The function f in this example is not integrable, meaning that

$$\int_0^1 \int_0^1 |f(x,y)|\, dxdy = \infty.$$

First, we define the product of two σ-algebras.

Definition 12.38 Let $(X, \mathcal{A})$ and $(Y, \mathcal{B})$ be measurable spaces. The *product σ-algebra* $\mathcal{A} \otimes \mathcal{B}$ is the σ-algebra on $X \times Y$ that is generated by the collection of sets

$$\{A \times B \mid A \in \mathcal{A},\ B \in \mathcal{B}\}. \tag{12.6}$$

The collection of sets in (12.6) does not form a σ-algebra, since the union of two such sets is not, in general, another such set. Next, we state a theorem which defines the product of two σ-finite measures.

Theorem 12.39 Suppose that $(X, \mathcal{A}, \mu)$ and $(Y, \mathcal{B}, \nu)$ are σ-finite measure spaces. There is a unique product measure $\mu \otimes \nu$, defined on $\mathcal{A} \otimes \mathcal{B}$, with the property that for every $A \in \mathcal{A}$ and $B \in \mathcal{B}$

$$(\mu \otimes \nu)(A \times B) = \mu(A)\nu(B).$$

Example 12.40 Suppose that $X = \mathbb{R}^m$ and $Y = \mathbb{R}^n$ are equipped with the Borel σ-algebras $\mathcal{A} = \mathcal{R}^m$ and $\mathcal{B} = \mathcal{R}^n$ and the Lebesgue measures $\mu = \lambda^m$ and $\nu = \lambda^n$. The product σ-algebra is $\mathcal{R}^m \otimes \mathcal{R}^n = \mathcal{R}^{m+n}$, and the product measure is $\lambda^m \otimes \lambda^n = \lambda^{m+n}$. Thus, Lebesgue measure on $\mathbb{R}^n$ is the n-fold product of Lebesgue measure on $\mathbb{R}$.

Let $f : X \times Y \to \overline{\mathbb{R}}$. We denote by $f^y : X \to \overline{\mathbb{R}}$ and $f_x : Y \to \overline{\mathbb{R}}$ the functions

$$f^y(x) = f(x,y), \qquad f_x(y) = f(x,y).$$

If $(X, \mathcal{A}, \mu)$ and $(Y, \mathcal{B}, \nu)$ are σ-finite measure spaces and $f : X \times Y \to \overline{\mathbb{R}}$ is an $\mathcal{A} \otimes \mathcal{B}$-measurable function, then one can prove that f^y is $\mathcal{A}$-measurable for every $y \in Y$ and f_x is $\mathcal{B}$-measurable for every $x \in X$. Furthermore, the function $I(x) = \int_Y f_x(y) \, d\nu(y)$ is μ-measurable, and the function $J(y) = \int_X f^y(x) \, d\mu(x)$ is ν-measurable. All the integrals appearing in the following statement of Fubini's theorem are therefore well defined.

Theorem 12.41 (Fubini) Let $(X, \mathcal{A}, \mu)$ and $(Y, \mathcal{B}, \nu)$ be σ-finite measure spaces. Suppose that $f : X \times Y \to \overline{\mathbb{R}}$ is an $(\mathcal{A} \otimes \mathcal{B})$-measurable function.

(a) The function f is integrable, meaning that

$$\int_{X \times Y} |f| \, d\mu \otimes d\nu < \infty,$$

if and only if either of the following iterated integrals is finite:

$$\int_X \left(\int_Y |f_x(y)| \, d\nu(y) \right) d\mu(x),$$

$$\int_Y \left(\int_X |f^y(x)| \, d\mu(x) \right) d\nu(y).$$

(b) If f is integrable, then

$$\int_{X \times Y} f(x,y) \, d(\mu(x) \otimes \nu(y)) = \int_X \left(\int_Y f_x(y) \, d\nu(y) \right) d\mu(x),$$

$$\int_{X \times Y} f(x,y) \, d(\mu(x) \otimes \nu(y)) = \int_Y \left(\int_X f^y(x) \, d\mu(x) \right) d\nu(y).$$

To apply this theorem, we usually check that one of the iterated integrals of $|f|$ is finite, and then compute the double integral of f by evaluation of an iterated integral.

Example 12.42 Suppose that x_{mn} is a doubly-indexed sequence of real or complex numbers, with $m, n \in \mathbb{N}$. The application of Fubini's theorem to counting measure on $\mathbb{N}$ implies that if

$$\sum_{m=1}^{\infty} \left(\sum_{n=1}^{\infty} |x_{mn}| \right) < \infty,$$

then

$$\sum_{m=1}^{\infty} \left(\sum_{n=1}^{\infty} x_{mn} \right) = \sum_{n=1}^{\infty} \left(\sum_{m=1}^{\infty} x_{mn} \right).$$

The product of two complete measures is not necessarily complete, and this leads to some technical complications in connection with Lebesgue measure.

Example 12.43 Suppose that $X = \mathbb{R}^m$ and $Y = \mathbb{R}^n$, with $m, n \geq 1$, are equipped with the Lebesgue σ-algebras $\mathcal{L}^m = \overline{\mathcal{R}^m}$ and $\mathcal{L}^n = \overline{\mathcal{R}^n}$, respectively. It is not true that $\mathcal{L}^m \otimes \mathcal{L}^n = \mathcal{L}^{m+n}$. Rather, we have

$$\mathcal{L}^{m+n} = \overline{\mathcal{L}^m \otimes \mathcal{L}^n}.$$

For example, if $E \subset \mathbb{R}^m$ is any non-Lebesgue measurable set (which cannot be a subset of a set of m-dimensional Lebesgue measure zero) and $y \in \mathbb{R}^n$, then the set $E \times \{y\} \subset \mathbb{R}^{m+n}$ does not belong to $\mathcal{L}^m \otimes \mathcal{L}^n$. It is, however, an $(\mathcal{L}^{m+n})$-measurable set, since it is a subset of $\mathbb{R}^m \times \{y\}$ which has $(m+n)$-dimensional Lebesgue measure zero.

The following version of Fubini's theorem applies in this context.

Theorem 12.44 Suppose that $(X, \mathcal{A}, \mu)$ and $(Y, \mathcal{B}, \nu)$ are complete σ-finite measure spaces, and let $(X \times Y, \mathcal{L}, \lambda)$ be the completion of the product space. If $f : X \times Y \to \overline{\mathbb{R}}$ is a nonnegative or integrable $\mathcal{L}$-measurable function, then f_x is $\mathcal{B}$-measurable μ-a.e. in $x \in X$ and f^y is $\mathcal{A}$-measurable ν-a.e. in $y \in Y$. Furthermore, $I(x) = \int f_x(y)\, d\nu(y)$ and $J(y) = \int f^y(x)\, d\mu(x)$ are measurable, and

$$\int_{X \times Y} f\, d\lambda = \int_X \left(\int_Y f_x(y)\, d\nu(y) \right) d\mu(x) = \int_Y \left(\int_X f^y(x)\, d\mu(x) \right) d\nu(y).$$

12.6   The L^p spaces

The L^p-spaces consist of functions whose pth powers are integrable. The space L^∞ requires a separate definition.

Definition 12.45 Let $(X, \mathcal{A}, \mu)$ be a measure space and $1 \leq p < \infty$. The space $L^p(X, \mu)$ is the space of equivalence classes of measurable functions $f : X \to \mathbb{C}$, with respect to the equivalence relation of a.e.-equality, such that

$$\int |f|^p\, d\mu < \infty.$$

The L^p-norm of f is defined by

$$\|f\|_p = \left(\int_X |f|^p\, d\mu \right)^{1/p}. \tag{12.7}$$

The space $L^\infty(X, \mu)$ consists of equivalence classes of functions $f : X \to \mathbb{C}$ such that there is a finite constant M with

$$|f(x)| \le M \qquad \mu\text{-a.e.}$$

The L^∞-norm is defined by

$$\begin{aligned}
\|f\|_\infty &= \text{ess sup } \{|f(x)| \mid x \in X\} \\
&= \inf\{M \mid |f(x)| \le M \ \mu\text{-a.e.}\}.
\end{aligned}$$

We are mainly interested in the case when X is a Lebesgue measurable subset of $\mathbb{R}^n$, but most results below apply to arbitrary measure spaces. When the measure μ is understood, we often abbreviate $L^p(X, \mu)$ to $L^p(X)$, or simply L^p. For example, we write $L^p(\mathbb{R}^n)$ for $L^p(\mathbb{R}^n, \lambda)$, where λ is Lebesgue measure.

Theorem 12.46 If $(X, \mathcal{A}, \mu)$ is a measure space and $1 \le p \le \infty$, then $L^p(X)$ is a Banach space.

Proof.   We will only prove the result for $1 \le p < \infty$. We abbreviate $L^p(X)$ to L^p and $\|\cdot\|_p$ to $\|\cdot\|$. The verification that L^p is a linear space and that $\|\cdot\|$ is a norm is straightforward, with the exception of the triangle inequality, which we prove in Theorem 12.56 below. We therefore just have to show that L^p is complete.

From Exercise 1.20, a normed linear space is complete if and only if every absolutely convergent series converges. Suppose that $f_n \in L^p$ with $n = 1, 2, \ldots$ is a sequence of functions such that

$$\sum_{n=1}^{\infty} \|f_n\| = M,$$

where $M < \infty$. We need to show that there is a function $f \in L^p$ such that

$$\lim_{N \to \infty} \left\| f - \sum_{n=1}^{N} f_n \right\| = 0.$$

First, we consider the sequence of nonnegative functions g_N defined by

$$g_N = \sum_{n=1}^{N} |f_n|.$$

The sequence (g_N) is monotone increasing, so it converges pointwise to a nonnegative, measurable extended real-valued function $g : X \to [0, \infty]$. We have $\|g_N\| \le M$ for every $N \in \mathbb{N}$, so

$$\int |g_N|^p d\mu = \|g_N\|^p \le M^p.$$

The monotone convergence theorem, Theorem 12.33, implies that

$$\|g\|^p = \lim_{N\to\infty} \|g_N\|^p \le M^p.$$

In particular, $\|g\| < \infty$, so $g \in L^p$. Therefore, g is finite μ-a.e., which means that the sum $\sum_{n=1}^{\infty} f_n(x)$ is absolutely convergent μ-a.e. We can then define a function f pointwise-a.e. by

$$f(x) = \sum_{n=1}^{\infty} f_n(x).$$

The partial sums of this series satisfy

$$\left|\sum_{n=1}^{N} f_n(x)\right|^p \le g(x)^p.$$

Since $g \in L^p$ and $|f(x)| \le g(x)$, we have that $f \in L^p$. Furthermore, $g^p \in L^1$ and

$$\left|f(x) - \sum_{n=1}^{N} f_n(x)\right|^p \le (2g(x))^p.$$

The dominated convergence theorem, Theorem 12.35, implies that

$$\lim_{N\to\infty} \int \left|f - \sum_{n=1}^{N} f_n\right|^p d\mu = 0,$$

so the series converges to f in L^p, which proves that L^p is complete. $\quad\square$

The following example shows that the L^p-convergence of a sequence does not imply pointwise-a.e. convergence. One can prove, however, that if $f_n \to f$ in L^p, then there is a subsequence of (f_n) that converges pointwise-a.e. to f.

Example 12.47 For $2^k \le n \le 2^{k+1}-1$, where $k = 0,1,2,\ldots$, we define the interval I_n by

$$I_n = \left[(n - 2^k)/2^k, (n + 1 - 2^k)/2^k\right],$$

and the function $f_n : [0,1] \to \mathbb{R}$ by

$$f_n = \chi_{I_n}.$$

The sequence (f_n) consists of characteristic functions of intervals of width 2^{-k} that sweep across the interval $[0,1]$. We have $f_n \to 0$ in $L^p([0,1])$, for $1 \le p < \infty$, but $f_n(x)$ does not converge for any $x \in [0,1]$. The subsequence (f_{2^k}) converges pointwise-a.e. to zero.

As we have seen, it is often useful to approximate an arbitrary element in some space as the limit of a sequence of elements with special properties. Every L^p-function may be approximated by simple functions.

Theorem 12.48 Suppose that $(X, \mathcal{A}, \mu)$ is a measure space and $1 \leq p \leq \infty$. If $f \in L^p(X)$, then there is a sequence (φ_n) of simple functions $\varphi_n : X \to \mathbb{C}$ such that

$$\lim_{n \to \infty} \|f - \varphi_n\|_p = 0.$$

Proof. It is sufficient to prove the result for nonnegative functions, since we may approximate a general function in L^p by approximating its positive and negative parts. We consider only the case $1 \leq p < \infty$ for simplicity. If $f \geq 0$, then from Theorem 12.26 there is a monotone increasing sequence of nonnegative simple functions (φ_n) that converges pointwise to f. The sequence (g_n) defined by

$$g_n = f^p - (f - \varphi_n)^p$$

is a monotone increasing sequence of nonnegative functions. The monotone convergence theorem implies that

$$\int_X g_n \, d\mu \to \int_X f^p \, d\mu$$

as $n \to \infty$, from which it follows that $\varphi_n \to f$ in L^p. $\qquad\square$

As an application of this theorem, we prove that $L^p(\mathbb{R}^n)$ is separable for $p < \infty$.

Theorem 12.49 If $1 \leq p < \infty$, then $L^p(\mathbb{R}^n)$ is a separable metric space.

Proof. We have to show that $L^p(\mathbb{R}^n)$ contains a countable dense subset. The set S of simple functions whose values are complex numbers with rational real and imaginary parts is dense in the space of simple functions. Hence, Theorem 12.48 implies that S is dense in $L^p(\mathbb{R}^n)$. The set S is not countable because there are far too many measurable sets, but we can approximate every simple function by a simple function of the form (12.3) in which each set A_i is chosen from a suitable countable collection $\mathcal{F}$ of measurable sets. For example, we can use the collection of cubes of the form $[a_1, b_1] \times [a_2, b_2] \times \cdots \times [a_n, b_n]$, where $a_1, b_1, a_2, b_2, \ldots, a_n, b_n \in \mathbb{Q}$. We omit a detailed proof. The result then follows. $\qquad\square$

This theorem also applies to $L^p(\Omega)$, with $1 \leq p < \infty$, where Ω is an arbitrary measurable subset of $\mathbb{R}^n$. The space $L^\infty(\mathbb{R}^n)$ is not separable (see Exercise 12.13).

For $p < \infty$, we can approximate functions in $L^p(\mathbb{R}^n)$ by continuous functions with compact support.

Theorem 12.50 The space $C_c(\mathbb{R}^n)$ of continuous functions with compact support is dense in $L^p(\mathbb{R}^n)$ for $1 \leq p < \infty$.

Proof. If $f \in L^p(\mathbb{R}^n)$, then the Lebesgue dominated convergence theorem implies that the sequence (f_n) of compactly supported functions,

$$f_n(x) = \begin{cases} f(x) & \text{if } |x| \leq n, \\ 0 & \text{if } |x| > n, \end{cases}$$

converges to f in L^p as $n \to \infty$. Each f_n may be approximated by simple functions, which are a finite linear combination of characteristic functions. It therefore suffices to prove that the characteristic function χ_A of a bounded, measurable set A in $\mathbb{R}^n$ may be approximated by continuous functions with compact support. From Theorem 12.10, for every $\epsilon > 0$ there is an open set G and a compact set $K \subset G$ such that $K \subset A \subset G$ and $\lambda(G \setminus K) < \epsilon^p$. By Urysohn's lemma (see Exercise 1.16), there is a continuous, real-valued function f such that $0 \le f(x) \le 1$, $f = 1$ on K, and $f = 0$ on G^c. Then

$$\|f - \chi_A\|_p = \left(\int_{G \setminus K} |f - \chi_A|^p \, dx \right)^{1/p} \le \lambda(G \setminus F)^{1/p} < \epsilon.$$

$\square$

We can use an approximate identity to smooth out C_c-approximations, thus obtaining C_c^∞-approximations.

Theorem 12.51 If $1 \le p < \infty$, then $C_c^\infty(\mathbb{R}^n)$ is a dense subspace of $L^p(\mathbb{R}^n)$.

Proof. For $\epsilon > 0$, let $\varphi_\epsilon \in C_c^\infty(\mathbb{R}^n)$ be an approximate identity. If $f \in C_c(\mathbb{R}^n)$, we define $f_\epsilon = \varphi_\epsilon * f$. Then $f_\epsilon \in C_c^\infty(\mathbb{R}^n)$ for every $\epsilon > 0$. Moreover, $f_\epsilon \to f$ uniformly, and hence in L^p, as $\epsilon \to 0^+$. Since $C_c(\mathbb{R}^n)$ is dense in $L^p(\mathbb{R}^n)$, the result follows. $\square$

12.7 The basic inequalities

It has been said that analysis is the art of estimating. In this section, we give the basic inequalities of L^p theory.

Many inequalities are based on convexity arguments. We first prove Jensen's inequality, which states that the mean of the values of a convex function is greater than or equal to the value of the convex function at the mean. We recall from Definition 8.47 that a function $\varphi : C \to \mathbb{R}$ on a convex set C is convex if

$$\varphi\left(tx + (1-t)y\right) \le t\varphi(x) + (1-t)\varphi(y)$$

for every $x, y \in C$ and $t \in [0, 1]$. We define the *mean* of an integrable function f on a finite measure space (X, μ) by

$$\langle f \rangle_\mu = \frac{1}{\mu(X)} \int_X f \, d\mu. \tag{12.8}$$

Theorem 12.52 (Jensen) Let (X, μ) be a finite measure space. If $\varphi : \mathbb{R} \to \mathbb{R}$ is convex and $f : X \to \mathbb{R}$ belongs to $L^1(X, \mu)$, then

$$\varphi\left(\langle f \rangle_\mu\right) \le \langle \varphi \circ f \rangle_\mu. \tag{12.9}$$

Proof. From Exercise 12.9, there is a constant $c \in \mathbb{R}$ such that

$$\varphi(y) \geq \varphi(\langle f \rangle_\mu) + c(y - \langle f \rangle_\mu) \qquad \text{for every } y \in \mathbb{R}. \qquad (12.10)$$

Setting $y = f(x)$ in this inequality and integrating the result over X, we obtain that

$$\int_X \varphi \circ f \, d\mu \;\geq\; \varphi(\langle f \rangle_\mu) \int_X d\mu + c \left(\int_X f \, d\mu - \langle f \rangle_\mu \int_X d\mu \right)$$
$$= \; \varphi(\langle f \rangle_\mu) \, \mu(X).$$

Dividing this equation by $\mu(X) < \infty$, we obtain (12.9). $\qquad\qquad\qquad$ □

The right-hand side of (12.9) may be infinite. Although $\varphi \circ f$ need not be in L^1, its negative part is integrable from (12.10), so its integral is well defined.

Example 12.53 Suppose that $\{x_1, x_2, \ldots, x_n\}$ is a finite subset of $\mathbb{R}$, and μ is a discrete probability measure on $\mathbb{R}$ with $\mu(\{x_i\}) = \lambda_i$, where $0 \leq \lambda_i \leq 1$ and

$$\lambda_1 + \lambda_2 + \cdots + \lambda_n = 1.$$

Then Jensen's inequality, with $f(x) = x$, implies that for any convex function $\varphi : \mathbb{R} \to \mathbb{R}$ we have

$$\varphi(\lambda_1 x_1 + \lambda_2 x_2 + \ldots + \lambda_n x_n) \leq \lambda_1 \varphi(x_1) + \lambda_2 \varphi(x_2) + \ldots + \lambda_n \varphi(x_n).$$

Hölder's inequality is one of the most important inequalities for proving estimates in L^p-spaces. We say that two numbers $1 \leq p, p' \leq \infty$ are *Hölder conjugates* or *conjugate exponents* if they satisfy

$$\frac{1}{p} + \frac{1}{p'} = 1, \qquad (12.11)$$

with the convention that $1/\infty = 0$. For example, $p = 1$ and $p' = \infty$ are Hölder conjugates, and $p = 2$ is conjugate to itself. Hölder's inequality applies to a pair of functions, one in L^p and one in $L^{p'}$.

Theorem 12.54 (Hölder) Let $1 \leq p, p' \leq \infty$ satisfy (12.11). If $f \in L^p(X, \mu)$ and $g \in L^{p'}(X, \mu)$, then $fg \in L^1(X, \mu)$ and

$$\left| \int fg \, d\mu \right| \leq \|f\|_p \|g\|_{p'}. \qquad (12.12)$$

Proof. For the conjugate pair of exponents $(p, p') = (1, \infty)$, Hölder's inequality is the obvious inequality

$$\left| \int_X fg \, d\mu \right| \leq \|g\|_\infty \int_X |f| \, d\mu.$$

We therefore assume that $1 < p, p' < \infty$.

The set $Y = \{x \in X \mid g(x) \neq 0\}$ is measurable, and

$$\left| \int_X f(x)g(x) \, d\mu(x) \right| = \left| \int_Y f(x)g(x) \, d\mu(x) \right| \leq \int_Y |f(x)| \, |g(x)| \, d\mu(x).$$

We also have

$$\|f\|_p^p = \int_X |f|^p \, d\mu \geq \int_Y |f|^p \, d\mu.$$

Therefore, replacing X by Y if necessary, it is sufficient to prove (12.12) under the assumption that $f(x) \geq 0$ and $g(x) > 0$ for every $x \in X$.

We define a new measure ν on X by

$$\nu(A) = \int_A g^{p'} \, d\mu.$$

The function $\varphi : \mathbb{R} \to \mathbb{R}$ defined by $\varphi(x) = |x|^p$ is convex for $p \geq 1$. An application of Jensen's inequality (12.9) to the function $h : X \to \mathbb{R}$ defined by

$$h = \frac{f}{g^{p'/p}}$$

implies that

$$\varphi(\langle h \rangle_\nu) \leq \langle \varphi \circ h \rangle_\nu.$$

Writing out this equation explicitly, we obtain

$$\left| \frac{1}{\nu(X)} \int_X \frac{f}{g^{p'/p}} \, d\nu \right|^p \leq \frac{1}{\nu(X)} \int_X \frac{f^p}{g^{p'}} \, d\nu.$$

Rewriting the integrals with respect to ν as integrals with respect to μ, and using the assumption that p and p' are dual exponents, we obtain that

$$\left| \frac{\int_X fg \, d\mu}{\int_X g^{p'} \, d\mu} \right|^p \leq \frac{1}{\int_X g^{p'} \, d\mu} \int_X f^p \, d\mu.$$

Taking the pth root of this equation, rearranging the result, and using the fact that p and p' are conjugate exponents again, we get (12.12). Since the right-hand side of this inequality is finite, it follows that $h \in L^1(X, \nu)$ and $fg \in L^1(X, \mu)$. $\square$

In the special case when $p = p' = 2$, Hölder's inequality is the Cauchy-Schwarz inequality for L^2-spaces. As an application of Hölder's inequality, we prove a result about the inclusion of L^p-spaces.

Proposition 12.55 Suppose that (X, μ) is a finite measure space, meaning that $\mu(X) < \infty$, and $1 \leq q \leq p \leq \infty$. Then

$$L^1(X, \mu) \supset L^q(X, \mu) \supset L^p(X, \mu) \supset L^\infty(X, \mu).$$

Proof. We define $1 \leq r \leq \infty$ by

$$\frac{1}{p} + \frac{1}{r} = \frac{1}{q}.$$

Then p/q and r/q are Hölder conjugates. Moreover, if $f \in L^p$, then $|f|^{p/q} \in L^q$ and $1 \in L^{r/q}$, since $\mu(X) < \infty$. Hölder's inequality therefore implies that

$$
\begin{aligned}
\|f\|_q^q &= \int_X |f|^q \cdot 1 \, d\mu \\
&\leq \left(\int_X (|f|^q)^{p/q} \, d\mu \right)^{q/p} \left(\int_X d\mu \right)^{q/r} \\
&= (\mu(X))^{p/(p-q)} \|f\|_p^q.
\end{aligned}
$$

Hence, $\|f\|_q$ is finite for every $f \in L^p$, which proves the claimed inclusion. $\quad\square$

For these inclusions to hold, it is crucial that $\mu(X) < \infty$, as illustrated in Exercise 12.14. Minkowski's inequality is the triangle inequality for the L^p-norm.

Theorem 12.56 (Minkowski) If $1 \leq p \leq \infty$, and $f, g \in L^p(X, \mu)$, then $f + g \in L^p(X, \mu)$, and

$$\|f + g\|_p \leq \|f\|_p + \|g\|_p. \tag{12.13}$$

Proof. We have

$$
\begin{aligned}
|f + g|^p &\leq (|f| + |g|)^p \\
&\leq 2^p \max (|f|^p, |g|^p) \\
&\leq 2^p (|f|^p + |g|^p).
\end{aligned}
$$

Hence, $f + g \in L^p$ if $f, g \in L^p$.

If $f + g \in L^p$, then $|f + g|^{p-1} \in L^{p'}$, where p' is the Hölder conjugate of p, and

$$\left\| |f + g|^{p-1} \right\|_{p'} = \|f + g\|_p^{p-1}.$$

Hence, using Hölder's inequality followed by this result, we find that

$$
\begin{aligned}
\|f + g\|_p^p &= \int |f + g| |f + g|^{p-1} \, d\mu \\
&\leq \int |f| |f + g|^{p-1} \, d\mu + \int |g| |f + g|^{p-1} \, d\mu \\
&\leq \|f\|_p \left\| |f + g|^{p-1} \right\|_{p'} + \|g\|_p \left\| |f + g|^{p-1} \right\|_{p'} \\
&\leq \left(\|f\|_p + \|g\|_p \right) \|f + g\|_p^{p-1}.
\end{aligned}
$$

If $\|f + g\|_p \neq 0$, then division of this inequality by $\|f + g\|_p^{p-1}$ gives (12.13). If $\|f + g\|_p = 0$, then the result is trivial. $\quad\square$

For $0 \leq p \leq 1$, nonnegative functions satisfy the reverse triangle inequality

$$\|f + g\|_p \geq \|f\|_p + \|g\|_p,$$

which explains why L^p is only a normed linear space for $p \geq 1$.

Tchebyshev's inequality is an elementary inequality that is especially useful in probability theory.

Theorem 12.57 (Tchebyshev) Suppose that $f \in L^p(X)$, where $0 < p < \infty$. For every $\epsilon > 0$, we have

$$\mu\left(\{x \in X \mid |f(x)| > \epsilon\}\right) \leq \frac{1}{\epsilon^p}\|f\|_p^p.$$

Proof. Define $S_\epsilon \in \mathcal{A}$ by

$$S_\epsilon = \{x \in X \mid |f(x)| > \epsilon\}.$$

Then

$$\|f\|_p^p = \int_X |f|^p \, d\mu \geq \int_{S_\epsilon} |f|^p \, d\mu \geq \epsilon^p \mu\left(S_\epsilon\right),$$

which is what we had to prove. $\square$

The following inequality for the L^p-norm of convolutions is called *Young's inequality*. This inequality shows that convolution is a continuous operation when defined on an appropriate choice of Banach spaces.

Theorem 12.58 (Young) Suppose that $1 \leq p, q, r \leq \infty$ satisfy

$$\frac{1}{p} + \frac{1}{q} = 1 + \frac{1}{r}. \tag{12.14}$$

If $f \in L^p(\mathbb{R}^n)$ and $g \in L^q(\mathbb{R}^n)$, then $f * g \in L^r(\mathbb{R}^n)$, and

$$\|f * g\|_r \leq \|f\|_p \|g\|_q.$$

Proof. We leave it to the reader to check that it is sufficient to prove the result for nonnegative functions f, g such that $\|f\|_p = \|g\|_q = 1$.

We first consider the special case $p = q = r = 1$. Using Fubini's theorem to exchange the order of integration, we have

$$
\begin{aligned}
\|f * g\|_1 &= \int \left[\int f(y) g(x - y) \, dy \right] dx \\
&= \left[\int f(y) \, dy \right] \left[\int g(z) \, dz \right] \\
&= \|f\|_1 \|g\|_1, \tag{12.15}
\end{aligned}
$$

which proves the result in this case.

For general values of p, q, we observe from (12.14) and the definition of the Hölder conjugate that

$$\frac{1}{p'} + \frac{1}{q'} + \frac{1}{r} = 1.$$

An application of the generalized Hölder inequality in Exercise 12.12 therefore implies that

$$f * g(x) = \int \left[f(y)^{p/r} g(x-y)^{q/r} \right] \left[f(y)^{1-p/r} \right] \left[g(x-y)^{1-q/r} \right] dy$$

$$\leq \left[\int f(y)^p g(x-y)^q \, dy \right]^{1/r} \left[\int f(y)^{(1-p/r)q'} \, dy \right]^{1/q'}$$

$$\left[\int g(x-y)^{(1-q/r)p'} \, dy \right]^{1/p'}.$$

Since $(1-p/r)q' = p$, $(1-q/r)p' = q$, and $\|f\|_p = \|g\|_q = 1$, it follows that

$$f * g(x)^r \leq \int f(y)^p g(x-y)^q \, dy,$$

meaning that $(f*g)^r \leq f^p * g^q$. The use of this inequality and (12.15) then implies that

$$\|f * g\|_r^r = \|(f*g)^r\|_1 \leq \|f^p * g^q\| = \|f^p\|_1 \|g^q\|_1 = \|f\|_p^p \|g\|_q^q = 1,$$

which proves the theorem. $\square$

Two common special cases of this result are:

$$\|f * g\|_1 \leq \|f\|_1 \|g\|_1, \qquad \|f * g\|_\infty \leq \|f\|_2 \|g\|_2.$$

12.8   The dual space of L^p

In Section 5.6, we gave the general definition of the dual space of a Banach space. In this section, we describe the dual space $L^p(X)^*$ of bounded, linear functionals on $L^p(X)$, where X is a measure space equipped with a measure μ.

Suppose that $1 \leq p \leq \infty$ and $g \in L^{p'}(X)$, where p' is the Hölder conjugate of p. We define $\varphi_g : L^p(X) \to \mathbb{C}$ by

$$\varphi_g(f) = \int_X fg \, d\mu \qquad \text{for every } f \in L^p(X). \tag{12.16}$$

Hölder's inequality implies that φ_g is a bounded linear functional on L^p, with

$$\|\varphi_g\|_{(L^p)^*} \leq \|g\|_{L^q}.$$

Here, $\|\cdot\|_{(L^p)^*}$ is the norm of a bounded linear function defined in (5.23). The next theorem states that if $1 \leq p < \infty$, then all linear functionals on L^p are of the form (12.16).

Theorem 12.59 If $1 < p < \infty$, then every $\varphi \in L^p(X)^*$ is of the form

$$\varphi(f) = \int_X fg \, d\mu$$

for some $g \in L^{p'}(X)$, where $1/p + 1/p' = 1$. If μ is σ-finite, the same conclusion holds when $p = 1$ and $p' = \infty$. Moreover,

$$\|\varphi\|_{(L^p)^*} = \|g\|_{L^{p'}}.$$

We will not give the proof. According to Theorem 12.59, we may identify the dual of L^p with $L^{p'}$. When $p = p' = 2$, we recover the result of the Riesz representation theorem that the dual of the Hilbert space L^2 may be identified with itself. The dual of L^1 is L^∞, but the dual of L^∞ is strictly larger than L^1 (except in trivial cases, such as when X is a finite set). The full description of $(L^\infty)^*$ is complicated and rarely useful, so we will not give it here. If $1 < p < \infty$, then $(L^p)^{**} = L^p$ and L^p is reflexive, but L^1 and L^∞ are not reflexive.

The continuous linear functionals define the weak topology. From Definition 5.59, Definition 5.60, and Theorem 12.59, we have the following definition of weak L^p-convergence.

Definition 12.60 Suppose that $1 \leq p < \infty$. A sequence (f_n) *converges weakly* to f in L^p, written $f_n \rightharpoonup f$, if

$$\lim_{n \to \infty} \int f_n g \, d\mu = \int fg \, d\mu \qquad \text{for every } g \in L^{p'}, \tag{12.17}$$

where p' is the Hölder conjugate of p. When $p = \infty$ and $p' = 1$, the condition in (12.17) corresponds to weak-$*$ convergence $f_n \overset{*}{\rightharpoonup} f$ in L^∞.

As in the case of Hilbert spaces, discussed in Section 8.6, weak L^p-convergence does not imply strong L^p-convergence, meaning convergence in the L^p-norm. The following example illustrates three typical ways in which a weakly convergent sequence of functions can fail to be strongly convergent.

Example 12.61 Let $g \in L^p(\mathbb{R})$ be a fixed nonzero function, where $1 < p < \infty$. For each of the following three sequences, we have $f_n \rightharpoonup 0$ weakly as $n \to \infty$, but not $f_n \to 0$ strongly, in $L^p(\mathbb{R})$.

(a) $f_n(x) = g(x) \sin nx$ (oscillation);
(b) $f_n(x) = n^{1/p} g(nx)$ (concentration);
(c) $f_n(x) = g(x - n)$ (escape to infinity).

The Banach-Alaoglu theorem, in Theorem 5.61, leads to the following weak compactness result for L^p.

Theorem 12.62 Suppose that (f_n) is a bounded sequence in $L^p(X)$, meaning that there is a constant M such that $\|f_n\| \leq M$ for every $n \in \mathbb{N}$. If $1 < p < \infty$, then there is a subsequence (f_{n_k}) and a function $f \in L^p(X)$ with $\|f\| \leq M$ such that $f_{n_k} \rightharpoonup f$ as $k \to \infty$ weakly in $L^p(X)$.

12.9 Sobolev spaces

Many problems in applied analysis involve differentiable functions. *Sobolev spaces* are Banach spaces of functions whose weak derivatives belong to L^p spaces. They provide the simplest and most useful setting for the application of functional analytic methods to the study of differential equations. We have already discussed several special cases of Sobolev spaces in the chapters on Fourier series and unbounded linear operators. Here, we give more general definitions of Sobolev spaces, and describe some of their main properties. We use the multi-index notation introduced in Section 11.1, and consider real-valued functions for simplicity.

We will define Sobolev spaces of functions whose domain is an open subset Ω of $\mathbb{R}^n$, equipped with n-dimensional Lebesgue measure. In particular, we could have $\Omega = \mathbb{R}^n$. As usual, $L^p(\Omega)$ denotes the space of Lebesgue measurable functions $f : \Omega \to \mathbb{R}$ whose pth powers are integrable. We also introduce the *local L^p spaces*, denoted by $L^p_{\mathrm{loc}}(\Omega)$. A function f belongs to $L^p_{\mathrm{loc}}(\Omega)$ if it is measurable and

$$\int_K |f|^p \, dx < \infty$$

for every compact subset K of Ω. For example, $1/x$ belongs $L^1_{\mathrm{loc}}((0,1))$, but not to $L^1((0,1))$ or $L^1_{\mathrm{loc}}(\mathbb{R})$. For every $1 \leq p \leq \infty$, we have the inclusions

$$L^1_{\mathrm{loc}}(\Omega) \supset L^p_{\mathrm{loc}}(\Omega) \supset L^p(\Omega).$$

Thus, $L^1_{\mathrm{loc}}(\Omega)$ is the largest space of integrable functions. We adopt the following definition of a test function for the purposes of this chapter.

Definition 12.63 A *test function* $\varphi : \Omega \to \mathbb{R}$ on an open subset Ω of $\mathbb{R}^n$ is a function with continuous partial derivatives of all orders whose support is a compact subset of Ω. We denote the set of test functions on Ω by $C_c^\infty(\Omega)$.

Definition 12.64 If $f, g_\alpha \in L^1_{\mathrm{loc}}(\Omega)$ are such that

$$\int_\Omega g_\alpha \varphi \, dx = (-1)^{|\alpha|} \int_\Omega f \partial^\alpha \varphi \, dx \qquad \text{for all } \varphi \in C_c^\infty(\Omega),$$

then we say that $g_\alpha = \partial^\alpha f$ is the αth *weak partial derivative* of f.

The weak derivative is only defined pointwise up to a set of measure zero.

Example 12.65 The function

$$f(x) = \frac{1}{|x|^a}$$

belongs to $L^1_{\text{loc}}(\mathbb{R}^n)$ if and only if $a < n$. The weak derivative of f with respect to x_i is given by

$$g_i(x) = -a\frac{x_i}{|x|}\frac{1}{|x|^{a+1}}$$

provided that g_i is locally integrable which is the case when $a < n-1$. For example, in one space dimension, if $f' \in L^p_{\text{loc}}(\mathbb{R})$ for some $p > 1$, then $a < 0$ and f is continuous. In fact, the Sobolev embedding theorem below implies that any function on $\mathbb{R}$ whose weak derivative belongs to $L^p_{\text{loc}}(\mathbb{R})$ for some $p > 1$ is continuous. In higher space dimensions, a function may be weakly differentiable but discontinuous. The strength of an allowable singularity in a weakly differentiable function increases with the number of space dimensions n.

Definition 12.66 Let k be a positive integer, $1 \leq p \leq \infty$, and Ω an open subset of $\mathbb{R}^n$. The *Sobolev space* $W^{k,p}(\Omega)$ consists of all functions $f : \Omega \to \mathbb{R}$ such that $\partial^\alpha f \in L^p(\Omega)$ for all weak partial derivatives of order $0 \leq |\alpha| \leq k$. We define a norm on $W^{k,p}(\Omega)$ by

$$\|f\|_{W^{k,p}(\Omega)} = \left(\sum_{0\leq|\alpha|\leq k} \int_\Omega |\partial^\alpha f|^p \, dx \right)^{1/p}$$

when $1 \leq p < \infty$, and by

$$\|f\|_{W^{k,\infty}(\Omega)} = \max_{0\leq|\alpha|\leq k} \left\{ \sup_{x\in\Omega} |\partial^\alpha f(x)| \right\}$$

when $p = \infty$. Here, the supremum is to be interpreted as an essential supremum. For $p = 2$, corresponding to the case of square integrable functions, we write $W^{k,2}(\Omega) = H^k(\Omega)$, and define an inner product on $H^k(\Omega)$ by

$$\langle f, g \rangle_{H^k(\Omega)} = \sum_{0\leq|\alpha|\leq k} \int_\Omega \partial^\alpha f \partial^\alpha g \, dx.$$

The space $W^{k,p}(\Omega)$ is a Banach space and $H^k(\Omega)$ is a Hilbert space.

Next, we define a Sobolev space of functions that "vanish on the boundary of Ω."

Definition 12.67 The closure of $C^\infty_c(\Omega)$ in $W^{k,p}(\Omega)$ is denoted by

$$W^{k,p}_0(\Omega) = \overline{C^\infty_c(\Omega)}.$$

We also define

$$H_0^k(\Omega) = W_0^{k,2}(\Omega).$$

Informally, we can think of $W_0^{k,p}(\Omega)$ as the $W^{k,p}(\Omega)$-functions whose derivatives of order less than or equal to $k-1$ vanish on the boundary $\partial\Omega$ of Ω. Compactly supported functions are dense in $W^{k,p}(\mathbb{R}^n)$, so that $W_0^{k,p}(\mathbb{R}^n) = W^{k,p}(\mathbb{R}^n)$.

The definition of the boundary values of Sobolev functions which do not vanish on the boundary is non-trivial. The boundary of a smooth set has measure zero, but Sobolev functions are not necessarily continuous and they are defined pointwise only up to sets of measure zero. The trace theorem, in Theorem 12.76 below, gives a way to assign boundary values to Sobolev functions.

Sobolev spaces of negative orders may be defined by duality.

Definition 12.68 Let k be a positive integer, $1 \leq p < \infty$, and p' the Hölder conjugate of p. The Sobolev space $W^{-k,p}(\Omega)$ is the dual space of $W_0^{k,p'}(\Omega)$. That is, $f \in W^{-k,p}(\Omega)$ is a continuous linear map

$$f: W_0^{k,p'}(\Omega) \to \mathbb{R}, \qquad f: u \mapsto \langle f, u \rangle.$$

We define a norm on $W^{-k,p}(\Omega)$ by

$$\|f\|_{W^{-k,p}} = \sup_{\substack{u \in W_0^{k,p'} \\ u \neq 0}} \frac{\langle f, u \rangle}{\|u\|}.$$

In particular, $H^{-k}(\Omega)$ is the dual space of $H_0^k(\Omega)$. Elements of $W^{-k,p}(\Omega)$ are distributions whose action on test functions extends continuously to an action on functions in $W_0^{k,p'}(\Omega)$. The dual space of $W^{k,p'}(\Omega)$ is not a space of distributions, because the action of a continuous linear functional on functions in $W^{k,p'}(\Omega)$ depends on the values of the $W^{k,p'}(\Omega)$-functions on the boundary $\partial\Omega$, and therefore it is not determined by its action on compactly supported test functions. It is possible to show that any distribution $f \in W^{-k,p}(\Omega)$ may be written (nonuniquely) as

$$f = \sum_{|\alpha| \leq k} \partial^\alpha g_\alpha$$

where $g_\alpha \in L^p(\Omega)$. The action of this distribution f on a $W_0^{k,p'}$-function u is given by

$$\langle f, u \rangle = \sum_{|\alpha| \leq k} (-1)^{|\alpha|} \int_\Omega g_\alpha \partial^\alpha u \, dx.$$

More generally, it is possible to define Sobolev spaces $W^{s,p}$ of fractional order for any $s \in \mathbb{R}$ and $p \in [1, \infty]$. These spaces arise naturally in connection with the trace theorem below, but we will not describe them here. For $p = 2$, the Hilbert spaces $H^s(\mathbb{R}^n)$ may be defined by use of the Fourier transform (see Definition 11.38).

12.10   Properties of Sobolev spaces

In this section, we summarize the main properties of Sobolev spaces without proof. These properties include the approximation of Sobolev functions by smooth functions (density theorems), the integrability or continuity properties of Sobolev functions (embedding theorems), compactness conditions (the Rellich-Kondrachov theorem), and the definition of boundary values of Sobolev functions (trace theorems). Depending on the context, there are many different regularity conditions that the domain Ω on which the Sobolev functions are defined must satisfy, and the differences between them are sometimes quite subtle. We will say that a domain is *regular* if it satisfies an appropriate regularity condition, without stating the precise condition that is required. Any domain bounded by a smooth hypersurface (meaning that the boundary is locally the zero set of a smooth function with nonzero derivative) satisfies the regularity conditions for all the results stated in this section. Domains with outward pointing cusps or needle-shaped protrusions are typical examples of domains with insufficient regularity, and some of the properties stated below do not hold on such domains.

We use $C(\overline{\Omega})$ to denote the space of uniformly continuous functions on Ω, and $C_0(\mathbb{R}^n)$ to denote the space of continuous functions on $\mathbb{R}^n$ that tend to zero as $x \to \infty$. This space is the closure of $C_c^\infty(\mathbb{R}^n)$ in $L^\infty(\mathbb{R}^n)$. The space $C^k(\overline{\Omega})$ consists of functions whose partial derivatives of order less than or equal to k are uniformly continuous in Ω, and $C^\infty(\overline{\Omega})$ consists of functions with uniformly continuous derivatives of all orders in Ω. If a function is uniformly continuous in the open set Ω, then it has a unique continuous extension to the closure $\overline{\Omega}$.

In the theorems stated below, we consider two types of domains: $\Omega = \mathbb{R}^n$; and Ω a regular, bounded, open subset of $\mathbb{R}^n$ with boundary $\partial\Omega$. It is frequently possible to consider more general domains, but this complicates the statements of some of the theorems. The order k is a positive integer and $1 \leq p \leq \infty$, unless stated otherwise.

Theorem 12.69 (Density) The space $C_c^\infty(\mathbb{R}^n)$ is dense in $W^{k,p}(\mathbb{R}^n)$. If Ω is an open subset of $\mathbb{R}^n$, then $C_c^\infty(\Omega)$ is dense in $W_0^{k,p}(\Omega)$, and if Ω is regular, then $C^\infty(\overline{\Omega})$ is dense in $W^{k,p}(\Omega)$.

Meyers and Serrin proved for general domains that $C^\infty(\Omega) \cap W^{k,p}(\Omega)$ is dense in $W^{k,p}(\Omega)$, but functions in $C^\infty(\Omega)$ need not be smooth up to the boundary.

Among the most important properties of Sobolev spaces are the embedding theorems, which provide information about the integrability or continuity of a function given information about the integrability of its derivatives. To motivate the embedding theorems, we first consider functions $u : \mathbb{R}^n \to \mathbb{R}$, and ask when it is possible to have an estimate of the form

$$\|u\|_{L^q} \leq C\|\nabla u\|_{L^p} \tag{12.18}$$

for a constant C that is independent of u. Here,

$$\nabla u = \left(\frac{\partial u}{\partial x_1}, \frac{\partial u}{\partial x_2}, \ldots, \frac{\partial u}{\partial x_n} \right)$$

is the gradient, or derivative, of u and

$$\|\nabla u\|_{L^p} = \left(\left\| \frac{\partial u}{\partial x_1} \right\|_{L^p}^p + \left\| \frac{\partial u}{\partial x_2} \right\|_{L^p}^p + \ldots + \left\| \frac{\partial u}{\partial x_n} \right\|_{L^p}^p \right)^{1/p}.$$

We also use the notation Du for ∇u.

For $t > 0$, we define the rescaled function

$$u_t(x) = u(tx).$$

A simple calculation shows that

$$\|u_t\|_{L^q} = t^{-n/q} \|u\|_{L^q}, \qquad \|\nabla u_t\|_{L^p} = t^{1-n/p} \|\nabla u\|_{L^p}. \tag{12.19}$$

These norms must scale according to the same exponent if the estimate in (12.18) is to hold. This occurs if and only if $p < n$ and $q = p^*$, where

$$p^* = \frac{np}{n-p}.$$

This equation may also be written as

$$\frac{1}{p^*} = \frac{1}{p} - \frac{1}{n}. \tag{12.20}$$

We call p^* the *Sobolev conjugate* of p. The inequality in (12.18) does in fact hold for every $u \in C_c^\infty(\mathbb{R}^n)$ when $q = p^*$, and it follows by a density argument that every function in $W^{1,p}(\mathbb{R}^n)$ belongs to $L^{p^*}(\mathbb{R}^n)$ when $p < n$.

The inclusion $W^{1,p}(\mathbb{R}^n) \subset L^{p^*}(\mathbb{R}^n)$ is equivalent to the existence of an embedding

$$J : W^{1,p}(\mathbb{R}^n) \to L^{p^*}(\mathbb{R}^n),$$

where $Jf = f$. The estimate (12.18) implies the continuity of J.

If $\Omega \subset \mathbb{R}^n$ is a bounded domain, then we have $W^{1,p}(\Omega) \subset L^{p^*}(\Omega)$ and $L^{p^*}(\Omega) \subset L^q(\Omega)$ for $1 \leq q \leq p^*$. Thus, there is a continuous embedding $J : W^{1,p}(\Omega) \to L^q(\Omega)$. Summarizing these results, we get the following theorem.

Theorem 12.70 (Embedding) Suppose that Ω is a regular, bounded, open set in $\mathbb{R}^n$, $p < n$, and p^* is the Sobolev conjugate of p, defined in (12.20).

(a) If $u \in W^{1,p}(\mathbb{R}^n)$, then $u \in L^{p^*}(\mathbb{R}^n)$. There is a constant $C = C(p, n)$ such that

$$\|u\|_{L^{p^*}(\mathbb{R}^n)} \leq C \|\nabla u\|_{L^p(\mathbb{R}^n)}.$$

(b) If $u \in W^{1,p}(\Omega)$ and $1 \le q \le p^*$, then $u \in L^q(\Omega)$. There is a constant $C = C(p, q, \Omega)$ such that

$$\|u\|_{L^q(\Omega)} \le C\|u\|_{W^{1,p}(\Omega)}.$$

To prove this theorem, one uses the Hölder inequality to show that the estimate holds for test functions. The result follows for arbitrary Sobolev functions by the density of test functions in Sobolev spaces (see Adams [1] for complete proofs).

The above embedding theorem applies if $p < n$. If $p > n$, then functions in $W^{1,p}$ are continuous and one can estimate their uniform norm in terms of their $W^{1,p}$-norm.

Theorem 12.71 (Embedding) Suppose that $n < p < \infty$, and Ω is a regular bounded open subset of $\mathbb{R}^n$.

(a) If $u \in W^{1,p}(\mathbb{R}^n)$, then $u \in C_0(\mathbb{R}^n)$. There exists a constant $C = C(p, n)$ such that

$$\|u\|_{L^\infty(\mathbb{R}^n)} \le C\|\nabla u\|_{L^p(\mathbb{R}^n)}.$$

(b) If $u \in W^{1,p}(\Omega)$ then $u \in C(\overline{\Omega})$ and there exists a constant $C = C(p, \Omega)$ such that

$$\|u\|_{L^\infty(\Omega)} \le C\|u\|_{W^{1,p}(\Omega)}.$$

A more refined version of this embedding theorem states that the functions are *Hölder continuous*.

Definition 12.72 A function $u : \Omega \to \mathbb{R}$ is *Hölder continuous* in the open set Ω, with exponent $0 < r \le 1$, if

$$\sup_{\substack{x,y\in\Omega \\ x\ne y}} \frac{|u(x) - u(y)|}{|x - y|^r} < \infty.$$

If u is Hölder continuous with exponent $r = 1$, then u is Lipschitz continuous. Any Hölder continuous function is continuous, but not conversely. The Banach space $C^{0,r}(\overline{\Omega})$ consists of all bounded Hölder continuous functions in Ω with the norm

$$\|u\|_{C^{0,r}(\overline{\Omega})} = \sup_{x\in\Omega} |u(x)| + \sup_{\substack{x,y\in\Omega \\ x\ne y}} \frac{|u(x) - u(y)|}{|x - y|^r}.$$

For each positive integer k and $0 < r \le 1$, we define $C^{k,r}(\overline{\Omega})$ to be the space of functions that are k times continuously differentiable in Ω, with uniformly continuous derivatives whose kth-order derivatives are Hölder continuous with exponent

r. This space is a Banach space with the norm

$$\|u\|_{C^{k,r}(\overline{\Omega})} = \max_{0 \le |\alpha| \le k} \left\{ \sup_{x \in \Omega} |\partial^\alpha u(x)| \right\} + \max_{|\alpha|=k} \sup_{\substack{x,y \in \Omega \\ x \ne y}} \frac{|\partial^\alpha u(x) - \partial^\alpha u(y)|}{|x - y|^r}.$$

Theorem 12.73 (Morrey) Suppose that $n < p < \infty$. Let

$$r = 1 - \frac{n}{p}.$$

(a) If $u \in W^{1,p}(\mathbb{R}^n)$, then $u \in C^{0,r}(\mathbb{R}^n)$ and there exists a constant $C = C(p, n)$ such that

$$\|u\|_{C^{0,r}(\mathbb{R}^n)} \le C \|\nabla u\|_{L^p(\mathbb{R}^n)}.$$

(b) If $u \in W^{1,p}(\Omega)$, then $u \in C^{0,r}(\Omega)$ and there exists a constant $C = C(p, \Omega)$ such that

$$\|u\|_{C^{0,r}(\Omega)} \le C \|u\|_{W^{1,p}(\Omega)}.$$

Spaces of continuous functions form an algebra with respect to the pointwise product, since the pointwise product of continuous functions is continuous, but the L^p-spaces do not form an algebra; for example, the product of two L^2-functions belongs to L^1, but not in general to L^2. The Sobolev spaces form an algebra when they consist of continuous functions.

Theorem 12.74 Suppose that Ω is an open subset of $\mathbb{R}^n$, including the possibility $\Omega = \mathbb{R}^n$. If $kp > n$, then $W^{k,p}(\Omega)$ is an algebra and there is a constant C such that

$$\|uv\|_{W^{k,p}} \le C \|u\|_{W^{k,p}} \|v\|_{W^{k,p}} \qquad \text{for all } u, v \in W^{k,p}(\Omega).$$

It is a general principle that a set of functions whose derivatives are uniformly bounded is compact. The Sobolev-space version of this principle is the Rellich-Kondrachov theorem which states that $W^{k,p}(\Omega)$ is compactly embedded in $L^q(\Omega)$ for $q < p^*$. The boundedness of the domain Ω and the condition that q is strictly less than the Sobolev conjugate p^* of p are both essential for compactness. In the critical case, $q = p^*$, the embedding is continuous but not compact.

Theorem 12.75 (Rellich-Kondrachov) Let Ω be a regular bounded domain in $\mathbb{R}^n$.

(a) Suppose that $1 \le p < n$ and $1 \le q < p^*$. Then bounded sets in $W^{1,p}(\Omega)$ are precompact in $L^q(\Omega)$.

(b) Suppose that $p > n$. Then bounded sets in $W^{1,p}(\Omega)$ are precompact in $C(\overline{\Omega})$.

In particular, suppose that (u_k) is a sequence of functions in $W^{1,p}(\Omega)$ such that

$$\|u_k\|_{W^{1,p}} \leq C$$

for a constant C that is independent of k. If $p < n$ and $1 \leq q < p^*$, then there is a subsequence of (u_k) that converges strongly in $L^q(\Omega)$. If $p > n$, then there is a uniformly convergent subsequence.

If $p > n$ and $0 < r < 1 - n/p$, then the embedding of $W^{1,p}(\Omega)$ into $C^{0,r}(\overline{\Omega})$ is compact. General compactness theorems follow by repeated application of this result. For example, if $kp < n$ then $W^{k,p}(\Omega)$ is compactly embedded in $L^q(\Omega)$ for any $1 \leq q < np/(n - kp)$, while if $kp > n$ then $W^{k,p}(\Omega)$ is compactly embedded in $C(\overline{\Omega})$.

There is no sensible way to assign boundary values $u|_{\partial\Omega}$ to a general function $u \in L^p(\Omega)$. Functions in L^p are defined only pointwise-a.e., and the boundary $\partial\Omega$ of a regular domain has measure zero. The situation is different for Sobolev functions. If $u \in W^{k,p}(\Omega)$, then one can assign boundary values to the derivatives of u of order less than or equal to $k - 1/p$. It is not possible to define boundary values of kth order derivatives, however, since they are just L^p functions.

Theorem 12.76 (Trace) There is a surjective bounded linear operator

$$\gamma : W^{1,p}(\Omega) \to W^{1-1/p,p}(\partial\Omega)$$

such that

$$\gamma u = u|_{\partial\Omega} \qquad \text{if } u \in W^{1,p}(\Omega) \cap C(\overline{\Omega}).$$

There is a "loss of $1/p$ derivatives" in restricting a Sobolev function to the boundary. For example, the boundary values of a function in $H^1(\Omega)$ belong to $H^{1/2}(\partial\Omega)$. Conversely given an element of $H^{1/2}(\partial\Omega)$, there is a function in $H^1(\Omega)$ which takes those boundary values.

Our last result is the Poincaré inequality, which has many variants. The common theme is that, after removing nonzero constant functions, one can estimate the L^p-norm of a function in terms of the L^p-norm of its derivative.

Theorem 12.77 (Poincaré) Suppose that Ω is a bounded domain. Then there is a constant C such that

$$\|u\|_{L^p} \leq C\|\nabla u\|_{L^p}$$

for every $u \in W_0^{1,p}(\Omega)$.

More generally, this estimate holds if Ω is bounded in one direction. The Poincaré estimate is false for nonzero constant functions, so the assumption that $u \in W_0^{1,p}$, rather than $u \in W^{1,p}$, is essential. A useful consequence of this estimate is that $\|\nabla u\|_{L^p}$ provides an equivalent norm on $W_0^{1,p}(\Omega)$. When $p = 2$, it follows that we can use

$$\langle u, v \rangle = \int_\Omega \nabla u(x) \cdot \nabla v(x)\, dx$$

as an inner product on $H_0^1(\Omega)$. Another Poincaré inequality is the following.

Theorem 12.78 (Poincaré) Suppose that Ω is a smooth connected bounded domain. There exists a constant C such that

$$\|u - \langle u \rangle\|_{L^p} \le C\|\nabla u\|_{L^p} \tag{12.21}$$

for every $u \in W^{1,p}(\Omega)$, where $\langle u \rangle$ is the mean of u over Ω,

$$\langle u \rangle = \frac{1}{|\Omega|} \int_\Omega u(x)\, dx,$$

and $|\Omega|$ is the volume of Ω.

12.11   Laplace's equation

The Dirichlet problem for the Laplacian is

$$-\Delta u = f \qquad x \in \Omega, \tag{12.22}$$
$$u(x) = 0 \qquad x \in \partial\Omega.$$

Here, $f : \Omega \to \mathbb{R}$ is a given function (or distribution) and Ω is a smooth bounded open set in $\mathbb{R}^n$. We assume homogeneous boundary conditions for simplicity; nonhomogeneous boundary conditions may be transferred to the PDE in the usual way.

To formulate any PDE problem in a precise way, we have to specify what function space solutions should belong to. We also have to specify how the derivatives are defined and in what sense the solution satisfies the boundary conditions and any other side conditions. There is often a great deal of choice in how this is done. A *classical solution* of (12.22) is a twice-continuously differentiable function u that satisfies the PDE pointwise, wheras a *weak solution* satisfies it in a distributional sense.

To motivate the definition of a weak solution, we suppose that u is a smooth solution. Let $\varphi \in C_c^\infty(\Omega)$ be any test function. Then multiplication of (12.22) by φ and an integration by parts imply that

$$\int_\Omega \nabla u(x) \cdot \nabla \varphi(x)\, dx = \int_\Omega f(x)\varphi(x)\, dx. \tag{12.23}$$

Conversely, if u is a smooth function that vanishes on $\partial\Omega$ and satisfies (12.23) for all test functions φ, then u is a classical solution of the original boundary value problem.

Let us require that the solution u and the test function φ belong to the same space. Then ∇u and $\nabla\varphi$ must both be square-integrable, so it is natural to look for solutions in the space $H_0^1(\Omega)$. Since $\varphi \in H_0^1(\Omega)$, we can make sense of the right-hand side of (12.23) provided that $f \in H^{-1}(\Omega)$. This leads to the following definition.

Definition 12.79 Given a distribution $f \in H^{-1}(\Omega)$, we say that u is a *weak solution* of (12.22) if $u \in H_0^1(\Omega)$ and

$$\int_\Omega \nabla u \cdot \nabla\varphi\, dx = \langle f, \varphi\rangle \qquad \text{for every } \varphi \in H_0^1(\Omega),$$

where $\langle \cdot, \cdot\rangle$ denotes the duality pairing between $H^{-1}(\Omega)$ and $H_0^1(\Omega)$.

If we define a quadratic functional $I : H_0^1(\Omega) \to \mathbb{R}$ by

$$I(u) = \frac{1}{2}\int_\Omega |\nabla u|^2\, dx - \langle f, u\rangle,$$

then, as we will see in Section 13.9, a function u that minimizes I is a weak solution of (12.22). As a result of this connection, the present approach to the study of the Laplacian is often called the variational method. The boundary condition $u = 0$ on $\partial\Omega$ is replaced by the condition that $u \in H_0^1(\Omega)$. The precise sense in which weak solutions satisfy boundary conditions or initial conditions often requires careful attention. Definition 12.79 is not the most general definition of weak solutions. For example, we could consider distributional solutions of (12.22) when $f \notin H^{-1}(\Omega)$. The definition given is the natural one for the following existence theorem.

Theorem 12.80 There is a unique weak solution $u \in H_0^1(\Omega)$ of (12.22) for every $f \in H^{-1}(\Omega)$. There is a constant $C = C(\Omega)$ such that

$$\|u\|_{H^1} \le C\|f\|_{H^{-1}} \qquad \text{for all } f \in H^{-1}(\Omega).$$

Proof. By the Poincaré inequality, we can use

$$(u, v) = \int_\Omega \nabla u \cdot \nabla v\, dx$$

as an inner product on $H_0^1(\Omega)$. Since $f \in H^{-1}(\Omega) = H_0^1(\Omega)^*$, and $H_0^1(\Omega)$ is a Hilbert space, the Riesz representation theorem implies that there is a unique $u \in H_0^1(\Omega)$ such that

$$(u, \varphi) = \langle f, \varphi\rangle$$

for every $\varphi \in H_0^1(\Omega)$. This function u is the unique weak solution of (12.22). Moreover, we have $\|u\|_{H_0^1} = \|f\|_{H^{-1}}$. $\qquad\square$

This theorem implies that

$$-\Delta : H_0^1(\Omega) \to H^{-1}(\Omega)$$

is a Hilbert space isomorphism; in fact, it is the isomorphism between $H_0^1(\Omega)$ and its dual space $H^{-1}(\Omega)$, in which the dual space is represented concretely as a space of distributions.

The proof of Theorem 12.80 gives a solution u of (12.22) that belongs to $H_0^1(\Omega)$. This is the best regularity one can hope for in the case of a general right hand side $f \in H^{-1}$. If, however, $f \in H^k$ is smooth, then elliptic regularity theory shows that the solution $u \in H^{k+2}$ and that

$$\|u\|_{H^{k+2}} \leq C\|f\|_{H^k}.$$

This gain of derivatives is typical of elliptic equations. One can estimate the L^2-norm of all second derivatives of u in terms of the L^2-norm of the single combination of second derivatives Δu. If $f \in H^k(\Omega)$ with $k > n/2$, then it follows from the Sobolev embedding theorem that $u \in H^{k+2}(\Omega) \subset C^2(\overline{\Omega})$, so u is a classical solution, and if $f \in C^\infty(\overline{\Omega})$, then $u \in C^\infty(\overline{\Omega})$.

If $f \in C(\overline{\Omega})$, then it is not necessarily true that $u \in C^2(\overline{\Omega})$. If, however, $f \in C^{k,r}(\overline{\Omega})$ is Hölder continuous, where $0 < r \leq 1$, then there is a unique Hölder continuous solution $u \in C^{k+2,r}(\overline{\Omega})$, and one can estimate the Hölder norms of the second derivatives of u in terms of the Hölder norm of f. Analogous existence, uniqueness, and regularity results hold in $L^p(\Omega)$ for $1 < p < \infty$, but not for $p = 1$ or $p = \infty$.

The idea in Theorem 12.80 of using the Riesz representation theorem to prove the existence and uniqueness of a weak solution applies to more general linear PDEs. Consider a linear equation that can be written in the abstract form

$$Au = f, \tag{12.24}$$

where $A : \mathcal{H} \to \mathcal{H}^*$ is a bounded linear operator from a Hilbert space $\mathcal{H}$ to its dual space $\mathcal{H}^*$, and $f \in \mathcal{H}^*$. In the case of Laplace's equation, we had $A = -\Delta$, $\mathcal{H} = H_0^1(\Omega)$, and $\mathcal{H}^* = H^{-1}(\Omega)$. Evaluation of (12.24) on a test function $v \in \mathcal{H}$, gives an equivalent weak formulation:

$$a(u, v) = \langle f, v \rangle \qquad \text{for all } v \in \mathcal{H},$$

where $a : \mathcal{H} \times \mathcal{H} \to \mathbb{R}$ is defined by

$$a(u, v) = \langle Au, v \rangle.$$

This bilinear form a is called the *Dirichlet form* associated with A. In the case of Laplace's equation, we have

$$a(u, v) = \int_\Omega \nabla u \cdot \nabla v \, dx.$$

The Dirichlet form of a Sturm-Liouville ordinary differential operator is given in (10.38). If a is a symmetric, positive definite, sesquilinear form, and there exists a constant $\alpha > 0$ such that

$$a(u, u) \geq \alpha \|u\|^2 \qquad \text{for all } u \in \mathcal{H},$$

then the *energy norm*

$$\|u\|_A = \sqrt{a(u, u)}$$

is equivalent to the original norm on $\mathcal{H}$, and the Riesz representation theorem implies the existence and uniqueness of a weak solution u of (12.24) for every $f \in \mathcal{H}^*$.

If a is not symmetric, then it does not define an inner product on $\mathcal{H}$, and the Riesz representation cannot be used directly to establish the existence of a weak solution. Nevertheless, a similar result, called the *Lax-Milgram lemma*, still applies. The proof is outlined in Exercise 12.23.

Theorem 12.81 (Lax-Milgram) Suppose that $a : \mathcal{H} \times \mathcal{H} \to \mathbb{R}$ is a sesquilinear form on a Hilbert space $\mathcal{H}$, and there are constants $\alpha > 0$, $\beta > 0$ such that

$$\alpha \|x\|^2 \leq |a(x, x)|, \qquad |a(x, y)| \leq \beta \|x\| \|y\|,$$

for all $x, y \in \mathcal{H}$. Then for every bounded linear functional $F : \mathcal{H} \to \mathbb{C}$ there is a unique element $x \in \mathcal{H}$ such that

$$a(x, y) = F(y) \qquad \text{for all } y \in \mathcal{H}.$$

Finally, we mention that (12.23) is a useful starting point for numerical methods of solving Laplace's equation, especially the *finite element method*.

12.12   References

Jones [25] gives a clear and well-motivated introduction to the Lebesgue integral. For a detailed account of measure theory, see Folland [12]. A concise, concrete introduction to the subject, including a discussion of L^p-spaces, is in Lieb and Loss [32]. For a detailed account of Sobolev spaces, see Adams [1]. For Sobolev spaces and elliptic PDEs, see Evans [11] and Gilbarg and Trudinger [15]. An extensive discussion of Sobolev spaces, variational problems, and related analysis of linear PDEs is contained in Dautry and Lions [7].

12.13   Exercises

Exercise 12.1 Prove that the Borel σ-algebra on $\mathbb{R}$ is generated by the following families of sets:

$$\{(a,b] \mid a < b\}, \qquad \{[a,b) \mid a < b\}, \qquad \{[a,b] \mid a < b\}, \qquad \{(a,\infty) \mid a \in \mathbb{R}\}.$$

Exercise 12.2 Let $\mathcal{A}$ be a σ-algebra of subsets of Ω, and suppose μ is a measure on Ω. Prove the following properties:

    (a) If $A, B \in \mathcal{A}$, then $A \setminus B \in \mathcal{A}$;
    (b) If $A, B \in \mathcal{A}$, and $A \subset B$, then $\mu(A) \leq \mu(B)$;
    (c) If $A, B \in \mathcal{A}$, then $\mu(A \cup B) \leq \mu(A) + \mu(B)$.

Exercise 12.3 If (A_i) is an increasing sequence of measurable sets, meaning that

$$A_1 \subset A_2 \subset \ldots \subset A_i \subset A_{i+1} \subset \ldots,$$

then prove that

$$\mu\left(\bigcup_{i=1}^{\infty} A_i\right) = \lim_{i \to \infty} \mu(A_i).$$

If (A_i) is a decreasing sequence of measurable sets, meaning that

$$A_1 \supset A_2 \supset \ldots \supset A_i \supset A_{i+1} \supset \ldots,$$

and $\mu(A_1) < \infty$, prove that

$$\mu\left(\bigcap_{i=1}^{\infty} A_i\right) = \lim_{i \to \infty} \mu(A_i).$$

Give a counterexample to show that this result need not be true if $\mu(A_i)$ is infinite for every i.

Exercise 12.4 Give an example of a monotonic decreasing sequence of nonnegative functions converging pointwise to a function f such that the equality in Theorem 12.33 does not hold.

Exercise 12.5 Check that the counting measure defined in Example 12.6 is a measure.

Exercise 12.6 Use the dominated convergence theorem to prove Corollary 12.36 for differentiation under an integral sign.

Exercise 12.7 Prove that $f \sim g$ if and only if $f = g$ pointwise-a.e. defines an equivalence relation on the space of all measurable functions.

Exercise 12.8 Let $f_n : X \to \mathbb{C}$ be a sequence of measurable functions converging to f pointwise-a.e. Suppose there exists $g \in L^p(X)$ such that $|f_n| \le g$ a.e. Prove that $f_n \to f$ in the L^p-norm.

Exercise 12.9 Let $\varphi : \mathbb{R} \to \mathbb{R}$ be a convex function. Prove the following properties:

(a) for all $x \in \mathbb{R}$ the left- and right-derivatives, $\varphi'_-(x)$ and $\varphi'_+(x)$ exist and satisfy

$$\varphi'_-(x) \le \varphi'_+(x);$$

(b) φ is continuous on $\mathbb{R}$;

(c) for all $x \in \mathbb{R}$, there exists a constant $c \in \mathbb{R}$ such that

$$\varphi(y) \ge \varphi(x) + c(y - x) \qquad \text{for all } y \in \mathbb{R}. \tag{12.25}$$

The graph of the function $y \mapsto c(y - x)$ satisfying (12.25) is called a *support line* of φ at x.

Exercise 12.10 If $x, y \ge 0$ and $\epsilon > 0$ is any positive number, show that

$$xy \le \frac{\epsilon}{2}x^2 + \frac{1}{2\epsilon}y^2.$$

This estimate is sometimes called the *Peter-Paul inequality*.

Exercise 12.11 Let $p_1, p_2, \ldots, p_n$ be positive numbers whose sum is equal to one. Prove that for any nonnegative numbers $x_1, x_2, \ldots, x_n$, we have the inequality

$$x_1^{p_1} x_2^{p_2} \ldots x_n^{p_n} \le p_1 x_1 + p_2 x_2 + \ldots + p_n x_n.$$

Exercise 12.12 Prove the following generalization of Hölder's inequality: if $1 \le p_i \ge \infty$, where $i = 1, \ldots, n$, satisfy

$$\sum_{i=1}^{n} \frac{1}{p_i} = 1,$$

and $f_i \in L^{p_i}(X, \mu)$, then $f_1 \cdots f_n \in L^1(X, \mu)$ and

$$\left| \int f_1 \cdots f_n \, d\mu \right| \le \|f_1\|_{p_1} \cdots \|f_n\|_{p_n}.$$

Exercise 12.13 Prove that $L^\infty([0,1])$ is not separable, and that $C([0,1])$ is not dense $L^\infty([0,1])$.

Exercise 12.14 Prove that for any pair of distinct exponents $1 \le p, q \le \infty$, we have $L^p(\mathbb{R}) \not\subset L^q(\mathbb{R})$. Show that the function

$$f(x) = \frac{1}{|x|^{1/2}\sqrt{1 + \log^2|x|}}$$

belongs to $L^2(\mathbb{R})$, but not to $L^p(\mathbb{R})$ for any $1 \le p \le \infty$ that is different from 2.

Exercise 12.15 If $f \in L^p(\mathbb{R}^n) \cap L^q(\mathbb{R}^n)$, where $p < q$, prove that $f \in L^r(\mathbb{R}^n)$ for any $p < r < q$, and show that

$$\|f\|_r \leq (\|f\|_p)^{\frac{1/r-1/q}{1/p-1/q}} (\|f\|_q)^{\frac{1/p-1/r}{1/p-1/q}}.$$

This result is one of the simplest examples of an *interpolation inequality*.

Exercise 12.16 A function $f : \mathbb{R}^n \to \mathbb{C}$ is said to be *L^p-continuous* if $\tau_h f \to f$ in $L^p(\mathbb{R}^n)$ as $h \to 0$ in $\mathbb{R}^n$, where $\tau_h f(x) = f(x - h)$ is the translation of f by h. Prove that, if $1 \leq p < \infty$, every $f \in L^p(\mathbb{R}^n)$ is L^p-continuous. Give a counter-example to show that this result is not true when $p = \infty$.
HINT. Approximate an L^p-function by a C_c-function.

Exercise 12.17 Prove that the unit ball in $L^p([0,1])$, where $1 \leq p \leq \infty$, is not strongly compact.

Exercise 12.18 Give an example of a bounded sequence in $L^1([0,1])$ that does not have a weakly convergent subsequence. Why doesn't this contradict the Banach-Alaoglu theorem?

Exercise 12.19 Let $1 \leq p \leq \infty$. Prove that if $f \in L^p(\mathbb{R})$, and its weak derivative is identically zero, then f is a constant function.

Exercise 12.20 Which of the following functions belongs to $H^1([-1,1])$?

  (a) $f(x) = |x|$;
  (b) $g(x) = \operatorname{sgn} x$;
  (c) $h(x) = \sum_{n=1}^{\infty}(1/n)^{3/2} \sin nx$.

Exercise 12.21 Prove a Poincaré inequality of the form (12.21) for periodic functions defined on the n-dimensional torus $\mathbb{T}^n$.

Exercise 12.22 Use the Riesz representation theorem to prove that there is a unique weak solution $u \in H^1(\mathbb{R}^n)$ of the equation

$$-\Delta u + u = f$$

for every $f \in H^{-1}(\mathbb{R}^n)$. Show that $(-\Delta + I) : H^1(\mathbb{R}^n) \to H^{-1}(\mathbb{R}^n)$ is an isomorphism. Is $-\Delta : H^1(\mathbb{R}^n) \to H^{-1}(\mathbb{R}^n)$ an isomorphism?

Exercise 12.23 Suppose that $a : \mathcal{H} \times \mathcal{H} \to \mathbb{R}$ is a sesquilinear form on a Hilbert space $\mathcal{H}$ that satisfies the hypotheses of the Lax-Milgram lemma in Theorem 12.81.

  (a) Show that there is a bounded linear map $J : \mathcal{H} \to \mathcal{H}$ such that Jx is the unique element satisfying

$$a(x,y) = \langle Jx, y \rangle \qquad \text{for all } y \in \mathcal{H},$$

  where $\langle \cdot, \cdot \rangle$ denote the inner product on $\mathcal{H}$.

(b) Show that $\alpha\|x\| \leq \|Jx\|$. Deduce that J is one-to-one and onto.

HINT. Show that ran J is closed and use the projection theorem to show that J is onto.

(c) Show that there is a unique solution x of the equation

$$\langle Jx, y \rangle = F(y) \qquad \text{for all } y \in \mathcal{H},$$

and prove the Lax-Milgram lemma.

Chapter 13

Differential Calculus and Variational Methods

In many of the preceding chapters, we studied linear spaces of functions and linear operators acting on them. This theory provides a natural framework for the majority of linear equations that arise in applied mathematics. Many problems lead to nonlinear equations that may be formulated in terms of nonlinear maps acting on Banach spaces. There is no general theory of nonlinear maps that is as powerful as, for example, the spectral theory of linear operators. If, however, we can approximate a nonlinear map locally by a linear map, then we can reduce various questions about nonlinear problems to ones about associated linear problems. The linearization of nonlinear maps is one of the most useful and widely applicable methods for the study of nonlinear problems, and accounts for the importance of linear analysis in nonlinear settings.

13.1 Linearization

Linearization is closely connected with differentiation: the central idea of differentiation is the local approximation of a nonlinear map by a linear map. A map $f : X \to Y$ between Banach spaces X, Y is *differentiable* at $x \in X$ if there is a bounded linear map $f'(x) : X \to Y$ such that

$$f(x + \epsilon h) = f(x) + \epsilon f'(x)h + o(\epsilon) \tag{13.1}$$

as $\epsilon \to 0$ for every $h \in X$. Here, $o(\epsilon)$ stands for a term that approaches zero faster than ϵ as $\epsilon \to 0$. We call the linear map $f'(x)$ the *derivative* of f at x.

In this section, we describe some problems where the linearization of a nonlinear map is useful. First, suppose that we want to find solutions $x \in X$ of a nonlinear equation of the form

$$f(x) = y, \tag{13.2}$$

where $f : X \to Y$ is a map between Banach spaces X, Y, and $y \in Y$ is given. If we know a solution $x_0 \in X$ for a particular $y_0 \in Y$, meaning that

$$f(x_0) = y_0, \tag{13.3}$$

then we can try to solve (13.2) when y is close to y_0 by looking for a solution x that is close to x_0. We write y as

$$y = y_0 + \epsilon y_1, \tag{13.4}$$

where ϵ is a small real or complex parameter. If f is differentiable at x_0, we may look for a solution $x(\epsilon)$ of the form

$$x(\epsilon) = x_0 + \epsilon x_1 + o(\epsilon). \tag{13.5}$$

Then, using (13.1), (13.4), and (13.5) in (13.2), we get

$$f(x_0) + \epsilon f'(x_0)x_1 = y_0 + \epsilon y_1 + o(\epsilon).$$

From (13.3), the leading order terms in ϵ are equal. Cancelling the leading order terms, dividing the equation by ϵ, and letting $\epsilon \to 0$, we find that x_1 and y_1 satisfy

$$f'(x_0)x_1 = y_1. \tag{13.6}$$

If $f'(x_0)$ is nonsingular, then we can solve (13.6) for x_1. It is then reasonable to expect that we can also solve the nonlinear equation (13.2) when y is sufficiently close to y_0. The *inverse function theorem* states that this expectation is correct, provided that f is continuously differentiable at x_0.

A second application of linearization concerns the stability of solutions of a nonlinear evolution equation of the form

$$x_t = f(x), \qquad x(0) = x_0, \tag{13.7}$$

where the solution $x : [0, \infty) \to X$ takes values in a Banach space X, and $f : X \to X$ is a vector field on X. For some equations, such as partial differential equations, the vector field f may only be defined on a dense subspace of X. A point $\overline{x} \in X$ is an *equilibrium solution*, or *stationary solution*, or *fixed point* of (13.7) if

$$f(\overline{x}) = 0.$$

In that case, the constant function $x(t) = \overline{x}$ is a solution of (13.7). Even though an equilibrium solution is an exact solution of (13.7), it may not be observed in practice if it is *unstable*, meaning that a small perturbation of the equilibrium grows in time.

To study the effect of a small perturbation on the equilibrium state, we look for solutions of (13.7) of the form $x(t) = \overline{x} + \epsilon y(t)$, where ϵ is small, and linearize the right hand side of (13.7) about $x = \overline{x}$. Neglecting higher order terms, we find that y satisfies the linear evolution equation

$$y_t = f'(\overline{x})\,y. \tag{13.8}$$

It is reasonable to expect that, under suitable conditions on f, the study of solutions of (13.8) will provide information about the stability of the equilibrium solution $\bar{x}$ of (13.7). Similar ideas may be used to study the stability of time-dependent solutions — for example, time-periodic solutions — of (13.7), but we will not describe them here.

As a third example of linearization, we consider the minimization of a functional $I : X \to \mathbb{R}$. We suppose that I is bounded from below and look for a minimizer $\bar{x} \in X$ such that

$$I(\bar{x}) = \inf_{x \in X} I(x). \tag{13.9}$$

We have already discussed the *direct method* for solving variational problems, in which we choose a minimizing sequence and attempt to show that it has a subsequence that converges to a minimizer. An alternative approach, called the *indirect method*, is to look for *critical points* of I, which are solutions of the equation

$$I'(x) = 0. \tag{13.10}$$

If I is differentiable, then any minimizer that is an interior point of the domain of I is a critical point of I. A point x at which $I'(x) \neq 0$ is called a *regular point*. Conversely, if an equation $f(x) = 0$ can be written in the form (13.10) for some functional I, then we may be able to use the associated variational principle (13.9) to construct solutions. When applicable, variational methods are one of the most powerful methods for analyzing equations.

The above examples illustrate the need for a notion of the derivative of a map between Banach spaces. There are many different definitions of the derivative. The most important is the Fréchet derivative, which generalizes the notion of the derivative or differential of a vector-valued function of several variables. We will also introduce the Gâteaux derivative, which generalizes the notion of the directional derivative.

13.2   Vector-valued integrals

In this section, we define the derivative and the Riemann integral of a vector-valued function of a real variable, and prove some of their basic properties. The definitions are essentially identical to the ones in elementary calculus for a real-valued function. We will need these tools, especially the estimate in Theorem 13.4, to prove results about differentiable functions, such as the inverse function theorem.

A vector-valued function of a real variable is a mapping from a subset of the real numbers into a Banach space, which could be finite or infinite-dimensional. Geometrically, such a function defines a parametrized curve in the Banach space. The derivative of the function is the tangent, or velocity, vector of the curve.

Definition 13.1 A function $f : (a, b) \to X$ from an open interval (a, b) into a Banach space X is *differentiable* at $a < t < b$, with derivative $f'(t) \in X$, if the following limit exists in X:

$$f'(t) = \lim_{h \to 0} \frac{f(t + h) - f(t)}{h}.$$

The function f is differentiable in (a, b) if it is differentiable at each point in (a, b), and continuously differentiable in (a, b) if $f' : (a, b) \to X$ is continuous.

Next, we define the *Riemann integral* of a vector-valued function $f : [a, b] \to X$ defined on a closed, bounded interval $[a, b]$. We say that $\varphi : [a, b] \to X$ is a *step function* if there is a partition $a = t_0 < t_1 < \ldots < t_n = b$ of the interval $[a, b]$, and constant vectors $c_i \in X$, with $i = 1, \ldots, n$, such that

$$\varphi(t) = c_i \qquad \text{for } t_{i-1} < t < t_i. \tag{13.11}$$

We denote the space of step functions from $[a, b]$ into X by $S([a, b])$, and regard it as a subspace of the Banach space $B([a, b])$ of bounded functions $f : [a, b] \to X$ equipped with the sup-norm

$$\|f\|_\infty = \sup_{a \le t \le b} \|f(t)\|.$$

We define a linear map

$$\Lambda : S([a, b]) \to X$$

that takes a step function φ, defined in (13.11), to its Riemann integral by

$$\Lambda\varphi = \sum_{i=1}^{n} (t_i - t_{i-1}) \, c_i.$$

Thus, the integral of a step function is a finite linear combination of the values c_i of the step function. The vectors c_i need not be parallel, and the integral need not be parallel to any of the c_i's, but it does lie in the linear subspace spanned by $\{c_1, \ldots, c_n\}$. The map Λ is well defined, since its value does not depend on how the step-function is represented. From the triangle inequality, we have

$$\|\Lambda\varphi\| \le \left(\max_{1 \le i \le n} \|c_i\| \right) \sum_{i=1}^{n} (t_i - t_{i-1}) = \|\varphi\|_\infty (b - a),$$

so Λ is bounded. We denote the closure of $S([a, b])$ in $B([a, b])$ by $R([a, b])$. Elements in the space $R([a, b])$ of uniform limits of step functions are sometimes called *regulated functions*. Theorem 5.19 implies that there is a unique bounded linear extension of Λ to $R([a, b])$, which we also denote by $\Lambda : R([a, b]) \to X$. For $f \in R([a, b])$, we call Λf the *Riemann integral* of f, and write it as

$$\Lambda f = \int_a^b f(t) \, dt.$$

The uniform continuity of a continuous function on the compact interval $[a, b]$ implies that $R([a, b])$ contains the space $C([a, b])$ of continuous functions from $[a, b]$ to X, so every continuous function is Riemann integrable (see Exercise 13.11). The space $R([a, b])$ also contains the piecewise continuous functions, which have a finite number of jump discontinuities in $[a, b]$, meaning that the left and right limits of the function exist at its points of discontinuity. In fact, it is possible to show [9] that a function f is in $R([a, b])$ if and only if the left and right hand limits,

$$\lim_{h \to 0^-} f(t + h), \qquad \lim_{h \to 0^+} f(t + h),$$

exist at every point of $[a, b]$ (except, of course, for the left limit at a and the right limit at b). The Riemann integral can be defined on a larger class of functions than the regulated functions, but once one has to deal with functions that are less regular than continuous or piecewise continuous functions, it is preferable to use the Lebesgue integral. One can also define integrals of functions taking values in an infinite-dimensional Banach space for which the Riemann sums, or the integrals of approximating simple functions, converge weakly instead of strongly. We will not consider such integrals in this book.

The estimate

$$\left\| \int_a^b f(t)\, dt \right\| \leq \int_a^b \|f(t)\|\, dt \leq M(b - a), \tag{13.12}$$

where

$$M = \sup_{a \leq t \leq b} \|f(t)\|,$$

follows from the continuity of Λ and the corresponding estimate for step functions. The usual algebraic properties of the Riemann integral also follow by continuity from the corresponding properties for step functions.

Next, we prove that if the derivative of a function is zero, then the function is constant. To do this, we use linear functionals on X to reduce to the real-valued case.

Proposition 13.2 If $f : (a, b) \to X$ is differentiable in (a, b) and $f' = 0$, then f is a constant function.

Proof. Let $\varphi : X \to \mathbb{R}$ be a bounded linear functional on X. We define $f_\varphi : (a, b) \to \mathbb{R}$ by

$$f_\varphi(t) = \varphi\left(f(t)\right).$$

The chain rule (see Theorem 13.8 below) implies that f_φ is differentiable and has zero derivative. The mean value theorem of elementary calculus (see Exercise 1.14) implies that f_φ is constant. Hence, for every $\varphi \in X^*$ and $s, t \in (a, b)$ we have

$$\varphi\left(f(s) - f(t)\right) = f_\varphi(s) - f_\varphi(t) = 0.$$

It follows from the Hahn-Banach theorem and Exercise 5.6 that $f(s) = f(t)$ for all $s, t \in (a, b)$, so f is constant. $\qquad\square$

The fundamental theorem of calculus holds for vector-valued maps.

Theorem 13.3 (Fundamental theorem of calculus) Suppose that X is a Banach space.

(a) If $f : [a, b] \to X$ is continuous, then

$$F(t) = \int_a^t f(s)\, ds$$

is continuously differentiable in (a, b) and $F' = f$.

(b) If f is continuously differentiable in an open interval containing $[a, b]$, then

$$f(b) - f(a) = \int_a^b f'(t)\, dt. \qquad (13.13)$$

Proof. To prove the first part, suppose that $a < t < b$ and h is sufficiently small. Then

$$F(t + h) - F(t) = \int_t^{t+h} f(s)\, ds.$$

It follows that

$$\frac{F(t + h) - F(t)}{h} - f(t) = \frac{1}{h} \int_t^{t+h} [f(s) - f(t)]\, ds.$$

Taking the norm of this equation, and using (13.12), we obtain that

$$\left\| \frac{F(t + h) - F(t)}{h} - f(t) \right\| \leq \sup \{ \|f(s) - f(t)\| \mid t \leq s \leq t + h \} \to 0$$

as $h \to 0$, by the continuity of f. Thus F is differentiable in (a, b), with continuous derivative f.

To prove the second part, suppose that f is continuously differentiable, and define

$$g(t) = \int_a^t f'(s)\, ds.$$

Then, from the first part, we have that g is continuously differentiable and $g' = f'$, so that the derivative of $(f - g)$ is zero. Since $g(a) = 0$, Proposition 13.2 implies that $f(t) - g(t) = f(a)$. Evaluation of this equation at $t = b$ gives (13.13). $\qquad\square$

The mean value theorem for real-valued functions does not hold for vector-valued functions, but the following estimate substitutes for the mean value theorem in many contexts.

Theorem 13.4 (Mean value) If f is continuously differentiable in an open interval that contains the closed, bounded interval $[a, b]$, with values in a Banach space, then

$$\|f(b) - f(a)\| \le M\,(b - a) \quad \text{where} \quad M = \sup_{a \le t \le b} \|f'(t)\|.$$

Proof. Using (13.12) and (13.13), we have

$$\|f(b) - f(a)\| = \left\| \int_a^b f'(t)\,dt \right\| \le \int_a^b \|f'(t)\|\,dt \le M(b - a).$$

$\square$

As an application of vector-valued integrals, we briefly consider the solution of a linear evolution equation

$$x_t = Ax, \qquad x(0) = x_0, \tag{13.14}$$

where $A : X \to X$ is a bounded linear operator on a Banach space X. For example, if $X = \mathbb{R}^n$, then (13.14) is an $n \times n$ system of ODEs. Similar ideas apply to PDEs, where $A : \mathcal{D}(A) \subset X \to X$ is an unbounded linear operator that generates a C_0-semigroup. Let

$$y(\lambda) = \int_0^\infty e^{-\lambda t} x(t)\,ds$$

denote the *Laplace transform* of $x : [0, \infty) \to X$. Then, taking the Laplace transform of (13.14), and integrating by parts, we find that

$$(\lambda I - A)\,y = x_0.$$

Thus, for $\lambda \in \rho(A)$, we have

$$y(\lambda) = R(\lambda)x_0,$$

where $R(\lambda) = (\lambda I - A)^{-1}$ is the resolvent operator of the generator A. The resolvent operator is related to the solution operator $T(t) = e^{At}$ by

$$R(\lambda) = \int_0^\infty e^{-\lambda s} T(s)\,ds,$$

meaning that the solution operator is the inverse Laplace transform of the resolvent. A nonhomogeneous linear evolution equation may be solved in terms of the solution operator of the homogeneous equation (see Exercise 13.12).

13.3 Derivatives of maps on Banach spaces

In this section, we define the derivative of a map between Banach spaces. The dimensions of the Banach spaces play little role in what follows, and a geometric understanding of the derivative is essential for a clear understanding of multivariable calculus on $\mathbb{R}^n$.

In order to generalize the notion of the derivative of a function of a real variable to a function defined on a Banach space, it is important to view the derivative in a slightly different way than is usual in elementary calculus. There, the derivative of a differentiable function $f : (a, b) \to \mathbb{R}$ is typically thought of as another function $f' : (a, b) \to \mathbb{R}$. Instead, we think of the derivative $f'(x)$ of f at a point x as a linear map that approximates f near x. For real-valued functions, this linear map is just multiplication by the value of the derivative at x.

Definition 13.5 A map $f : U \subset X \to Y$ whose domain U is an open subset of a Banach space X and whose range is a Banach space Y is *differentiable* at $x \in U$ if there is a bounded linear map $A : X \to Y$ such that

$$\lim_{h \to 0} \frac{\|f(x + h) - f(x) - Ah\|}{\|h\|} = 0.$$

This definition of the derivative is sometimes called the *Fréchet derivative*, to distinguish it from the directional, or Gâteaux, derivative in Definition 13.9 below. When we refer to the derivative of a function, without other qualifications, we will mean the Fréchet derivative, but there is little consistency in the literature in the usage of the words "derivative," "differential," and "differentiable."

We can restate the definition using the following o-notation. Suppose that

$$r : U \subset X \to Y$$

is a function whose domain U is a neighborhood of the origin in a Banach space X, with values in a Banach space Y. We write

$$r(h) = o(h^n) \qquad \text{as } h \to 0,$$

pronounced r is "small oh" of h^n, if

$$\lim_{h \to 0} \frac{\|r(h)\|}{\|h\|^n} = 0,$$

meaning that $\|r(h)\|$ approaches zero as $h \to 0$ faster than $\|h\|^n$. We also write

$$r(h) = O(h^n) \qquad \text{as } h \to 0,$$

pronounced r is "big oh" of h^n, if there are constants $\delta > 0$ and $C > 0$ such that

$$\|r(h)\| \leq C\|h\|^n \qquad \text{when } \|h\| < \delta.$$

If $f, g : U \subset X \to Y$, we write

$$f(h) = g(h) + o(h^n) \quad \text{if} \quad f(h) - g(h) = o(h^n) \quad \text{as } h \to 0,$$
$$f(h) = g(h) + O(h^n) \quad \text{if} \quad f(h) - g(h) = O(h^n) \quad \text{as } h \to 0.$$

Thus, $o(h^n)$ denotes a term that approaches zero faster than $\|h\|^n$, and $O(h^n)$ denotes a term that is bounded by a constant factor of $\|h\|^n$ near 0.

The function f is differentiable at x if and only if there is a bounded linear map $A : X \to Y$ such that

$$f(x + h) = f(x) + Ah + o(h) \quad \text{as } h \to 0.$$

If such a linear map exists, then it is unique (see Exercise 13.1), and we write it as $A = f'(x)$. If f is differentiable at each point of U, then

$$f' : U \to \mathcal{B}(X, Y)$$

is the map that assigns to each point $x \in U$ the bounded linear map $f'(x) : X \to Y$ that approximates f near x. We say that f is *continuously differentiable* at x if the map f' is continuous at x, where the domain U is equipped with the norm on X and the range $\mathcal{B}(X, Y)$ is equipped with the operator norm. We say that f is continuously differentiable in U if it is continuously differentiable at each point $x \in U$. Other common notations for the derivative are df, Df, and f_x.

Example 13.6 Suppose that $f : U \subset \mathbb{R}^n \to \mathbb{R}^m$. We use coordinates $(x_1, x_2, \ldots, x_n)$ on $\mathbb{R}^n$ and $(y_1, y_2, \ldots, y_m)$ on $\mathbb{R}^m$. Then the component expression for f is

$$
\begin{aligned}
y_1 &= f_1(x_1, x_2, \ldots, x_n), \\
y_2 &= f_2(x_1, x_2, \ldots, x_n), \\
&\vdots \\
y_m &= f_m(x_1, x_2, \ldots, x_n),
\end{aligned}
$$

where $f_i : \mathbb{R}^n \to \mathbb{R}$. We assume that the partial derivatives of the coordinate functions f_i exist and are continuous in U. Then it follows from the remark below Theorem 13.11 that f is differentiable, and the matrix of $f' : U \to \mathcal{B}(\mathbb{R}^n, \mathbb{R}^m)$ is the *Jacobian matrix* of f:

$$
f' = \begin{pmatrix}
\partial f_1/\partial x_1 & \partial f_1/\partial x_2 & \cdots & \partial f_1/\partial x_n \\
\partial f_2/\partial x_1 & \partial f_2/\partial x_2 & \cdots & \partial f_2/\partial x_n \\
\vdots & \vdots & \ddots & \vdots \\
\partial f_m/\partial x_1 & \partial f_m/\partial x_2 & \cdots & \partial f_m/\partial x_n
\end{pmatrix}.
$$

Example 13.7 Let Ω be a smooth domain in $\mathbb{R}^n$, $X = H^1(\Omega)$, and $Y = L^1(\Omega)$. We consider real-valued functions $u : \Omega \to \mathbb{R}$ for simplicity. We will show that the quadratic map $f : X \to Y$ defined by

$$f(u) = \frac{1}{2}|\nabla u|^2 + \frac{1}{2}u^2$$

is differentiable, and find its derivative. We have

$$f(u + h) = f(u) + A(u)h + f(h),$$

where $A(u) : H^1(\Omega) \to L^1(\Omega)$ is defined by

$$A(u)h = \nabla u \cdot \nabla h + uh.$$

From the Cauchy-Schwarz inequality, we have

$$\|A(u)h\|_{L^1} \leq \|u\|_{H^1}\|h\|_{H^1},$$

so $A(u)$ is a bounded linear map. The term $f(h) = o(h)$ as $h \to 0$, since

$$\|f(h)\|_{L^1} = \frac{1}{2}\int_\Omega \left(|\nabla h|^2 + h^2\right)\,dx = \frac{1}{2}\|h\|_{H^1}^2,$$

so

$$\frac{\|f(h)\|_{L^1}}{\|h\|_{H^1}} = \|h\|_{H^1} \to 0 \qquad \text{as } h \to 0.$$

Thus, f is differentiable in $H^1(\Omega)$, and $f'(u) = A(u)$.

One of the most important results concerning derivatives is the *chain rule*. Geometrically, the chain rule states that the linear approximation of the composition of two differentiable maps is the composition of their linear approximations.

Theorem 13.8 (Chain rule) Suppose that X, Y, Z are Banach spaces, and

$$f : U \subset X \to Y, \qquad g : V \subset Y \to Z,$$

where U and V are open subsets of X and Y, respectively. If f is differentiable at $x \in U$ and g is differentiable at $f(x) \in V$, then $g \circ f$ is differentiable at x and

$$(g \circ f)'(x) = g'(f(x))\,f'(x).$$

Proof. By the differentiability of f, we have

$$f(x + h) = f(x) + f'(x)h + r(h),$$

where $r(h)/\|h\| \to 0$ as $h \to 0$. Let $y = f(x)$ and $k = f'(x)h + r(h)$. Then

$$g(y + k) = g(y) + g'(y)k + s(k),$$

where $s(k)/\|k\| \to 0$ as $k \to 0$. Hence,

$$g(f(x + h)) = g(y) + g'(y)\,f'(x)h + t(h),$$

where

$$t(h) = g'(y) r(h) + s(k)$$
$$= g'(f(x)) r(h) + s(f'(x)h + r(h)).$$

Since $f'(x)$ and $g'(y)$ are bounded linear maps, we have

$$\frac{\|t(h)\|}{\|h\|} \leq \frac{\|g'(y)\| \|r(h)\|}{\|h\|} + \frac{\|s(k)\|}{\|k\|} \left(\frac{\|f'(x)\| \|h\| + \|r(h)\|}{\|h\|} \right).$$

It follows that $\|t(h)\|/\|h\| \to 0$ as $h \to 0$, which proves the result. $\qquad\square$

A useful way to compute the derivative of a function is in terms of its *directional derivative*, or *Gâteaux derivative*. For example, the matrix of the derivative of a map on $\mathbb{R}^n$ is the Jacobian matrix of its partial derivatives.

Definition 13.9 Let X and Y be Banach spaces, and $f : U \subset X \to Y$, where U is an open subset of X. The *directional derivative* of f at $x \in U$ in the direction $h \in X$ is given by

$$\delta f(x; h) = \lim_{t \to 0} \frac{f(x + th) - f(x)}{t}. \tag{13.15}$$

If this limit exists for every $h \in X$, and $f'_G(x) : X \to Y$ defined by $f'_G(x)h = \delta f(x; h)$ is a linear map, then we say that f is *Gâteaux differentiable* at x, and we call f'_G the *Gâteaux derivative* of f at x.

The directional derivative may also be written as

$$\delta f(x; h) = \left. \frac{d}{dt} f(x + th) \right|_{t=0}.$$

If f is Fréchet differentiable at x, then it is Gâteaux differentiable at x (Exercise 13.4), and the Fréchet derivative $f'(x)$ is given by

$$f'(x)h = \delta f(x; h).$$

The converse is not true. Even for functions defined on $\mathbb{R}^2$, the existence and linearity of directional derivatives does not imply the differentiability, or even the continuity, of the function (see Exercise 13.5). To give a sufficient condition for the existence of directional derivatives to imply differentiability, we first prove the following immediate consequence of the mean value theorem in Theorem 13.4.

Theorem 13.10 Suppose that $f : U \subset X \to Y$ is a Gâteaux differentiable function from an open subset U of a Banach space X to a Banach space Y. If $x, y \in U$ and the line segment $\{tx + (1 - t)y \mid 0 \leq t \leq 1\}$ connecting x and y is contained in U, then

$$\|f(x) - f(y)\| \leq M\|x - y\| \quad \text{where} \quad M = \sup_{0 \leq t \leq 1} \|f'_G(tx + (1 - t)y)\|.$$

Proof. The definition of the Gâteaux derivative implies that the function

$$h(t) = f\left(tx + (1-t)y\right)$$

is differentiable in an open interval that contains $[0,1]$ and

$$h'(t) = f'_G\left(tx + (1-t)y\right)(x-y).$$

The result then follows from an application of Theorem 13.4 to h. □

Theorem 13.11 Suppose that $f : U \subset X \to Y$ is a Gâteaux differentiable function from an open subset U of a Banach space X to a Banach space Y. If the Gâteaux derivative $f'_G : U \subset X \to \mathcal{B}(X,Y)$ is continuous at $x \in U$, then f is Fréchet differentiable at x and $f'(x) = f'_G(x)$.

Proof. For sufficiently small $\|h\|$, we define

$$r(h) = f(x+h) - f(x) - f'_G(x)h. \tag{13.16}$$

The Gâteaux differentiability of f implies that r is Gâteaux differentiable, and

$$r'_G(h) = f'_G(x+h) - f'_G(x).$$

From Theorem 13.10, we have that

$$\|r(h)\| \le M(h)\|h\|,$$

where

$$M(h) = \sup_{0 \le t \le 1} \|r'_G(th)\|.$$

The continuous Gâteaux differentiability of f implies that $M(h) \to 0$ as $h \to 0$, so $r(h) = o(h)$ as $h \to 0$. Equation (13.16) implies that f is Fréchet differentiable at x, and $f'(x) = f'_G(x)$. □

A more refined argument shows that f is Fréchet differentiable at a point if its directional derivatives $\delta f(x; h)$ exist in a neighborhood of the point and are uniformly continuous functions of x and continuous functions of h (see Lusternik and Sobolev [33], for example).

Next, we consider some examples of directional derivatives.

Example 13.12 Let $X = L^p(\Omega)$, where $1 \le p < \infty$. We will compute the Gâteaux derivative of the L^p-norm. We define $F : X \to \mathbb{R}$ by $F(u) = \|u\|_p^p$. Then

$$\delta F(u; h) = \frac{d}{dt} \int |u + th|^p \bigg|_{t=0}.$$

First, we show that we can interchange the derivative and the integral by the dominated convergence theorem (Theorem 12.35). For each $x \in X$ we have

$$\frac{d}{dt}|u(x) + th(x)|^p\bigg|_{t=0} = \lim_{t \to 0} \frac{1}{t}\left(|u(x) + th(x)|^p - |u(x)|^p\right).$$

Since the function $x \mapsto |x|^p$ is convex for $p \geq 1$, we have

$$|u(x) + th(x)|^p \leq t|u(x) + h(x)|^p + (1-t)|u(x)|^p \qquad \text{for } 0 < t < 1.$$

Hence

$$\frac{|u(x) + th(x)|^p - |u(x)|^p}{t} \leq |u(x) + h(x)|^p - |u(x)|^p,$$

and similarly

$$|u(x)|^p - |u(x) - h(x)|^p \leq \frac{|u(x) + th(x)|^p - |u(x)|^p}{t}.$$

These two inequalities imply that

$$\left|\frac{|u(x) + th(x)|^p - |u(x)|^p}{t}\right| \leq |u(x)|^p + |u(x) + h(x)|^p + |u(x) - h(x)|^p. \qquad (13.17)$$

The left-hand side of (13.17) converges pointwise a.e. as $t \to 0$, since

$$\lim_{t \to 0^+} \frac{1}{t}\left\||u(x) + th(x)|^p - |u(x)|^p\right\| = \frac{d}{dt}\left\||u(x) + th(x)|^p - |u(x)|^p\right\|\bigg|_{t=0}$$

$$= \frac{p}{2}|u(x)|^{p-2}\left(\overline{u(x)}h(x) + u(x)\overline{h(x)}\right),$$

where we write $|z|^p$ as $(z\overline{z})^{p/2}$ before differentiating. The right-hand side of (13.17) is in L^1. Therefore, the dominated convergence theorem implies that

$$\lim_{t \to 0} \frac{1}{t}\int \left\||u(x) + th(x)|^p - |u(x)|^p\right\| = \frac{p}{2}\int |u(x)|^{p-2}\left(\overline{u(x)}h(x) + u(x)\overline{h(x)}\right).$$

Hence, the directional derivative is given by

$$\delta F(u; h) = \frac{p}{2}\int |u|^{p-2}\left(\overline{u}h + u\overline{h}\right).$$

Since $|u|^{p-1} \in L^{p'}$ when $u \in L^p$ where p' is the Hölder conjugate of p, Hölder's inequality implies that $\delta F(u; \cdot)$ is a bounded linear functional on L^p. It follows that F is Gâteaux differentiable.

Example 13.13 Suppose that $\varphi : \mathbb{R} \to \mathbb{R}$ is a continuous function such that

$$|\varphi(t)| \leq a + b|t|^{p/q} \qquad (13.18)$$

for suitable constants $a, b > 0$ and $p, q \geq 1$, and Ω is a bounded, measurable subset of $\mathbb{R}^n$. We define a nonlinear map $N_\varphi : L^p(\Omega) \to L^q(\Omega)$ by

$$(N_\varphi u)(x) = \varphi(u(x)).$$

Thus, N_φ is the operation of composition with the function φ, regarded as a map on L^p. Such an operator is called a *Nemitski operator*. It follows from (13.18) that N_φ is bounded, meaning that it maps bounded sets in L^p into bounded sets in L^q. It is also possible to show that if $u_n \to u$ in L^p, then $N_\varphi(u_n) \to N_\varphi(u)$ in L^q, so N_φ is continuous. This continuity does not follow from the boundedness of N_φ because N_φ is nonlinear.

Now suppose that $\varphi : \mathbb{R} \to \mathbb{R}$ is continuously differentiable. The pointwise calculation

$$\frac{d}{d\epsilon}\varphi(u(x) + \epsilon h(x))\Big|_{\epsilon=0} = \varphi'(u(x)) h(x)$$

suggests that, when it exists, the derivative of N_φ at u is multiplication by the function $\varphi'(u)$. To give conditions under which this is true, suppose that $p > 2$ and

$$|\varphi'(t)| \leq a + b|t|^{p-2}. \tag{13.19}$$

Then, if $u, h \in L^p(\Omega)$, we have $\varphi'(u) h \in L^q(\Omega)$, where q is the Hölder conjugate of p. Thus the map $h \mapsto \varphi'(u) h$ is a bounded linear map from L^p to L^q. It is possible to show that, in this case, the Nemitski operator $N_\varphi : L^p \to L^q$ is Fréchet differentiable, and

$$(N_\varphi)'(u)h = \varphi'(u) h.$$

In the limiting case, when $p = 2$ and $|\varphi'(t)| \leq a$, the Nemitski operator $N_\varphi : L^2 \to L^2$ is Gâteaux differentiable, but not Fréchet differentiable, unless $\varphi(t) = a + bt$ is a linear function of t. The proof of these facts requires some measure-theoretic arguments which we omit.

If $f : X \times Y \to Z$ is a differentiable map on the product of two Banach spaces, then we have

$$f(x + h, y + k) = f(x, y) + Ah + Bk + o(h, k)$$

for suitable linear maps $A : X \to Z$ and $B : Y \to Z$. We call A and B the *partial derivatives* of f with respect to x and y, repectively, and denote them by

$$A = D_x f(x, y), \qquad B = D_y f(x, y).$$

Other common notations for the partial derivatives $(D_x f, D_y f)$ are

$$(d_x f, d_y f), \quad (D_1 f, D_2 f), \quad (f_x, f_y).$$

We may define higher order derivatives as multilinear maps in a similar way to the definition of the first derivative as a linear map. For example, we say that $f : U \subset X \to Y$ is twice differentiable at $x \in U$ if there is a continuous, bilinear map $f''(x) : X \times X \to Y$ such that

$$f(x + h) = f(x) + f'(x)h + \frac{1}{2}f''(x)(h, h) + o(h^2) \qquad \text{as } h \to 0.$$

If $f : U \subset X \to Y$ is k-times continuously differentiable at each point of U, then we say that f belongs to $C^k(U)$.

Example 13.14 If $f = (f_1, f_2, \ldots, f_m) : \mathbb{R}^n \to \mathbb{R}^m$, where $f_i : \mathbb{R}^n \to \mathbb{R}$ is twice continuously differentiable, and $h = (h_1, h_2, \ldots, h_n) \in \mathbb{R}^n$, then

$$[f''(x)(h, h)]_i = \sum_{j,k=1}^{n} \frac{\partial f_i}{\partial x_j \partial x_k}(x) h_j h_k.$$

Just as we defined the equivalence of the topological properties of two spaces in terms of homeomorphisms, and the equivalence of metric space properties in terms of metric space isomorphisms, we may define the equivalence of the smoothness properties of two spaces in terms of diffeomorphisms.

Definition 13.15 If $f : U \subset X \to V \subset Y$ is a one-to-one, onto map from an open subset U of a Banach space X to an open subset V of a Banach space Y such that $f \in C^k(U)$ and $f^{-1} \in C^k(V)$, where $k \geq 1$, then f is called a C^k-*diffeomorphism*, or a *diffeomorphism*. Two open sets $U \subset X$, $V \subset Y$ are *diffeomorphic* if there is a diffeomorphism $f : U \to V$.

13.4 The inverse and implicit function theorems

In this section, we prove the inverse function theorem, which states that a continuously differentiable function is locally invertible if its derivative is invertible.

Theorem 13.16 (Inverse function) Suppose that $f : U \subset X \to Y$ is a differentiable map from an open subset U of a Banach space X to a Banach space Y. If f is continuously differentiable in U and $f'(x)$ has a bounded inverse at $x \in U$, then there are open neighborhoods $V \subset U$ of x and $W \subset Y$ of $f(x)$ such that $f : V \to W$ is a one-to-one, onto continuous map with continuous inverse $f^{-1} : W \to V$. Moreover, the local inverse is continuously differentiable at $f(x)$ and

$$\left(f^{-1}\right)'(f(x)) = [f'(x)]^{-1}. \tag{13.20}$$

Proof. We want to show that for a given sufficiently small $k \in Y$ there is a solution $h \in X$ of the equation

$$f(x + h) = f(x) + k. \tag{13.21}$$

The idea of the proof is to use the contraction mapping theorem to show that there is a solution of the nonlinear equation close to the solution of the linearized equation. We write

$$f(x + h) = f(x) + f'(x)h + r(h),$$

where $r(h) = o(h)$ as $h \to 0$. Since f is continuously differentiable at x, we see that r is continuously differentiable at 0, and $r'(0) = 0$. Since $f'(x)$ is invertible, we may rewrite (13.21) as a fixed point equation

$$h = T(h) \quad \text{where} \quad T(h) = [f'(x)]^{-1}(k - r(h)). \tag{13.22}$$

The vector k occurs in this equation as a parameter.

First, we show that T contracts distances when $\|h\|$ is sufficiently small. From (13.22), we have

$$\|T(h_1) - T(h_2)\| \le \left\|[f'(x)]^{-1}\right\| \|r(h_1) - r(h_2)\|.$$

Theorem 13.10 implies that

$$\|r(h_1) - r(h_2)\| \le \sup_{0 \le t \le 1} \|r'(th_1 + (1-t)h_2)\| \|h_1 - h_2\|.$$

Since $r'(0) = 0$ and r is continuously differentiable at 0, there is a $\delta > 0$ such that

$$\|r'(h)\| \le \frac{1}{2\left\|[f'(x)]^{-1}\right\|} \quad \text{for } \|h\| \le \delta.$$

We denote the closed ball in X of radius δ and center zero by

$$B_\delta = \{h \in X \mid \|h\| \le \delta\}.$$

It follows that

$$\|r(h_1) - r(h_2)\| \le \frac{\|h_1 - h_2\|}{2\left\|[f'(x)]^{-1}\right\|} \quad \text{for } h_1, h_2 \in B_\delta, \tag{13.23}$$

and therefore that

$$\|T(h_1) - T(h_2)\| \le \frac{1}{2}\|h_1 - h_2\| \quad \text{for } h_1, h_2 \in B_\delta.$$

To apply the contraction mapping theorem, we need to show that T maps B_δ into itself when k is sufficiently small. Taking the norm of $T(h)$ in (13.22), we get

$$\|T(h)\| \le \|[f'(x)]^{-1}\| (\|k\| + \|r(h)\|). \tag{13.24}$$

Equation (13.23), with $h_1 = h$ and $h_2 = 0$, implies that

$$\|r(h)\| \le \frac{\|h\|}{2\left\|[f'(x)]^{-1}\right\|} \quad \text{for } h \in B_\delta.$$

It therefore follows from (13.24) that if $\|h\| \le \delta$ and

$$\|k\| \le \eta \qquad \text{where} \quad \eta = \frac{\delta}{2\| [f'(x)]^{-1} \|}, \tag{13.25}$$

then $\|T(h)\| \le \delta$. Thus $T : B_\delta \to B_\delta$ is a contraction on the complete set B_δ when $k \in B_\eta$, where B_η is the closed ball in Y of radius η and center zero. The contraction mapping theorem implies that T has a unique fixed point in B_δ. We may therefore define $g : B_\eta \to B_\delta$ by the requirement that $h = g(k)$ is the unique solution of (13.22) belonging to B_δ. From (13.21), the function g provides a local inverse of f, with

$$f^{-1}(f(x) + k) = x + g(k). \tag{13.26}$$

To complete the proof, we need to show that f^{-1} is continuously differentiable.
  From (13.22), if $h = g(k)$, then

$$h = [f'(x)]^{-1}(k - r(h)). \tag{13.27}$$

Subtracting the equations corresponding to (13.27) for $h_1 = g(k_1)$ and $h_2 = g(k_2)$, taking the norm of the result, and using (13.23), we find that

$$
\begin{aligned}
\|h_1 - h_2\| &\le \left\| [f'(x)]^{-1} \right\| (\|k_1 - k_2\| + \|r(h_1) - r(h_2)\|) \\
&\le \left\| [f'(x)]^{-1} \right\| \|k_1 - k_2\| + \frac{1}{2}\|h_1 - h_2\|.
\end{aligned}
$$

Rewriting this inequality, we obtain that

$$\|g(k_1) - g(k_2)\| \le 2\| [f'(x)]^{-1} \| \|k_1 - k_2\| \qquad \text{for } k_1, k_2 \in B_\eta. \tag{13.28}$$

Thus, g is Lipschitz continuous in B_η.
  Setting $h = g(k)$ in (13.27), we find that

$$g(k) = [f'(x)]^{-1} k + s(k), \tag{13.29}$$

where the remainder $s : B_\eta \to Y$ is defined by

$$s(k) = -[f'(x)]^{-1} r(g(k)).$$

From (13.28), with $k_1 = k$ and $k_2 = 0$, we have

$$\|g(k)\| \le 2\|k\| \| [f'(x)]^{-1} \| \qquad \text{for } k \in B_\eta.$$

Hence, $s(k) = o(k)$ as $k \to 0$ because

$$\frac{\|s(k)\|}{\|k\|} = \frac{\|h\|}{\|k\|} \frac{\| [f'(x)]^{-1} r(h)\|}{\|h\|} \le 2\| [f'(x)]^{-1} \|^2 \frac{\|r(h)\|}{\|h\|} \to 0 \qquad \text{as } k \to 0.$$

Equation (13.29) therefore implies that g is differentiable at $k = 0$ with $g'(0) = [f'(x)]^{-1}$. It follows from (13.26) that f^{-1} is differentiable at $f(x)$, and its derivative is given by (13.20).

The continuous differentiability of f^{-1} follows from the continuous differentiability of f, and the continuity of inversion on the set of bounded, nonsingular linear operators.   □

The fact that the derivative of the inverse is the inverse of the derivative may also be deduced from the chain rule. Exercise 13.7 shows that the continuity requirement on the derivative of f in the hypotheses of the inverse function theorem cannot be dropped.

Example 13.17 The map $s : \mathbb{R} \to \mathbb{R}$ given by $s(x) = x^2$ is locally invertible at every $x \neq 0$. If $x > 0$, a local inverse is $r : (0, \infty) \to (0, \infty)$ where $r(y) = \sqrt{y}$. If $x < 0$, a local inverse is $r : (0, \infty) \to (-\infty, 0)$ where $r(y) = -\sqrt{y}$. The map s is not locally invertible at 0 where its derivative vanishes. The map $c : \mathbb{R} \to \mathbb{R}$ given by $c(x) = x^3$ is globally invertible on $\mathbb{R}$, with continuous inverse $c^{-1} : \mathbb{R} \to \mathbb{R}$ where $c^{-1}(x) = x^{1/3}$. The inverse function is not differentiable at $x = 0$ where the derivative of c vanishes. Thus, $c : \mathbb{R} \to \mathbb{R}$ is a homeomorphism, but not a diffeomorphism.

Example 13.18 Consider the map $\exp : \mathbb{C} \to \mathbb{C}$ defined by $\exp z = e^z$. This map may also be regarded as a map $\exp : \mathbb{R}^2 \to \mathbb{R}^2$ defined by $\exp(x, y) = (u, v)$, where

$$u = e^x \cos y, \qquad v = e^x \sin y.$$

The derivative of exp is nonsingular at every point, so it is locally invertible. The map is not globally invertible, however, since $\exp(z + 2\pi i n) = \exp z$ for every $n \in \mathbb{Z}$.

Example 13.19 Suppose that $f : U \subset \mathbb{R}^n \to \mathbb{R}^n$ is a continuously differentiable map. From Example 13.6, the matrix of the derivative f' is the Jacobian matrix of f. The determinant of this matrix, $J : U \subset \mathbb{R}^n \to \mathbb{R}$,

$$J = \det \left(\frac{\partial f_i}{\partial x_j} \right)$$

is called the *Jacobian* of f. The inverse function theorem implies that the map f is locally invertible near x if its Jacobian is nonzero at x. Moreover, the local inverse f^{-1} is differentiable, and its Jacobian matrix is the inverse of the Jacobian matrix of f.

Example 13.20 The *hodograph method* is a method for linearizing certain nonlinear PDEs by exchanging the role of independent and dependent variables. As an example, we consider the *transonic small disturbance equation*, which provides a simplified model of the equations for steady fluid flows near the speed of sound (such as the flow around an aircraft flying at a speed close to the speed of sound):

$$uu_x + v_y = 0,$$
$$u_y - v_x = 0, \tag{13.30}$$

where $u = u(x, y)$, $v = v(x, y)$. If the Jacobian

$$J = \begin{vmatrix} u_x & u_y \\ v_x & v_y \end{vmatrix} = u_x v_y - u_y v_x$$

is nonzero, then we may locally invert the map $(x, y) \mapsto (u, v)$ and write $x = x(u, v)$, $y = y(u, v)$. Moreover, we have

$$\begin{pmatrix} u_x & u_y \\ v_x & v_y \end{pmatrix} = \begin{pmatrix} x_u & x_v \\ y_u & y_v \end{pmatrix}^{-1} = \frac{1}{j} \begin{pmatrix} y_v & -x_v \\ -y_u & x_u \end{pmatrix},$$

where

$$j = x_u y_v - x_v y_u = \frac{1}{J}.$$

Hence,

$$u_x = \frac{y_v}{j}, \quad u_y = \frac{-x_v}{j}, \quad v_x = \frac{-y_u}{j}, \quad v_y = \frac{x_u}{j}.$$

The use of these equations in (13.30) implies that

$$u y_v + x_u = 0,$$
$$x_v - y_u = 0.$$

The Jacobian j cancels, because all terms are linear in a first order derivative of u or v, and consequently the resulting system for $x = x(u, v)$, $y = y(u, v)$ is linear. From the second equation, we may write $x = \varphi_u$ and $y = \varphi_v$ for some function φ. The first equation then implies that $\varphi = \varphi(u, v)$ satisfies

$$\varphi_{uu} + u \varphi_{vv} = 0.$$

This PDE is called the *Tricomi equation*. It is one of the simplest equations of mixed type, being elliptic when $u > 0$ (corresponding to subsonic flow) and hyperbolic when $u < 0$ (corresponding to supersonic flow). Despite the greater simplicity of the linear equations for (x, y) than the nonlinear equations for (u, v), the hodograph method has a significant drawback: solutions may contain curves or regions where the Jacobians j or J vanish, and then the local invertibility between (x, y) and (u, v) is lost.

Example 13.21 Neglecting friction, the angle of $u(t)$ of a forced pendulum satisfies

$$\ddot{u} + \sin u = h, \tag{13.31}$$

where $h(t)$ is a given forcing function. We suppose that h is a T-periodic function, where $T > 0$, and ask if (13.31) has T-periodic solutions. When $h = 0$, (13.31) has the trivial T-periodic solution $u = 0$, and we can use the implicit function theorem to prove the existence of T-periodic solutions for small, nonzero h. Let

$$X = \left\{ u \in C^2(\mathbb{R}) \mid u(t + T) = u(t) \right\}, \quad Y = \left\{ u \in C(\mathbb{R}) \mid u(t + T) = u(t) \right\}.$$

Then we may write (13.31) as $f(u) = h$ where $f : X \to Y$ is defined by

$$f(u) = \ddot{u} + \sin u.$$

The map f is continuously differentiable, and its derivative $f'(0) : X \to Y$ is given by

$$f'(0)v = \ddot{v} + v.$$

The linear map $f'(0) : X \to Y$ is nonsingular if and only if $T \neq 2n\pi$ for some $n \in \mathbb{N}$. In that case, there is a unique T-periodic solution of (13.31) when $\|h\|_\infty$ is sufficiently small. The case $T = 2n\pi$ corresponds to a *resonance* of the external forcing with the linearized oscillator.

The implicit function theorem is a generalization of the inverse function theorem.

Theorem 13.22 (Implicit function theorem) Suppose that X, Y, Z are Banach spaces, and $F : U \subset X \times Y \to Z$ is a continuously differentiable map defined on an open subset U of $X \times Y$. If $(x_0, y_0) \in U$ is a point such that $F(x_0, y_0) = 0$, and $D_y F(x_0, y_0) : Y \to Z$ is a one-to-one, onto, bounded linear map, then there is an open neighborhood $V \subset X$ of x_0, an open neighborhood $W \subset Y$ of y_0, and a unique function $f : V \to W$ such that

$$F(x, f(x)) = 0 \qquad \text{for all } x \in V.$$

The function f is continuously differentiable, and

$$f'(x) = -[D_y F(x, f(x))]^{-1} D_x F(x, f(x)).$$

The proof of this theorem is similar to the proof of the inverse function theorem, so we will omit it. The implicit function theorem reduces to the inverse function theorem when $F(x, y) = x - f(y)$.

Example 13.23 If $F = (F_1, F_2, \ldots, F_m) : \mathbb{R}^n \times \mathbb{R}^m \to \mathbb{R}^m$ is a continuously differentiable function, then the matrix of the partial derivative $D_y F$ is

$$\begin{pmatrix} \partial F_1/\partial y_1 & \partial F_1/\partial y_2 & \cdots & \partial F_1/\partial y_m \\ \partial F_2/\partial y_1 & \partial F_2/\partial y_2 & \cdots & \partial F_2/\partial y_m \\ \vdots & \vdots & \ddots & \vdots \\ \partial F_m/\partial y_1 & \partial F_m/\partial y_2 & \cdots & \partial F_m/\partial y_m \end{pmatrix}. \qquad (13.32)$$

The $m \times m$ system of nonlinear equations

$$F(x, y) = 0$$

has a unique local solution for y in terms of x near any point where $F(x, y) = 0$ and the determinant of the Jacobian matrix in (13.32) is nonzero.

Many problems lead to an equation that depends on a parameter $\mu \in \mathbb{R}$. For example, μ may be a dimensionless parameter characteristic of the system being modeled by the equation. For a time-independent problem, we may write such an equation in the abstract form

$$F(x, \mu) = 0, \qquad (13.33)$$

where $F : X \times \mathbb{R} \to Y$ and X, Y are Banach spaces. The study of how the solution set of (13.33) varies as μ varies is part of *bifurcation theory*. We assume that F is a smooth function. A *solution branch* of (13.33) is a smooth map $\overline{x} : I \subset \mathbb{R} \to X$ from an open interval I in $\mathbb{R}$ into X such that

$$F\left(\overline{x}(\mu), \mu\right) = 0 \qquad \text{for all } \mu \in I.$$

We say that μ_* is a *bifurcation point* of (13.33) from the solution branch $\overline{x}$, if there is a sequence of solutions (x_n, μ_n) in $X \times I$ such that

$$F(x_n, \mu_n) = 0, \qquad x_n \neq \overline{x}(\mu_n),$$
$$x_n \to \overline{x}(\mu_*), \quad \mu_n \to \mu_* \qquad \text{as } n \to \infty.$$

The implicit function theorem implies that if the derivative

$$D_x F\left(\overline{x}(\mu), \mu\right) \qquad (13.34)$$

is a nonsingular, bounded linear map from X to Y, then there is a unique local solution branch. Thus, a necessary condition for μ_* to be a bifurcation point is that the derivative in (13.34) is singular at $\mu = \mu_*$.

Example 13.24 Consider the equation

$$x^3 - \mu x = 0,$$

where $x, \mu \in \mathbb{R}$, corresponding to $F(x, \mu) = x^3 - \mu x$. The zero solution $x = 0$ is a solution branch. We have $F_x(0, \mu) = -\mu$, so the only possible bifurcation point from the zero solution is at $\mu = 0$. The solutions in this case are $x = 0$, and $x = \pm\sqrt{\mu}$ when $\mu > 0$. Thus, a new branch of solutions appears at $\mu = 0$. This bifurcation is called a *pitchfork bifurcation*.

The next example shows that the singularity of the derivative in (13.34) is a necessary but not sufficient condition for a bifurcation to occur.

Example 13.25 Consider the following system of equations for $(x, y) \in \mathbb{R}^2$:

$$y^3 - \mu x = 0,$$
$$x^3 + \mu y = 0.$$

The zero solution $(x, y) = (0, 0)$ satisfies this system for all $\mu \in \mathbb{R}$. The derivative of the left-hand side with respect to (x, y) has the matrix

$$\begin{pmatrix} -\mu & 3y^2 \\ 3x^2 & \mu \end{pmatrix}.$$

This matrix is singular at $x = y = 0$ if and only if $\mu = 0$, in which case it has a two-dimensional null space. Elimination of μ from the original system of equations implies that $x^4 + y^4 = 0$. Therefore, the zero solution is the only solution, and $\mu = 0$ is not a bifurcation point.

The same ideas apply to bifurcation problems for equations on infinite-dimensional spaces.

Example 13.26 Consider the following nonlinear Dirichlet problem on a smooth, bounded domain $\Omega \subset \mathbb{R}^n$:

$$-\Delta u = \mu \sin u \qquad \text{in } \Omega,$$
$$u = 0 \qquad \text{on } \partial\Omega.$$

We write this equation in the form (13.33), where $F : X \times \mathbb{R} \to Y$ with

$$F(u, \mu) = -\Delta u + \mu \sin u,$$
$$X = \left\{ u \in C^{2,r}\left(\overline{\Omega}\right) \mid u = 0 \text{ on } \partial\Omega \right\}, \qquad \cdot Y = C^{0,r}\left(\overline{\Omega}\right).$$

Here $0 < r \leq 1$, and $C^{k,r}\left(\overline{\Omega}\right)$ denotes a space of Hölder continuous functions. One can show that F is differentiable at $u = 0$, and

$$D_u F(0, \mu)h = -\Delta h - \mu h.$$

The theory of elliptic PDEs implies that $D_u F(0, \mu) : X \to Y$ is a bounded, nonsingular map unless μ is an eigenvalue of $-\Delta$. Thus, the possible bifurcation points from the trivial solution $u = 0$ are the eigenvalues of the Laplacian operator on Ω. It is possible to show that a bifurcation must occur at a simple eigenvalue, but need not occur at a multiple eigenvalue.

13.5 Newton's method

Newton's method is an iterative method for the solution of a finite or infinite-dimensional system of nonlinear equations,

$$f(x) = 0, \tag{13.35}$$

where f is a smooth mapping between Banach spaces. Suppose that x_n is an approximate solution. As $x \to x_n$, we have

$$f(x) = f(x_n) + f'(x_n)(x - x_n) + o(x - x_n).$$

If $f'(x_n)$ is nonsingular and x_n is sufficiently close to a solution of (13.35), then it is reasonable to expect that the solution $x = x_{n+1}$ of the linearized equation,

$$f(x_n) + f'(x_n)(x - x_n) = 0,$$

is a better approximation to the solution of the nonlinear equation than x_n. The resulting iteration scheme, called *Newton's method*, is given by

$$x_{n+1} = x_n - [f'(x_n)]^{-1} f(x_n) \qquad \text{for } n \geq 0. \tag{13.36}$$

After the choice of a starting point, x_0, Newton's method generates a sequence (x_n) of iterates, provided that $f'(x_n)$ is nonsingular for every n. The Newton iterates may be obtained from an iteration of the fixed point problem

$$x = x - [f'(x)]^{-1} f(x),$$

which is clearly equivalent to (13.35) when $f'(x)$ is nonsingular. A basic question concerning Newton's method is: When does the sequence of Newton iterates converge to a solution of (13.35)?

There are many variants of Newton's method. One of the simplest is the *modified Newton's method*:

$$x_{n+1} = x_n - [f'(x_0)]^{-1} f(x_n) \qquad \text{for } n \geq 0. \tag{13.37}$$

This method has the numerical advantage that the derivative $f'(x_0)$ only has to be computed and inverted once, at the starting point x_0; the sequence of approximations, however, does not converge as rapidly as the sequence obtained from Newton's method. The modified Newton's method is simpler to analyze than Newton's method, and, following Lusternik and Sobolev [33], we will prove a convergence result for it here. A proof of the convergence of Newton's method, under suitable assumptions on f and x_0, may be found in Kantorovich and Akilov [27].

In order to prove convergence, we will assume that f' satisfies a Lipschitz condition. If $f : U \subset X \to Y$ is a differentiable function, then we say that the derivative $f' : U \subset X \to \mathcal{B}(X, Y)$ is *Lipschitz continuous* in U if there is a constant C, called a *Lipschitz constant*, such that

$$\|f'(x) - f'(y)\| \leq C\|x - y\| \qquad \text{for all } x, y \in U. \tag{13.38}$$

Theorem 13.27 Let $f : U \subset X \to Y$ be a differentiable map from an open subset U of a Banach space X into a Banach space Y such that f' is Lipschitz continuous in U with Lipschitz constant C. Suppose that $x_0 \in U$, $f'(x_0)$ is nonsingular, and

$$h = C \left\| [f'(x_0)]^{-1} \right\| \left\| [f'(x_0)]^{-1} f(x_0) \right\| \leq \frac{1}{4}. \tag{13.39}$$

Define $\delta \geq 0$ by

$$\delta = \left\| [f'(x_0)]^{-1} f(x_0) \right\| \left(\frac{1 - \sqrt{1 - 4h}}{2h} \right), \tag{13.40}$$

and suppose further that the closed ball B_δ of radius δ centered at x_0,

$$B_\delta = \{x \in X \mid \|x - x_0\| \leq \delta\}, \tag{13.41}$$

is contained in U. Then there is a unique solution of the equation $f(x) = 0$ in B_δ, and the sequence (x_n) of modified Newton iterates, defined in (13.37), converges to the solution of $f(x) = 0$ in B_δ as $n \to \infty$.

Proof.   The modified Newton iterates are obtained from the fixed point iteration $x_{n+1} = T(x_n)$, where

$$T(x) = x - [f'(x_0)]^{-1} f(x).$$

First, we show that $T : B_\delta \to B_\delta$. We may write

$$T(x) - x_0 = -[f'(x_0)]^{-1} [r(x) + f(x_0)], \tag{13.42}$$

where

$$r(x) = f(x) - f(x_0) - f'(x_0)(x - x_0). \tag{13.43}$$

Taking the norm of (13.42), we find that

$$\|T(x) - x_0\| \leq M \|r(x)\| + \eta, \tag{13.44}$$

where

$$M = \left\| [f'(x_0)]^{-1} \right\|, \qquad \eta = \left\| [f'(x_0)]^{-1} f(x_0) \right\|. \tag{13.45}$$

Computing the derivative of r, and using the Lipschitz condition (13.38) for f', we obtain that

$$\|r'(x)\| = \|f'(x) - f'(x_0)\| \leq C \|x - x_0\|.$$

Since $r(x_0) = 0$, the mean value theorem implies that

$$\|r(x)\| = \|r(x) - r(x_0)\| \leq \sup_{0 \leq t \leq 1} \|r'(tx + (1-t)x_0)\| \, \|x - x_0\| \leq C \|x - x_0\|^2.$$

Using this result in (13.44), we find that

$$\|T(x) - x_0\| \leq CM \|x - x_0\|^2 + \eta.$$

Hence, T maps the ball $\{x \mid \|x - x_0\| \leq \epsilon\}$ into itself provided that

$$CM\epsilon^2 + \eta \leq \epsilon. \tag{13.46}$$

This inequality can be satisfied for some $\epsilon > 0$ if

$$h = CM\eta \leq \frac{1}{4}. \tag{13.47}$$

Using (13.45), we see that this is the condition in (13.39). In that case, the smallest value δ of ϵ for which (13.46) holds is

$$\delta = \eta\tau, \tag{13.48}$$

where τ is the smallest root of the equation $h\tau^2 - \tau + 1 = 0$, or

$$\tau = \frac{1 - \sqrt{1 - 4h}}{2h}. \tag{13.49}$$

Using (13.45) and (13.49) in (13.48), we find that δ is given by (13.40). This proves that $T : B_\delta \to B_\delta$.

Next, we prove that T is a contraction on B_δ. Differentiating (13.42) and (13.43), we find that

$$T'(x) = -[f'(x_0)]^{-1}[f'(x) - f'(x_0)].$$

Hence, using (13.45) and the Lipschitz condition on f' in (13.38), we have

$$\|T'(x)\| \le M\|f'(x) - f'(x_0)\| \le CM\|x - x_0\| \le CM\delta \qquad \text{for all } x \in B_\delta.$$

It follows from (13.47), (13.48), and (13.49) that

$$CM\delta = \frac{1 - \sqrt{1 - 4h}}{2} \le \frac{1}{2}.$$

We therefore have $\|T'(x)\| \le 1/2$ in B_δ, so from the mean value theorem

$$\|T(x) - T(y)\| \le \frac{1}{2}\|x - y\| \qquad \text{for all } x, y \in B_\delta.$$

The theorem now follows from the contraction mapping theorem. $\square$

Note that the conditions (13.39) and $B_\delta \subset U$ in the hypotheses of the theorem are satisfied when x_0 is sufficiently close to a solution of $f(x) = 0$ at which the derivative of f is nonsingular.

A significant practical difficulty in the implementation of Newton's method, and its modifications, is that the iterates may diverge unless the starting point is very close to the solution. For this reason, Newton's method is often used in conjunction with continuation methods, in which one slowly varies a parameter in the equation, and uses the solution for a previous parameter value as an initial guess for the Newton iterations for the next parameter value.

13.6 Linearized stability

We consider an equilibrium of the evolution equation

$$u_t = f(u), \tag{13.50}$$

where $f : \mathcal{D}(f) \subset X \to X$ is a vector field on a Banach space X. We assume that the initial value problem for (13.50) is well posed, meaning that there is a unique solution $u : [0, \infty) \to X$ for every initial condition $u(0) = u_0 \in X$, and $u(t)$ depends continuously on u_0 with respect to the normed topology on X. A state $\overline{u} \in X$ is an *equilibrium* of (13.50) if $f(\overline{u}) = 0$. There are many inequivalent ways to define the stability of an equilibrium. We only consider the two most common ways here.

Definition 13.28 An equilibrium $\overline{u}$ of (13.50) is *stable* if for every neighborhood U of $\overline{u}$ there is a neighborhood V of $\overline{u}$ such that if $u(0) \in V$, then $u(t) \in U$ for all $t \geq 0$. If $\overline{u}$ is not stable, then it is *unstable*. An equilibrium $\overline{u}$ is *asymptotically stable* if it is stable and there is a neighborhood W of $\overline{u}$ such that $u(t) \to \overline{u}$ as $t \to \infty$ whenever $u(0) \in W$.

Thus, if a stable equilibrium is perturbed, the perturbation remains small, and if an asymptotically stable equilibrium is perturbed, then the perturbation remains small and eventually dies out.

The linearization of (13.50) about $\overline{u}$ is

$$v_t = Av, \qquad \text{where } A = f'(\overline{u}). \tag{13.51}$$

We define the linear stability of an equilibrium $\overline{u}$ of (13.50) in terms of the stability of the equilibrium $v = 0$ of (13.51).

Definition 13.29 The equilibrium $u = \overline{u}$ of (13.50) is *linearly stable*, or *linearly asymptotically stable*, if $v = 0$ is a stable, or asymptotically stable, equilibrium of (13.51), respectively.

In the case of ODEs, we have the following result.

Theorem 13.30 Suppose that $f : \mathbb{R}^n \to \mathbb{R}^n$ is continuously differentiable and $f(\overline{u}) = 0$. If all the eigenvalues of $f'(\overline{u})$ have a strictly negative real part, then $\overline{u}$ is an asymptotically stable equilibrium of the system of ODEs $\dot{u} = f(u)$. If one of the eigenvalues of $f'(\overline{u})$ has a strictly positive real part, then $\overline{u}$ is unstable.

The stability part of this theorem is proved by the construction of a suitable quadratic *Liapunov function* $V : X \to \mathbb{R}$ for (13.50) that has a minimum at $\overline{u}$ and the property that $V_t \leq 0$ on solutions of (13.50) in a neighborhood of $\overline{u}$ (see [21]).

Example 13.31 Consider a 2×2 system of ODEs for $(u, v) \in \mathbb{R}^2$:

$$\dot{u} = f(u, v), \qquad \dot{v} = g(u, v). \tag{13.52}$$

The eigenvalues of the derivative

$$A = \begin{pmatrix} f_u(\overline{u}, \overline{v}) & f_v(\overline{u}, \overline{v}) \\ g_u(\overline{u}, \overline{v}) & g_v(\overline{u}, \overline{v}) \end{pmatrix} \tag{13.53}$$

have strictly negative real parts if and only if

$$\text{tr } A = f_u\left(\overline{u}, \overline{v}\right) + g_v\left(\overline{u}, \overline{v}\right) < 0,$$
$$\det A = f_u\left(\overline{u}, \overline{v}\right) g_v\left(\overline{u}, \overline{v}\right) - f_v\left(\overline{u}, \overline{v}\right) g_u\left(\overline{u}, \overline{v}\right) > 0. \tag{13.54}$$

Thus, an equilibrium $(\overline{u}, \overline{v})$ of (13.52) is asymptotically stable when the condition in (13.54) holds.

If the spectrum of $f'\left(\overline{u}\right)$ touches the imaginary axis, meaning that all points in the spectrum have nonpositive real parts and the real part of at least one point is equal to zero, then the equilibrium $\overline{u}$ may be linearly stable, but we cannot draw conclusions about the nonlinear stability of $\overline{u}$ from its linearized stability alone.

Example 13.32 Consider the following 2×2 system of ODEs for $(u, v) \in \mathbb{R}^2$:

$$\dot{u} = \mu u - v - \alpha\left(u^2 + v^2\right) u,$$
$$\dot{v} = u + \mu v - \alpha\left(u^2 + v^2\right) v,$$

where α, μ are real parameters. It is convenient to write this equation in complex form for $w = u + iv \in \mathbb{C}$ as

$$\dot{w} = (\mu + i) w - \alpha|w|^2 w. \tag{13.55}$$

This equation may be solved by writing it in the polar form

$$\dot{r} = \mu r - \alpha r^3, \qquad \dot{\theta} = 1,$$

where

$$w = r e^{i\theta}, \quad r = \sqrt{u^2 + v^2}, \quad \tan\theta = \frac{v}{u}.$$

If $\mu < 0$, then the equilibrium $w = 0$ is asymptotically stable, and if $\mu > 0$ it is unstable. If $\mu = 0$, then the eigenvalues of the linearization are purely imaginary, and the equilibrium $w = 0$ is linearly stable. It is asymptotically stable if $\alpha > 0$, stable if $\alpha = 0$, and unstable if $\alpha < 0$.

When α and μ have the same sign, (13.55) has a periodic *limit cycle* solution

$$w(t) = \sqrt{\frac{\mu}{\alpha}} e^{it}.$$

As μ increases through zero, the equilibrium $w = 0$ becomes unstable. If $\alpha > 0$, then a stable limit cycle appears for $\mu > 0$, while if $\alpha < 0$, then an unstable limit cycle shrinks down to the equilibrium $w = 0$ and disappears for $\mu > 0$. This type of bifurcation is called a *Hopf bifurcation*. The Hopf bifurcation is said to be *supercritical* if $\alpha > 0$ and *subcritical* if $\alpha < 0$.

For PDEs, the relationship between different types of stability, and between linear and nonlinear stability, can be rather subtle. It is usually true that if the spectrum of $f'(\overline{u})$ is contained in a left-half plane $\{\lambda \in \mathbb{C} \mid \operatorname{Re}\lambda \leq \omega\}$ for some $\omega < 0$, then $\overline{u}$ is linearly asymptotically stable (see the discussion of (9.24)), and that linear asymptotic stability implies nonlinear asymptotic stability, but there are exceptions. Moreover, the linearized equation may have continuous or residual spectrum in addition to, or instead of, the pure point spectrum that occurs for ODEs.

We will illustrate the linearization of nonlinear PDEs by considering an important class of examples called *reaction-diffusion equations*. These nonlinear PDEs describe the dynamics of spatially dependent chemical concentrations and temperature in the presence of chemical reactions and the diffusion of reactants and heat. They also model the population of spatially distributed species in ecology. The general form of a reaction-diffusion equation for $u(x,t) \in \mathbb{R}^m$, where $x \in \mathbb{R}^n$ and $t \geq 0$, is

$$u_t = D\Delta u + f(u). \tag{13.56}$$

The effect of diffusion is described by the linear term $D\Delta u$, where D is a positive definite, symmetric $m \times m$ matrix, called the diffusion matrix. In most applications, D is diagonal, and, for simplicity, we assume that D is constant. The effect of reactions is described by the term $f(u)$, where $f : \mathbb{R}^m \to \mathbb{R}^m$ is a given nonlinear function. For models of chemical reactions, f is often a polynomial in the chemical concentrations, as follows from the *law of mass action*, with coefficients that depend exponentially on the temperature. To be specific, we consider a reaction-diffusion equation for a function u defined on a regular, bounded domain $\Omega \subset \mathbb{R}^n$ subject to Dirichlet boundary conditions. In that case, we supplement (13.56) with the initial and boundary conditions

$$u(x,t) = 0 \qquad \text{for } x \in \partial\Omega \text{ and } t > 0,$$
$$u(x,0) = u_0(x) \qquad \text{for } x \in \Omega.$$

An equilibrium solution $\overline{u} : \Omega \subset \mathbb{R}^n \to \mathbb{R}^m$ of (13.56) satisfies the elliptic system of PDEs

$$D\Delta\overline{u} + f(\overline{u}) = 0,$$
$$\overline{u}(x) = 0 \qquad \text{for } x \in \partial\Omega.$$

To study the linear stability of $\overline{u}$, we have to compute the Fréchet derivative of the map F given by

$$F(u) = D\Delta u + f(u). \tag{13.57}$$

The derivative of the linear term is trivial to compute, so we only need to compute the derivative of the nonlinear term, defined on a suitable space of functions.

Lemma 13.33 Suppose that $\Omega \subset \mathbb{R}^n$ is a regular, bounded open domain, and $f : \mathbb{R}^m \to \mathbb{R}^m$ is a twice continuously differentiable function with derivative $f' :$ $\mathbb{R}^m \to \mathbb{R}^{m \times m}$. Then the map $N_f : H^k(\Omega) \to L^2(\Omega)$ defined by $N_f(u) = f(u)$ is differentiable for every $k > n/2$, and its derivative $N_f'(u) : H^k(\Omega) \to L^2(\Omega)$ at $u \in H^k(\Omega)$, is given by

$$N_f'(u)(v) = f'(u)v.$$

Proof. Let $|u|$ denote the Euclidean norm of $u \in \mathbb{R}^m$. Since f is twice continuously differentiable, for each $R > 0$ there is constant $C(R)$ such that

$$|f(u+h) - f(u) - f'(u)v| \leq C(R)|v|^2$$

for all $u, v \in \mathbb{R}^n$ such that $|u| \leq R$ and $|u + v| \leq R$. By the Sobolev embedding theorem (Theorem 12.70) and the assumption that $k > n/2$, there is a constant M such that

$$\|u\|_\infty \leq M \|u\|_{H^k}.$$

By combining these inequalities and integrating the result over Ω, we obtain for $\|v\|_{H^k} \leq 1$ that

$$
\begin{aligned}
\|f(u+v) - f(u) - f'(u)v\|_{L^2} &\leq C \|v^2\|_{L^2} \\
&\leq C' \|v\|_\infty \|v\|_{L^2} \\
&\leq C'' \|v\|_{H^k}^2
\end{aligned}
$$

where $C = C(R)$ with $R = \|u\|_\infty + M$, C', and C'' are constants depending on the bounded domain Ω and u, but not on v. The result follows by dividing this equation by $\|v\|_{H^k}$ and taking the limit of the result as $\|v\|_{H^k} \to 0$. $\square$

The Laplacian maps H^k into L^2 if $k \geq 2$. Thus, it follows that if $k \geq 2$ and $k > n/2$, then the map F defined in (13.57), where

$$F : \mathcal{D}(F) \subset H^k(\Omega) \to L^2(\Omega), \qquad \mathcal{D}(F) = \{u \in H^k(\Omega) \mid u = 0 \text{ on } \partial\Omega\},$$

is differentiable, and the derivative of F at $\overline{u} \in \mathcal{D}(F)$ is given by

$$F'(\overline{u})\,v = D\Delta v + Av,$$

where $A = f'(\overline{u})$. The linearization of (13.56) is therefore

$$v_t = D\Delta v + Av. \tag{13.58}$$

As for the Laplacian on a regular bounded domain, the spectrum of $F'(\overline{u})$ consists entirely of eigenvalues. The matrix A need not be symmetric, and if it is not, then $F'(\overline{u})$ is not self-adjoint, so its eigenvalues need not be real. The equilibrium $\overline{u}$ is linearly asymptotically stable if every eigenvalue of $F'(\overline{u})$ has strictly negative real part.

In general, the equilibrium $\overline{u}$ is a function of x, so that $F'(\overline{u})$ is a variable coefficient elliptic differential operator. It is usually not possible to compute its eigenvalues explicitly, although the eigenvalues with the largest real parts can be computed numerically. If, however, $\overline{u}$ is a constant state and the reaction-diffusion equation is posed on $\mathbb{R}^n$, or $\mathbb{T}^n$, then we can use Fourier analysis to study the spectrum of the linearization. We illustrate this procedure with a discussion of the *Turing instability*, which was proposed by Turing in 1952 as a mechanism for the development of spatial patterns from a spatially uniform state, and in particular as a possible mechanism for morphogenesis.

The state $\overline{u} \in \mathbb{R}^m$ is a spatially uniform equilibrium of (13.56) if and only if $f(\overline{u}) = 0$. We look for solutions of the linearization (13.58) of (13.56) at $\overline{u}$ of the form

$$v(x,t) = e^{ik\cdot x + \lambda t}\widehat{v}, \qquad (13.59)$$

where $k \in \mathbb{R}^n$, $\lambda \in \mathbb{C}$, and $\widehat{v} \in \mathbb{C}^m \setminus 0$. General solutions of the IVP for (13.58) may be obtained from these solutions by use of the Fourier transform. The solution (13.59) grows exponentially in time if $\operatorname{Re}\lambda > 0$. Thus, $\overline{u}$ is linearly unstable if $\operatorname{Re}\lambda > 0$ for some $k \in \mathbb{R}^n$. The use of (13.59) in (13.58) implies that

$$\left(-|k|^2 D + A\right)\widehat{v} = \lambda\widehat{v}. \qquad (13.60)$$

It follows that λ is an eigenvalue of $-|k|^2 D + A$. Turing observed that $-|k|^2 D + A$ may have an eigenvalue with positive real part for some $|k| > 0$ even though all the eigenvalues of A have negative real part. In that case, $\overline{u}$ is an asymptotically stable equilibrium of the reaction equations $\dot{u} = f(u)$, and is therefore stable to spatially uniform perturbations, but is unstable to spatially nonuniform perturbations. The growth of spatially unstable perturbations, and the possible saturation of the growth by nonlinear effects, leads to the formation of spatial patterns from a spatially uniform state. This instability is called a *Turing instability* or a *diffusion-driven instability*.

The simplest system that exhibits Turing instability is a 2×2 system of reaction diffusion equations for $(u,v) \in \mathbb{R}^2$ with a diagonal diffusion matrix $D = \operatorname{diag}(\mu,\nu)$, where $\mu,\nu > 0$:

$$\begin{aligned} u_t &= \mu\Delta u + f(u,v), \\ v_t &= \nu\Delta v + g(u,v). \end{aligned} \qquad (13.61)$$

Here, $f, g : \mathbb{R}^2 \to \mathbb{R}$ are given functions that describe the reaction equations. The eigenvalue problem (13.60) for (13.61) is

$$\begin{pmatrix} -|k|^2\mu + f_u & f_v \\ g_u & -|k|^2\nu + g_v \end{pmatrix} \begin{pmatrix} \widehat{u} \\ \widehat{v} \end{pmatrix} = \lambda \begin{pmatrix} \widehat{u} \\ \widehat{v} \end{pmatrix},$$

where we do not indicate explicitly that the derivatives of f, g are evaluated at $u = \overline{u}$, $v = \overline{v}$.

The eigenvalues λ are solutions of the quadratic equation

$$\lambda^2 - \left[f_u + g_v - |k|^2 \left(\mu + \nu\right)\right] \lambda + h\left(|k|^2\right) = 0, \tag{13.62}$$

where

$$h\left(|k|^2\right) = \mu\nu|k|^4 - \left(\mu g_v + \nu f_u\right)|k|^2 + f_u g_v - f_v g_u.$$

From (13.54), the equilibrium state is stable to spatially uniform perturbations with $k = 0$ if

$$f_u + g_v < 0, \qquad f_u g_v - f_v g_u > 0. \tag{13.63}$$

The quadratic equation (13.62) for λ then has a root with positive real part if and only if $h\left(|k|^2\right) < 0$, and this can occur for some $|k| > 0$ only if

$$\mu g_v + \nu f_u > 0. \tag{13.64}$$

In that case, $h\left(|k|^2\right)$ has a minimum at $|k| = \kappa$, where

$$\kappa^2 = \frac{\mu g_v + \nu f_u}{2\mu\nu},$$

and $h\left(\kappa^2\right) < 0$ if and only if

$$\mu g_v + \nu f_u > 2\sqrt{\mu\nu\left(f_u g_v - f_v g_u\right)}. \tag{13.65}$$

Thus, the conditions (13.63)–(13.65) imply that a Turing instability occurs in the system defined by (13.61).

At first sight, it may seem surprising that diffusion can cause instability in a state that is stable to spatially uniform perturbations, but there is a simple explanation. It follows from (13.63) and (13.64) that f_u and g_v have opposite signs and $\mu \neq \nu$. This difference between the diffusivities is essential for the Turing instability. Exchanging u and v, if necessary, we may assume that $f_u < 0$ and $g_v > 0$, when $\mu > \nu$. Furthermore, replacing v by $-v$, if necessary, we find that the sign structure of the entries in the matrix A of the derivative of the reaction terms in a 2×2 system that is subject to a Turing instability can always be put in the form

$$A = \begin{pmatrix} - & + \\ - & + \end{pmatrix}.$$

In this case, we call v an *activator* because, in the absence of diffusion, it grows exponentially in time when $u = 0$, and we call u an *inhibitor* because positive values of u reduce the growth of v. If $v = 0$, then the inhibitor u decays exponentially to its equilibrium state. The equilibrium is a stable state of the reaction equations because a positive perturbation in the activator from its equilibrium value causes a growth in the inhibitor, and this in turn stabilizes the activator. We have seen that $\mu > \nu$, which means that the diffusivity of the inhibitor is greater than the diffusivity of the activator. If a spatially nonuniform perturbation in the activator begins to

grow at some point, then the inhibitor diffuses away faster than the activator, and
as a result the generation of inhibitor may not be sufficient to prevent the continued
growth of the activator.

Morphogenesis is too complex to be explained by a Turing instability. It has been
suggested, however, that some biological patterns, such as coat pigmentations, are
the result of a Turing instability. Turing instability has been observed in chemically
reacting systems in gels, although the first successful experimental observations took
place almost forty years after Turing's original theoretical work.

13.7 The calculus of variations

The calculus of variations is an enormous subject, with applications to physics, geometry, and optimization theory, among many other areas. The following discussion
is therefore only a brief, and incomplete, introduction.

A basic problem in the calculus of variations is the minimization of a functional.
Suppose that $I : X \to \mathbb{R}$ is a real-valued functional defined on a Hilbert or Banach
space X. If I has a local minimum at $\overline{x}$, then for each $h \in X$, the function $I(\overline{x}+\epsilon h)$ of
the scalar parameter ϵ has a local minimum at $\epsilon = 0$. Therefore, if I is differentiable
at $\overline{x}$, then

$$\frac{d}{d\epsilon}I(\overline{x} + \epsilon h)\bigg|_{\epsilon=0} = I'(\overline{x})h = 0,$$

so the derivative of I at $\overline{x}$ is zero. We call a point where the derivative of a functional
I vanishes a *critical point* or *stationary point* of I. Thus, a necessary condition for a
differentiable functional I to have a minimum at an interior point $\overline{x}$ of its domain is
that $\overline{x}$ is a critical point of I. In searching for the minimizers of I, we may therefore
restrict attention to the critical points of I, as well as any boundary points of the
domain of I, and points where I fails to be differentiable. A critical point need not
be a local minimum. For example, it could be a local maximum or a saddle point.
We will not discuss here sufficient conditions for a critical point to be a minimum,
but critical points of functionals are often of interest in their own right.

In this section, we introduce the fundamental ideas in the calculus of variations
by a study of functionals of the form

$$I(u) = \int_0^1 L\left(x, u(x), Du(x)\right) \, dx, \tag{13.66}$$

where $u : [0, 1] \to \mathbb{R}^m$ is a continuously differentiable function of one variable, and

$$L : [0, 1] \times \mathbb{R}^m \times \mathbb{R}^m \to \mathbb{R}$$

is a given smooth function, called the *Lagrangian*. It is convenient to use the
notation D for the derivative with respect to x, so $Du = u'$.

We will derive a differential equation, called the *Euler-Lagrange* equation, that is satisfied by any sufficiently smooth critical point of I. Abusing notation slightly, we write

$$L_x\left(x, u(x), Du(x)\right) = \left.\frac{\partial}{\partial x}L(x, y, v)\right|_{y=u(x), v=Du(x)},$$

$$L_u\left(x, u(x), Du(x)\right) = \left.\frac{\partial}{\partial y}L(x, y, v)\right|_{y=u(x), v=Du(x)},$$

$$L_{Du}(x, u(x), Du(x)) = \left.\frac{\partial}{\partial v}L(x, y, v)\right|_{y=u(x), v=Du(x)}.$$

If $u = (u_1, u_2, \ldots, u_m)$ takes values in $\mathbb{R}^m$, then these partial derivatives denote the gradient, and

$$L_u = (L_{u_1}, L_{u_2}, \ldots, L_{u_m}), \qquad L_{Du} = (L_{Du_1}, L_{Du_2}, \ldots, L_{Du_m})$$

also take values in $\mathbb{R}^m$. If $\varphi = (\varphi_1, \varphi_2, \ldots, \varphi_m) : [0, 1] \to \mathbb{R}^m$, then we write

$$L_u \cdot \varphi = \sum_{i=1}^{m} L_{u_i}\varphi_i, \qquad L_{Du} \cdot \varphi = \sum_{i=1}^{m} L_{Du_i}\varphi_i.$$

For example, the chain rule implies that

$$\begin{aligned} DL\left(x, u(x), Du(x)\right) = \ & L_x\left(x, u(x), Du(x)\right) + L_u\left(x, u(x), Du(x)\right) \cdot Du(x) \\ & + L_{Du}(x, u(x), Du(x)) \cdot D^2u(x). \end{aligned}$$

There are many possible choices for the space X of admissible functions on which I is defined, and the "correct" space in which to look for a minimizer, if one exists at all, depends in general on the functional. For definiteness, we first suppose that u is a continuously differentiable function that satisfies the boundary conditions $u(0) = u(1) = 0$. Then $I : X \to \mathbb{R}$, where

$$X = \left\{u \in C^1([0, 1]) \mid u(0) = u(1) = 0\right\}.$$

The functional I is differentiable on X, and $I'(u) : X \to \mathbb{R}$ is given by

$$\begin{aligned} I'(u)\varphi &= \left.\frac{d}{d\epsilon}I(u + \epsilon\varphi)\right|_{\epsilon=0} \\ &= \left.\frac{d}{d\epsilon}\int_0^1 L(x, u + \epsilon\varphi, Du + \epsilon D\varphi)\, dx\right|_{\epsilon=0} \\ &= \int_0^1 \left\{L_u(x, u, Du) \cdot \varphi + L_{Du}(x, u, Du) \cdot D\varphi\right\} dx. \end{aligned}$$

Thus, if u is a critical point of I, we have

$$\int_0^1 \left\{L_u(x, u, Du) \cdot \varphi + L_{Du}(x, u, Du) \cdot D\varphi\right\} dx = 0 \qquad \text{for all } \varphi \in X.$$

Suppose that the critical point u belongs to $C^2([0,1])$. We may then integrate by parts in this equation to obtain

$$\int_0^1 \{L_u(x,u,Du) - D[L_{Du}(x,u,Du)]\} \cdot \varphi \, dx = 0 \qquad \text{for all } \varphi \in X. \qquad (13.67)$$

The boundary terms vanish because φ is zero at the endpoints. To obtain the differential equation satisfied by u, we use the following *fundamental lemma of the calculus of variations*, or *du Bois-Reymond lemma*.

Lemma 13.34 (Fundamental) If $f : [a,b] \to \mathbb{R}$ is a continuous function such that

$$\int_a^b f(x)\varphi(x) \, dx = 0 \qquad \text{for every } \varphi \in C_c^\infty((a,b)),$$

then $f(x) = 0$ for every $a \le x \le b$.

Proof. Suppose that f is not identically zero. Then there is an $x_0 \in (a,b)$ such that $f(x_0) \ne 0$. Multiplying f by -1, if necessary, we may assume that $f(x_0) > 0$. Since f is continuous, there is an interval $I \subset (a,b)$ such that

$$|f(x) - f(x_0)| \le \frac{1}{2}f(x_0) \qquad \text{for every } x \in I,$$

which implies that $f(x) \ge f(x_0)/2$ for every $x \in I$. Let $\varphi \in C_c^\infty((a,b))$ be a nonnegative function with integral equal to one and support contained in I. Then

$$\int_a^b f(x)\varphi(x) \, dx \ge \frac{1}{2}f(x_0) > 0.$$

This contradiction proves the lemma. $\qquad\qquad\qquad\qquad\qquad\qquad\qquad\square$

Applying the fundamental lemma componentwise to (13.67), we see that every C^2-critical point u of the functional I defined in (13.66) satisfies the following *Euler-Lagrange equation*:

$$-DL_{Du} + L_u = 0. \qquad (13.68)$$

If $u = (u_1, u_2, \ldots, u_m)$ is an m-vector-valued function, where $u_i : [0,1] \to \mathbb{R}$, then the component form of the Euler-Lagrange equation (13.68) is

$$-DL_{Du_i} + L_{u_i} = 0, \qquad i = 1, 2, \ldots, m.$$

Using the chain rule to expand the derivative, we may write this equation as

$$-\sum_{j=1}^m \{L_{Du_i Du_j} D^2 u_j + L_{Du_i u_j} Du_j\} - L_{Du_i x} + L_{u_i} = 0, \qquad i = 1, 2, \ldots, m.$$

If the second derivative L_{DuDu} of L with respect to Du, with matrix

$$\left(L_{Du_i Du_j}\right)_{i,j=1}^m,$$

is nonsingular, then we may solve this equation for $D^2 u$ to obtain a second-order system of equations of the form $D^2 u = f(x, u, Du)$.

Exactly the same argument applies if we minimize I over the affine space of functions u that satisfy the nonhomogeneous boundary conditions

$$u(0) = a, \qquad u(1) = b, \qquad\qquad\qquad (13.69)$$

since $u + \epsilon \varphi$ satisfies the same boundary conditions as u if and only if $\varphi(0) = \varphi(1) = 0$. A C^2-critical point of I on the space of functions $u \in C^1([0,1])$ such that $u(0) = a$ and $u(1) = b$ therefore satisfies the ODE (13.68) and the boundary conditions (13.69).

Example 13.35 Suppose that a curve $y = u(x)$ connects the origin $(0,0)$ and a point (a, b) in the (x, y)-plane. The length $I(u)$ of the curve is given by the arclength integral

$$I(u) = \int_0^a \sqrt{1 + (Du)^2}\, dx.$$

The corresponding Euler-Lagrange equation is

$$-D\left(\frac{Du}{\sqrt{1 + (Du)^2}}\right) = 0,$$

which simplifies to

$$D^2 u = 0.$$

The solution is a linear function of x, and the shortest curve connecting two points is a straight line.

Example 13.36 One of the original problems in the calculus of variations was the *brachistochrone problem*, first formulated by Galileo in 1638, who suggested incorrectly that the solution is a circular arc. The problem was formulated independently and solved correctly by Johann Bernoulli in 1697. Suppose that a frictionless particle, or bead, slides along a curve $y = u(x)$ under the influence of gravity. We choose the y coordinate downwards, so that gravity acts in the positive y-direction. If the particle starts at the origin $O = (0,0)$ with zero velocity, then conservation of energy implies that after it has dropped a vertical distance y, the velocity v of the particle satisfies

$$\frac{1}{2}v^2 = gy,$$

where g is the acceleration due to gravity. Therefore, $v = \sqrt{2gy}$. The time $I(u)$ taken by the particle to move from the origin O to a point $P = (a, b)$ on the curve is given by the integral of arclength divided by velocity, or

$$I(u) = \int_0^a \sqrt{\frac{1 + (Du)^2}{2gu}}\, dx.$$

The brachistochrone problem is to find the curve connecting given points O and P such that a particle starting at rest slides from O to P along the curve in the shortest possible time. The curve should be steep initially, so the particle accelerates rapidly, but it should not be too steep, because this increases its arclength. The required curve satisfies the Euler-Lagrange equation associated with I, which is

$$-D\left(\frac{Du}{\sqrt{2gu\left[1 + (Du)^2\right]}}\right) - \frac{1}{2}\sqrt{\frac{1 + (Du)^2}{2gu^3}} = 0. \qquad (13.70)$$

In order to solve (13.70), we will show that the Euler-Lagrange equation has a first integral whenever the Lagrangian does not depend explicitly on the independent variable x. This result is one of the simplest instances of *Noether's theorem*, which connects symmetries of the Lagrangian with conservation laws of the Euler-Lagrange equation.

Proposition 13.37 If $L = L(u, Du)$ is independent of x, then any solution u of the Euler-Lagrange equation (13.68) satisfies the conservation law

$$L_{Du}(u, Du) \cdot Du - L(u, Du) = \text{constant}.$$

Proof. We define $H : \mathbb{R}^m \times \mathbb{R}^m \to \mathbb{R}$ by

$$H(u, Du) = L_{Du}(u, Du) \cdot Du - L(u, Du).$$

Then, using the chain rule, we find that

$$DH = (DL_{Du} - L_u) \cdot Du.$$

Hence, if u satisfies (13.68), then H is constant. $\qquad \square$

Example 13.38 The Lagrangian for brachistochrone problem in Example 13.36,

$$L(u, Du) = \sqrt{\frac{1 + (Du)^2}{2gu}},$$

is independent of x. Proposition 13.37 therefore implies that (13.70) has the first integral

$$\frac{(Du)^2}{\sqrt{2gu\left[1 + (Du)^2\right]}} - \sqrt{\frac{1 + (Du)^2}{2gu}} = c,$$

where c is a constant of integration. This equation simplifies to

$$2gc^2u\left[1 + (Du)^2\right] = 1.$$

Writing the curve $y = u(x)$ in the parametric form $x = x(t)$, $y = y(t)$, where $y(t) = u(x(t))$ so $Du = \dot{y}/\dot{x}$, and setting $y(t) = k(1 - \cos t)$ where $k = 1/(4gc^2)$, we find that

$$\dot{x} = k(1 - \cos t).$$

The solution through the points $(0,0)$ and (a, b) is therefore the cycloid

$$x(t) = k(t - \sin t), \qquad y(t) = k(1 - \cos t)$$

for $0 \le t \le T$. The constants of integration k, T are chosen so that

$$k(T - \sin T) = a, \qquad k(1 - \cos T) = b.$$

In view of the importance of the expression on the left-hand side of the Euler-Lagrange equation (13.68), we make the following definition.

Definition 13.39 The *variational derivative*, or *functional derivative*, of the functional I in (13.66) at a smooth function u is the function

$$\frac{\delta I}{\delta u} = -DL_{Du} + L_u.$$

The *Euler operator* L_I of I is the operator

$$L_I(u) = \frac{\delta I}{\delta u}$$

that maps a function u to the variational derivative of I at u.

Using this notation, we may write the Euler-Lagrange equation for I as

$$\frac{\delta I}{\delta u} = 0.$$

If u and φ are sufficiently smooth, and φ is compactly supported in $(0, 1)$, then

$$\frac{d}{d\epsilon}I(u + \epsilon\varphi)\bigg|_{\epsilon=0} = \int_0^1 \frac{\delta I}{\delta u}(x) \cdot \varphi(x)\, dx. \tag{13.71}$$

We may think of the L^2-inner product

$$\langle u, v \rangle = \int_0^1 u(x) \cdot v(x)\, dx$$

as a continuous analog of the Euclidean inner product on $\mathbb{R}^n$,

$$\langle u, v \rangle = \sum_{i=1}^{n} u_i v_i,$$

in which an integral over the continuous index x replaces a sum over the discrete index i. Equation (13.71) is the continuous analog of the formula for the directional derivative of a function $I : \mathbb{R}^n \to \mathbb{R}$:

$$\frac{d}{d\epsilon}I(u + \epsilon\varphi)\bigg|_{\epsilon=0} = \sum_{i=1}^{n} \frac{\partial I}{\partial u_i}\varphi_i.$$

Thus, the variational derivative of a functional defined on a suitable subspace of L^2 is a formal continuous analog of the gradient of a function defined on $\mathbb{R}^n$. From (13.71), we may write the value of the variational derivative at x formally as

$$\frac{\delta I}{\delta u}(x) = \frac{d}{d\epsilon}I(u + \epsilon\delta_x)\bigg|_{\epsilon=0},$$

where δ_x is the delta-function supported at x. Thus, heuristically, the value of $\delta I/\delta u$ at the point x measures the sensitivity of I to changes in u at x.

In the above analysis, we looked for critical points of the functional I in (13.66), defined on a restricted class of admissible functions that satisfy prescribed Dirichlet conditions at the endpoints. Suppose, instead, we look for critical points of $I(u)$ without imposing any boundary conditions on the admissible functions u, so that $I : X \to \mathbb{R}$ where $X = C^1([0,1])$. If a critical point u belongs to $C^2([0,1])$, then exactly the same argument as before implies that

$$\int_0^1 \{L_u(x, u, Du) - DL_{Du}(x, u, Du)\} \cdot \varphi\, dx = -[L_{Du}(x, u, Du) \cdot \varphi]_0^1, \qquad (13.72)$$

for all $\varphi \in C^1([0,1])$. The boundary terms arising from the integration by parts need not vanish, since φ is not required to vanish at the endpoints. If, however, we first consider (13.72) for functions φ that do vanish at the endpoints, then we see, as before, that u must satisfy the Euler-Lagrange equation (13.68). It then follows from (13.72) that

$$[L_{Du}(x, u, Du) \cdot \varphi]_0^1 = 0 \qquad \text{for all } \varphi \in C^1([0,1]).$$

Choosing a smooth function $\varphi : [0,1] \to \mathbb{R}^m$ such that $\varphi_i(0) = 1$, $\varphi_i(1) = 0$, or $\varphi_i(0) = 0$, $\varphi_i(1) = 1$, with all other components zero at both endpoints, we see that the critical point u must satisfy the boundary conditions

$$L_{Du}(x, u, Du) = 0 \qquad \text{when } x = 0, 1.$$

These boundary conditions are called *natural boundary conditions* or *free boundary conditions* for I, since they are the ones picked out automatically by the variational principle.

A function $N(x, y, v)$ is called a *null Lagrangian* if the functional

$$F(u) = \int N(x, u, Du)\, dx$$

has the property that $L_F(u) = 0$ for all smooth functions u. If L and $\tilde{L}$ are two Lagrangians that differ by a null Lagrangian, then the associated variational principles have the same Euler-Lagrange equations, but they may have different natural boundary conditions. Null Lagrangians are also of interest in other contexts.

Example 13.40 If $u : \mathbb{R} \to \mathbb{R}$, the Lagrangian $N = Du$ is a null Lagrangian, since the Euler operator of the associated functional $F = \int N \, dx$ is

$$L_F(u) = -DN_{Du} = -D1 = 0.$$

The Euler-Lagrange equation of the functional

$$I_\alpha(u) = \int_0^1 \left(\frac{1}{2}(Du)^2 + \alpha Du - fu \right) dx$$

is independent of α:

$$-D^2 u - f = 0;$$

but the natural boundary conditions do depend on α:

$$Du + \alpha = 0 \qquad \text{when } x = 0, 1.$$

13.8 Hamilton's equation and classical mechanics

If the Lagrangian $L(x, y, v)$ is a convex function of v, we may use a *Legendre transform* to rewrite the second-order Euler-Lagrange equation (13.68) as a first-order, Hamiltonian system of ODEs. We begin by describing the Legendre transform.

Let $f : \Omega \subset \mathbb{R}^n \to \mathbb{R}$ be a twice continuously differentiable function defined on a convex, open set Ω. We say that f is *uniformly convex* if the second derivative

$$f''(x) : \mathbb{R}^n \times \mathbb{R}^n \to \mathbb{R}$$

is positive definite for every $x \in \Omega$, meaning that

$$f''(x)(h, h) > 0 \qquad \text{for all } h \in \mathbb{R}^n \setminus \{0\}.$$

This condition is equivalent to the positive definiteness of the *Hessian matrix* of f,

$$\left(\frac{\partial^2 f}{\partial x_i \partial x_j} \right).$$

A uniformly convex function is strictly convex (see Exercise 13.16). We define the *gradient mapping* $\varphi : \Omega \to \Omega^*$ associated with f by

$$\varphi(x) = f'(x), \quad \Omega^* = \{x^* \in \mathbb{R}^n \mid x^* = \varphi(x) \text{ for some } x \in \Omega\}. \tag{13.73}$$

Geometrically, $f'(x) : \mathbb{R}^n \to \mathbb{R}$, so x^* belongs to the dual space of $\mathbb{R}^n$. Here, we will use the Euclidean inner product $\cdot$ to identify $\mathbb{R}^n$ with its dual.

Lemma 13.41 If $f : \Omega \to \mathbb{R}$ is a uniformly convex function on a convex, open subset Ω of $\mathbb{R}^n$ that belongs to $C^2(\Omega)$, then the gradient map $\varphi : \Omega \to \Omega^*$ defined in (13.73) is a C^1-diffeomorphism of Ω onto Ω^*.

Proof. Since $\varphi' = f''$ is nonsingular, the inverse function theorem implies that φ is a locally invertible C^1-map. By the definition of Ω^*, the gradient map φ is onto, so we only have to show that φ is globally one-to-one. Suppose that $x, y \in \Omega$. Then, since Ω is convex, we have

$$[\varphi(x) - \varphi(y)](x - y) = \left[\int_0^1 \frac{d}{dt} \varphi(tx + (1-t)y) \, dt \right] (x - y)$$

$$= \int_0^1 f''(tx + (1-t)y)(x - y, x - y) \, dt.$$

Using the positive definiteness and continuity of f'', we see that

$$[\varphi(x) - \varphi(y)](x - y) > 0 \quad \text{if } x \neq y.$$

Hence $\varphi(x) \neq \varphi(y)$ if $x \neq y$, so φ is globally invertible. □

It follows from this lemma that $\varphi^{-1} : \Omega^* \to \Omega$ is a C^1-diffeomorphism. The following Legendre transform is therefore well defined.

Definition 13.42 Let $f : \Omega \to \mathbb{R}$ be a uniformly convex function on a convex, open subset Ω of $\mathbb{R}^n$ that belongs to $C^2(\Omega)$. The *Legendre transform* of f is the map $f^* : \Omega^* \to \mathbb{R}$ defined by

$$f^*(x^*) = x \cdot x^* - f(x), \quad \text{where } x = \varphi^{-1}(x^*).$$

Here, $x \cdot x^*$ denotes the Euclidean inner product of x and x^* in $\mathbb{R}^n$, and $\varphi : \Omega \to \Omega^*$ is the gradient map associated with f defined in (13.73).

We call x, x^* *dual variables* or *conjugate variables*, and f, f^* *dual functions* or *conjugate functions*.

Example 13.43 If $f : \mathbb{R} \to \mathbb{R}$ is uniformly convex, then $x^* = \varphi(x)$ is the slope of the graph of f at x, which is a strictly increasing function of x. The value of f^* at x^* is the difference between the values of the linear function whose graph is a line through the origin of slope x^* and f at the point where the slope of f is equal to x^*.

We now return to the variational principle for the functional I in (13.66). We assume that the Lagrangian $L(x, y, v)$ is a uniformly convex function of v, and we define the *Hamiltonian* $H(x, y, p)$ to be the Legendre transform of $L(x, y, v)$ with respect to v, meaning that

$$H(x, y, p) = p \cdot v - L(x, y, v), \quad p = L_v(x, y, v). \tag{13.74}$$

It follows from (13.74) that $L_u = -H_u$. The Euler-Lagrange equation (13.68) may therefore be written as a first-order system of ODEs:

$$Du = H_p(x, u, p), \qquad Dp = -H_u(x, u, p).$$

We call such a system a *Hamiltonian system.*

Variational principles provide a general formulation of the laws of classical mechanics, which may be written in either the Lagrangian or Hamiltonian forms. This formulation is essential in understanding the connection between classical and quantum mechanics. As an example, we consider the equations for a particle of mass m moving in $\mathbb{R}^n$, acted on by a conservative force field $F = -\nabla V$, where $V : \mathbb{R}^n \to \mathbb{R}$ is a smooth potential energy function. We change notation, and write the independent variable as t, instead of x, and the dependent variable as q, instead of u. We use a dot to denote the derivative with respect to t. A particle path for times $0 \le t \le T$ is given by a function $q : [0, T] \to \mathbb{R}^n$. We define the *action* $S(q)$ of a path q to be the time-integral of the difference between the particle's kinetic and potential energies along the path:

$$S(q) = \int_0^T \left(\frac{1}{2} m \dot{q}^2 - V(q) \right) dt.$$

Here, $\dot{q}^2 = \dot{q} \cdot \dot{q}$. The corresponding Lagrangian is

$$L(q, \dot{q}) = \frac{1}{2} m \dot{q}^2 - V(q).$$

Thus, the action $S : C^1([0, T]) \to \mathbb{R}$ is a functional defined on the space of possible particle paths. Hamilton's *principle of stationary action* states that the actual path traveled by a particle with given positions at $t = 0$ and $t = T$ is a stationary point of the action. The path therefore satisfies the Euler-Lagrange equation associated with S, which is

$$m \ddot{q} = -\nabla V. \tag{13.75}$$

This equation is Newton's second law.

The Lagrangian L is independent of t, and Proposition 13.37 implies that

$$\frac{1}{2} m \dot{q}^2 + V(q) = \text{constant} \tag{13.76}$$

on a solution, which expresses the conservation of energy. The correspondence between the invariance of the Lagrangian under time translations and the conservation of energy is a very general one.

Conservation of energy may be verified directly from (13.75). Taking the scalar product of (13.75) with $\dot{q}$, we obtain that

$$m \dot{q} \cdot \ddot{q} + \nabla V(q) \cdot \dot{q} = 0.$$

Using the chain rule, we find that

$$\frac{d}{dt}\left[\frac{1}{2}m\dot{q}^2 + V(q)\right] = 0,$$

which implies (13.76).

The reason why classical mechanics is given by a principle of stationary action is not at all clear at the classical level, but the principle may be derived from quantum mechanics. For example, in Feynman's path-integral formulation of quantum mechanics, the action is the phase of the quantum-mechanical amplitude of a particle path, and the classical paths are paths of stationary phase.

The Legendre transform (13.74) implies that the momentum $p = mv$ is the dual variable to the velocity $v = \dot{q}$, and the Hamiltonian H is the total energy of the particle:

$$H(q,p) = \frac{1}{2m}p^2 + V(q).$$

The Hamiltonian form of the Euler-Lagrange equation is

$$\dot{q} = \frac{1}{m}p, \qquad \dot{p} = -\nabla_q V.$$

Hamilton's equation may itself be given a variational formulation, as the Euler-Lagrange equation of the functional

$$\widetilde{S}(q,p) = \int_0^T \{p \cdot \dot{q} - H(p,q)\}\, dt.$$

The Lagrangian function is not a uniformly convex function of the derivatives, since it is a linear function of $\dot{q}$ and is independent of $\dot{p}$. This explains why the associated Euler-Lagrange equation is first-order, rather than second-order.

13.9   Multiple integrals in the calculus of variations

The Euler-Lagrange equation for a functional of functions of several variables is a PDE, rather than an ODE. Suppose that

$$I(u) = \int_\Omega L\left(x, u, Du\right)\, dx, \qquad (13.77)$$

where Ω is a smooth bounded domain in $\mathbb{R}^n$, $u : \Omega \subset \mathbb{R}^n \to \mathbb{R}^m$, and

$$Du = (D_1 u, D_2 u, \ldots, D_n u)$$

is the derivative of u, where D_i is the partial derivative with respect to x_i keeping x_j fixed for $j \neq i$. A similar calculation to the one in Section 13.7 shows that the

Euler-Lagrange equation for I is

$$-\sum_{i=1}^{n} D_i L_{D_i u} + L_u = 0.$$

Example 13.44 The Euler-Lagrange equation associated with the functional

$$I(u) = \int_{\Omega} \left\{ \frac{1}{2} u_t^2 - \frac{1}{2} |Du|^2 - (1 - \cos u) \right\} \, dx dt$$

is the *sine-Gordon equation*,

$$-u_{tt} + \Delta u - \sin u = 0.$$

We will consider the variational principle for Laplace's equation. Similar ideas apply to variational principles of the form (13.77) in which L is a convex function of Du. We define a quadratic functional $I : H_0^1(\Omega) \to \mathbb{R}$ by

$$I(u) = \frac{1}{2} \int_{\Omega} |Du|^2 \, dx - f(u),$$

where $f : H_0^1(\Omega) \to \mathbb{R}$ is a bounded linear functional on $H_0^1(\Omega)$, meaning that $f \in H^{-1}(\Omega)$. From the Poincaré inequality in Theorem 12.77, we may use

$$\langle u, v \rangle = \int_{\Omega} Du \cdot Dv \, dx$$

as the H_0^1-inner product. Then

$$I(u) = \frac{1}{2} \|u\|^2 - f(u).$$

It follows from Theorem 8.50 (see Exercise 8.20) that I has a unique minimizer on $H_0^1(\Omega)$. This minimizer is a critical point of I, so that $I'(u)(\varphi) = 0$ for all $\varphi \in H_0^1(\Omega)$, meaning that

$$\int_{\Omega} Du \cdot D\varphi \, dx + f(\varphi) = 0 \qquad \text{for all } \varphi \in H_0^1(\Omega).$$

From Definition 12.79, the minimizer u is a weak solution of Laplace's equation,

$$-\Delta u = f,$$

providing another proof of the existence of weak solutions.

In many problems, we are interested in minimizing a functional $I : X \to \mathbb{R}$ subject to a nonlinear constraint $J(x) = 0$, where $J : X \to \mathbb{R}$. A constrained minimization problem may often be replaced by an unconstrained problem by the introduction of a *Lagrange multiplier* λ. We define $F : X \times \mathbb{R} \to \mathbb{R}$ by

$$F(x, \lambda) = I(x) - \lambda J(x).$$

If $(\overline{x}, \overline{\lambda})$ is a critical point of F, then taking the partial derivatives of F with respect to x and λ, we see that

$$I'(\overline{x}) - \overline{\lambda} J'(\overline{x}) = 0, \qquad J(\overline{x}) = 0.$$

It follows that $\overline{x}$ is a critical point of I on the constraint manifold $J(x) = 0$ (see Giaquinta and Hildebrandt [14] for a detailed discussion).

Example 13.45 Consider the problem of minimizing $I : H_0^1(\Omega) \to \mathbb{R}$ given by

$$I(u) = \frac{1}{2} \int_\Omega |Du|^2 \, dx$$

subject to the constraint

$$\frac{1}{2} \int_\Omega u^2 \, dx = 1.$$

A constraint on the value of an integral of an admissible function is called an *isoperimetric constraint*. Introducing a Lagrange multiplier $\lambda \in \mathbb{R}$, we consider critical points of the functional

$$F(u, \lambda) = \frac{1}{2} \int_\Omega |Du|^2 \, dx - \lambda \left(\frac{1}{2} \int_\Omega u^2 \, dx - 1 \right).$$

Taking the derivative of F with respect to u, we find that a smooth critical point satisfies

$$-\Delta u = \lambda u.$$

Thus, u is an eigenfunction of the Laplacian, and the Lagrange multiplier λ is an eigenvalue.

Example 13.46 Consider a function

$$u = (u_1, u_2, \ldots, u_{m+1}) : \Omega \subset \mathbb{R}^n \to \mathbb{S}^m \subset \mathbb{R}^{m+1}$$

from a subset Ω of n-dimensional Euclidean space into an m-dimensional sphere. We use the notation

$$|u|^2 = \sum_{j=1}^{m+1} u_j^2, \quad |Du|^2 = \sum_{i=1}^{n} \sum_{j=1}^{m+1} \left(\frac{\partial u_j}{\partial x_i} \right)^2.$$

A function u that minimizes the functional

$$I(u) = \frac{1}{2} \int_\Omega |Du|^2 \, dx, \tag{13.78}$$

subject to the constraint that $u(x) \in \mathbb{S}^m$, meaning that

$$|u(x)|^2 = 1, \tag{13.79}$$

is called a *harmonic map* from $\mathbb{R}^n$ into $\mathbb{S}^m$. A constraint on the pointwise values of an admissible function and its derivative is called a *holonomic constraint*.

In this case, the Lagrange multiplier is a function $\lambda : \Omega \subset \mathbb{R}^n \to \mathbb{R}$. We consider critical points of the functional

$$F(u, \lambda) = \frac{1}{2} \int_\Omega |Du|^2 \, dx - \int_\Omega \lambda \, |u|^2 \, dx.$$

The Euler-Lagrange equation is

$$-\Delta u = \lambda u. \tag{13.80}$$

Differentiating the constraint $|u|^2 = 1$, we find that $u \cdot Du = 0$. Hence, taking the scalar product of (13.80) with u, and rearranging the result, we find that

$$\lambda = -u \cdot \Delta u = -D \cdot (u \cdot Du) + |Du|^2 = |Du|^2.$$

Thus the Euler-Lagrange equation for harmonic maps from Euclidean space into a sphere is the following nonlinear elliptic system of PDEs:

$$-\Delta u = |Du|^2 u.$$

Such harmonic maps provide a simple model for the steady state configuration of systems with orientational order, such as liquid crystal director fields. An interesting feature of the solutions is the possible presence of topological defects in the field u of unit vectors.

13.10 References

Newton's method is discussed in Kantorovich and Akilov [27]. For a discussion of evolution equations and Liapunov functions, see Walker [55]. For more on symmetries and variational principles, see Olver [42]. For classical mechanics, see Arnold [2] and Gallavotti [13]. The Legendre transformation in convex analysis is described in Rockafellar [46]. The classical calculus of variations is discussed in much more depth in Giaquinta and Hildebrandt [14]. An indication of the extent of the subject is that, despite the fact that these two volumes have a total length of over $1,000$ pages, the authors state that their account is an introduction to the subject, and is not encyclopaedic!

13.11 Exercises

Exercise 13.1 Prove that the derivative of a differentiable map is unique.

Exercise 13.2 Prove that if $A : X \to Y$ is a bounded linear map, then A is differentiable in X, with constant Fréchet derivative equal to A itself.

Exercise 13.3 Suppose that $f, g : X \to Y$ are two differentiable maps between Banach spaces X, Y. Show that $f + g$ is differentiable, and $(f + g)' = f' + g'$.

Exercise 13.4 Prove that a Fréchet differentiable map is Gâteaux differentiable.

Exercise 13.5 Define the function $f : \mathbb{R}^2 \to \mathbb{R}$ by

$$f(x, y) = \frac{xy^2}{x^2 + y^4} \qquad \text{for } (x, y) \neq (0, 0),$$

and $f(0, 0) = 0$. Show that the directional derivatives of f at the origin exist in every direction, but f is not continuous or Fréchet differentiable at the origin.

Exercise 13.6 Let $k : \mathbb{R} \times [0, 1] \to \mathbb{R}$ be a continuous function such that for each $t \in [0, 1]$, the function $k(\cdot, t)$ is in $C^1(\mathbb{R})$. Define a functional $f : C([0, 1]) \to \mathbb{R}$ by

$$f(u) = \int_0^1 k(u(t), t) \, dt.$$

Determine the differentiability properties of f.

Exercise 13.7 Let $f : \mathbb{R} \to \mathbb{R}$ be an increasing function such that

$$f(x) = \begin{cases} 1/n & \text{if } 1/n - 1/(4n^2) \leq x \leq 1/n + 1/(4n^2), \\ x + O(x^2) & \text{as } x \to 0, \end{cases}$$

where $n \in \mathbb{N}$. Show that $f'(0) \neq 0$, but f is not locally invertible at 0. Why doesn't this example contradict the inverse function theorem?

Exercise 13.8 Consider the BVP

$$u'' = \mu^2 \sin v, \quad v'' = \mu^2 u \cos v,$$
$$u'(0) = u(1) = 0, \quad v(0) = v'(1) = 0.$$

Show that there are no solutions that bifurcate off the trivial solution $u = v = 0$ unless $\mu \in \mathbb{R}$ is a solution of

$$1 + \cos \mu \cosh \mu = 0.$$

Exercise 13.9 Suppose that $\mathcal{H}$ is a Hilbert space and $F : \mathcal{H} \times \mathbb{R} \to \mathcal{H}$ is a continuously differentiable operator such that $F(0, \mu) = 0$ for all $\mu \in \mathbb{R}$, so $u = 0$ is a solution branch of the equation $F(u, \mu) = 0$. Suppose that $D_u F(0, 0) : \mathcal{H} \to \mathcal{H}$ is a singular Fredholm operator (see Definition 8.22), assumed self-adjoint for simplicity. Then $\mathcal{H} = \mathcal{M} \oplus \mathcal{N}$ where $\mathcal{M} = \operatorname{ran} D_u F(0, 0)$ and $\mathcal{N} = \ker D_u F(0, 0)$. Let P denote the orthogonal projection of $\mathcal{H}$ onto $\mathcal{M}$ and Q the orthogonal projection onto $\mathcal{N}$. Prove that there are open neighborhoods $U \subset \mathcal{H}$, $V \subset \mathcal{N}$, and $I \subset \mathbb{R}$ of 0 and a continuously differentiable function $\varphi : V \times I \to \mathcal{M}$ such that $(u, \mu) \in U \times I$

is a solution of $F(u, \mu) = 0$ if and only if $u = \varphi(v, \mu) + v$ where $v \in V$ is a solution of $G(v, \mu) = 0$ with $G : V \times \mathbb{R} \to \mathcal{N}$ defined by

$$G(v, \mu) = QF \left(\varphi(v, \mu) + v, \mu \right).$$

The finite-dimensional system of equations $G(v, \mu) = 0$ for $v \in \ker D_u F(0, 0)$ is called the *bifurcation equation* associated with the original, possibly infinite-dimensional, system of equations $F(u, \mu) = 0$. This procedure is called *Liapunov-Schmidt reduction*. With appropriate modifications, a similar procedure applies to a continuously differentiable map $F : X \times \mathbb{R} \to Y$ between Banach spaces X, Y.

Exercise 13.10 Suppose that $f : \mathcal{H} \to \mathbb{R}$ is a differentiable functional on a Hilbert space $\mathcal{H}$. Show that there is a function $\nabla f : \mathcal{H} \to \mathcal{H}$, called the *gradient* of f, such that

$$f'(x)h = \langle \nabla f(x), h \rangle.$$

Compute the gradient of the function $f(x) = \|x\|^2$.

Exercise 13.11 Prove that the closure $R([a, b])$ of the space $S([a, b])$ of step functions in the space $B([a, b])$ of bounded functions $f : [a, b] \to X$ on a compact interval $[a, b]$ into a Banach space X, equipped with the sup-norm, includes all continuous functions. Show that the characteristic function of the rationals in $[a, b]$ does not belong to $R([a, b])$.

Exercise 13.12 Let $A : X \to X$ be a bounded linear operator on a Banach space X, and $f : \mathbb{R} \to X$ a continuous, vector-valued function. Show that the solution of the nonhomogeneous linear evolution equation

$$x_t = Ax + f, \qquad x(0) = x_0$$

is given by

$$x(t) = T(t)x_0 + \int_0^t T(t - s)f(s) \, ds,$$

where $T(t) = e^{tA}$ is the solution operator of the homogeneous equation. This result is called *Duhamel's formula*.

Exercise 13.13 Suppose that $T > 0$ and $T \neq 2n\pi$ for any $n \in \mathbb{N}$. Write out the iteration scheme of the modified Newton method for finding T-periodic solutions of the forced pendulum,

$$\ddot{u} + \sin u = h,$$

where h is a given T-periodic function. Assume that the initial point for the modified Newton's method is $u_0 = 0$. Find an estimate on $\|h\|_\infty$ that is sufficient to ensure convergence of the modified Newton iterates, and estimate the norm $\|u\|_{C^2}$ of the corresponding T-periodic solution.

Exercise 13.14 Derive the Euler-Lagrange equation satisfied by C^4-critical points of a functional $I : C^2([0,1]) \to \mathbb{R}$ defined by

$$I(u) = \int_0^1 L\left(x, u, Du, D^2 u\right) dx,$$

where the Lagrangian $L : \mathbb{R} \times \mathbb{R}^m \times \mathbb{R}^m \times \mathbb{R}^m \to \mathbb{R}$ is a twice continuously differentiable function.

Exercise 13.15 The area of a surface obtained by revolving the graph $y = u(x)$ about the x-axis, where $0 \le x \le 1$, is given by

$$I(u) = 2\pi \int_0^1 u(x)\sqrt{1 + [Du(x)]^2}\, dx.$$

Write out the Euler-Lagrange equation, and the first integral that follows from the independence of the Lagrangian of x. Show that the curve with smallest surface area of revolution connecting given endpoint $u(0) = a$, $u(1) = b$ is a catenary.

Exercise 13.16 Prove that a uniformly convex function $f : \Omega \subset \mathbb{R}^n \to \mathbb{R}$ on an open, convex set Ω is strictly convex.

Exercise 13.17 Compute the Legendre transform of:

(a) $f(x) = e^x - 1$, where $f : \mathbb{R} \to \mathbb{R}$;
(b) $f(x) = x^T A x / 2$ where A is an $n \times n$ positive definite matrix, and $f : \mathbb{R}^n \to \mathbb{R}$.

Exercise 13.18 Compute the Euler-Lagrange equation of the quadratic functional

$$I(u) = \int_\Omega \left\{ \sum_{i,j=1}^n \frac{1}{2} a_{ij}(x) \frac{\partial u}{\partial x_i} \frac{\partial u}{\partial x_j} - f(x) u \right\} dx,$$

where $a_{ij} = a_{ji}$ without loss of generality. Show that the resulting linear PDE is formally self-adjoint.

Exercise 13.19 Let Ω be a regular, bounded open subset of $\mathbb{R}^n$. Show that

$$\lambda = \inf_{\substack{u \in H_0^1(\Omega) \\ u \neq 0}} \left\{ \frac{\int_\Omega |\nabla u|^2\, dx}{\int_\Omega |u|^2\, dx} \right\}$$

is the smallest eigenvalue of the Dirichlet problem for the Laplacian on Ω, and that the infimum is attained at the corresponding eigenfunctions. Use the trial function $u(x,y) = xy(1 - x - y)$ to obtain an upper bound on the lowest eigenvalue of the Dirichlet Laplacian on the triangle $\Omega = \{(x,y) \in \mathbb{R}^2 \mid 0 < x < 1, 0 < y < 1 - x\}$.

Bibliography

[1] R. A. Adams, *Sobolev Spaces*, Academic Press, New York, 1975.

[2] V. I. Arnold, *Mathematical Methods of Classical Mechanics*, 2nd ed., Springer-Verlag, New York, 1989.

[3] R. Beals, *Advanced Mathematical Analysis*, Graduate Texts in Mathematics Vol. 12, Springer-Verlag, New York, 1973.

[4] A. Bellini-Morante, and A. C. McBride, *Applied Nonlinear Semigroups*, John Wiley & Sons, Chichester, 1998.

[5] R. Courant, and D. Hilbert, *Methods of Mathematical Physics*, Vol. 1, Interscience, New York, 1953.

[6] E. A. Coddington, and N. Levinson, *Theory of Ordinary Differential Equations*, McGraw-Hill, New York, 1955.

[7] R. Dautry, and J.-L. Lions, *Mathematical Analysis and Numerical Methods for Science and Technology*, Vol 1–6, Springer-Verlag, New York, 1988–93.

[8] R. Devaney, *An Introduction to Chaotic Dynamical Systems*, 2nd ed., Addison-Wesley, Redwood City, 1989.

[9] J. Dieudonné, *Foundations of Modern Analysis*, Academic Press, New York, 1960.

[10] H. Dym, and H. P. McKean, *Fourier Series and Integrals*, Academic Press, New York, 1972.

[11] L. C. Evans, *Partial Differential Equations*, Graduate Studies in Mathematics, Volume 19, American Mathematical Society, Providence, 1998.

[12] G. B. Folland, *Real Analysis*, 2nd ed., John Wiley & Sons, New York, 1999.

[13] G. Gallavotti, *The Elements of Mechanics*, Springer-Verlag, New York, 1983.

[14] M. Giaquinta, and S. Hildebrandt, *Calculus of Variations I, II*, Springer-Verlag, New York, 1996.

[15] D. Gilbarg, and N. Trudinger, *Elliptic Partial Differential Equations of Second Order*, Springer-Verlag, New York, 1977.

[16] I. S. Gradshteyn, and I. M. Ryzhik, *Table of Integrals, Series, and Products*, 6th ed., Academic Press, Boston, 2000.

[17] G. R. Grimmett, and D. R. Stirzaker, *Probability and Random Processes*, 2nd ed., Clarendon Press, Oxford, 1992.

[18] J. Guckenheimer, and P. Holmes, *Nonlinear Oscillations, Dynamical Systems, and Bifurcations of Vector Fields*, Corr. 5th print., Springer-Verlag, New York, 1997.

[19] P. R. Halmos, *Finite-dimensional Vector Spaces*, 2nd ed., D. Van Nostrand Company, Princeton, 1958.

[20] P. R. Halmos, *A Hilbert Space Problem Book*, 2nd ed., Springer-Verlag, New York,

1982.

[21] M. W. Hirsch, and S. Smale, *Differential Equations, Dynamical Systems and Linear Algebra*, Academic Press, New York, 1974.

[22] H. Hochstadt, *Integral Equations*, John Wiley & Sons, New York, 1973.

[23] H. Hochstadt, *The Functions of Mathematical Physics*, Dover, New York, 1986.

[24] R. A. Horn, and C. R. Johnson, *Matrix Analysis*, Cambridge University Press, Cambridge, 1985.

[25] F. Jones, *Lebesgue Integration on Euclidean Space*, Jones and Bartlett, Boston, 1993.

[26] T. Kato, *Perturbation Theory for Linear Operators*, 2nd ed., Springer-Verlag, Berlin, 1980.

[27] L. V. Kantorovich, and G. P. Akilov, *Functional Analysis*, 2nd ed., Pergamon Press, New York, 1982.

[28] J. L. Kelley, *General Topology*, Springer-Verlag, New York, 1955.

[29] T. W. Körner, *Fourier Analysis*, Cambridge University Press, Cambridge, 1988.

[30] P. D. Lax, *Linear Algebra*, John Wiley & Sons, New York, 1997.

[31] N. N. Lebedev, *Special Functions and their Applications*, Dover, New York, 1972.

[32] E. H. Lieb and M. Loss, *Analysis*, Graduate Studies in Mathematics volume 14, American Mathematical Society, 1997.

[33] L. A. Lusternik, and V. J. Sobolev, *Elements of Functional Analysis*, 3rd ed., Hindustan Publishing Corporation, Delhi, and John Wiley & Sons, New York, 1974.

[34] D. G. Luenberger, *Optimization by Vector Space Methods*, John Wiley & Sons, New York, 1969.

[35] S. Mallet, *A Wavelet Tour of Signal Processing*, Academic Press, San Diego, 1998.

[36] J. E. Marsden, and M. J. Hoffman, *Basic Complex Analysis*, 3rd ed., W. H. Freeman and Company, New York, 1999.

[37] J. E. Marsden, and M. J. Hoffman, *Elementary Classical Analysis*, 2nd ed., W. H. Freeman and Company, New York, 1993.

[38] S. G. Mikhlin, *Mathematical Physics; an Advanced Course*, North-Holland Pub. Co., Amsterdam, 1970.

[39] P. M. Morse, and H. Feshbach, *Methods of Theoretical Physics, Parts I–II*, McGraw-Hill, New York, 1953.

[40] A. W. Naylor, and G. R. Sell, *Linear Operator Theory in Engineering and Science*, Springer-Verlag, New York, 1982.

[41] B. Øksendal, *Stochastic Differential Equations*, 5th ed., Springer-Verlag, New York, 1998.

[42] P. J. Olver, *Applications of Lie Groups to Differential Equations*, 2nd ed., Springer-Verlag, New York, 1993.

[43] W. H. Press et al., *Numerical Recipes in C*, 2nd ed., Cambridge University Press, Cambridge, 1992.

[44] J. Rauch, *Partial Differential Equations*, Springer-Verlag, New York, 1991.

[45] M. Reed, and B. Simon, *Methods of Modern Mathematical Physics*, Vol. 1: Functional Analysis, 2nd ed., Academic Press, New York, 1980.

[46] R. T. Rockafellar, *Convex Analysis*, Princeton University Press, Princeton, 1970.

[47] W. Rudin, *Principles of Mathematical Analysis*, 2nd ed., McGraw-Hill, New York, 1964.

[48] W. Rudin, *Functional Analysis*, 2nd ed., McGraw-Hill, New York, 1991.

[49] W. Rudin, *Real and Complex Analysis*, 3rd ed., McGraw-Hill, New York, 1986.

[50] G. F. Simmons, *Introduction to Topology and Modern Analysis*, McGraw-Hill, New York, 1963.

[51] M. Schroeder, *Fractals, Chaos, Power Laws: Minutes from an Infinite Paradise*,

W. H. Freeman, New York, 1991.

[52] I. Stakgold, *Green's Functions and Boundary Value Problems*, 2nd ed., John Wiley & Sons, New York, 1998.

[53] J. C. Strikwerder, *Finite Difference Schemes and Partial Differential Equations*, Wadsworth & Brooks/Cole, Pacific Grove, CA, 1989.

[54] L. Trefethen, and D. Bai, *Numerical Linear Algebra*, Society for Industrial and Applied Mathematics, Philadelphia, 1997.

[55] J. A. Walker, *Dynamical Systems and Evolution Equations*, Plenum, New York, 1980.

[56] G. B. Whitham, *Linear and Nonlinear Waves*, John Wiley & Sons, New York, 1974.

[57] E. Zauderer, *Partial Differential Equations of Applied Mathematics*, 2nd ed., John Wiley & Sons, New York, 1998.

Index